Lecture Notes in Computer Scie

Commenced Publication in 1973
Founding and Former Series Editors:
Gerhard Goos, Juris Hartmanis, and Jan van Leeuwen

Andrew Ravenscroft
Stefanie Lindstaedt
Carlos Delgado Kloos
Davinia Hernández-Leo (Eds.)

21st Century Learning for 21st Century Skills

7th European Conference
on Technology Enhanced Learning, EC-TEL 2012
Saarbrücken, Germany, September 18-21, 2012
Proceedings

 Springer

Volume Editors

Andrew Ravenscroft
University of East London (UEL)
CASS School of Education and Communities
Stratford Campus, Water Lane, London E15 4LZ, UK
E-mail: a.ravenscroft@uel.ac.uk

Stefanie Lindstaedt
Graz University of Technology (TUG)
Knowledge Management Institute and Know-Center GmbH
Inffeldgasse 21a, 8010 Graz, Austria
E-mail: lindstaedt@tugraz.at

Carlos Delgado Kloos
Universidad Carlos III de Madrid
Departamento Ingeniería Telemática
Avenida Universidad 30, 28911 Leganés (Madrid), Spain
E-mail: cdk@it.uc3m.es

Davinia Hernández-Leo
Universitat Pompeu Fabra
Departament de Tecnologies de la Informació i les Comunicacions
Roc Boronat 138, 08018 Barcelona, Spain
E-mail: davinia.hernandez@upf.edu

ISSN 0302-9743   e-ISSN 1611-3349
ISBN 978-3-642-33262-3   e-ISBN 978-3-642-33263-0
DOI 10.1007/978-3-642-33263-0
Springer Heidelberg Dordrecht London New York

Library of Congress Control Number: 2012946459

CR Subject Classification (1998): H.4, H.3, I.2, C.2, H.5, J.1

LNCS Sublibrary: SL 2 – Programming and Software Engineering

Typesetting: Camera-ready by author, data conversion by Scientific Publishing Services, Chennai, India

Printed on acid-free paper

Springer is part of Springer Science+Business Media (www.springer.com)

Preface

The European Conferences on Technology Enhanced Learning (EC-TEL) are now established as a main reference point for the state of the art in Technology Enhanced Learning (TEL) research and development, particularly within Europe and also worldwide. The seventh conference took place in Saarbrucken in Germany, and was hosted by CeLTech – Centre for e-Learning Technology (Saarland University/DFKI) – during 18–21 September 2012. This built upon previous conferences held in Palermo, Italy (2011), Barcelona, Spain (2010), Nice, France (2009), Maastricht, The Netherlands (2008), and Crete, Greece (2006 and 2007). EC-TEL 2012 provided a unique opportunity for researchers, practitioners, and policy makers to address current challenges and advances in the field. This year the conference addressed the pressing challenge facing TEL and education more widely, namely how to support and promote *21st century learning for the 21st century skills*. This theme is a key priority within the European Union and constituent countries and also worldwide, as research needs to address crucial contemporary questions such as:

- How can schools prepare young people for the technology-rich workplace of the future?
- How can we use technology to promote informal and independent learning outside traditional educational settings?
- How can we use next generation social and mobile technologies to promote informal and responsive learning?
- How does technology transform education?

Our programme tackled the theme and key questions comprehensively through its related activities, namely: 4 world leading keynote speakers; 38 high-quality long and short scientific papers; 9 pre-conference workshops and 2 tutorials; an industrial track; a doctorial consortium; and interactive demonstrations and posters. 'Interactivity' was a key feature of the conference, which encouraged the provision of demonstrations linked to scientific articles, continually running video sequences of demonstrations throughout the conference venue and holding a competitive 'TEL Shootout' where delegates voted on the best demonstration.

The four keynote speakers provided exciting and complementary perspectives on the conference theme and sub-themes. Mary Lou Maher (Design Lab, University of Maryland) emphasized the role and importance of designing for diversity and creativity during her talk about *Technology Enhanced Innovation and Learning: Design Principles for Environments that Mediate and Encourage Diversity and Creativity*. Richard Noss (Director, London Knowledge Lab and UK Teaching and Learning Research Programme) provided an insightful examination of the precepts and implications of the conference theme in his address *21st Century Learning for 21st Century Skills: What Does It Mean, and How Do*

We Do It? Wolfgang Wahlster (Director, German Research Centre for Artificial Intelligence) provided an innovative perspective on the essential links between situated learning and industry requirements and practices in his talk *Situated Learning and Assistance Technologies for Industry 4.0*. A further international perspective was provided by Prof. Ruimin Shen (Shanghai Jiao Tong University) who gave a keynote on *Technology Enhanced Learning in China: The example of the SJTU E-Learning Lab*.

In addition, in what has become a tradition for EC-TEL conferences, the delegates were addressed by Marco Marsella, Deputy Head of the Unit for eContent and Safer Internet from the European Commission, so that ongoing and future research and development could be clearly articulated with the priorities for research funding within Europe. A key overarching objective of EC-TEL 2012 was to examine and improve the transitions between research, practice and industry. This was reflected through the co-sponsors of the conference, which included the European Association of Technology Enhanced Learning (EATEL), TELspain, eMadrid, Springer and IMC information multimedia communication AG.

This year saw 130 submissions from 35 countries for consideration as full papers at EC-TEL 2012. After intense scrutiny in the form of some 300 reviews, the programme committee selected just 26 papers, or 20% of those submitted. Also selected for inclusion in the proceedings were 12 short papers; 16 papers from demonstration sessions and 11 poster papers.

Specifically, the conference programme was formed through the themes: learning analytics and retrieval; academic learning and context; personalized and adaptive learning; learning environments; organizational and workplace learning; serious and educational games; collaborative learning and semantic means; and, ict and learning. Collectively, these themes embraced a key feature of the conference and EC-TEL research and practice. This is that TEL research and development needs to embrace the increasing interconnectedness of learning technologies and the contextualized formal and informal practices for learning and education.

Finally, in introducing these proceedings to you we hope that the high-quality, rich and varied articles that are included take TEL research and thinking forward in ways that address the changing and technology-rich landscape in which we think, learn and work in the 21st century.

September 2012

Andrew Ravenscroft
Stefanie Lindstaedt
Carlos Delgado Kloos
Davinia Hernández-Leo

Conference Organisation

Executive Committee

General Chair

Carlos Delgado Kloos — eMadrid/TELSpain/Universidad Carlos III de Madrid, Spain

Programme Chairs

Andrew Ravenscroft — CASS School of Education and Communities, University of East London (UEL), UK

Stefanie Lindstaedt — Graz University of Technology & Know-Center, Austria

Workshop Chair

Tobias Ley — Talinn University, Estonia

Poster and Demonstration Chair

Davinia Hernández-Leo — Universitat Pompeu Fabra, Spain

Dissemination Chair

Sergey Sosnovsky — CeLTech – Centre for e-Learning Technology (Saarland University / DFKI), Germany

Industrial Session Chair

Volker Zimmermann — imc, Germany

Local Organization Chair

Christoph Igel — CeLTech – Centre for e-Learning Technology (Saarland University / DFKI), Germany

Doctoral Consortium Chairs

Katherine Maillet — Institut Mines-Telecom, France

Mario Allegra — Istituto per le Tecnologie Didattiche, Italy

Programme Committee

Heidrun Allert	Christian-Albrechts-Universität zu Kiel
Charoula Angeli	University of Cyprus
Luis Anido Rifon	Universidade de Vigo
Inmaculada Arnedillo-Sanchez	Trinity College Dublin
Nicolas Balacheff	CNRS
Francesco Bellotti	University of Genoa
Adriana Berlanga	Open University of the Netherlands
Katrin Borcea-Pfitzmann	Technische Universität Dresden
Francis Brouns	Open University of the Netherlands
Peter Brusilovsky	University of Pittsburgh
Daniel Burgos	International University of La Rioja
Lorenzo Cantoni	Università della Svizzera italiana
Linda Castañeda	University of Murcia
Manuel Castro	UNED
Mohamed Amine Chatti	RWTH Aachen University
Giuseppe Chiazzese	Italian National Research Council
Agnieszka Chrzaszcz	AGH-UST
Audrey Cooke	Curtin University
Raquel M. Crespo	Universidad Carlos III de Madrid
Ulrike Cress	Knowledge Media Research Center
Alexandra I. Cristea	University of Warwick
Paul de Bra	Eindhoven University of Technology
Carlos Delgado Kloos	Universidad Carlos III de Madrid
Christian Depover	Université de Mons
Michael Derntl	RWTH Aachen University
Philippe Dessus	LSE, Grenoble
Darina Dicheva	Winston-Salem State University
Stefan Dietze	L3S Research Center
Yannis Dimitriadis	University of Valladolid
Vania Dimitrova	University of Leeds
Hendrik Drachsler	Open University of the Netherlands
Jon Dron	Athabasca University
Benedict Du Boulay	University of Sussex
Erik Duval	K.U. Leuven
Martin Ebner	University of Graz
Alfred Essa	Desire2Learn
Dieter Euler	University of St. Gallen
Baltasar Fernandez-Manjon	Universidad Complutense de Madrid
Carmen Fernández-Panadero	Universidad Carlos III de Madrid
Christine Ferraris	Université de Savoie
Katharina Freitag	imc AG
Muriel Garreta	Universitat Oberta de Catalunya
Dragan Gasevic	Athabasca University

Denis Gillet   Swiss Federal Institute of Technology in
                                 Lausanne
Fabrizio Giorgini   eXact learning solutions
Christian Glahn   Swiss Federal Institute of Technology Zurich
Monique Grandbastien   LORIA, UHP Nancy1
David Griffiths   University of Bolton
Begona Gros   Universitat Oberta de Catalunya
Christian Guetl   Graz University of Technology
Joerg Haake   FernUniversitaet in Hagen
Andreas Harrer   Catholic University Eichstätt-Ingolstadt
Davinia Hernández-Leo   Universitat Pompeu Fabra
Knut Hinkelmann   University of Applied Sciences Northwestern
                                 Switzerland
Patrick Hoefler   Know-Center
Ulrich Hoppe   University Duisburg-Essen
Christoph Igel   CeLTech – Centre for e-Learning Technology
Sanna Järvelä   University of Oulu
Julia Kaltenbeck   Know-Center
Petra Kaltenbeck   Know-Center
Marco Kalz   Open University of the Netherlands
Nuri Kara   Middle East Technical University
Nikos Karacapilidis   University of Patras
Michael Kickmeier-Rust   University of Graz
Barbara Kieslinger   Centre for Social Innovation
Ralf Klamma   RWTH Aachen University
Joris Klerkx   Katholieke Universiteit Leuven
Tomaz Klobucar   Jozef Stefan Institute
Milos Kravcik   RWTH Aachen University
Mart Laanpere   Tallinn University
Lydia Lau   University of Leeds
Effie Law   University of Leicester
Tobias Ley   Tallinn University
Stefanie Lindstaedt   Know-Center
Andreas Lingnau   University of Strathclyde
Martin Llamas-Nistal   University of Vigo
Rose Luckin   The London Knowledge Lab
George Magoulas   Birkbeck College
Katherine Maillet   Institut Mines-Télécom
Allegra Mario   Italian National Research Council
Russell Meier   Milwaukee School of Engineering
Martin Memmel   DFKI GmbH
Doris Meringer   Know-Center
Riichiro Mizoguchi   University of Osaka
Paola Monachesi   Utrecht University
Pablo Moreno-Ger   Universidad Complutense de Madrid
Pedro J. Muñoz Merino   Carlos III University of Madrid

Mario Muñoz-Organero	Carlos III University of Madrid
Rob Nadolski	Open University of the Netherlands
Ambjorn Naeve	Royal Institute of Technology
Roger Nkambou	Université du Québec à Montréal
Viktoria Pammer	Know-Center
Abelardo Pardo	Carlos III University of Madrid
Kai Pata	Tallinn University
Jermann Patrick	Ecole Polytechnique Fédérale de Lausanne
Jan Pawlowski	University of Jyväskylä
Asensio Perez Juan	University of Valladolid
Eric Ras	Public Research Centre Henri Tudor
Andrew Ravenscroft	University of East London (UEL)
Uwe Riss	SAP Research
Miguel Rodriguez Artacho	UNED University
Vicente Romero	Atos Origin
J.I. Schoonenboom	VU University Amsterdam
Britta Seidel	CeLTech – Centre for e-Learning Technology
Mike Sharples	The Open University
Bernd Simon	WU, Vienna
Peter Sloep	Open University of the Netherlands
Sergey Sosnovsky	CeLTech – Centre for e-Learning Technology
Marcus Specht	Open University of the Nethderlands
Slavi Stoyanov	Open University of the Netherlands
Deborah Tatar	VirginiaTech
Pierre Tchounikine	University of Grenoble
Stefaan Ternier	Open University of the Netherlands
Stefan Trausan-Matu	"Politehnica" University of Bucharest
Martin Valcke	Ghent University
Christine Vanoirbeek	Ecole Polytechnique Fédérale de Lausanne
Katrien Verbert	K.U. Leuven
Wim Westera	Open University of the Netherlands
Peter Wetz	Know-Center
Fridolin Wild	The Open University of the UK
Martin Wolpers	Fraunhofer Institute of Applied Information Technology
Volker Zimmermann	imc AG

Additional Reviewers

Alario-Hoyos, Carlos
De La Fuente Valentín, Luis
Falakmasir, Mohammad Hassan
Gutiérrez Rojas, Israel
Pérez-Sanagustín, Mar
Rodríguez Triana, María Jesús

Ruíz Calleja, Adolfo
Seta, Luciano
Simon, Bernd
Taibi, Davide
Voigt, Christian

Table of Contents

Part III: Short Papers

Part IV: Demonstration Papers

Part V: Poster Papers

Part I
Invited Paper

21st Century Learning for 21st Century Skills: What Does It Mean, and How Do We Do It?

Richard Noss

London Knowledge Lab and Technology Enhanced Learning (TEL) Research Programme,
Institute of Education, University of London, London, WC1N 3QS, United Kingdom
r.noss@ioe.ac.uk

Abstract. I want to argue in this lecture, that life – especially educational life – is never that simple. What exactly are 21st century skills? How, for example, do they differ from 'knowledge'? And once we know what they are, does there follow a strategy – or at least a set of principles – for what learning should look like, and the roles we ascribe to technology? Most importantly, if 21st century knowledge is qualitatively different from the 19th and 20th century knowledge that characterises much of our existing curricula, we will need to consider carefully just how to make that knowledge learnable and accessible through the design of digital technologies and their evaluation.

Keywords: pedagogy, technology, teaching, learning.

Abstract

21st Century Learning for 21st Century Skills. What's not to like? We know, it seems, that the newish century demands new, process-oriented skills like teamwork, flexibility, problem solving, to take account of the shift from material labour to immaterial, weightless production. We can take for granted, at least in a gathering like this, that 21c. learning is learning with digital technology. And we can surely agree that we are gaining with impressive speed, understanding of the technology's potential to enable a new kind of pedagogy.

I want to argue in this lecture, that life – especially educational life – is never that simple. What exactly are 21st century skills? How, for example, do they differ from 'knowledge'? And once we know what they are, does there follow a strategy – or at least a set of principles – for what learning should look like, and the roles we ascribe to technology? Most importantly, if 21st century knowledge is qualitatively different from the 19th and 20th century knowledge that characterises much of our existing curricula, we will need to consider carefully just how to make that knowledge learnable and accessible through the design of digital technologies and their evaluation.

The problem is this. The needs of the 21st century are seen as broadly dichotomised. Much of the discussion about who needs to know what, is predicated on the assumption that technology has created the need for fewer and fewer people really to understand the way the world works; and for more and more merely to

A. Ravenscroft et al. (Eds.): EC-TEL 2012, LNCS 7563, pp. 3–5, 2012.

respond to what technology demands of them. There is partial truth here: very few people need to know how derivatives work (it seems that the bankers don't either); and the supermarket checkout operator no longer needs to calculate change. So this gives rise to the belief that there is stuff that the elite need to know; and stuff that everyone needs to know – and that these have very little in common. Inevitably, the latter is reduced to process-oriented skills, denuded of real knowledge that can help individuals engage as empowered agents in their own lives. And the gap between these two poles is widening, despite the best intentions of educators and policymakers. So the danger is real: Knowledge for the top of the pyramid; skills and processes for the bottom.

Of course, the imperatives of the workplace should not be the only driver of educational policy or practice. But they cannot be ignored, and if they are going to inform or even direct it, it would be helpful if we were clear about what we are trying to do. This is all the more important as we are at something of a crossroads in our thinking about technology (more fashionably, a 'tipping point').

The first 30 years of educational computing were dominated by a commercial paradigm borrowed from business and industry. When educators and policy makers thought of technology for schools, colleges and universities, they were guided with reference to a social niche nicely occupied by *Windows,* the all-pervasive metaphor of the *office,* the *desktop,* the *filing system,* and so on. It worked fine in many respects, except one: it pretty much guaranteed that the existing practices of teaching and learning institutions remained more or less intact, lubricated by the application of technology, but not changed fundamentally by it. The technology beautifully legitimised the commercial/business paradigm of learning – think, for example, how the interactive whiteboard has been, for the most part, the technological end of a pedagogy based on eyes-front, teacher-led practice.

I don't want to future-gaze too much, and certainly do *not* want to stand accused of technocentrism, which I've been pretty vocal about over these last thirty years[1]. But technology *does* shape the ambient culture, as well as being shaped by it, and understanding how that works is an important part of how we should respond. It is hard not to notice a change in the ways technology is impacting people's lives; and again, without attributing magical powers to this or that passing platform, I think that the sudden ubiquity of the i-Pad/smartphone paradigm – a paradigm quite different from the commercial paradigm that preceded it - should give us pause for thought. Until now, technology has been seen as institutional; but now, we have reached the point where it has moved from the institution to the home, the pocket and the street – it has become personal.

There is a lot to say about this, and I'll save it for the lecture. But one thing is clear: this change is double-edged. i-Pads are wonderful machines for viewing photos, organising playlists, and providing a platform for the exponentially increasing number of apps, all just a click away. That click is attractive for schools and colleges – no

[1] Seymour Papert describes technocentrism as *"the fallacy of referring all questions to the technology"*.
(http://www.papert.org/articles/ACritiqueofTechnocentrism.html)

need for training, cumbersome networks, and above all, a convergence between what learners already do and what they might be encouraged to do. But as we all know, ease of access comes at the price of invisibility – the same digital natives who know how to use their phones for mashups and facebook, are digital immigrants when it comes to engaging with the technology in any deep way: and for educators that's a very expensive price to pay for simplicity.

So the challenge I want to take up in this lecture is this: how can we design, implement and exploit technology, so that it recognises the diversity of what we are trying to teach, and to whom we are trying to teach it? And, just as important, can technology help us to achieve what seems, at first sight, to be the impossible: to help all learners, across the social and economic spectrum, to learn about *their* agency in a world where, increasingly, agency is at best invisible and at worst, non-existent. For that they will need knowledge, not 'knowledge about knowledge' or 'learning about learning'.[2]

The structure of the lecture will be as follows. First, I'll take a look at what is known about the needs of 'knowledge economies', and the gap that has opened up between the knowledge rich and the knowledge poor. Second, I want to review what we know about technology, share some research findings of the Technology Enhanced Learning Research Programme, and show how our state of knowledge can be put to use in bridging the gap. Third, I want to future gaze just a little, in terms of developments in technology, and how they focus our attention on the question of *what* to teach, rather than merely *how* to teach it. And finally, I want to return to the first theme, and show that by focusing on the new things we can learn with technology (things that are essentially unlearnable without it), we can address the problem this conference has set itself by somewhat adapting the title – to understand *21st Century Learning for 21st Century Knowledge*.

References

1. Young, M.: Bringing Knowledge Back. In: From Social Constructivism to Social Realism in the Sociology of Education. Routledge, Oxford (2008)
2. Papert, S.: A Critique of Technocentrism in Thinking About the School of the Future (2012), http://www.papert.org/articles/ACritiqueofTechnocentrism.html (accessed July 16, 2012)

---

[2] Michael Young's book, *Bringing Knowledge Back In: From Social Constructivism to Social Realism in the Sociology of Education,* extends these ideas.

Part II
Full Paper

Exploiting Semantic Information
for Graph-Based Recommendations
of Learning Resources

Mojisola Anjorin, Thomas Rodenhausen,
Renato Domínguez García, and Christoph Rensing

Multimedia Communications Lab,
Technische Universität Darmstadt, Germany
{mojisola.anjorin,thomas.rodenhausen,renato.dominguez.garcia,
christoph.rensing}@kom.tu-darmstadt.de
http://www.kom.tu-darmstadt.de

Abstract. Recommender systems in e-learning have different goals as
compared to those in other domains. This brings about new requirements
such as the need for techniques that recommend learning resources beyond their similarity. It is therefore an ongoing challenge to develop recommender systems considering the particularities of e-learning scenarios
like CROKODIL. CROKODIL is a platform supporting the collaborative
acquisition and management of learning resources. It supports collaborative semantic tagging thereby forming a folksonomy. Research shows that
additional semantic information in extended folksonomies can be used
to enhance graph-based recommendations. In this paper, CROKODIL's
folksonomy is analysed, focusing on its hierarchical activity structure.
Activities help learners structure their tasks and learning goals. AScore
and AInheritScore are proposed approaches for recommending learning
resources by exploiting the additional semantic information gained from
activity structures. Results show that this additional semantic information is beneficial for recommending learning resources in an application
scenario like CROKODIL.

Keywords: ranking, resource recommendation, folksonomy, tagging.

1 Introduction

Resources found on the Web ranging from multimedia websites to collaborative
web resources, become increasingly important for today's learning. Learners appreciate a learning process in which a variety of resources are used [9]. This shows
a shift away from instructional-based learning to resource-based learning [17].
Resource-based learning is mostly self-directed [3] and the learner is often confronted, in addition to the actual learning process, with an overhead of finding
relevant high quality learning resources amidst the huge amount of information available on the Web. In learning scenarios, recommender systems support
learners by suggesting relevant learning resources [15]. An effective ranking of

A. Ravenscroft et al. (Eds.): EC-TEL 2012, LNCS 7563, pp. 9–22, 2012.

learning resources would reduce the overhead when learning with resources found on the Web.

Social bookmarking applications, in which users collaboratively attach tags to resources, offer support to the user during the search, annotation and sharing tasks involved in resource-based learning [3]. Tagging helps to quickly retrieve a resource later via search, or navigation, or to give an overview about the resource's content. Through the collaborative tagging of resources, a structure called a folksonomy is created. Promising results using additional semantic information to improve the ranking of resources in extended folksonomies have been made [1]. It is therefore of great interest to investigate how semantic information can benefit the ranking of learning resources in an e-learning scenario such as CROKODIL [4]. CROKODIL[1] is a platform supporting the collaborative acquisition and management of learning resources. It offers support to the learner in all tasks of resource-based learning [3]. CROKODIL is based on a pedagogical concept which focuses on *activities* as the main concept for organizing learning resources [3]. *Activities* aim to support the learner during his learning process by organizing his tasks in a hierarchical activity structure. Relevant knowledge resources found on the Web are then attached to these activities. The resulting challenge is now how best to exploit these activity structures in order to recommend relevant learning resources to other users working on related activities.

In this work, we consider the hierarchical activity structures available in the CROKODIL application scenario [4] as additional semantic information which can be used for ranking resources. We therefore propose the algorithms AScore and AInheritScore which exploit the activity structures in CROKODIL to improve the ranking of resources in an extended folksonomy for the purpose of recommending relevant learning resources.

The extended folksonomy of the CROKODIL application scenario is defined in Sect. 2. Related work is summarized in Sect. 3. Proposed approaches are implemented in Sect. 4 and evaluated in Sect. 5. This paper concludes with a brief summary and an outlook on possible future work.

2   Analysis of Application Scenario: CROKODIL

CROKODIL supports the collaborative semantic tagging [5] of learning resources thereby forming a folksonomy structure consisting of users, resources and tags [3]. Tags can be assigned tag types such as topic, location, person, event or genre. Activities as mentioned in Sect.1 are created describing learning goals or tasks to be accomplished by a learner or group of learners. Resources needed to achieve these goals are attached to these activities. In addition, CROKODIL offers social network functionality to support the learning community [3]. Groups of learners working on a common activity can be created, as well as friendship relations between two learners. In the following a folksonomy and CROKODIL's extended folksonomy are defined.

---

[1] http://www.crokodil.de/, http://demo.crokodil.de(retrieved 06.07.2012)

A **folksonomy** is described as a system of classification derived from collaboratively creating and managing tags to annotate and categorize content[16]. This is also known as a social tagging system or a collaborative tagging system. A folksonomy can also be represented as a folksonomy graph G_F as defined in Sect. 4.

Definition 1 (Folksonomy). *A folksonomy is defined as a quadruple [11]: $F := (U, T, R, Y)$ where:*

- *U is a finite set of **users***
- *T is a finite set of **tags***
- *R is a finite set of **resources***
- *$Y \subseteq U \times T \times R$ is a **tag assignment** relation over these sets*

E.g., user *thomas* $\in U$ attaches a tag *London* $\in T$ to the resource *olympic.org* $\in R$, thus forming a tag assignment *(thomas, London, olympic.org)* $\in Y$.

An **extended folksonomy** is a folksonomy enhanced with additional semantic information [1]. CROKODIL is an extended folksonomy where the semantic information gained from activities, semantic tag types, learner groups and friendships extend the folksonomy. These additional semantic information can also be seen as giving a context to elements in the folksonomy [4] [1]. For example, resources belonging to the same activity, can be seen as belonging to the same context of this activity.

Definition 2 (CROKODIL's Extended Folksonomy). *CROKODIL's extended folksonomy is defined as: $F_C := (U, T_{typed}, R, Y_T, (A, <), Y_A, Y_U, G, friends)$ where:*

- *U is a finite set of **learners***
- *T_{typed} is a finite set of **typed tags** consisting of pairs (t, type), where t is an arbitrary tag and type $\in \{topic, location, event, genre, person, other\}$*
- *R is a finite set of **learning resources***
- *$Y_T \subseteq U \times T_{typed} \times R$ is a **tag assignment** relation over the set of users, typed tags and resources*
- *$(A, <)$ is a finite set of **activities** with a partial order $<$ indicating sub-activities*
- *$Y_A \subseteq U \times A \times R$ is an **activity assignment** relation over the set of users, activities and resources*
- *$Y_U \subseteq U \times A$ is an **activity membership assignment** relation over the set of users and activities*
- *$G \subseteq \mathcal{P}(U)$ is the finite set of subsets of learners called **groups of learners***
- *friends $\subseteq U \times U$ is a symmetric binary relation which indicates a **friendship relation** between two learners*

E.g., *thomas* is preparing for a quiz about the olympic games. He therefore creates an activity *prepare quiz about the olympics* having a sub-activity *collect historical facts*. This means $A = \{prepare\ quiz\ about\ the\ olympics,\ collect\ historical\ facts\}$ and *collect historical facts* $<$ *prepare quiz about the olympics*.

In addition, (*thomas, prepare quiz about the olympics*) $\in Y_U$ and (*thomas, collect historical facts*) $\in Y_U$. He finds the website *olympic.org*, to which he attaches the tag *London* with tag type *location*, (*thomas, (London, location), olympic.org*) $\in Y_T$. He then attaches this resource to the activity *prepare quiz about the olympics*, (*thomas, prepare quiz about the olympics, olympic.org*) $\in Y_A$. Thomas creates a group *olympic experts* $\in G$ and invites *moji* $\in U$ and his friend *renato* $\in U$ to help him gather facts about the olympic games.

In this paper, we will be focusing on the additional semantic information gained from the activities in CROKODIL's extended folksonomy and investigating how this can improve the ranking of learning resources.

3   Related Work

Recommender systems have shown to be very useful in e-learning scenarios [15]. Collaborative filtering approaches use community data such as feedback, tags or ratings from learners to make recommendations e.g.[8] whereas content-based approaches make recommendations based on the similarity between learning resources e.g. [18]. Recommender systems in e-learning have different information retrieval goals as compared to other domains thus leading to new requirements like recommending items beyond their similarity [15]. It is therefore increasingly important to develop recommender systems that consider the particularities of the e-learning domain. Graph-based recommendation techniques can be classified as neighborhood-based collaborative filtering approaches, having the advantage of avoiding the problems of sparsity and limited coverage [7]. Graph-based recommender systems e.g. [1,6] consider the graphical structure when recommending items in a folksonomy. The data is represented in the form of a graph where nodes are users, tags or resources and edges the transactions or relations between them. One of the most popular approaches is FolkRank [12] which is based on the PageRank computation on a graph created from a folksonomy. FolkRank can be used to recommend users, tags or resources in social bookmarking systems. The intuition is that a resource tagged with important tags by important users becomes important itself. The same holds for tags and users.

Furthermore, it is of interest for recommender systems in e-learning to take advantage of additional semantic information such as context awareness which includes pedagogical aspects like learning goals [15]. Abel [1] shows it is worth exploiting additional semantic information which are found in extended folksonomies to improve ranking strategies. Approaches, for example GFolkRank [1], are introduced which extend FolkRank to a context-sensitive ranking algorithm exploiting the additional semantic information gained from the grouping of resources in GroupMe!2. Groups in GroupMe! allow resources e.g. belonging to a common topic to be semantically grouped together. Groups can also contain other groups [2]. GFolkRank, an extension of FolkRank [12] is a ranking algorithm that leverages groups available in GroupMe! for ranking. Groups are interpreted as tags i.e. if a user adds a resource r to a group g then GFolkRank

---

2 http://groupme.org/, retrieved 06/07/2012

translates this as a tag (group) assignment. The folksonomy graph is therefore extended with additional group nodes and group assignments. In addition, other approaches are proposed such as GRank [1]. GRank is designed for ranking resources with a tag as input. It computes a ranking for all resources, which are related to the input tag with respect to the group structure in GroupMe!

The concept of groups in the GroupMe! application is similar to the concept of activities in the CROKODIL application. Therefore, this opportunity to exploit the semantic information gained from activities in CROKODIL will be investigated in the following sections.

4 Concept and Implementation

Given a certain user u as input, the resource recommendation task is to find a resource r which is relevant to this user. This recommendation task is also seen as a ranking task. A ranking algorithm computes for an input user u a score vector that contains the score values $score(r)$ for each resource r in the graph. These scored resources are then ordered forming a ranked list according to their score values with the highest scored resource at the top of the list. The top ranked resources are then recommended to the user u. For example, the scores $score(r_1) = 5$ and $score(r_2) = 7$ and $score(r_3) = 3$ create a ranked list: r_2, r_1 and r_3. Therefore the top recommendation to user u will be resource r_2.

We propose two ranking algorithms, AScore and AInheritscore. Both algorithms compute a folksonomy graph G_F considering not only activities when ranking resources but also including activity hierarchies and users assigned to work on these activities in the graph structure.

In the following, three sets are defined that will be used in Definition 3 to determine the weights of the edges in the folksonomy graph G_F. For a given user $u \in U$, tag $t \in T$ and resource $r \in R$:

- Let $U_{t,r} = \{ u \in U \mid (u,t,r) \in Y \} \subseteq U$ be the set of all users that have assigned resource r a tag t
- Let $T_{u,r} = \{ t \in T \mid (u,t,r) \in Y \} \subseteq T$ be the set of all tags that user u assigned to resource r
- Let $R_{u,t} = \{ r \in R \mid (u,t,r) \in Y \} \subseteq R$ be the set of all resources that user u assigned a tag t

Definition 3 (Folksonomy Graph). *Given a folksonomy F, the folksonomy graph G_F [1] is defined as an undirected, weighted graph $G_F := (V_F, E_F)$ where:*

- *$V_F = U \cup T \cup R$ is the set of nodes*
- *$E_F = \{ \{u,t\}, \{t,r\}, \{u,r\} \mid u \in U, t \in T, r \in R, (u,t,r) \in Y \} \subseteq V_F \times V_F$ is the set of undirected edges*
- *Each of these edges is given a weight $w(e), e \in E_F$ according to their frequency within the set of tag assignments:*
  - *$w(u,t) = |R_{u,t}|$ the number of resources that user u assigned the tag t*
  - *$w(t,r) = |U_{t,r}|$ the number of users who assigned tag t to resource r*
  - *$w(u,r) = |T_{u,r}|$ the number of tags that user u assigned to resource r*

4.1   AScore

AScore is an algorithm based on GFolkRank [1] as described in Sect. 3. AScore extends the folksonomy graph G_F in a similar way with activity nodes and activity assignments. However, in addition, AScore extends the folksonomy graph with activity hierarchy relations between activities (4) as well as with users belonging to an activity (3). A user u is said to belong to an activity a, when the user u is working on the activity a. This is represented as an edge in the graph between u and a. Furthermore, AScore considers the hierarchical activity structure when determining the weights of the newly introduced edges. The AScore algorithm is described below:

- Let $G_C = (V_C, E_C)$ be the folksonomy graph of the extended folksonomy F_C
- $V_C = V_F \cup A$
- E_C is a combination of edges (1) from the folksonomy graph E_F with E_A (2), which are all activity assignments where a user u added a resource r to an activity a. Additionally, E_U (3) is added, which comprises all assignments of a user u to an activity a. Finally, the activity hierarchies E_H (4) are added as edges between a sub-activity a_{sub} and a super-activity a_{super}.

$$E_C = E_F \cup E_A \cup E_U \cup E_H . \tag{1}$$

$$E_A = \{\{u, a\}, \{a, r\}, \{u, r\} \mid u \in U, r \in R, a \in A, (u, a, r) \in Y_A\} . \tag{2}$$

$$E_U = \{\{u, a\} \mid u \in U, a \in A, (u, a) \in Y_U\} . \tag{3}$$

$$E_H = \{\{a_{sub}, a_{super}\} \mid a_{sub}, a_{super} \in A, a_{sub} < a_{super}\} . \tag{4}$$

The newly introduced edges are now given weights. The edges in E_A are given all the same weight $activityAssign(u, r, a)$ (5) because, similar to GFolkRank [1], a resource can only be added once to an activity. Attaching additional semantic information to a resource (like assigning it to a group in GroupMe! or to an activity in CROKODIL) is seen as more valuable than simply tagging it [1], therefore $activityAssign(u, r, a)$ is assigned the maximum number of users who assigned tag t to resource r (5). Similarly, the edges between a user u and an activity a are given the weight $w_{Membership}(u, a)$ (6) which is the maximum number of resources assigned with tag t by user u, who is working on activity a. The edges between activities of the same hierarchy are given the weight $w_{Hierarchy}(a_{sub}, a_{super})$. These edges are seen to be at least as strong as the connections between an activity and other nodes in the graph, therefore in (7), the maximum weight is assigned.

$$w(u, a) = w(a, r) = w(u, r) = activityAssign(u, r, a)$$
$$\text{where } activityAssign(u, r, a) = max(\mid U_{t,r} \mid) . \tag{5}$$

$$w_{Membership}(u, a) = max(\mid R_{u,t} \mid) . \tag{6}$$

$$w_{Hierarchy}(a_{sub}, a_{super}) = max(activityAssign(u, r, a_{sub}), w_{Membership}(u, a_{sub})) \ . \tag{7}$$

After the folksonomy graph G_C has been created and the weights of the edges determined, any graph-based ranking algorithm for folksonomies e.g. FolkRank can now be applied to calculate the scores of each node.

4.2   AInheritScore

AInheritScore is an algorithm based on GRank [1] as described in Sect. 3. AInheritscore computes for an input user u a score vector that contains the score values $score(r)$ for each resource r. The input user u however needs to be transformed into input tags t_q, depending upon how many tags the user u has. Each of these input tags t_q is weighted according to its frequency of usage by user u. The parameters d_a, d_b, d_c are defined to emphasize the "inherited" scores gained by relations in the hierarchy. The values of these parameters are set in Sect. 5 for the evaluations.

1. d_a for resources having the input tag directly assigned to them
2. d_b for resources in the activity hierarchy having a resource that is tagged with the input tag
3. d_c for users in the activity hierarchy having assigned the input tag

Additionally, an activity distance $activityDist(a_1, a_2)$ between two activities is calculated as the number of hops from activity a_1 to activity a_2. However, it is also possible to calculate a lesser distance for sub-activities, or include the fan-out in the computation. AInheritscore contrasts to GRank in the following points:

1. Activities are not considered to be resources and cannot be assigned a tag.
2. AInheritscore considers activity hierarchies as well as users assigned to activities when computing the scores.
3. Activity hierarchies are leveraged by the inheritance of scores. These scores are emphasized by considering the connections in the activity hierarchy. The distance between activities in the hierarchy are considered as well.

AInheritscore algorithm is described in the following steps:

1. For each input tag t_q
2. Let $score = 0$ be the score vector
3. Determine $R_q = R_a \cup R_b \cup R_c$ where:
   (a) R_a contains all resources with the input tag t_q directly assigned to them $w(t_q, r) > 0$.
   (b) R_b contains all resources belonging to the same activity hierarchy as another resource r, that has the input tag t_q directly assigned to it: $w(t_q, r) > 0$
   (c) R_c contains all resources belonging to the same activity hierarchy as a user u, who has tagged a resource with the input tag tq: $w(u, t_q) > 0$

4. For all $r \in R_q$ belonging to activity a do
   (a) increase the score value of r:

$$score(r)+ = w(t_q, r) \cdot d_a \qquad (8)$$

   (b) for each $r' \in R_q$ belonging to activity a', where a' and a are in the same activity hierarchy, increase again the score of r:

$$score(r)+ = \frac{w(t_q, r')}{activityDist(a, a')} \cdot d_b \qquad (9)$$

   (c) for each $u \in U_q$ working on activity a', where a' and a are in the same activity hierarchy, increase again the score of r:

$$score(r)+ = \frac{w(u, t_q)}{activityDist(a, a')} \cdot d_c \qquad (10)$$

5. Output: *score*

5   Evaluation

The goal of this paper is to investigate how the implicit semantic information contained in activity hierarchies can be exploited to improve the ranking of resources in an extended folksonomy such as CROKODIL. As the CROKODIL data set has not yet attained a sufficient size for significant evaluation, a data set with an extended folksonomy containing similar concepts to those of activities in CROKODIL was sought.

5.1   Corpus

The GroupMe! data set was chosen as the concept of groups in GroupMe! is a similar concept to the activities and activity hierarchies in CROKODIL as mentioned in Sect. 3. There are however differences and a mapping of the concepts is necessary to be able to use the data set:

- The aim of groups in GroupMe! is to provide a collection of related resources. In CROKODIL however, activities are based on a pedagogical concept to help learners structure their learning goals in a hierarchical structure. Learning resources needed to achieve these goals are attached to these activities. Therefore, the assignment of a resource to a group in GroupMe! is interpreted as attaching a resource to an activity in CROKODIL.
- Groups in GroupMe! are considered resources and can therefore belong to other groups. These groups of groups or hierarchies of groups are interpreted as activity hierarchies in CROKODIL.
- Tags can be assigned to groups in GroupMe!. In contrast however, tags can not be assigned to activities in CROKODIL. These tags on groups in GroupMe! are therefore not considered in the data set.

Groups of groups or group hierarchies are unfortunately sparse in the GroupMe! data set. A p-core extraction [13] would reduce these hierarchies even more, therefore no p-core extraction is made. The data set has the characteristics described in Table 1.

Table 1. The extended folksonomy GroupMe! data set

Users	Tags	Resources	Groups	Posts	Tag Assignments
649	2580	1789	1143	1865	4366

5.2   Evaluation Methodology

The evaluation methodology **LeavePostOut** [13] is used for the evaluations of AScore and AInheritscore. In addition, we propose an evaluation methodology **LeaveRTOut** which is inspired from LeavePostOut. A post $P_{u,r}$ is defined in [11] as all tag assignments of a specific user u to a specific resource r. Leave-PostOut as shown in Fig.1 removes the post $P_{u,r}$, thereby ensuring that no information in the folksonomy remains that could connect the user u directly to resource r [13]. LeaveRTOut as shown in Fig. 2 eliminates the connection in the folksonomy between a tag t and a resource r instead of eliminating the connection between a user u and a resource r. LeaveRTOut therefore sets a different task to solve as LeavePostOut. For the evaluations, the user u of a post is used as input. LeavePostOut is used to determine adequate parameters for the algorithms. AInheritScore takes the values of GRank's parameters which according to a sensitivity analysis in [1] shall be set to $d_a = 10$, $d_b = 2$. d_c is set as well as $d_b = d_c = 2$.

For the evaluations, the metrics *Mean Average Precision (MAP)* and *Precision at k* [14] are used. MAP is used to determine the overall ranking quality while Precision at k determines the ranking quality in the top k positions. Precision at k is extended to *Mean Normalized Precision (MNP) at k* to obtain a

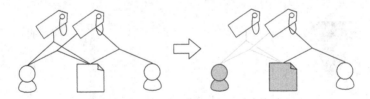

Fig. 1. LeavePostOut evaluation methodology

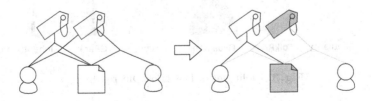

Fig. 2. LeaveRTOut evaluation methodology

single measure over a number of information needs Q as well as to be more suitable for the evaluation methodology, i.e. in respect to the maximal achievable $Precision_{max}(k)$. Mean Normalized Precision at k is defined as follows:

$$MNP(Q, k) = \frac{1}{|Q|} \cdot \sum_{j=1}^{|Q|} \frac{Precision(k)}{Precision_{max}(k)} \qquad (11)$$

For the statistical significant tests, *Average Precision* [14] is used for a single information need q, applying the *Wilcoxon signed-rank tests*[3].

5.3   Results

LeavePostOut and LeaveRTOut results from AScore and AInheritScore are compared to those of GRank, GFolkRank, FolkRank and Popularity. Popularity is calculated as the number of tags and users a resource is connected to. The results are visualized as a violin plot [10] in Fig.3 and Fig.4. The distribution of the data values are shown along the y-axis. The width of the violin plot is proportional to the estimated density at that point. As can be seen, most of the algorithms have most items ranked in positions < 500, whereas popularity still has too many items ranked in further positions.

The MAP results for LeaveRTOut are presented in Table 3. GFolkRank and AScore perform best with a MAP of 0.20, followed by FolkRank, GRank, AInheritScore and last Popularity. The results of the Mean Normalized Precision at k for $k \in [1, 10]$ for both LeavePostOut (left) and LeaveRTOut (right) are shown in Fig.5.

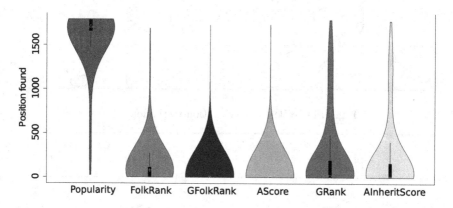

Fig. 3. Violinplot of LeavePostOut results

---

[3] http://stat.ethz.ch/R-manual/R-patched/library/stats/html/wilcox.test. html, retrieved 20/03/2012

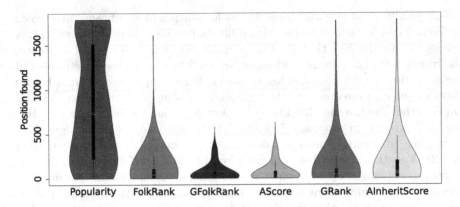

Fig. 4. Violinplot of LeaveRTOut results

Table 2. Mean Average Precision (MAP) results for LeavePostOut

Popularity	FolkRank	GFolkRank	AScore	GRank	AInheritscore
0,00	0,19	0,70	0,70	0,38	0,47

Table 3. Mean Average Precision (MAP) results for LeaveRTOut

Popularity	FolkRank	GFolkRank	AScore	GRank	AInheritscore
0.02	0.18	0.20	0.20	0.14	0.11

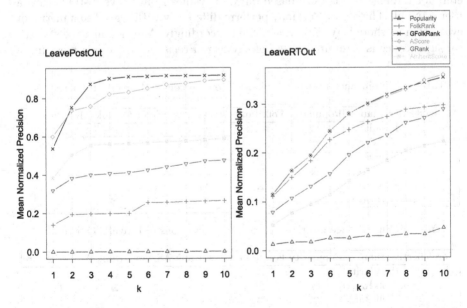

Fig. 5. Mean Normalized Precision at k: LeavePostOut (left) and LeaveRTOut(right)

The results of all pairwise comparisons for statistical significance are shown in Table 4 and Table 5. The LeavePostOut results differ from the LeaveRTOut results due to the fact that they set a differently hard task to solve. Hence, the results from the two methodologies are useful to assess the effectiveness of the algorithms in different ranking scenarios. For example, results from Leave-PostOut show on the one hand, that GFolkRank is more effective than AScore. On the other hand, results from LeaveRTOut show that AScore is more effective than GFolkRank. In summary, LeavePostOut results show that the algorithms leveraging additional semantic information are overall more effective than FolkRank as these algorithms designed for the extended folksonomy have the advantage of being able to leverage the additional information gained from activities to recommend relevant resources. The selection of an algorithm for ranking learning resources will therefore depend upon its application scenario and what is important for ranking. For example, AScore would be the choice when activity hierarchies are particularly important for ranking learning resources such as in the CROKODIL application scenario or GFolkRank if this is not the case in other scenarios.

Limitations. The proposed algorithms AScore and AIhneritscore are fundamentally based on the concept of activity hierarchies from the CROKODIL application scenario. The results achieved with the GroupMe! data set thus may not be representative as the group hierarchies from the GroupMe! data set modeled as the CROKODIL activity hierarchies were very sparse. Furthermore, the parameters for the algorithms were based on MAP values from LeavePostOut with a user as input. The algorithms may perform differently with regard to a metric or evaluation methodology, if parameterized accordingly. Additionally, the statistical significance is computed based on Average Precision, which is a measure of

Table 4. Significance matrix of pair-wise comparisons of LeavePostOut results

More effective than →	Popularity	FolkRank	GFolkRank	AScore	GRank	AInheritScore
Popularity	□	□	□	□	□	□
FolkRank	⊠	□	□	□	□	□
GFolkRank	⊠	⊠	□	⊠	⊠	⊠
AScore	⊠	⊠	□	□	⊠	⊠
GRank	⊠	⊠	□	□	□	□
AInheritScore	⊠	⊠	□	□	⊠	□

Table 5. Significance matrix of pair-wise comparisons of LeaveRTOut results

More effective than →	Popularity	FolkRank	GFolkRank	AScore	GRank	AInheritScore
Popularity	□	□	□	□	□	□
FolkRank	⊠	□	□	□	⊠	⊠
GFolkRank	⊠	⊠	□	□	⊠	⊠
AScore	⊠	⊠	⊠	□	⊠	⊠
GRank	⊠	□	□	□	□	⊠
AInheritScore	⊠	□	□	□	□	□

the overall ranking quality. If the statistical significance is to be compared based on the effectiveness of ranking in top positions, a different series of significance tests needs to be conducted.

6 Conclusion

Resource-based learning is mostly self-directed and the learner is often confronted with an overhead of finding relevant high quality learning resources on the Web. Graph-based recommender systems that recommend resources beyond their similarity can reduce the effort of finding relevant learning resources. We therefore propose in this paper two approaches AScore and AInheritScore that exploit the hierarchical activity structures in CROKODIL to improve the ranking of resources in an extended folksonomy for the purpose of recommending learning resources. Evaluation results show that this additional semantic information is beneficial for recommending learning resources in an application scenario such as CROKODIL. The algorithms leveraging additional semantic information are overall more effective than FolkRank as these algorithms designed for the extended folksonomy have the advantage of being able to leverage the additional information gained from activities and activity hierarchies to recommend relevant resources.

Future work will be to evaluate these algorithms with a data set from the CROKODIL application scenario. Additionally, a user study in the CROKODIL application scenario is planned to determine the true relevance of recommendations of learning resources based on human judgement in a live evaluation.

Acknowledgments. This work is supported by funds from the German Federal Ministry of Education and Research and the mark 01 PF 08015 A and from the European Social Fund of the European Union (ESF). The responsibility for the contents of this publication lies with the authors.

Special thanks to Fabian Abel for making the most current data set from GroupMe! available for the evaluations.

References

1. Abel, F.: Contextualization, User Modeling and Personalization in the Social Web. PhD Thesis, Gottfried Wilhelm Leibniz Universitst Hannover (2011)
2. Abel, F., Frank, M., Henze, N., Krause, D., Plappert, D., Siehndel, P.: GroupMe! - Where Semantic Web Meets Web 2.0. In: Aberer, K., Choi, K.-S., Noy, N., Allemang, D., Lee, K.-I., Nixon, L.J.B., Golbeck, J., Mika, P., Maynard, D., Mizoguchi, R., Schreiber, G., Cudré-Mauroux, P. (eds.) ASWC 2007 and ISWC 2007. LNCS, vol. 4825, pp. 871–878. Springer, Heidelberg (2007)
3. Anjorin, M., Rensing, C., Bischoff, K., Bogner, C., Lehmann, L., Reger, A., Faltin, N., Steinacker, A., Lüdemann, A., Domínguez García, R.: CROKODIL - A Platform for Collaborative Resource-Based Learning. In: Kloos, C.D., Gillet, D., Crespo García, R.M., Wild, F., Wolpers, M. (eds.) EC-TEL 2011. LNCS, vol. 6964, pp. 29–42. Springer, Heidelberg (2011)

4. Anjorin, M., Rensing, C., Steinmetz, R.: Towards ranking in folksonomies for personalized recommender systems in e-learning. In: Proc. of the 2nd Workshop on Semantic Personalized Information Management: Retrieval and Recommendation. CEUR-WS, vol. 781, pp. 22–25 (October 2011)
5. Böhnstedt, D., Scholl, P., Rensing, C., Steinmetz, R.: Collaborative Semantic Tagging of Web Resources on the Basis of Individual Knowledge Networks. In: Houben, G.-J., McCalla, G., Pianesi, F., Zancanaro, M. (eds.) UMAP 2009. LNCS, vol. 5535, pp. 379–384. Springer, Heidelberg (2009)
6. Cantador, I., Konstas, I., Jose, J.: Categorising Social Tags to Improve Folksonomy-Based Recommendations. Web Semantics: Science, Services and Agents on the World Wide Web 9, 1–15 (2011)
7. Desrosiers, C., Karypis, G.: A Comprehensive Survey of Neighborhood-Based Recommendation Methods. In: Ricci, F., Rokach, L., Shapira, B., Kantor, P. (eds.) Recommender Systems Handbook, pp. 107–144. Springer (2011)
8. Drachsler, H., Pecceu, D., Arts, T., Hutten, E., Rutledge, L., van Rosmalen, P., Hummel, H.G.K., Koper, R.: ReMashed – Recommendations for Mash-Up Personal Learning Environments. In: Cress, U., Dimitrova, V., Specht, M. (eds.) EC-TEL 2009. LNCS, vol. 5794, pp. 788–793. Springer, Heidelberg (2009)
9. Hannafin, M., Hill, J.: Resource-Based Learning. In: Handbook of Research on Educational Communications and Technology. pp. 525–536 (2008)
10. Hintze, J., Nelson, R.: Violin plots: A box plot-density trace synergism. The American Statistician 52(2), 181–184 (1998)
11. Hotho, A., Jäschke, R., Schmitz, C., Stumme, G.: BibSonomy: A Social Bookmark and Publication Sharing System. In: Proc. of the Conceptual Structures Tool Interoperability Workshop (2006)
12. Hotho, A., Jäschke, R., Schmitz, C., Stumme, G.: Information Retrieval in Folksonomies: Search and Ranking. In: Sure, Y., Domingue, J. (eds.) ESWC 2006. LNCS, vol. 4011, pp. 411–426. Springer, Heidelberg (2006)
13. Jäschke, R., Marinho, L., Hotho, A., Schmidt-Thieme, L., Stumme, G.: Tag Recommendations in Folksonomies. In: Kok, J.N., Koronacki, J., Lopez de Mantaras, R., Matwin, S., Mladenič, D., Skowron, A. (eds.) PKDD 2007. LNCS (LNAI), vol. 4702, pp. 506–514. Springer, Heidelberg (2007)
14. Manning, C., Raghavan, P., Schÿtze, H.: Introduction to Information Retrieval. Cambridge University Press (2008)
15. Manouselis, N., Drachsler, H., Vuorikari, R., Hummel, H., Koper, R.: Recommender Systems in Technology Enhanced Learning. In: Ricci, F., Rokach, L., Shapira, B., Kantor, P. (eds.) Recommender Systems Handbook, pp. 387–415. Springer (2011)
16. Peters, I.: Folksonomies: Indexing and Retrieval in Web 2.0. De Gruyter Saur (2010)
17. Rakes, G.: Using the Internet as a Tool in a Resource-Based Learning Environment. Educational Technology 36, 52–56 (1996)
18. Romero Zaldivar, V.A., Crespo García, R.M., Burgos, D., Kloos, C.D., Pardo, A.: Automatic Discovery of Complementary Learning Resources. In: Kloos, C.D., Gillet, D., Crespo García, R.M., Wild, F., Wolpers, M. (eds.) EC-TEL 2011. LNCS, vol. 6964, pp. 327–340. Springer, Heidelberg (2011)

An Initial Evaluation of Metacognitive Scaffolding for Experiential Training Simulators

Marcel Berthold[1], Adam Moore[2], Christina M. Steiner[1], Conor Gaffney[3],
Declan Dagger[3], Dietrich Albert[1], Fionn Kelly[4], Gary Donohoe[4],
Gordon Power[3], and Owen Conlan[2]

[1] Knowledge Management Institute, Graz University of Technology,
Rechbauerstr. 12, 8010 Graz, Austria
{marcel.berthold,christina.steiner,dietrich.albert}@tugraz.at
[2] Knowledge, Data & Engineering Group, School of Computer Science and Statistics,
Trinity College, Dublin, Ireland
{mooread,oconlan}@scss.tcd.ie
[3] EmpowerTheUser, Trinity Technology & Enterprise Campus,
The Tower, Pearse Street, Dublin, Ireland
{conor.gaffney,declan.dagger,gordon.power}@empowertheuser.com
[4] Department of Psychiatry, School of Medicine, Trinity College, Dublin, Ireland
fkelly@stpatsmail.com, DONOGHUG@tcd.ie

Abstract. This paper elaborates on the evaluation of a Metacognitive Scaffolding Service (MSS), which has been integrated into an already existing and mature medical training simulator. The MSS is envisioned to facilitate self-regulated learning (SRL) through thinking prompts and appropriate learning hints enhancing the use of metacognitive strategies. The MSS is developed in the European ImREAL (Immersive Reflective Experience-based Adaptive Learning) project that aims to augment simulated learning environments throughout services that are decoupled from the simulation itself. Results comparing a baseline evaluation of the 'pure' simulator (N=131) and a first user trial including the MSS (N=143) are presented. The findings indicate a positive effect on learning motivation and perceived performance with consistently good usability. The MSS and simulator are perceived as an entity by medical students involved in the study. Further steps of development are discussed and outlined.

Keywords: self-regulated learning, metacognitive scaffolding, training simulator, augmentation.

1 Introduction

Self-regulated learning (SRL) and especially metacognition, is currently a prominent topic in technology-enhanced learning (TEL) research. Many studies provide evidence of the effectiveness of SRL in combination with metacognitive scaffolding (cf. [1, 2]). Self-regulated learning refers to learning experiences that are directed by the learner and describes the ways in which individuals regulate their own cognitive and metacognitive processes in educational settings (e.g. [3, 4]). An important aspect

A. Ravenscroft et al. (Eds.): EC-TEL 2012, LNCS 7563, pp. 23–36, 2012.
© Springer-Verlag Berlin Heidelberg 2012

of self-regulated learning is therefore the learners' use of different cognitive and metacognitive strategies, in order to control and direct their learning [5]. These strategies include cognitive learning strategies, self-regulatory strategies to control cognition (i.e. metacognitive strategies) and resource management strategies. Self-regulated learning also involves motivational processes and motivational beliefs [4]. It has been shown that good self-regulated learners perform better and are more motivated to learn [6] than weak self-regulated learners. TEL environments provide opportunities to support and facilitate metacognitive skills, but most learners need additional help and guidance [7] to perform well in such environments.

In the EU project, ImREAL[1] (Immersive Reflective Experience-based Adaptive Learning), intelligent services are being developed to augment and improve experiential simulated learning environments – including one to scaffold metacognitive processes. The development of the scaffolding service focuses on the salient and timely support of learners in their metacognitive processes and self-regulated learning in the context of a simulation environment. Herein we report a concrete study examining the medical training simulator provided by EmpowerTheUser[2] augmented with the ImREAL Metacognitive Scaffolding Service (MSS). The service will provide prompts and suggestions adapted to a learner's needs and traits of metacognition and aiming at enhancing motivation towards the learning activity in the simulation. While the aspect of supporting metacognition needs to be integrated in the learning process, the according service will be technically decoupled from the specific learning system itself. Overall, the research presented investigates the effectiveness and appropriateness of the service and the scaffolding it provides. To allow a more detailed examination of the issues, we address four sub questions:

1. **Is self-regulated learning supported?** For the evaluation and analysis of self-regulated learning we distinguish between the *general learning approach* (i.e. application of cognitive, metacognitive strategies), and the *metacognitive and specific learning processes in the simulation* (i.e. cognitive, metacognitive strategies or actions within simulator context); thereby, learning and metacognitive scaffolding in the simulation may optimally, and on a long-term basis, influence the general learning approach of a learner. That means learning in the simulation in combination with metacognitive scaffolding may optimally influence the general learning approach. If this approach is successful it may have also an influence on SRL on a long-term basis.

2. **Does the simulator augmentation through the service lead to better *learning performance*?** The learning performance refers to the *(objective or subjective/perceived) learners' knowledge/competence acquisition and performance* in the learning situation and to the *transfer* of acquired *knowledge* to other situations.

3. **Does the simulator augmentation through the service increase *motivation*?** The aspect of motivation addresses the motivation to learn, i.e. the structures and processes explaining learning actions and the effects of learning [8].

---

[1] http://www.imreal-project.eu
[2] http://www.empowertheuser.ie

4. **Is the service *well integrated* in the simulation and learning experience?** This refers to the question whether the scaffolding interventions provided during the simulation via the MSS are perceived by learners as appropriate and useful – in terms of their content, context and timing.

In order to answer these evaluation questions the paper is organized in the following structure. Section 2 gives an overview of the MSS and outlines related work. Section 3 presents the simulator and its normal usage. Section 3 gives an overview of the MSS and outlines related work. In section 4 the experimental design of the study is introduced and section 5 includes the according results. These results are discussed in section 5. Section 6 provides a conclusion and an outlook to further research.

2   Metacognitive Scaffolding – Background and Technology

Scaffolding is an important part of the educational process, supporting learners in their acquisition of knowledge and developing their learning skills. Scaffolding has been a major topic of research since the pioneering work of Vygotsky (e.g. 1978 [9]) and the key work of Bruner, Wood and colleagues (cf. [10]). Bruner [11] identified several aspects which should be considered when providing feedback to students such as form and timing.

Work on the use of scaffolding with the help of computer-based learning environments has been extensive (cf. [12]). Originally, the emphasis was on cognitive scaffolding which has many forms (cf. [13]). In the last ten years there has been a move towards research in metacognitive scaffolding (e.g. [14–17]) as well as in the use of metacognitive scaffolding in adaptive learning environments (e.g. [18–21]).

Other forms of scaffolding have also been explored both in educational and technology enhanced learning contexts – such as affective scaffolding and conative scaffolding. Van de Pol et al. [14] sought to develop a framework for the analysis of different forms of scaffolding. In the technology enhanced learning community, Porayska-Pomsta and Pain [22] explored affective and cognitive scaffolding through a form of face theory (the affective scaffolding also included an element of motivational scaffolding). Aist et al. [23] examined the notion of emotional scaffolding and found different kinds of emotional scaffolding had an effect on children's persistence using a reading tutoring system.

There are different forms of metacognitive scaffolding. Molenaar et al. [2] investigated the distinction between structuring and problematizing forms of metacognitive scaffolding and found that problematizing scaffolding seemed to have a significant effect on learning the required content. They used Orientation, Planning, Monitoring, Evaluation and Reflection as subcategories of metacognitive scaffolding.

Sharma and Hannafin [24] reviewed the area of scaffolding in terms of the implications for technology enhanced learning systems. They point out the need to balance metacognitive and "procedural" scaffolds since only receiving one kind can lead to difficulties – with only procedural scaffolding students take a piecemeal approach, and with only metacognitive scaffolding students tend to fail to complete their work. They also argue for systems that are sensitive to the needs of individuals.

Boyer et al. [25] examined the balance between motivational and cognitive scaffolding through tutorial dialogue and found evidence that cognitive scaffolding supported learning gains while motivational scaffolding supported increase in self-efficacy.

The aim of the ImREAL project is to bring simulators closer to the 'real world'. As part of training for a diagnostic interview, in the 'Real World' a mentor sits at back observing and providing occasional input / interventions as necessary. The MSS has been developed to integrate into the simulator learning experience as an analogue of a mentor, sitting alongside the simulator to provide scaffolding. The ETU simulator supports meta-comprehension and open reflection via note taking.

For this trial metacognitive scaffolding was provided using calls to a RESTful [26] service developed as part of the ImREAL project. The service utilises technology initially developed for the ETTHOS model [27] and presents Items from the Metacognitive Awareness Inventory [28] according to an underlying cognitive activity model, matched to Factors in the MAI. In this way the importance of the tasks being undertaken by the learner is clear scaffolding is developed in order to match a learners' cognitive activity to metacognitive support.

The scaffolding service supplements the pre-existing ETU note-taking tool, both of which are illustrated in Figure 1 below. The text of the thinking prompt item is phrased in order to elicit a yes/no response. If additional context / rephrasing has been added by the instructional experts that is displayed before the open text response area. A link that activates an explanatory text occurs underneath the text input area, as well as a "Like button" which can be selected and the submit action.

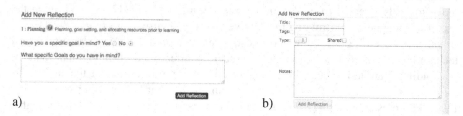

Fig. 1. a) MSS Interface b) ETU Note-taking tool

3   Overview of Simulator and Normal Usage

For this research the ETU Talent Development Platform was used, with training for medical interview situations. The user plays the role of a clinical therapist and selects interview questions from a variety of possible options to ask the patient. When a question is selected a video is presented that shows the verbal interaction of the therapist with the patient (close up of the patient, voice of the therapist) and the verbal and non-verbal reaction of the patient (close up of the patient). Starting the simulation, users can choose between two types of scenarios (Depression and Mania), which offer the same types of subcategories: Introduction and negotiating the agenda, eliciting information, outlining a management plan and closing the interview.

After a scenario is chosen, the user may simulate the interview as long as they prefer or until the interview is "naturally" finished. Furthermore, the users could have as many runs of the simulation as they want and could choose a different scenario in the following attempts. When going through the simulator the student obtain scores. The simulator performance scores are a measure of the students' potential to perform effectively in a real interview. In this study we focused only on the Depression interview scenario. A screenshot of a typical interaction within the ETU system is show below in Figure 2.

Fig. 2. Screen shot of the EmpowerTheUser Simulator. The scenario of diagnosing a patient with clinical depression is just beginning.

4   Experimental Design

4.1   Cohort

143 medical students participated in the study and performed the simulation as part of their second year (2011/2012) medical training at Trinity College, Dublin (TCD). They were on average approximately 22 years old (40% male vs. 60% female, 80% Irish). In addition, these results are compared, as far as they have been assessed at both time points, to a baseline evaluation based on using the simulator without ImREAL services. In the baseline evaluation, 131 TCD medical students from the previous year group (2010/2011) participated (cross-section design).

4.2   Measurement Instrument

ETU Simulator. Within the simulation learning performance is assessed by tracking scores for each of the 4 subsections, as well as dialogue scores and notes are recorded that were written in a note pad for reflections.

Questionnaire on Self-Regulated Learning. Self-regulated learning skills were measured by the *Questionnaire for Self-Regulated Learning* (*QSRL*; [29]). The QSRL consists of 54 items, which belong to six main scales (Memorizing / Elaboration / Organization / Planning / Self-monitoring / Time management) and three subscales (Achievement Motivation / Internal attribution / Effort). In the online version of the questionnaire, respondents indicate their agreement to an item by moving a slider on an analogue scale between the positions "strongly disagree" to "strongly agree". The possible score range is 0-100 in each case, with higher values indicating a better result.

Survey Questions on Use of ETU, Experience in Performing Clinical Interviews and Relevance of the ETU Simulator. In order to control possible influences of prior experience with the respective simulated learning environment or real world medical interviews, their experiences were assessed through three survey questions. The following survey question assessed the relevance of the simulator with answer options ranging from "not relevant at all" to "very relevant".

Thinking Prompts. Triggers that made calls to the MSS were inserted into the practice phase of the simulator but not available during 'live/scored usage'. The triggers were created using the ETU authoring platform and made a call to the MSS requesting a prompt of a particular Factor (Planning, Information Management, Comprehension, Debugging or Evaluation). As explained above, each Factor consisted of a number of Items or *Thinking Prompts*. An item was not redisplayed once a reflection had been entered with it.

Motivation. Motivation was assessed with four survey questions referring to learning more about clinical interviews, improve own interview skills, performing a good interview during the simulation and applying what has been learned in a real interview.

Workload. Measures of workload were assessed by six subscales of the NASA-TX [30, 31] with a score range of 0-100. In this case higher values indicate a higher workload. An overall workload score was calculated based on the subscales by computing a mean of all item contributions. In contrast to the original NASA-TLX, students did not mark their answers to an analogue scale, but entered digits between 0-100 into a text field.

Usability and Service Specific Integration. The Short Usability Scale (SUS, [32]) consists of ten items with answer options of a five-point-Likert-scale ranging from "strongly disagree" to "strongly agree". The raw data were computed to an overall SUS score. The overall SUS score ranged from 0-100 with higher values indicating higher usability. Additionally to the SUS questions, three service specific usability questions were administered regarding the relation of the prompts to the rest of the simulation and obvious differences. The answer options were the same as for the SUS.

Learning Experience with MSS. Learning experience with MSS was measured by 10 questions referring to helpfulness and appropriateness of the MSS thinking prompts within the simulator with answer options on a 5-point-Likert-scale ranging from "not at all" to "very much". In addition, a free text comment field was provided.

Procedure. The baseline evaluation, using the pure simulator, was conducted in mid-February and beginning of March of 2011; the first user trial was carried out in Dublin from mid-February until the beginning of March of 2012. The TCD medical students used the ETU medical training simulator.

Data collection was carried out during the simulation (e.g. ETU scores, MSS data) and after learning with the ETU simulator (questionnaire data). At first the students worked on the simulation for as long as they wanted and could choose between two scenarios: Mania or Depression. After they were finished they were directed to the online questionnaires. In this stage they filled in the survey questions on relevance and on motivation, NASA-TLX, SUS, questions on prompts, learning experience and the QSRL.

After working on the simulation in the TCD course students still had access to the ETU simulator via the internet for approximately two weeks. It was not mandatory to use the simulation in the medical course at TCD or to participate in the evaluation.

5   Experimental Results

PAWS Statistics, version 18.0 [33] and Microsoft Excel (2010) were used for statistical analyses and graphical presentations. If not explicitly mentioned, statistical requirements for inference statistical analyses and procedures were fulfilled. For all analyses the alpha level was $\alpha=.05$. Due to an unbalanced number of participants in the samples in regard to comparisons of the first user trial and baseline evaluation appropriate pre-tests have been performed and the according values are presented.

This section focuses mainly on the first user trial evaluation based on using the ETU simulator with the integrated MSS ImREAL services.

5.1   Log-Data

ETU Simulator – Descriptive Data. All students of the first user trial reported that they have never used the ETU medical training simulator before. Nonetheless, they were quite experienced in conducting clinical interviews, since 97% reported to have already performed at least one, but only 15 % had experienced interviewing a psychiatric patient.

A comparison of the first user trial and the baseline evaluation showed that duration time in minutes ($M_{base}=17.89$, $SD_{base}=11.15$; $M_{1UT}=15.45$, $SD_{1UT}=6.81$) and scored points ($M_{base}=31.34$, $SD_{base}=6.33$; $M_{1UT}=27.61$, $SD_{1UT}=5.91$) in the simulation decreased from baseline evaluation to the first user trial (duration time: $t_{211,49}=2.17$, $p=.031$; score: $t_{272}=5.10$, $p<.001$). These results show that students spent on average less time in the simulator and reached lower scores. This is rather surprising, because, students of the baseline cohort and first user trial cohort were similarly experienced whereas the participants of the first user trial worked with the additional MSS. In this case longer duration time was expected for the cohort of the first user trial.

Metacognitive Scaffolding Service (MSS) Comments. 10 comments have been collected by MSS learning experience questionnaire free text comment field. The participants provided interesting comments, which however referred more to the simulator than to the MSS. This implies that the MSS seems to be perceived as well integrated in the simulation, because students do not seem to differentiate between the additional service and the simulation itself. The participants pointed to sometimes inappropriate prompts in combination with the simulator in situations, especially, when only one answering option was available in the dialogue with the patient and they were asked to think about their strategy. Nevertheless, one learner recorded that *"I am learning a lot actually, it is amazing how much you can miss just by asking a question in a slightly different way! I keep going back a step and looked through the other options to see where the scenario goes. Usually I've picked the most suitable one, but not always. Sometimes I am surprised about how much I would have missed!!"*.

Prompts Analysis. Five different types of prompts were presented according to the five MAI phases described in section 3. In total 2001 prompts (Planning: 469, Information Management: 752, Monitoring: 425, Debugging: 301 and Reflection: 54) were shown to 50 students. The other students ignored the up-popping prompts. Every student who used the practice facility in the simulator was presented with a pop-up suggesting they reflected. Clicking on that pop-up would move the simulation to the MSS screen (Figure 2a). The relative frequency of the prompts was compared to the expected frequency based on the probability of available prompts for each phase. The results indicate that the learners were scaffolded more often in the second phase "Information Management" and were less scaffolded in the reflection phase as could have been expected ($\chi^2_{(4,0.95)}$= 314.55, p<.001, Figure 3). On the one hand, learners seem to need more assistance in effectively processing information by hints to use more organizational, elaborative, summarizing or selective learning strategies. On the other hand they are rather confident in the reflection phase and wave the offer of scaffolds.

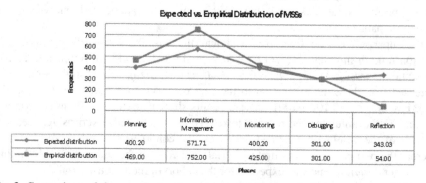

Expected vs. Empirical Distribution of MSSs

Phases	Planning	Information Management	Monitoring	Debugging	Reflection
Expected distribution	400.20	571.71	400.20	301.00	343.03
Empirical distribution	469.00	752.00	425.00	301.00	54.00

Fig. 3. Comparison of the expected and empirical distribution of metacognitive scaffolds for the five phases of Schraw's Metacognitive Awareness Inventory

5.2   Questionnaire Data

QSRL. The quantitative results of the MSS are a little surprising, because students estimated the use of cognitive learning strategies, especially elaboration strategies, relatively high. In general, all SRL scores are located above the center point of the score range and indicate positive results for all cognitive and metacognitive strategies. However, a stronger use of elaboration strategies is reported (t_{20} = 3.34, $p=.003$) in the first user trial. It needs to be explicitly stated that this is not an unfavorable result as such as elaboration strategies are strategies of deeper learning [34], which should be further supported by scaffolding services.

A comparison to the baseline study shows no significant increase in any of the usage of reported learning strategies.

Motivation. 38 participants filled in the motivation questions. The results show that the scores were on a high motivation level around 3.16-3.49 on a 4-point-Likert rating scale. This implies that the students were very motivated to learn about the clinical interview during the simulation, to improve their interview skills, perform a good interview during the simulation and to apply what they have just learnt in the simulation in a real world clinical interview context. Furthermore, a comparison of the overall motivation scores assessed immediately after the simulation of the first user trail and immediately after the baseline evaluation reveals significant higher motivation scores for the MSS trial ($M=3.35$, $SD=.4.14$) compared to the baseline ($M=2.48$, $SD=.73$; $t_{118.47}=-8.64$, $p<.001$).

Workload. A moderate overall workload could be observed. It has to be noted at this stage that for a learning environment it should not be aimed at reducing the workload to a minimum; rather, the challenge should be at an appropriate, medium level of challenge – in an optimal case adapted to the individual learner. Participants reported the highest, but still moderately pronounced, load for effort. This subscale refers to the mental and physical resources that had to be mobilized to accomplish the task. Consequently, the result for effort can be relegated rather to mental than physical demand. Yet the simulation is a complex program that supports and requires active learning processes; a reduction of mental demand is somewhat challenging, but could possibly be realized by improving the MSS and addressing the challenge to reduce repetitions and provide only appropriate scaffolds. Furthermore, the second highest score was observed for performance (see Figure 4), showing that the students felt they successfully accomplished what they were supposed to do. Performance scores (referring to subjective/perceived learning outcome) even increased for the first user trial ($M=54.83$, $SD=17.90$) compared to the baseline evaluation ($M=43.00$, $SD=23.68$) significantly ($t_{109}=-2.63$, $p=.01$). A t-test for independent groups remains insignificant comparing overall workload scores for the first user trial ($M=44.19$, $SD=10.86$) and baseline evaluation ($M=44.81$, $SD=12.011$; $t_{75.52}=.27$, ns.).

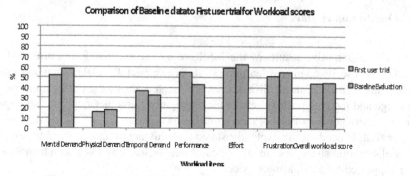

Fig. 4. Comparison of Baseline and first user trial data for workload

Usability and Service Specific Integration. No differences could be observed between the rather high usability overall scores for the first user trial ($M=62.50$, $SD=17.90$) and the baseline evaluation ($M=62.80$, $SD=16.08$; $F_{57.90}=.90$, ns.).

With respect to service specific integration with the ETU medical training and MSS prompts the majority of the students ratings were positive with 21 out of 33 stating they felt supported during their learning process by the MSS and that the service was well integrated in the system ($M=3.26$, $SD=.40$).

Learning Experience with MSS. The learning experience with the MSS was relatively positive. More than 63% of the participants perceived the MSS learning experience overall as very much helpful and appropriate. The high score for the individual items were all above the center point of the scale, which underlies this encouraging impression.

6   Discussion

In this paper we examined the effectiveness and appropriateness of the MSS. Results of the first user trial have been reported, involving the ETU medical training simulator augmented with the MSS. These results have been compared to a baseline evaluation where the 'pure' simulator was administered without any additional ImREAL services. Addressing the evaluation questions stated in the introduction section:

Is Self-Regulated Learning Supported? Even though self-regulated learning and metacognitive scaffolding are closely connected, because the SRL process heavily relies on applying cognitive and especially metacognitive learning strategies and techniques, no changes in SRL profile could be observed comparing the first user trial to the baseline data. This is because influencing self-regulated learning aspects is rather a long-term process [35]. This result might also be explained by having a look at the usage frequency of the simulator. The students were confronted with the simulation only in the TCD course and no one had used the simulation before.

Furthermore, duration time of working with the simulator, which was on average less than half an hour, might not be too short to change a rather stable learning approach.

For future studies the application of a longitudinal-evaluation-design could be suggested instead of a cross-section evaluation, to better meet the requirements of a longer-term process. In addition, teachers' or supervisors' judgments on SRL performance could be included to assess their observations on potential changes in learners' daily learning behavior. However, the last point might be difficult to realize in an university setting with more than 140 students in a course.

In general, all SRL self-reports were positive, indicating a higher use of elaboration strategies compared to memorizing strategies. Elaboration strategies represent strategies of deeper learning [34]. Nevertheless, fostering memorizing/rehearsal strategies might be taken up as an idea for improving the MSS. Assuming that the participants in the evaluation trials constitute a representative sample of ETU simulator users, ImREAL could start from this result and aim at improving users' rehearsal strategies through the provision of appropriate scaffolding. Of course, this strategy type should not be the only one to be supported. Rehearsal strategies help the learner to select and remember important information, but may not represent very deep levels of cognitive processing [34]. As a result, ImREAL services should especially try to further support elaboration as well as organizational strategies. In the ImREAL pedagogical framework learning [36] is seen as a cyclic process of three phases: forethought, learning and reflection. These individual phases are already represented in the ETU system, but not covered comprehensively. As described above, medical students do not tend to use the ETU simulator very often and if they do they undertake the interview scenario only for a short period of time. Therefore, reflection and coverage of the SRL phases should be further extended and supported by the ImREAL MSS.

Does the Simulator Augmentation through the Service Lead to Better Learning Performance? Results concerning the learning performance draw a clear picture. The actual objective data collected by the ETU simulator demonstrates that overall scores decreased from the baseline evaluation to the first user trial. Accordingly also self-reports on performance decreased. A decrease in self-report scores is expected, if actual performance is lower and may have been influenced by the MSS encouraging learners to think about their learning process and therefore make an accurate estimation of their performance. Accurate self-estimation might be seen as a factor to regulate the own learning approach. One reason why overall scores decreased could be the fact that the students spend less time working with the simulator. This is due to a change in the curriculum

Does the Simulator Augmentation through the Service Increase Motivation? Motivation scores increased from the baseline evaluation to the first user trial.

In addition to the consideration of motivation as a state characteristic, motivational beliefs (motivational traits in terms of being more stable and outlasting than state motivation) can be further influenced by positive sounding scaffolds and hints to optimize learning. If students see the prompts as support of their learning approach a positive attitude to the whole learning process can be expected and could explain the current result, because these motivational beliefs are factors influencing the initial motivation of the learner [37].

Is the Service Well Integrated in the Simulation and Learning Experience? Results on usability of the whole system (simulator + MSS) and service specific integration provide evidence that the MSS is well integrated in the simulation and leads to real augmentation. This is not only demonstrated by the positive scores on the service specific integration questions, but also by user comments, which were overall quite positive. Such positive results may be attributed to the MSS operating in an appropriately timely and salient manner, with the pop-up triggers appearing at apposite times created by the instructional design experts. Also the RESTful interface allows calls to be made to an ETU simulator-specific interface for the MSS, ensuring there are no obvious presentational and interactional differences between the hosting simulator and the MSS.

7   Conclusions and Outlook

The results above demonstrate a clear advantage in providing a MSS to augment an experiential training simulator, leading to more engaged, motivated learners without overly burdening them or interrupting the flow of their learning experience. With respect to the actual learning performance no positive effect could be identified. This would be desirable to investigate in more detail in future studies. These further studies should optimally be realized in a longitudinal-evaluation-design, as well as an assessment of real-world performance on medical interviews (i.e. learning transfer).

The collecting and monitoring of the development of motivation throughout both evaluation runs is important, because in the next version of the ImREAL MSS there will be a strong focus on extending it by 'affective scaffolding'. As a result, the data from the first user trial evaluation (with metacognitive scaffolding 'only') will serve as benchmark for a comparison with evaluation outcomes for the affective metacognitive scaffolding, thus allowing to investigate the additional benefit of the affective part.

The MSS will be integrated within additional experiential simulators to investigate the service's capabilities for generalization and integration within different systems and usage cases and to further evaluate its effect on learning experience.

Acknowledgement. The research leading to these results has received funding from the European Community's Seventh Framework Program (FP7/2007-2013) under grant agreement no 257831 (ImREAL project) and could not be realized without the close collaboration between all ImREAL partners.

References

1. Veenman, M.V.J.: Learning to Self-Monitor and Self-Regulate. Regulation (2008)
2. Molenaar, I., Van Boxtel, C.A.M., Sleegers, P.J.C.: The effects of scaffolding metacognitive activities in small groups. Computers in Human Behavior 26, 1727–1738 (2010)
3. Puustinen, M., Pulkkinen, L.: Models of Self-regulated Learning: A review. Scandinavian Journal of Educational Research 45, 269–286 (2001)
4. Zimmerman, B.: Becoming a Self-Regulated Learner: An Overview. Theory Into Practice 41, 64–70 (2002)
5. Pintrich, P.R.: The role of motivation in promoting and sustaining self-regulated learning. International Journal of Educational Research 31, 459–470 (1999)
6. Veenman, M.V.J.: Learning to Self-Monitor and Self-Regulate. In: Mayer, R., Alexander, P. (eds.) Handbook of Research on Learning and Instruction, pp. 197–218. Routledge, New York (2011)
7. Manuela, B.: Effects of Reflection Prompts when Learning with Hypermedia. Journal of Educational Computing Research 35, 359–375 (2006)
8. Krapp, A.: Die Psychologie der Lernmotivation. Perspektiven der Forschung und Probleme ihrer pädagogischen Rezeption [The psychology of motivation to learn. Perspectives of research and their pedagogical reception]. Zeitschrift für Pädagogik 39, 187–206 (1993)
9. Vygotskiĭ, L.S., Cole, M.: Mind in society: the development of higher psychological processes. Harvard University Press, Cambridge (1978)
10. Wood, D., Bruner, J.S., Ross, G.: The role of tutoring in problem solving. Journal of Child Psychology and Psychiatry 17, 89–100 (1976)
11. Bruner, J.S.: Toward a theory of instruction. Harvard University Press, Cambridge (1966)
12. Lajoie, S.P.: Extending the Scaffolding Metaphor. Instructional Science 33, 541–557 (2005)
13. Clark, A.: Towards a science of the bio-technological mind. International Journal of Cognition and Technology 1, 21–33 (2002)
14. Pol, J., Volman, M., Beishuizen, J.: Scaffolding in Teacher–Student Interaction: A Decade of Research. Educational Psychology Review 22, 271–296 (2010)
15. Dinsmore, D.L., Alexander, P.A., Loughlin, S.M.: Focusing the Conceptual Lens on Metacognition, Self-regulation, and Self-regulated Learning. Educational Psychology Review 20, 391–409 (2008)
16. Greene, J., Azevedo, R.: The Measurement of Learners' Self-Regulated Cognitive and Metacognitive Processes While Using Computer-Based Learning Environments. Educational Psychologist 45, 203–209 (2010)
17. Azevedo, R.: Does adaptive scaffolding facilitate students' ability to regulate their learning with hypermedia? Contemporary Educational Psychology 29, 344–370 (2004)
18. Azevedo, R., Hadwin, A.F.: Scaffolding Self-regulated Learning and Metacognition – Implications for the Design of Computer-based Scaffolds. Instructional Science 33, 367–379 (2005)
19. Luckin, R., Hammerton, L.: Getting to Know Me: Helping Learners Understand Their Own Learning Needs through Metacognitive Scaffolding. In: Cerri, S.A., Gouardéres, G., Paraguaçu, F. (eds.) ITS 2002. LNCS, vol. 2363, pp. 759–771. Springer, Heidelberg (2002)

20. Roll, I., Aleven, V., McLaren, B.M., Koedinger, K.R.: Designing for metacognition—applying cognitive tutor principles to the tutoring of help seeking. Metacognition and Learning 2, 125–140 (2007)

21. Roll, I., Aleven, V., McLaren, B.M., Koedinger, K.R.: Improving students' help-seeking skills using metacognitive feedback in an intelligent tutoring system. Learning and Instruction 21, 267–280 (2011)

22. Porayska-Pomsta, K., Pain, H.: Providing Cognitive and Affective Scaffolding Through Teaching Strategies: Applying Linguistic Politeness to the Educational Context. In: Lester, J.C., Vicari, R.M., Paraguaçu, F. (eds.) ITS 2004. LNCS, vol. 3220, pp. 77–86. Springer, Heidelberg (2004)

23. Aist, G., et al.: Experimentally augmenting an intelligent tutoring system with human-supplied capabilities: adding human-provided emotional scaffolding to an automated reading tutor that listens. In: Proceedings Fourth IEEE International Conference on Multimodal Interfaces, pp. 483–490 (2002)

24. Sharma, P., Hannafin, M.J.: Scaffolding in technology-enhanced learning environments. Interactive Learning Environments 15, 27–46 (2007)

25. Boyer, K.E., Phillips, R., Wallis, M., Vouk, M.A., Lester, J.C.: Balancing Cognitive and Motivational Scaffolding in Tutorial Dialogue. In: Woolf, B.P., Aïmeur, E., Nkambou, R., Lajoie, S. (eds.) ITS 2008. LNCS, vol. 5091, pp. 239–249. Springer, Heidelberg (2008)

26. Fielding, R.T.: Architectural Styles and the Design of Network-based Software Architectures, PhD-thesis, Citeseer (2000)

27. Macarthur, V., Moore, A., Mulwa, C., Conlan, O.: Towards a Cognitive Model to Support Self-Reflection: Emulating Traits and Tasks in Higher Order Schemata. In: EC-TEL 2011 Workshop on Augmenting the Learning Experience with Collaborative Reflection, Palermo, Sicily, Italy (2011)

28. Schraw, G., Sperling Dennison, R.: Assessing metacognitive awareness. Contemporary Educational Psychology 19, 460–475 (1994)

29. Fill Giordano, R., Litzenberger, M., Berthold, M.: On the Assessment of strategies in self-regulated learning (SRL)–differences in adolescents of different age group and school type. Poster. 9. Tagung der Österreichischen Gesellschaft für Psychologie, Salzburg (2010)

30. Hart, S.G., Staveland, L.E.: Development of NASA-TLX (Task Load Index): Results of Empirical and Theoretical Research. In: Hancock, N.M.P.A. (ed.) Human Mental Workload, pp. 239–250. North Holland Press, Amsterdam (1988)

31. Hart, S.G.: NASA-Task Load Index (NASA-TLX); 20 Years Later. Proceedings of the Human Factors and Ergonomics Society Annual Meeting 50, 904–908 (2006)

32. Brooke, J.: SUS: A "quick and dirty" usability scale. In: Jordan, P.W., Thomas, B., Weerdmeester, B., McClelland, A.L. (eds.) Usability Evaluation in Industry, pp. 189–194. Taylor & Francis, London (1996)

33. SPSS Inc., PASW Statistics 18 Core System User's Guide. SPSS Inc. (2007)

34. Weinstein, C.E., Mayer, R.E.: The teaching of learning strategies. In: Wittrock, M. (ed.) Handbook of Research on Teaching, pp. 315–327. Macmillan, New York (1986)

35. Pressley, M.: More about the development of self-regulation: Complex, long-term, and thoroughly social. Educational Psychologist 30, 207–212 (1995)

36. Hetzner, S., Steiner, C., Dimitrova, V., Brna, P., Conlan, O.: Adult Self-regulated Learning through Linking Experience in Simulated and Real World: A Holistic Approach. In: Kloos, C.D., Gillet, D., Garcia, R.M.C., Wild, F., Wolpers, M. (eds.) EC-TEL 2011. LNCS, vol. 6964, pp. 166–180. Springer, Heidelberg (2011)

37. Vollmeyer, R., Rheinberg, F.: Motivational effects on self-regulated learning with different tasks. Educational Psychology Review 18, 239–253 (2006)

Paper Interfaces for Learning Geometry

Quentin Bonnard, Himanshu Verma, Frédéric Kaplan, and Pierre Dillenbourg

CRAFT, École Polytechnique Fédérale de Lausanne,
RLC D1 740, Station 20, 1015 Lausanne, Switzerland
{quentin.bonnard,h.verma,frederic.kaplan,pierre.dillenbourg}@epfl.ch

Abstract. Paper interfaces offer tremendous possibilities for geometry education in primary schools. Existing computer interfaces designed to learn geometry do not consider the integration of conventional school tools, which form the part of the curriculum. Moreover, most of computer tools are designed specifically for individual learning, some propose group activities, but most disregard classroom-level learning, thus impeding their adoption. We present an augmented reality based tabletop system with interface elements made of paper that addresses these issues. It integrates conventional geometry tools seamlessly into the activity and it enables group and classroom-level learning. In order to evaluate our system, we conducted an exploratory user study based on three learning activities: classifying quadrilaterals, discovering the protractor and describing angles. We observed how paper interfaces can be easily adopted into the traditional classroom practices.

Keywords: Paper interfaces, Sheets, Cards, Geometry learning, Tabletop.

1 Introduction

Geometry education in primary schools is a domain ripe for exploiting the possibilities of computers, as they allow for an easy exploration of the problem space. However, there are some constraints which make it difficult to effectively utilize computers in a classroom scenario. Particularly, they do not cover the entire curriculum, which is based on pen and paper. For example, the only way for children to learn how to draw an arc is by using a physical compass.

Paper interfaces can prove to be an effective solution to this dilemma, as paper is already situated and integrated in the classroom environment and its practices. In addition, paper is cheap to produce, yet persistent and malleable to adapt to the dynamics of the classroom. As a computer interface it can transform into a dynamic display capable of computing and processing data. Besides these benefits of paper interfaces, paper has different properties and affordances depending upon its material, shape and size. Also, many interface metaphors such as cut-copy-paste, files and folders, check-boxes etc. are actually inspired by practices involving paper. Effective identification of these properties followed by a proper utilization, might render the paper interface intuitive for the users to interact. We hypothesize that geometry education in primary schools can greatly

A. Ravenscroft et al. (Eds.): EC-TEL 2012, LNCS 7563, pp. 37–50, 2012.

benefit from the use of paper interfaces and their characteristics. For example, folding is a natural embodiment of axial symmetries, cutting can be the physical counterpart of recomposing figures in order to compute their areas.

In this article, we present an augmented reality based tabletop system to facilitate geometry learning for primary school children. Our system incorporates a camera-projector device which is capable of projecting content over sheets and cards placed on a table top. We also present three exploratory user studies to study the influence and feasibility of using paper interfaces in primary schools. We report on the observations related to these user studies with three different geometry learning activities concerning shapes and angles.

2 Related Work

The domain of paper interfaces is broad and not very well defined, just like the paper is used as an umbrella term for a variety of artefacts and practices. The archetype of using paper consists of writing with a pen (or a pencil) on a white rectangular sheet, but paper interfaces have been built for book reading [1], sticky notes [2], painting [3], presentation notes [4], trading cards [5], postcards [6] and even cover sheets [7]. However, in this section we would focus on the approaches addressing education, and start with the work related to the use of computers in geometry.

2.1 Computers in Geometry Education

Many researchers have tried to study the use of computers in geometry education involving software controlled by mouse and screen [8], augmented reality systems [9], or emulation of pen and paper [10]. Garcia et al. [8] identified that students appreciate the ability to repeat a geometrical construction (and playing it step-by-step) as allowed by a computer. Also, Dynamic Geometry Software (DGS) such as Cabri Géomètre [11], GeoGebra[1] enables learners to explore the dynamic behavior of a geometrical construction, i.e. what moves and what remains fixed under given constraints. Straesser [12] explains how DGS opens new possibilities in geometry education, by enabling geometric constructions not easily possible with pen and paper. However, the use of WIMP interfaces in teaching involves the risk of spending more time to learn the software than learning geometry, as these interfaces are completely different from the typical geometry tools frequently used in classrooms such as compass, ruler etc. [13].

Augmented reality interfaces aim at making the interaction more natural by integrating virtual elements in the real world. For example, Kaufmann and his colleagues [14] addressed some of the shortcomings of learning spatial geometry on a mouse/screen/keyboard system: with head mounted displays, the manipulation of 3 dimensional objects is more direct, and they allow for face-to-face collaboration. Martín-Gutiérrez [15] and his colleagues designed an augmented

---

[1] www.geogebra.org

book combined with a screen to develop spatial abilities in engineering students. They measured a positive impact on the spatial abilities, and the users found the system easy-to-use, attractive, and useful. Underkoffler and Ishii [9] made the reality augmenting device even less intrusive than head mounted displays or screens by using the so called *I/O bulbs* to simulate optics. *I/O bulbs* are camera/projector system above an interaction surface allowing students to manipulate tangible artefacts representing optical elements and see the effects on the trajectory of light.

Oviatt [10] and her colleagues bring forward this intent of making the interface as *quiet* as possible. They compared how student worked on geometrical problems using pen and paper, and interfaces approximating pen and paper with less and less exactitude: a smart pen using the microscopic pattern, a pen tablet, and a graphical tablet. They showed that the closer from the familiar work practice (i.e. pen and paper), the better is the performance.

To summarize, computers can add an essential dimension to geometry learning: dynamic information. However, existing educational interfaces for geometry are not adapted to classroom education, where paper prevails. Thus, paper interfaces can act as a bridge between computers and learning practices.

2.2   Paper-Based Interfaces in Education

We review the work related to paper interfaces in education based on the two aspects of paper that can be useful in the educational context. The first one, introduced by Wellner's seminal paper [16] on linking digital documents with their paper counterpart, presents paper as the support for working transparently on a digital document and its physical copy. This aspect is important for education, because it allows the researcher to study and extend the existing practice, in order to integrate the classroom more easily. Practices existing in the classroom that can be augmented include taking notes [17], reading textbooks [18], storytelling [19,20], or drawing schema [21].

The second aspect of paper useful for deployment in the classroom is its tangible aspect. It provides a cheap, easy way to attach virtual elements to reality. For example, Radu and MacIntyre [22] used cards for their tangible programming environment for pupils. Song and her colleagues used a cube covered by marked paper [23] to combined the advantages of digital and physical media. Millner and Resnick [24] even used a paper plate to prototype a steering wheel control, with printed buttons. Several frameworks [25,26,27] have already been proposed to study the design space of tangible interfaces [28,29]. For example, Hornecker and Dünser [30] showed that pupils expect the system to match the physical properties of the tangible interface.

In both aspects, it is important to identify the properties of paper. Regarding the work practices related to paper, McGee [31] analyzed the established usages in order to list the properties that natural interfaces should have. In their literature review [32], Klemmer and Landay classified the other approaches based on whether they were using a book, a document, a table, or a printer (among other things).

To sum-up, it has been identified that paper has two characteristics: it can be annotated, and it can be easily manipulated. These two features are associated to the established classroom practices that can be used to design intuitive paper interfaces. In this paper, we will investigate the most common forms allowing this: *sheets* and *cards*.

3   System Used

Our system for geometry education is built on the TinkerLamp [33], which is a tabletop environment developed at our lab. The TinkerLamp, shown in Figure 1, incorporates a camera and a projector directed to the tabletop surface via a mirror, which extends the augmented surface. The augmented surface is of dimension 70 × 40 cm. The camera and projector are connected to an embedded computer, so that the interaction with the hardware is minimum for the end user: switch ON or OFF. It only requires to be plugged into an electric outlet.

We use fiducial markers similar to ARTags[2] to identify and precisely track the various elements of the interface. Since the interface is projected from the top, it is possible to use interface elements (paper sheets and cards) as a projection surface in addition to the tabletop surface.

The different interface elements mainly consist of paper *sheets* and *cards*. The properties and behaviors of these interface elements are identified by the system using the fiducial markers printed over them. In addition to paper elements, we also use traditional geometry tools such as ruler and protractor as part of our system. We refer to this kind of interface as a *scattered interface* [34].

Fig. 1. Our camera-projector on a table, along with various types of objects which can be augmented: sheets, cards, tools and wooden blocks

4   Exploratory User Study

In order to study the influence and potential of paper interfaces in geometry education for primary school pupils, we used our system to design three learning

---

[2] http://www.artag.net

activities. These activities are based on *cards* and *sheets*, which are the two elements of our interface. Each activity was designed while keeping in mind the three circles of usability in the classroom - individual, group and classroom - as examined by Dillenbourg et al. [35]. This was done in order to integrate the system well in the conventional classroom curriculum.

Our analysis is based on observational field notes made during the experiment, the videos from a panoramic camera placed under the lamp filming the pupils, and the snapshots of the interaction surface taken every second by the camera of the lamp. We logged the position of every fiducial marker with a time stamp, which allowed us to replay the interaction with the system, since the fiducial markers are the only input of the projected augmentation. This way, we could generate any additional log from the software. From the information collected, it would be possible to conduct more detailed analyses, however this will be the topic of future work.

4.1   First Activity: Classifying Quadrilaterals

We designed the first activity as a pedagogical script to introduce the classification of quadrilaterals (squares, rhombuses, trapezoids, etc.) as shown in Figure 2a. The script consists of *sheets*, four *cards*, and a set of quadrilateral cardboard *shapes*. Each of these elements has a fiducial marker to identify them and they were produced with a regular printer. The cardboard shapes were numbered, so that they could be referenced from the sheets.

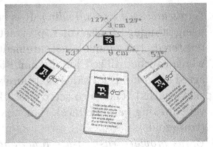

(a) The components of the activity about the classification of quadrilaterals: five cardboard quadrilaterals are classified into two groups on the instruction sheet, a card shows the measure of the angles of a rectangle, the feedback card displays the validation text.

(b) Configuration of the tool cards into a test bench, where cardboard shapes (a trapezoid here) are brought to display all their characteristics.

Fig. 2. First Activity: Classifying Quadrilaterals

The sheets, carrying instructions, are shown in the left part of Figure 2a. They consists of a short instructional text and two areas (marked with different colors - gray and white) denoting two different classes of quadrilaterals. The text instructs the learner to use the three cards shown on Figure 2b to find a common characteristic in a subset of shapes, and separate them into two classes. The cards have a small text describing their function. When a specific card is brought close to a shape, the system will display the given characteristic of the shape (such as side length, angle measures and parallel sides).

The learner is instructed to place a fourth card next to the current page once the shapes are placed in the classification areas (see the top right part of Figure 2a). If all shapes have not been placed in the areas, the learner will be reminded to do so. If the grouping is not the expected one, the learner will be invited to try again. If the grouping into areas is correct, the formulation of the answer will appear, e.g. *"Good job! Quadrilaterals with a pair of parallel sides are called trapezoids"*. Feedbacks are intentionally trivial; the cards are not meant to replace teachers.

Procedure and Discussion

This activity was deployed at two occasions in schools with pupils in the age group of 7–10 years. On the first occasion, 13 pupils in groups of 2-3 individuals worked on the first sheet of the activity for 5 minutes. On the second occasion, the study was performed with 12 pupils (in groups of 3) who worked on the complete activity for 40 minutes. In both cases a short presentation of the system was given to the whole class. Hereafter, we present the observed usage of various interface elements while identifying their characteristic behaviors.

Usage of the Cards

- *Cards are used as scaffolding.* It is crucial that pupils learn how to measure using standard tools (ruler, protractor etc.). However, once these skills are mastered, the manual measurement can become menial and wastes time which can be utilized for the main topic of the lesson. In this regard, cards acted as scaffolds for skills that pupils have already mastered well (measuring side lengths). They also acted as scaffolds for skills that pupils did not master yet (drawing parallel lines), that was necessary to introduce another concept (trapezoids).
- *Cards provided easy-to-use functionalities.* We observed that pupils had no difficulty in using the cards, thanks to the printed self-description and their simple, easy to try functionalities. Cards were used in two ways: either they were brought close to the shape to display properties, or the shape was brought close to them.
- *Cards allowed the composition of new functionalities.* One group provided an interesting example of appropriation of the interface. They created a test bench by placing the tool cards together, and bringing the cardboard shapes in the common neighborhood of all the cards so as to show all the related information at once, as shown in Figure 2b.

Usage of the Sheets

- *Sheets structured the activity in space. As opposed to the ephemeral workspaces that can emerge with cards as seen previously with the test bench, sheets* predefined a necessary workspace i.e. the two areas corresponding to the groups in which the cardboard shapes are to be placed.

- *Sheets structured the activity in time.* The sequence of exercises is also predefined by the sheets. We note that this structure is flexible in a sense that if the teacher wants to skip an exercise in the software, it is as simple as skipping a page in the sheets.

Usage of the Cardboard Shapes

- *Cardboard shapes are more concrete.* Cardboard shapes can be replaced by cards with a textual description of shape or an illustration of the shape represented. However, the lower level of abstraction provided by the visual match between a geometrical shape and its corresponding cardboard representation, assists in reducing the cognitive effort to discover common points between them.
- *Cardboard shapes are persistent.* The cardboard shapes in this activity have an existence of their own, and not only in the context of our system. Since they are made of cardboard and not altered, they could be reused between two experiments.

4.2   Second Activity: Discovering the Protractor

We designed the second activity as an exploratory activity for pupils, in order to learn to use the protractor, after the introduction of angles in the classroom by the teacher. This activity incorporates a deck of cards of two kinds - two *angle control cards* and ten *angle measure cards* (see Figure 3a).

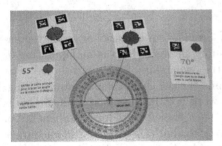

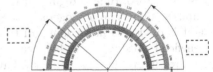

(a) The various elements of the task introducing angles: the two control cards and two of the angle measure. One is flipped and shows the measure of the angle constructed with the blue control card (70°).

(b) The drawing representing a protractor used on the pre- and post-test sheets for the task introducing angles is not necessarily associated with a real protractor by the pupils.

Fig. 3. Second Activity: Discovering the Protractor

These cards can be divided into two groups based on the orange or a blue icon printed on them. These two colors indicate the direction of measurement of angles - orange cards correspond to *clockwise* measurement, whereas blue cards denote *counter-clockwise* measurement. This distinction had been identified during our collaboration with the teachers as the main difficulty when learning to use a protractor. When a control card is shown to the system, an angle appears, with its origin in the centre of the projection area, an extremity on the centre of the control card, and the other extremity fixed horizontally on the left or right side of the origin (the X-axis), depending on whether the card is orange or blue (see Figure 3a).

Each angle measure card has a different angle value (in degrees) printed on them along with the instructions to construct an angle by using the corresponding control card. In order to check if the value of the produced angle (indicated on the angle measure card) is satisfying, the measure card is flipped and the current value of angle is displayed in a color depending on the degree of error (green for correct, yellow for close enough and red for otherwise).

Procedure and Discussion

This activity was conducted with 106 pupils (between 8–10 years) from 4 classes in a group of 2. Each group was required to go through 10 angle measure cards in 10 minutes, with individuals taking turns to measure angles. For the first 2 classes, the experimenters distributed the cards in a designated order one after the other. Whereas, for the other 2 classes the whole stack of cards was given to the group and no ordering was enforced. Also, the pupils were asked to take a pre-test and a post-test on paper where they were asked to identify and write down the angle measures next to a printed protractor as shown in Figure 3b. Next, we present our observations about how different interface elements were used.

Usage of Cards

- *Cards materialize roles.* This activity provides an example of group regulation via shared resources, as cards simply showed who was manipulating or checking. Also, time is regulated via the ownership of the control card. The pupils would try to homogenize the time each of them spends manipulating, as it is obvious who is doing all the work (i.e. having all the fun). This is beneficial since a lack of balance has been shown to reduce the benefits of learning in groups [36]. Similarly, having to share the control will encourage its negotiation, which has been shown to lead to greater learning gains [37].
- *Cards materialize progress.* Often, the measure cards were kept next to the pupil who managed to build the corresponding angle, acting as a trophy. Apart from the gain in engagement for the pupils, it is also a valuable help towards orchestration of the classroom, which refers to the teacher's responsibility to identify and manage the evolving learning opportunities and constraints, in real-time [35]. In this case, a teacher can easily get an instant summary of what each pupil did, and react accordingly.
- *Cards materialize the mode.* Cards also materialize even more ephemeral parts of the interaction, such as the current mode (building or checking). In this activity, it had a great implication on the engagement: all the groups preferred switching the feedback on and off for the sake of suspense rather

than continuously displaying it. When we told them that they can also display the variations of measure of the angle being built, one pupil answered: "I'm hiding it to see if [the pupil manipulating the control card] manages to build the angle".

- *The order of the cards did not matter.* This activity revealed the fact that the order of the cards is not important, as pupils often selected the measure card they were more comfortable with. For example, a group skipped all the cards corresponding to clockwise angles. Out of the eight groups for which the order of the cards was not enforced, only two followed the designated sequence of cards. Two groups skipped angles of a given orientation (one built only clockwise angles while the other only counter-clockwise angles).

Usage of Tools

- *Tools cannot be replaced.* During the study, we realized the importance of using a real protractor and not a printed representation. The pre-test and post-test did not give any statistically significant results due to a ceiling effect, but yielded an interesting anecdote. During the pre-test, one of the pupil counted each increment within the angle to measure the graduation instead of reading the measure directly. During the activity, she correctly read the measure directly on the protractor. Again during the post-test, she counted the increments. She clearly did not match the printed graduation with the one on the real protractor.

4.3   Third Activity: Describing Angles

Whereas the second activity was designed to introduce the concept of measuring angles, the third activity regards describing and communicating angles. In order to communicate an angle to someone, the pupil has to describe the angle measure, direction of measurement (clockwise or anti-clockwise) as well as the most convenient reference for measuring this angle (which axis to choose). In this direction, the third activity was designed as a game to get rid of space junk, non-functional satellites that continue to orbit around Earth. We consider that there are 3 laser guns deployed at 3 locations around Earth capable of destroying space junk (see the right side of Figure 4). This activity also allows for the use of protractor during the problem solving task.

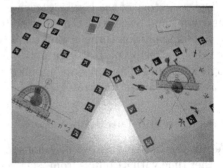

Fig. 4. The various elements of the problem using protractors

We divide a group of 4 pupils into 2 collaborating teams (of 2 pupils) and call them *observers* and *controllers*, with a physical separation between them (see Figure 5). The observers have a sheet (right side of Figure 4) with the view of

Fig. 5. The observer team measures the orientation to give to the laser (left), and communicates it to the controller team (right)

Earth along with all the satellites printed on them. Already destroyed satellites are highlighted in green and the next target in red. Also, the position of the three laser guns along with the baseline (axis) is also printed in the observer sheet. The observers are supposed to draw a line originating from one of the 3 laser guns to the target satellite (using a ruler). Next, they use the protractor to measure this angle with respect to the horizontal axis for this laser. Finally the observers have to describe this angle, direction of measurement and what laser gun to use to the controller team, by writing this information on a small piece of paper. This piece of paper is considered to be an *ammunition* for the laser gun.

The controllers are provided with 3 sheets corresponding to the 3 laser guns (see the left part of Figure 4). The controllers can change the inclination of the appropriate laser using the control card (similar to the one used in second activity). They reproduce the angle received from the observers using a protractor. Finally, the lasers can be activated by flipping this small paper received from the observers which contains a fiducial marker. Before firing, a yellow rectangle grows for 3 seconds over the ammunition (small paper), allowing to cancel the shot by flipping back or hiding the ammunition.

The trajectory of the laser is shown for 3 seconds on the sheets of controllers (laser gun) and observers (Earth view), with a fading blue line. If the satellite is hit, the ammunition turns green otherwise it turns red indicating a missed shot (see the centre top part of Figure 4). Each ammunition can only be used for a single shot, and the pupils are supplied with limited number of them, in order to avoid trial-and-error strategies.

Procedure and Discussion

We ran this activity on two occasions: once with 140 pupils from 7 classes and another time with 41 pupils from 2 classes. Groups of 4 pupils (2 observers and 2 controllers) were asked to complete this activity on a single system. Each group was given 25 minutes with this activity and they were asked to shoot as many satellites as they can. During the first study, we used 6 systems in a single room, while 2 systems were used in the second study. Next, we present our observations about the way sheets were used by groups during this activity.

Usage of Sheets

- *The workspace of a sheet was a stable referential.* Both observers and controllers placed their protractors on a sheet, which became a referential. All the groups but one kept the satellite view in the same orientation, even if it would have been easier to rotate the sheet before drawing the lines or measuring the angles.
- *Progress was written on the sheet.* While cards can act as ephemeral trophy, sheets durably store the progression with ink. The orchestration of a whole class was made a lot easier by the fact that the satellite view kept track of the intended trajectories in the form of lines between the location of the laser on Earth and the satellite. It helped to diagnose which part of the group (the observers or the controllers) was wrong in their measurement. The annotations on the ammunitions kept track of the progress of the group. The main difficulty in the activity is to establish a convention to describe and communicate an angle without seeing it. Giving the measurement was obvious, and the pupils would quickly realize that the origin of the shot (i.e. which laser to use) has to be communicated too. The more tricky part concerned the orientation of the angle. Since each shot has to be described on the ammunitions, it was easy to track when the pupils started to realize which information was needed.
- *Sheets do not restrict expressiveness.* When we explained the activity to the pupils, we intentionally remained vague on how to describe the angle, simply hinting them that there were several informations to provide. The angle measure and the laser gun to use were easily given as numbers. However, the pupils did not have an established convention for the orientation. This constructivist exploration paves the way for the teacher to explain the concept of *clockwise* and *counter-clockwise* measurement, since the need has been felt directly.

5   Conclusion

The tabletop system presented in this article was designed to facilitate geometry learning for primary school pupils. As existing classroom curriculum is based on paper and conventional geometry tools (ruler, compass, etc.), our system incorporates paper *sheets* and *cards* as the two main interface elements. We designed and conducted 3 exploratory user studies focusing on the different usages of sheets and cards, in order to study the impact and potential of paper interfaces in geometry learning. Our observations show very positive results regarding the adoption of paper interfaces by the pupils, as the use of sheets and cards was easily perceived and minimal effort was required to learn how to use the interface.

Our system takes into account the three circles of usability outlined by Dillenbourg et al. [35] - individual, group and classroom. On the individual level the pupils were highly engaged and participated actively in the activities, even in the classes that were less affected by the novelty effect in our subsequent visits.

This is a success given that the activities revolved around using a (boring) protractor, or classifying quadrilaterals. On the group level, the system naturally promoted collaboration, allowing pupils to help each other and learn in teams. At the classroom level, the paper interface enabled the teacher to monitor the progress of teams and thus orchestrate the classroom activities accordingly. This aims at facilitating smooth integration and adoption of computers in the entire curriculum.

In addition, our observations regarding the characteristics of sheets and cards provided insights about the affordances of the different paper elements. On one hand, we observed that sheets are important for their content. Sheets were used to organize the discourse on two levels. On the first level, the layout on a single page encodes the order in which to read the various information and proceed with the activity. On the second level, several sheets can be organized together in a sequence (by stapling or binding), which enables us to implement several lessons or exercises similar to a book. As the trace of a pen is persistent over sheets, they can act as a permanent memory, which can be used as a way to trace the performance of pupils during a learning activity, or display publicly the progress within the group.

On the other hand, cards are mostly used as a physical body. The position of the card is usually relative, and bringing one close to another element allows to show additional properties. Further, the side of a card is another useful property; it can be flipped to control a binary value. In general, cards can materialize the reversible and ephemeral pieces of interaction according to rules. For example, the presence of a card on the table or next to a pupil indicated its role in the group.

We believe that careful identification of these characteristics of paper interface elements might provide crucial design guidelines towards the development of paper interfaces for education in general. The affordances of different paper elements (depending on the shape, size and material) render the interfaces easy-to-use and highly intuitive.

In future, we would like to conduct a formal evaluation of the effects of paper interfaces on learning. Also, we would like to investigate the technological issues related to the predisposition of the system and to learning design. The aim would be to enable teachers to set up pedagogical experiences without assistance from researchers. This would naturally link the activities to specific mathematics learning theories.

Acknowledgement. The authors wish to sincerely thank the teachers for their precious collaboration, as well as Olivier Guédat for building the TinkerLamp; Michael Chablais, Chia-Jung Chan Fardel, and Carlos Sanchez Witt for their contribution to the activities during their internship.

References

1. Klemmer, S.R., Graham, J., Wolff, G.J., Landay, J.A.: Books with voices: paper transcripts as a physical interface to oral histories. In: CHI 2003, pp. 89–96. ACM, New York (2003)

2. Moran, T.P., Saund, E., Van Melle, W., Gujar, A.U., Fishkin, K.P., Harrison, B.L.: Design and technology for collaborage: collaborative collages of information on physical walls. In: UIST 1999, pp. 197–206. ACM, New York (1999)
3. Flagg, M., Rehg, J.: Projector-guided painting. In: UIST 2006, pp. 235–244. ACM (2006)
4. Nelson, L., Ichimura, S., Pedersen, E.R., Adams, L.: Palette: a paper interface for giving presentations. In: CHI 1999, pp. 354–361. ACM, New York (1999)
5. Lam, A.H.T., Chow, K.C.H., Yau, E.H.H., Lyu, M.R.: Art: augmented reality table for interactive trading card game. In: VRCIA 2006, pp. 357–360. ACM, New York (2006)
6. Cho, H., Jung, J., Cho, K., Seo, Y.H., Yang, H.S.: Ar postcard: the augmented reality system with a postcard. In: VRCIA 2011, pp. 453–454. ACM, New York (2011)
7. Hong, J., Price, M.N., Schilit, B.N., Golovchinsky, G.: Printertainment: printing with interactive cover sheets. In: CHI EA 1999, pp. 240–241. ACM, New York (1999)
8. García, R., Quirós, J., Santos, R., González, S., Fernanz, S.: Interactive multimedia animation with Macromedia Flash in Descriptive Geometry teaching. Computers & Education 49(3), 615–639 (2007)
9. Underkoffler, J., Ishii, H.: Illuminating light: a casual optics workbench. In: CHI EA 1999, pp. 5–6. ACM, New York (1999)
10. Oviatt, S., Arthur, A., Brock, Y., Cohen, J.: Expressive pen-based interfaces for math education. In: International Society of the Learning Sciences (CSCL), pp. 573–582 (2007)
11. Laborde, C., Keitel, Ruthven, K.: The computer as part of the learning environment: the case of geometry. In: Learning from Computers: Mathematics Education and Technology, pp. 48–67. Springer (1993)
12. Straesser, R.: Cabri-geometre: Does dynamic geometry software (dgs) change geometry and its teaching and learning? International Journal of Computers for Mathematical Learning 6(3), 319–333 (2002)
13. Kortenkamp, U., Dohrmann, C.: User Interface Design for Dynamic Geometry Software. Acta Didactica Napocensia 3 (2010)
14. Kaufmann, H., Dünser, A.: Summary of Usability Evaluations of an Educational Augmented Reality Application. In: Shumaker, R. (ed.) HCII 2007 and ICVR 2007. LNCS, vol. 4563, pp. 660–669. Springer, Heidelberg (2007)
15. Martín-Gutiérrez, J., Luís Saorín, J., Contero, M., Alcañiz, M., Pérez-López, D., Ortega, M.: Design and validation of an augmented book for spatial abilities development in engineering students. Computers & Graphics 34(1), 77–91 (2010)
16. Wellner, P.: Interacting with paper on the DigitalDesk. Communications of the ACM 36(7), 87–96 (1993)
17. Malacria, S., Pietrzak, T., Tabard, A., Lecolinet, É.: U-Note: Capture the Class and Access It Everywhere. In: Campos, P., Graham, N., Jorge, J., Nunes, N., Palanque, P., Winckler, M. (eds.) INTERACT 2011, Part I. LNCS, vol. 6946, pp. 643–660. Springer, Heidelberg (2011)
18. Asai, K., Kobayashi, H., Kondo, T.: Augmented instructions - a fusion of augmented reality and printed learning materials. In: ICALT, pp. 213–215. IEEE Computer Society (2005)
19. Portocarrero, E., Robert, D., Follmer, S., Chung, M.: The Never Ending Storytelling Machine a platform for creative collaboration using a sketchbook and everyday objects. In: Proc. PaperComp 2010 (2010)

20. Koike, H., Sato, Y., Kobayashi, Y., Tobita, H., Kobayashi, M.: Interactive textbook and interactive venn diagram: natural and intuitive interfaces on augmented desk system. In: CHI 2000, pp. 121–128. ACM, New York (2000)
21. Lee, W., de Silva, R., Peterson, E., Calfee, R., Stahovich, T.: Newton's Pen: A pen-based tutoring system for statics. Computers & Graphics 32(5), 511–524 (2008)
22. Radu, I., MacIntyre, B.: Augmented-reality scratch: a children's authoring environment for augmented-reality experiences. In: IDC 2009, pp. 210–213. ACM, New York (2009)
23. Song, H., Guimbretière, F., Ambrose, M.A., Lostritto, C.: CubeExplorer: An Evaluation of Interaction Techniques in Architectural Education. In: Baranauskas, C., Abascal, J., Barbosa, S.D.J. (eds.) INTERACT 2007, Part II. LNCS, vol. 4663, pp. 43–56. Springer, Heidelberg (2007)
24. Millner, A., Resnick, M.: Tools for creating custom physical computer interfaces. Demonstration presented at Interaction Design and Children, Boulder, CO (2005)
25. Ullmer, B., Ishii, H.: Emerging frameworks for tangible user interfaces. IBM Systems Journal 39(3.4), 915–931 (2000)
26. Ullmer, B., Ishii, H., Jacob, R.: Token+ constraint systems for tangible interaction with digital information. ACM Transactions on Computer-Human Interaction (TOCHI) 12(1), 81–118 (2005)
27. Fishkin, K.: A taxonomy for and analysis of tangible interfaces. Personal and Ubiquitous Computing 8(5), 347–358 (2004)
28. Fitzmaurice, G.: Graspable user interfaces. PhD thesis, Citeseer (1996)
29. Ishii, H., Ullmer, B.: Tangible bits: towards seamless interfaces between people, bits and atoms. In: CHI 1997, pp. 234–241. ACM, New York (1997)
30. Hornecker, E., Dünser, A.: Of pages and paddles: Children's expectations and mistaken interactions with physical-digital tools. Interacting with Computers 21(1-2), 95–107 (2009)
31. Mcgee, D.R.: Augmenting environments with multimodal interaction. PhD thesis, Oregon Health & Science University (2003) AAI3100651
32. Klemmer, S.R., Landay, J.A.: Toolkit support for integrating physical and digital interactions. Human-Computer Interaction 24(3), 315–366 (2009)
33. Zufferey, G., Jermann, P., Dillenbourg, P.: A tabletop learning environment for logistics assistants: activating teachers. In: Proceedings of the Third IASTED International Conference on Human Computer Interaction, HCI 2008, pp. 37–42. ACTA Press, Anaheim (2008)
34. Cuendet, S., Bonnard, Q., Kaplan, F., Dillenbourg, P.: Paper interface design for classroom orchestration. In: Proceedings of the 2011 Annual Conference Extended Abstracts on Human Factors in Computing Systems, CHI EA 2011, pp. 1993–1998. ACM, New York (2011)
35. Dillenbourg, P., Zufferey, G., Alavi, H.S., Jermann, P., Do, L.H.S., Bonnard, Q., Cuendet, S., Kaplan, F.: Classroom orchestration: The third circle of usability. In: Connecting Computer-Supported Collaborative Learning to Policy and Practice: CSCL 2011 Conference Proceedings. Volume I - Long Papers. International Society of the Learning Sciences, pp. 510–517 (2011)
36. Cohen, E.: Restructuring the classroom: Conditions for productive small groups. Review of Educational Research 64(1), 1 (1994)
37. Do-Lenh, S., Kaplan, F., Dillenbourg, P.: Paper-based concept map: the effects of tabletop on an expressive collaborative learning task. In: Proceedings of the 23rd British HCI Group Annual Conference on People and Computers: Celebrating People and Technology, BCS-HCI 2009, pp. 149–158. British Computer Society, Swinton (2009)

The European TEL Projects Community from a Social Network Analysis Perspective

Michael Derntl and Ralf Klamma

RWTH Aachen University
Advanced Community Information Systems (ACIS)
Informatik 5, Ahornstr. 55, 52056 Aachen, Germany
{derntl,klamma}@dbis.rwth-aachen.de

Abstract. In this paper we draw a community landscape of European Commission-funded TEL projects and organizations in the 6th and 7th Framework Programmes and *eContentplus*. The project metadata were crawled from the web and maintained as part of the TEL Mediabase, a large collection of data obtained from different web sources including blogs, bibliographies, and project fact sheets. We apply social network analysis and impact analysis on the project consortium progression graph and on the organizational collaboration graph to identify the most central TEL projects and organizations. The key findings are that networks of excellence and integrated projects have the strongest impact on the project network; that *eContentplus* was a funding bridge between FP6 and FP7; and that the tightly knit collaboration network may inhibit the assimilation of new organizations and ideas into the TEL community.

1 Introduction

For many years, technology enhanced learning (TEL) has been a well-funded thematic area in the work programmes of the European Community's (EC) Framework Programmes (FP). Current projects like the STELLAR network of excellence [http://stellarnet.eu] and the TEL-Map support action [http://telmap.org] take a supporting role for the EC and other TEL stakeholders to provide input to future challenges and roadmapping for the European TEL community. This is evidence that the TEL community has a genuine interest in its thematic and collaborative structures and dynamics. To facilitate this kind of self-reflection based on available data, we established the Mediabase [1] in 2006 as a collection of TEL related social media artifacts that have been crawled from the web, fed in to relational databases, and analyzed using web-based tools available to the TEL community for self-observation and self-modeling [2].

Following that spirit, this paper focuses on the macro level of funding and organizational collaboration in European TEL by analyzing the current status and historic evolution of the landscape of TEL projects and organizations using social network analysis (SNA) techniques. Specifically, we research these aspects:

A. Ravenscroft et al. (Eds.): EC-TEL 2012, LNCS 7563, pp. 51–64, 2012.

- *Project progression*: identify sustained organizational collaboration ties between consecutive projects to identify cliques of successful collaborators and a measure of project impact; identify the most central projects, and consortia that unite central organizations.
- *Inter-organizational collaboration*: identify central participating organizations in terms of SNA metrics (e.g. betweenness centrality, PageRank, and clustering in the organizational collaboration network) and EC funding.
- *Times series analysis*: analyze the development of SNA metrics in the project progression network and in the organizational collaboration networks since the start of FP6.

The paper is structured as follows. In the next section we discuss related work. In Section 3 we outline the TEL projects community data set and formal foundations of projects as social networks. In Section 4 we analyze project progression, i.e. the sustained collaboration of project partners in follow-up projects. In Section 5 we analyse the organizational collaboration network in TEL projects. In Section 6 we highlight the dynamics in these social networks since the start of FP6. Sections 7 and 8 sum up the key findings and conclude the paper.

2   Related Work

An analysis of consortium involvements of TEL partners in other projects, partners, and events was previously conducted and reported [3] by the STELLAR network of excellence. This analysis was centered on STELLAR as the central entity. A more general approach was adopted in [4], where the authors define a formal model for social network analysis of all projects funded by the EC in FP1 through FP4. They model projects and organizations as graphs and apply SNA algorithms to identify the overall characteristics of these R&D networks. They identify these networks as being typical of complex, scale-free networks with small diameter and high clustering. Similar findings are reported in [5] following a social network analysis of the first five FPs.

In a recent paper [6], the authors model the affiliation network of FP6 projects using an agent-event metaphor. In their model, organizations are agents, who participate in projects, which represent the events. This model was employed to obtain general insight into FP6 collaborations using SNA techniques, studying the effects of different network representations—one-mode network (actors and events separated) vs. two-mode network (actors and events unified)—on the analyses and results. A similar study that focused on the forming and evolving of these networks is reported in [7], finding the emergence of dense hierarchical networks resting on an "oligarchic core" of participants.

Previous work also includes the application of community detection in R&D project collaborations, e.g. in [8] the authors apply community detection on the FP6 organizational collaboration graph, with a main interest in the role of nationality and type of organizations. An analysis of FP5 collaboration with emphasis on the role of geographical or technological proximity of partners is reported in [9]. The authors found that both factors impact the collaboration network, with technological proximity having the greater effect.

In [10] an analysis of the collaboration networks in FP4 with particular emphasis on the differences of telecommunications and agro-industrial industries is performed. The paper focuses on specific thematic areas and their comparative characteristics in terms of consortium size, required funding, and the role of scientific vs. industry partners.

In this paper, we build on the ideas and formal foundations laid out in these previous research endeavors. Our locus of interest, however, is different and unique: we are interested in a particular thematic area, i.e. technology enhanced learning, and draw insights and findings of relevance to participants and stakeholders in that particular community. In a related presentation [11] we formulated first ideas about project impact based on social network analysis, however without proposing a formal measure of impact. To close this gap, the present paper reports analyses of the R&D networks in TEL from a temporal-dynamic perspective, going beyond previous work by proposing and applying a measure of impact of project consortia on the TEL collaboration landscape.

3 The European TEL Projects Community

Data Set. The TEL projects database includes details on TEL projects funded under FP6, FP7, and *eContentplus* programmes. In total, the metadata of 77 TEL projects (see Table 1) were collected and used for the analyses presented in this paper. The data includes detailed information on the projects like start and end dates, cost, EC funding, and consortium members (in total there are 604 distinct organizations participating in these 77 projects). The database was fed by a crawler that was deployed to collect and scrape project facts from the CORDIS website [12] for FP6 and FP7 projects, as well as from the respective *eContentplus* pages. As evident in Table 1, only projects funded under TEL related calls were included in the data set.[1]

In the CORDIS data we found that there were several typos and variants in the spelling of organizations and countries; the list of organizations was therefore manually post-processed to merge variants and correct spelling errors. Organizational name changes were not accounted for. For instance, Giunti Labs S.r.l. was rebranded to eXact Learning Solutions in 2010. In the data set, these—and all organizations with similar rebrandings—are represented as separate entities. Likewise, organizational mergers are not accounted for, e.g. ATOS Origin and Siemens Learning, which merged in 2011, and Aalto University in Finland, which was established in 2010 as a merger of three universities. Also, the CORDIS fact sheets expose some omissions; one that was discovered is the fact sheet of the EdReNe *eContentplus* project, which only includes the coordinator without any of the consortium members. Finally, a plethora of projects from other TEL related funding channels is not yet included.

---

[1] A browseable version of the complete TEL projects data set is avaialble on the Learning Frontiers portal at http://learningfrontiers.eu/?q=project_space

Table 1. TEL projects data set

Call	Projects	#
eContent*plus* Call 2005	CITER, JEM, MACE, MELT	4
eContent*plus* Call 2006	COSMOS, EdReNe, EUROGENE, eVip, Intergeo, KeyToNature, Organic.Edunet	7
eContent*plus* Call 2007	ASPECT, iCOPER, EduTubePlus	3
eContent*plus* Call 2008	LiLa, Math-Bridge, mEducator, OpenScienceResources, OpenScout	5
FP6 IST-2002-2.3.1.12	CONNECT, E-LEGI, ICLASS, KALEIDOSCOPE, LEACTIVEMATH, PROLEARN, TELCERT, UNFOLD	8
FP6 IST-2004-2.4.10	APOSDLE, ARGUNAUT, ATGENTIVE, COOPER, ECIRCUS, ELEKTRA, I-MAESTRO, KP-LAB, L2C, LEAD, PALETTE, PROLIX, RE.MATH, TENCOMPETENCE	14
FP6 IST-2004-2.4.13	ARISE, CALIBRATE, ELU, EMAPPS.COM, ICAMP, LOGOS, LT4EL, MGBL, UNITE, VEMUS	10
FP7 ICT-2007.4.1	80DAYS, GRAPPLE, IDSPACE, LTFLL, MATURE, SCY	6
FP7 ICT-2007.4.4	COSPATIAL, DYNALEARN, INTELLEO, ROLE, STELLAR, TARGET, XDELIA	7
FP7 ICT-2009.4.2	ALICE, ARISTOTELE, ECUTE, GALA, IMREAL, ITEC, METAFORA, MIROR, MIRROR, NEXT-TELL, SIREN, TEL-MAP, TERENCE	13

Projects as Social Networks. A collaborative project can be modeled as a social network [13,6]. A social network is modeled as a graph $G = (V, E)$ with V being the set of vertices (or nodes) and E being the set of edges connecting the vertices with one another [14]. We define P as the set of projects, and O as the set of organizations involved in these projects. Similar to [6], we define a function μ representing the membership of any organization $o \in O$ in the consortium of any TEL project $p \in P$ as follows to enable graph-based analyses of the TEL project and organization networks:

$$\mu : P \times O \to \begin{cases} true & \text{if } o \text{ is or was member of the consortium of } p \\ false & \text{otherwise .} \end{cases}$$

4   Project Consortium Progression

The project consortium progression graph $G_P = (V_P, E_P)$ contains projects as nodes ($V_P = P$) and their successor relationships as directed edges. A successor relationship between two projects is established if (1) at least k organizations have participated in both projects, and (2) the successor project started at least t time units after the predecessor project. Let $s : P \to \mathbb{R}$ map projects to their start points in time, represented for simplicity as real numbers, then we can define the set of edges as

$$E_P = \{(u, v) : u, v \in V_P \land s(v) - s(u) \geq t \land |\{o \in O : \mu(u, o) \land \mu(v, o)\}| \geq k\}.$$

The least restrictive parameter pair $k = 1, t = 0$ produces a graph including all 77 TEL projects and a total of 712 edges (note that $t = 0$ implies that the graph also includes edges between projects that started at the same time). Looking for a reasonable value for k to represent consortium progression in the sense of continued collaboration we require at least two consortium members present in

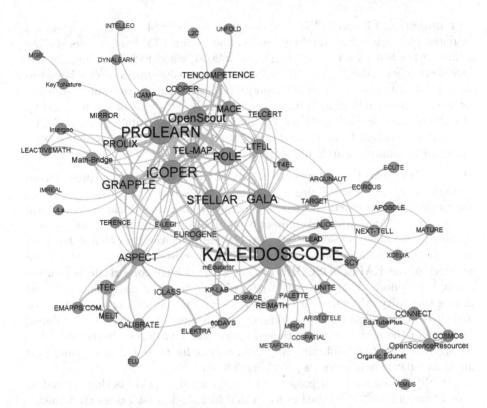

Fig. 1. Project progression graph spanning FP6, FP7, and *eContentplus* TEL projects

a successor project, i.e. $k = 2$. The time span parameter $t \in \mathbb{R}_{\geq 0}$ should allow for a time gap between two consecutive project starts to reasonably establish a predecessor-successor relationship. Of course the time span between call deadline and project start may vary, but the actual point in time when partners team up for a project proposal is not represented in our data. So we let months be the unit of time and chose $t = 3$ months as the lower threshold for the time between the start of two projects. With this threshold the younger project can reasonably considered as a successor of the older project.

The resulting graph for $k = 2, t = 3$ is illustrated in Fig. 1, including 68 nodes and 198 edges. The size of each node and its label in the figure is proportional to its weighted degree centrality, i.e. the number of adjacent projects weighted by consortium overlap. The graph layout was produced by applying the ForceAtlas layout in Gephi. Evidently, KALEIDOSCOPE—one of the two "inaugural" TEL Networks of Excellence in FP6—is the most degree-central node, boosted by its large consortium of 83 partners, which is more than five times the average consortium size of 14.5 in FP6. The graph also reveals that in addition to strong ties between FP6 and FP7 projects, several *eContentplus* projects (e.g. OpenScout, ICOPER, ASPECT) have central positions and strong connections with projects in FP6 and FP7. This can probably be explained by the fact that *eContentplus* bridged a "funding gap" in 2007 when FP6 funding was stalling following the last FP6 projects launched in 2006, while FP7 TEL funding was kicked off only in 2008. In fact, in 2007 only *eContentplus* projects were launched with EC funding in our data set (compare also the time series analysis of the project networks in Section 6, in particular Fig. 9). This kind of gap bridging by *eContentplus*, where a large share of organizations funded under FP6 and FP7 engaged in e-content focused R&D projects, could be interpreted as evidence for an organizational "research follows money" attitude.

Project Impact. One interesting aspect of G_P is the impact of project consortium members on sustaining and shaping the social TEL project ties after the project start (in fact, this shaping already commences during the proposal writing phase). Measuring this impact by applying graph centrality metrics (e.g. degree, betweenness, closeness, etc.) to G_P entails several problems. For one, these metrics will favor projects with very large consortia. In our data set the presence of the KALEIDOSCOPE project with its huge consortium is a prime example of a node that represents an "outlier" in terms of consortium size and degree centrality. Moreover, in G_P the projects' chances of having predecessor and successor projects vary depending on the distribution of project start dates. In terms of the social network, these factors bias a project's chance of having stronger incoming and outgoing edges. To control for these biases we conceived an impact measure of projects $p \in V_P$ as follows.

Let $S_p^{t,k}$ be the successor projects of p, i.e. the set of projects that started at least t time units after the start of p and that include at least k consortium members of p. Let D_p^t be the set of all potential successor projects, i.e. all successor projects that started at least t time units after the start of p, and let $C_p \subseteq O$ be the set of consortium members of project p. It holds that $S_p^{t,k} \subseteq D_p^t \subseteq P$. We define the impact δ of project p as

$$\delta_p = \frac{|S_p^{t,k}|}{|D_p^t|} \sum_{q \in S_p^{t,k}} \frac{|C_p \cap C_q|}{|C_p|}.$$

In this formula, the term $\frac{|S_p^{t,k}|}{|D_p^t|}$ accounts for the actual number of successor projects of p relative to opportunity, that is, the fraction of actual vs. potential successor projects. Essentially, this eliminates the potential (dis)advantages from a projects' position on the timeline. The term $\frac{|C_p \cap C_q|}{|C_p|}$ represents the weighted overlap of a project with other project consortia and thus accounts for the varying sizes of project consortia and the varying number of organizations that overlap between two projects. Actually the latter term is a shorthand of $\frac{|C_q \cap C_p|}{|C_q|} \cdot \frac{|C_q|}{|C_p|}$, in which the first term represents the share of q's consortium that was "fed" with members of project p, and the second term represents the ratio of the sizes of the two consortia. Consequently, summing up these ratios with $\sum_{q \in S_p^{t,k}} \frac{|C_q \cap C_p|}{|C_q|}$ essentially represents the cumulative share of successive project consortia in G_P filled up exclusively with p's consortium members.

Table 2. Top 15 projects by impact on the TEL projects landscape

#	Project p	Runtime	Funding	Programme	Type	$|C_p|$	$|S_p^{3,2}|$	$|D_p^3|$	IFM	δ_p
1	PROLEARN	2004–07	6.01	FP6	NoE	22	22	69	.19	1.14
2	KALEIDOSCOPE	2004–07	9.35	FP6	NoE	83	34	69	.07	.66
3	GRAPPLE	2008–11	3.85	FP7	STREP	15	8	28	.10	.38
4	iCOPER	2008–11	4.80	ECP	BPN	23	7	25	.07	.33
5	STELLAR	2009–12	4.99	FP7	NoE	16	5	18	.05	.23
6	ASPECT	2008–11	3.70	ECP	BPN	22	6	25	.06	.21
7	COOPER	2005–07	1.95	FP6	STREP	8	6	49	.10	.20
8	MACE	2006–09	3.15	ECP	CEP	13	7	41	.06	.20
9	LTFLL	2008–11	2.85	FP7	STREP	11	5	28	.06	.18
10	PROLIX	2005–09	7.65	FP6	IP	19	7	49	.02	.17
11	TELCERT	2004–06	1.80	FP6	STREP	9	7	69	.09	.16
12	ROLE	2009–13	6.60	FP7	IP	16	3	18	.01	.09
13	TENCOMPETENCE	2005–09	8.80	FP6	IP	15	5	49	.01	.09
14	CONNECT	2004–07	4.69	FP6	STREP	18	6	69	.02	.08
15	ICLASS	2004–08	12.59	FP6	IP	17	6	69	.01	.08

Funding ... European Commission contribution in million euro.

IFM ... Impact for Money = δ/Funding.

Type ... Network of Excellence (NoE), Integrated Project (IP), Content Enrichment Project (CEP), Specific Targeted Research Project (STREP), Best Practice Network (BPN).

The top 15 projects by impact for $t = 3, k = 2$ ordered by descending δ are displayed in Table 2. The table includes projects from all three programmes with FP6 represented strongest, and it prominently features Networks of Excellence and Integrated Projects, although these are by number among the rarest project types in the Framework Programmes. The best impact-for-money (IFM) ratio was achieved by PROLEARN, followed by COOPER and GRAPPLE. More data and further research is required to identify indicators for high-impact projects. The project proposals and deliverables would certainly help in this regard, since these typically contain information on and cross-references to work in other projects.

5   Organizational Collaboration

In addition to the project consortium progression network presented in the previous section, TEL projects can be viewed from another angle: the organizational collaboration graph $G_O = (V_O, E_O)$ contains organizations and their collaborations in the project consortia [13]. This graph shows organizations as nodes and an (undirected) edge between two nodes if there are at least k projects where both organizations have participated in, i.e. $V_O = O$ and

$$E_O = \{(u, v) : u, v \in V_O \land u \neq v \land |\{p \in P : \mu(p, u) \land \mu(p, v)\}| \geq k\}.$$

G_O is an undirected graph, which is visualized in Fig. 2. For $k = 1$ it includes 603 distinct partners and 9315 edges, of which only those representing at least two shared projects are displayed in Fig. 2. Apart from the core component, there are no strongly connected sub-networks, which means that in every project there is at least one partner who is involved in another project.

We calculated SNA metrics and funding for each participant in V_O; the resulting table of the top ten organizations is given in Table 3. The table is ordered

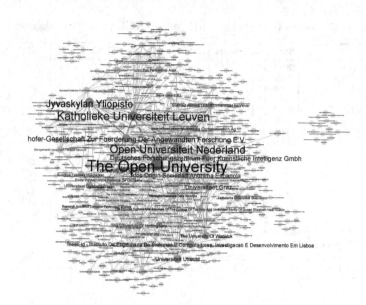

Fig. 2. Partner collaborations spanning FP6, FP7, and *eContentplus* TEL projects

by PageRank, a metric that not only takes into account the number of edges of each node, but also the "importance" of the adjacent neighbor nodes [15]. This means, an organization's importance depends on the number of collaborations with other organizations and on the importance of the organization's collaborators.

Table 3. Top 10 organizations by PageRank

#	Organization	PR	BC	LC	DC	Funding
1	The Open University, United Kingdom	.0125 [1]	.1185 [1]	.2151 [601]	219 [1]	3.55 [3]
2	Katholieke Universiteit Leuven, Belgium	.0090 [2]	.0752 [2]	.1716 [604]	148 [3]	2.56 [6]
3	Open Universiteit Nederland, Netherlands	.0086 [3]	.0414 [6]	.2161 [600]	133 [7]	3.45 [4]
4	Jyvaskylan Yliopisto, Finland	.0080 [4]	.0667 [3]	.3170 [588]	170 [2]	1.26 [39]
5	Fraunhofer-Gesellschaft zur Foerderung der Angewandten Forschung E.V., Germany	.0068 [5]	.0529 [4]	.1833 [603]	111 [22]	3.40 [5]
6	Deutsches Forschungszentrum fuer Kuenstliche Intelligenz Gmbh, Germany	.0066 [6]	.0390 [7]	.1916 [602]	106 [27]	3.68 [1]
7	Atos Origin Sociedad Anonima Espanola, Spain	.0064 [7]	.0236 [15]	.4316 [565]	142 [5]	1.33 [33]
8	Universitaet Graz, Austria	.0064 [8]	.0230 [18]	.4016 [573]	148 [3]	2.03 [10]
9	Universiteit Utrecht, Netherlands	.0061 [9]	.0203 [23]	.4323 [564]	139 [6]	1.62 [19]
10	INESC ID - Instituto de Engenharia de Sistemas e Computadores, Investigacao e Desenvolvimento em Lisboa, Portugal	.0061 [10]	.0368 [8]	.4741 [552]	130 [8]	1.68 [16]

PR... PageRank — **BC**... Betweenness centrality — **LC**... Local clustering coefficient — **DC**... Degree centrality — **Funding**... EC contribution to the project cost in million Euro. Note that CORDIS states the total funding for each project. The funding per consortium member for each project was computed by dividing the total EC contribution to that project by the number of consortium members. This should give a good estimate.

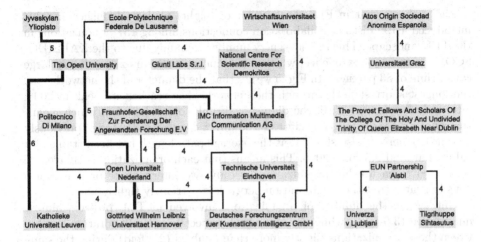

Fig. 3. Strongest organizational ties in FP6, FP7, and *eContentplus* TEL projects

The top partnership bonds across all TEL projects are displayed in Fig. 3. The figure shows the 22 collaboration pairs (edges) between organizations (nodes) that are based on at least four projects (the number of projects is displayed as a label for each edge). Assuming that partnership is only continued from successful previous collaborations, we can conjecture that those projects where the organization pairs displayed in Fig. 3 were involved can be flagged as having lasting impact, at least in terms of continuity in research collaborations. The most important of these projects, ordered by frequency of partnerships, are:

1. PROLEARN (FP6): 16 pairs,
2. ICOPER (*eContentplus*): 10 pairs,
3. OpenScout (*eContentplus*): 9 pairs,
4. GRAPPLE (FP7): 8 pairs,
5. STELLAR (FP7), ROLE (FP7), PROLIX (FP6): 5 pairs.

It is evident that the PROLEARN network of excellence that co-kicked off FP6 succeeded in creating and sustaining strong partnerships, while the KALEIDO-SCOPE network of excellence, which started at the same time as PROLEARN, did not achieve this despite its much larger consortium.

6 Time Series Analysis

The previous figures all took the current status of collaborations and projects as a basis for calculating social network metrics. To understand the dynamics of the projects and their consortium collaborations this section presents the development of SNA metrics of the collaboration network over time, starting from 2004 when the first FP6 projects were launched, up to the year 2010 (inclusive). The years 2011 and 2012 were omitted from the analyses since no new TEL projects were launched in FP7 and *eContentplus* after 2010 to date.

Fig. 4 shows that in FP6 the first set of (eight) projects launched in 2004 introduced 4,199 distinct collaboration connections among 157 organizations in the TEL landscape. This massive entry number is mainly due to the KALEIDO-SCOPE network of excellence, which was launched with an extraordinary large consortium of 83 partners. In Fig. 5 we see that the diameter of the network—i.e. the longest shortest path through the network—has reached its peak in 2006, after only 2 years; in 2010, the diameter shrunk to a value of 4, which means that one or more projects have introduced direct connections between previously distant partners. It also shows that the average path length has been stable at a value of around 2.5 since 2006. This means that each organization is on average connected to each other organization by only two intermediate organizations. This indicates that the collaboration network is extremely tightly knit.

Until 2010, the number of organizations involved in TEL projects almost quadrupled (3.9-fold), while the number of project-based collaboration ties between those organizations slightly more than doubled (2.2-fold) during the same time window (cf. Fig. 4). This gap can partly be explained by Fig. 6, which shows that although there has been a steady flow of new projects, these projects have added fewer and fewer new organizations to the picture, exposing a drop from 8.1 new organizations per new project in 2006 to a value of 4.8 in 2010.

Fig. 8 demonstrates that the average size of the consortia of newly launched projects has been relatively stable since 2005, ranging between 10.9 and 14.1. In contrast, the average share of newly introduced organizations per launched project has dropped from 66% in 2005 to 40% in 2010. The sharpest drop is evident for projects that started in the year 2008 (from 62% to 42%); this was the year when the first six FP7 projects plus three new *eContentplus* projects were launched (cf. Fig. 9). At the transition from FP6 to FP7 and *eContentplus*, the project consortia apparently resorted to an established core of members.

While Fig. 7 shows that new projects have introduced a relatively stable number of new collaboration ties to the landscape in recent years, Fig. 10 demonstrates that the average number of new collaboration ties created by each organization making its debut in TEL projects has, after an initial fall between 2004 and 2005, increased from 7.9 in 2005 to 17.7 in 2010. Hence, starting to participate in TEL projects has an increasingly positive effect in terms of new collaborations with other organizations involved in TEL.

The project participation data shows that of the 34 TEL projects launched between 2008 and 2010, 20% were coordinated by organizations which had not participated in any previous (or at that time running) TEL project. The development of this percentage over time is plotted in Fig. 11. The sharp increase in 2007 is likely due to *eContentplus*, where the focus shifted to e-content and metadata, and thus new organizations were introduced. The data shows that even for total "newbie organizations" in TEL it is absolutely feasible to write a successful project proposal in the coordinator role.

However, the tendency evident in most of the figures in this section points in another direction; it appears that there is less and less demand for new organizations in the TEL community. One the one hand, this is understandable: if an organization launches a new project it is likely to resort to partners it has already successfully collaborated with, particularly as more competing organizations are entering the community every year. On the other hand, it shows that project consortia and collaboration ties between organizations behave like an inertial mass, which impedes the involvement of new and fresh organizations, and likely also new ideas and research foci.

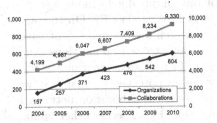

Fig. 4. Organizational collaboration network: Nodes and edges

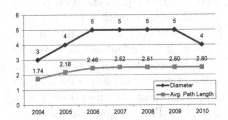

Fig. 5. Organizational collaboration network: Network measures

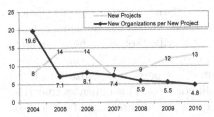

Fig. 6. New organizations introduced by newly launched projects

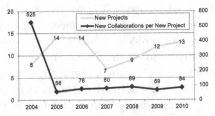

Fig. 7. New collaboration ties introduced by newly launched projects

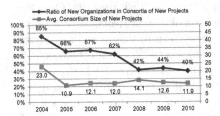

Fig. 8. New organizations in consortia of new projects

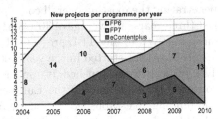

Fig. 9. Number of launched projects by year and programme

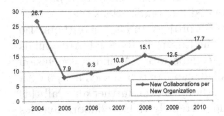

Fig. 10. New collaboration ties introduced by new organizations

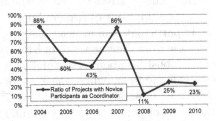

Fig. 11. Novice organizations as project coordinators

7 Key Findings

Funding Bridge. While there are strong ties between FP6 and FP7 in terms of participating organizations, it was demonstrated that *eContentplus* acted as a broker between FP6 and FP7 project consortia. Particularly some Best Practice Networks like ASPECT or ICOPER, and also Targeted Projects like OpenScout, have many strong consortium overlaps with both preceding FP6 projects and succeeding FP7 projects. This pattern is probably simply due to the fact that in 2007 there were neither new project launches in FP6 nor in FP7. On the other hand it could also be attributed to a plain "research follows money" attitude. That is, if there had not been funding from *eContentplus*, organizations would likely have looked for funding opportunities in TEL related programmes with different focus between 2006 and 2008. Anyway, *eContentplus* apparently was supportive and non-disruptive for the organizational collaboration network in European TEL. A question for future research would be whether other funding schemes that do not explicitly carry the "Technology Enhanced Learning" label have similar effects on the project and collaboration landscape.

Role of Project Type. Integrated Projects (IP) and Networks of Excellence (NoE) are prominently placed among those projects with the highest impact on successor projects, whereby this cannot solely be ascribed solely to the larger size compared to e.g. STREPs. For instance, these projects, along with some large *e-Contentplus* consortia, typically also include multiple pairs of organizations that appear in the network of the most frequent collaborators. This indicates that IPs and NoEs are very important not only for shaping the research agenda, but also for creating strong and sustained collaboration ties between TEL organizations.

TEL Family. With every new TEL project, relatively fewer organizations are penetrating the existing overall collaboration network in TEL projects. Over the last three years, an average of 40% of the consortia of new projects was not previously involved in any TEL projects. The sharpest drop in this number occurred for projects that started in the year 2008 (from 62% to 42%), when the first FP7 TEL projects were launched. It appears that at the transition to FP7, the project consortia—and ultimately the European Commission—resorted to building on and funding an established core of organizations, thus strengthening existing collaboration bonds. This has lead to a tightly knit family-like community of TEL organizations, an inertial mass that can impede the involvement of new organizations. This is strengthened by the fact that of the 34 launched TEL projects since 2008, four out of five are being coordinated by organizations that have already participated in at least one previous TEL project. In [4], the authors also conclude from their analyses of FPs 1–4 that the European Research Area builds on a "robust backbone structure" of frequent collaborators. A similar conclusion can be found in [5], where the authors identified a core of established actors with increasing integration over time. In [7] the authors go even further and call this backbone of partners with long-standing and extremely tight collaboration ties the "oligarchic core" of the FPs. In the light of these related studies, we can state that the TEL community exposes similar bonding characteristics as the complete Framework Programme networks. Of course, from the EC's viewpoint it seems reasonable to fund projects where a large share of the consortium have previous experience in EC-funded TEL projects. Still, this appears to be a policy issue that requires attention.

8 Conclusion

This paper has reported analyses, results and implications of the application of social network analysis on European Commission funded TEL projects to provide stakeholders with an overview on historic development and the current state. The three key findings we distilled are that organizations are resourceful in finding alternative funding opportunities; that integrated projects and networks of excellence have a central role in shaping the collaboration landscape; and that the collaboration ties within and across TEL projects in Europe expose characteristics of oligarchic structures.

There are several limitations in the current data sources and the analyses, which will have to be addressed in forthcoming work. Most importantly, the projects dataset currently exclusively contains TEL related projects from FP6, FP7 and *eContentplus*. There are many additional sources and projects that could be included, e.g. the Lifelong Learning Programme, additional projects from the EC's Policy Support Programme, the UK JISC funded projects, and many more. Additionally, several projects have strong associate partnership programmes and funded sub-projects (e.g. STELLAR theme teams) that could be integrated into the analyses. Also, we currently have descriptive project metadata only and do not consider project deliverables. These would significantly augment the potential analysis toolbox with text mining, topic modeling, and information on involved researchers. Finally, the funded projects analyzed in this paper likely represent only a small fraction of the actual collaboration network, since the competition in TEL calls is fierce with very low success rates. It would therefore be worthwhile to include unsuccessful project proposals in the analysis.

To keep interested stakeholders up-to-date with facts and figures from the TEL projects community we deployed a widget-based dashboard [16] for visual interaction with the Mediabase data sources in the Learning Frontiers portal[2]. The next update to the TEL projects data set will arrive later this year, when several new TEL projects will be funded from bids submitted to FP7 ICT Call 8. It remains to be seen how the results of this call will impact the project and collaboration networks. Projecting the past onto the future, we can expect that the new projects to be mainly composed of established organizations, with a few new ones hopefully entering the scene.

Acknowledgments. This work was funded by the European Commission through the 7th Framework Programme ICT Coordination and Support Action TEL-Map (FP7 257822).

---

[2] http://learningfrontiers.eu/?q=dashboard

References

1. Klamma, R., Spaniol, M., Cao, Y., Jarke, M.: Pattern-Based Cross Media Social Network Analysis for Technology Enhanced Learning in Europe. In: Nejdl, W., Tochtermann, K. (eds.) EC-TEL 2006. LNCS, vol. 4227, pp. 242–256. Springer, Heidelberg (2006)

2. Petrushyna, Z., Klamma, R.: No Guru, No Method, No Teacher: Self-classification and Self-modelling of E-Learning Communities. In: Dillenbourg, P., Specht, M. (eds.) EC-TEL 2008. LNCS, vol. 5192, pp. 354–365. Springer, Heidelberg (2008)

3. Voigt, C. (ed.): 4th Evaluation Report – Including Social Network Analysis. Deliverable D7.5, STELLAR Nework of Excellence (2011)

4. Barber, M., Krueger, A., Krueger, T., Roediger-Schluga, T.: Network of European Union–funded collaborative research and development projects. Physical Review E 73 (2006)

5. Roediger-Schluga, T., Barber, M.J.: R&D collaboration networks in the European Framework Programmes: data processing, network construction and selected results. International Journal of Foresight and Innovation Policy 4(3/4), 321–347 (2008)

6. Frachisse, D., Billand, P., Massard, N.: The Sixth Framework Program as an Affiliation Network: Representation and Analysis (2008), http://ssrn.com/abstract=1117966

7. Breschi, S., Cusmano, L.: Unveiling the texture of a European Research Area: emergence of oligarchic networks under EU Framework Programmes. International Journal of Technology Management 27(8), 747–772 (2004)

8. Lozano, S., Duch, J., Arenas, A.: Analysis of large social datasets by community detection. The European Physical Journal Special Topics 143(1), 257–259 (2007)

9. Scherngell, T., Barber, M.J.: Spatial interaction modelling of cross-region R&D collaborations: empirical evidence from the 5th EU framework programme. Papers in Regional Science 88(3), 531–546 (2009)

10. Roediger-Schluga, T., Dachs, B.: Does technology affect network structure? - A quantitative analysis of collaborative research projects in two specific EU programmes. UNU-MERIT Working Paper Series 041 (2006)

11. Derntl, M., Klamma, R.: Social Network Analysis of European Project Consortia to Reveal Impact of Technology-Enhanced Learning Projects. In: 12th IEEE Int. Conf. on Advanced Learning Technologies, ICALT 2012. IEEE (2012)

12. European Commission: Community Research and Development Information Service (CORDIS), http://cordis.europa.eu/home_en.html

13. Derntl, M., Renzel, D., Klamma, R.: Mapping the European TEL Project Landscape Using Social Network Analysis and Advanced Query Visualization. In: 1st Int. Workshop on Enhancing Learning with Ambient Displays and Visualization Techniques, ADVTEL 2011 (2011)

14. Brandes, U., Erlebach, T. (eds.): Network Analysis. LNCS, vol. 3418. Springer, Heidelberg (2005)

15. Brin, S., Page, L.: The anatomy of a large-scale hypertextual web search engine. Computer Networks and ISDN Systems 30, 107–117 (1998)

16. Derntl, M., Erdtmann, S., Klamma, R.: An Embeddable Dashboard for Widget-Based Visual Analytics on Scientific Communities. In: 12th Int. Conf. on Knowledge Management and Knowledge Technologies, I-KNOW 2012. ACM (2012)

TinkerLamp 2.0: Designing and Evaluating Orchestration Technologies for the Classroom

Son Do-Lenh, Patrick Jermann, Amanda Legge,
Guillaume Zufferey, and Pierre Dillenbourg

CRAFT, Ecole Polytechnique Fédérale de Lausanne (EPFL)
1015 Lausanne, Switzerland
{son.dolenh,patrick.jermann,pierre.dillenbourg}@epfl.ch,
{legge.amanda@gmail.com,guillaume.zufferey}@simpliquity.com

Abstract. Orchestration refers to the real-time classroom management of multiple activities and multiple constraints conducted by teachers. Orchestration emphasizes the classroom constraints, integrative scenarios, and the role of teachers in managing these technology-enhanced classrooms. Supporting orchestration is becoming increasingly important due to the many factors and activities involved in the classroom. This paper presents the design and evaluation of TinkerLamp 2.0, a tangible tabletop learning environment that was explicitly designed to support classroom orchestration. Our study suggested that supporting orchestration facilitates teachers' work and leads to improvements in both the classroom atmosphere and learning outcomes.

1   Introduction

Due to the technological evolution in schools, the learning process now involves multiple activities, resources, and constraints in the classroom. Teachers not only have to prepare lesson plans, accommodate curricula, and teach, but also understand and manage various technologies such as interactive whiteboards and computers, and improvise the lesson when appropriate. This real-time management of multiple activities with multiple constraints conducted by teachers, also known as *classroom orchestration*, is crucial for the materialization of learning.

Orchestration emphasizes the classroom constraints and the teachers' role in managing these technology-enhanced classrooms. Although occasionally mentioned in the literature [26,6], until recently, orchestration has not received much attention from the CSCL community [7,15,9]. It has been argued that orchestration is important for more technology adoption in authentic classrooms [8].

Orchestration technologies are tools that assist the teachers in their task of orchestrating integrated classroom activities. They aim to provide support for teachers, who will then be able to orchestrate and manage the class on-the-fly, intervening with students to adapt teaching plans and learning activities. While a few early examples of technologies designed to support orchestration have started to emerge [3,1], little work explores the requirements and guidelines for the design of such technologies in real classroom settings.

A. Ravenscroft et al. (Eds.): EC-TEL 2012, LNCS 7563, pp. 65–78, 2012.

Motivated by the increasing need for better orchestration support in classroom settings and the lack of design guidelines, we developed TinkerLamp 2.0, an interactive tabletop learning environment that explores the design space of orchestration technologies (Figure 1). TinkerLamp 2.0 draws upon TinkerLamp 1.0 [14,28,10], which was designed to support the training of vocational apprentices in logistics. TinkerLamp 2.0 introduces new and redesigned features that explicitly support classroom orchestration as well as new classroom practices that lead to improvements in both the classroom atmosphere and learning outcomes.

This paper presents the design, implementation, and evaluation of the TinkerLamp 2.0 learning environment and its supporting orchestration tools. Our study of the environment, which involved 6 classes and 93 vocational apprentices, showed that the system facilitated the teachers' work, making it easier for them to manage both the class and the learning resources in real-time. Importantly, it resulted in more opportunities for reflection, higher learning outcomes, better support for class-wide activities, and a more playful atmosphere, compared to two baseline conditions including an identical system without orchestration support.

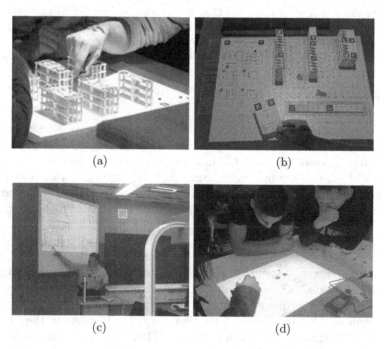

(a)   (b)

(c)   (d)

Fig. 1. The components of TinkerLamp 2.0: (a) Tangible model, (b) TinkerKey, (c) TinkerBoard, (d) TinkerQuiz

2   Related Work

As argued by [8], the success of a learning system in a technology-enhanced classroom environment increasingly depends on considering the technology in a broader context, including classroom orchestration, rather than just focusing on a positive learning outcome in a lab setting. Orchestration, which refers to the teachers' real-time classroom management of multiple activities with multiple constraints, promotes productive learning in a class ecosystem by using integrative scenarios and empowering teachers [7,15,9,22]. More specifically, it concerns the integration of activities at multiple social planes within the classroom, such as individual reading, team argumentation, and plenary sessions [8].

Recent studies have shown the benefits of considering orchestration in the classroom [22,15]. For example, [15] ran a study with a number of eighth-grade high school classrooms studying Biology. The results of this study demonstrated that orchestration, in this case alternating plenary, small group, and dyadic learning phases, led to higher levels of learning competence than having all activities at only one level.

How we can design learning environments and technologies that facilitate orchestration is still an open question. An early example of orchestration technologies is the One Mouse Per Child project [3]. It provides the teachers with a visualization display that shows simplified aggregated data about each of the 40 children in the room. This information is displayed permanently for the teachers to facilitate their awareness of the class progress and individual statuses without posing queries. Mischief [16] is a teaching system designed to enhance social awareness between collocated students and support classroom-wide interactions. Mischief enables the simultaneous interaction of up to 18 students in a classroom using a large shared display.

nQuire [18] is a system developed to guide personal inquiry learning, sharing the orchestration responsibility between the teachers and the students. It allows inquiries to be created, scripted, configured and used, all on-the-fly, by either role. nQuire incorporates different technological devices and promotes the support for inquiry activities across individual, group, and class levels at different parts of the inquiry. Some other research provides logistic support for teachers to monitor students' activities by connecting multi-touch tabletops in the classroom to the teacher's desk [1] in order to cope with different levels of student expertise [17], provide task-specific context to the teacher [23], and provide distributed awareness tools to the tutor [2].

Orchestration technologies are still in their infancy stage. As argued by [9], most previous research focuses more on the core pedagogical task, at the individual or group level of collaborative learning. Therefore, research is needed to explore guidelines specifically designed for the development of learning environments that explicitly support teacher orchestration and activities at multiple levels in the classroom. This paper presents such an attempt, using tabletop technology to develop our orchestration environment.

Tangible tabletops have been researched and used across many educational contexts [11,20,27]. Tangible tabletops can be effective in supporting co-located learning by providing a large shared workspace, increasing group members' awareness [13]. They also enable more participation and active learning thanks to their simultaneous interaction capabilities [25,21]. Tangible tabletops support building learning activities in which users can interact directly with their hands by touching and manipulating objects. This sensori-motor experience that tangible tabletops offer has been described as beneficial for learning [21], relying on the idea that they support an enactive mode of reasoning [4,19], and that they leverage metaphors of object usage and take advantage of the close inter-relation between cognition and perception of the physical world [12,24].

3   The TinkerLamp 2.0 Environment

TinkerLamp 2.0, an interactive tangible tabletop learning environment, is designed to help logistics apprentices understand theoretical concepts presented at school by letting them experiment with these concepts on an augmented small-scale model of a warehouse. In terms of hardware, TinkerLamp 2.0 consists of a projector and a camera, which are mounted in a metal box suspended above the tabletop.

The TinkerLamp 2.0 was the result of our evaluation and re-design of the TinkerLamp 1.0 system which was deployed in several vocational schools for two years. We conducted field evaluations and controlled experiments of this 1.0 system with nearly 300 students and 8 teachers in several separate studies from 2008 to 2010 [14,28,10,24].

While the TinkerLamp 1.0 focused on supporting group-level activities, TinkerLamp 2.0 introduces a whole new set of functionalities for explicitly supporting classroom-level activities and teacher orchestration, therefore allowing the continuity of learning throughout the entire classroom. These orchestration tools include a teacher-exclusive card-based interface, a public display, and a collection of interactive quizzes. In addition, we redesigned the learning scenario to include new learning activities not available in the TinkerLamp 1.0 system.

Following, we briefly present the features inherited from TinkerLamp 1.0 before describing the new orchestration and learning tools in TinkerLamp 2.0.

3.1   Inherited Features: Small-Scale Model and TinkerSheet

Apprentices interact with the TinkerLamp 2.0 through two interaction modalities inherited from TinkerLamp 1.0: a tangible warehouse model and a paper-based interface, called TinkerSheet (Figure 2).

Users interact with the warehouse model using miniature plastic shelves, docks, and offices. Each element of this small-scale warehouse is tagged with a fiducial marker that enables automatic camera object recognition. The model is augmented with visual feedback and information through a projector in the lamp's head. The apprentices can also run simulations on the models, which compute statistics related to the physical structure of the warehouse such as the areas used for storing goods, the distance between shelves, etc. The simulations use simple models of customers and suppliers that generate a flow of goods

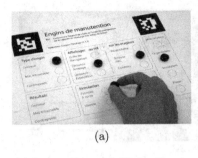 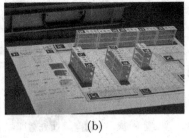

(a)   (b)

Fig. 2. (a) Interacting with TinkerSheet using tokens, (b) The small-scale model next to a TinkerSheet

entering and leaving the warehouse in real-time. This real-time simulated information (e.g. animation of how forklifts approach the shelves, statistics about the warehouse inventory, etc.) is displayed directly on top of the model and on the TinkerSheet.

A TinkerSheet is a piece of paper automatically tracked in real-time by fiducial markers that allow users to control the system (e.g. setting parameters for the simulation, changing the size of the forklift, etc.). It also serves as a visual feedback space on which textual or graphical summary information from the simulation is projected (e.g. the warehouse statistics such as surface areas, degree of use, etc.). Interaction with a TinkerSheet is primarily performed by using a small physical token.

3.2   TinkerKey: Orchestration Card for Teacher

We observed that the teacher needs to be empowered in a classroom equipped with multiple TinkerLamps. With TinkerLamp 1.0, the teacher's job was limited to walking around the class and discussing with students. His role was "weakened" by the TinkerLamps, in the sense that the students were too engaged in the simulation, often ignoring his instructions. This was not ideal for their learning because, without the presence and guidance of the teacher, the tangible interface sometimes tempted the apprentices to manipulate too much. This led to less intensive cognitive effort to understand the solutions, and less useful reflection and discussion for learning [5,10].

We aimed to support the teacher's orchestration to alleviate this problem. When properly supported, the teacher presense could notably increase the reflection level, and in turn, learning outcomes of his students with the lamps. For example, when present, the teacher could pose reflective questions to individual students and encourage group discussions and comparisons.

We developed TinkerKey, a small paper card used by the teacher to orchestrate the class (Figure 3). Its purpose is to prevent manipulation temptation by empowering the teachers. It provides them with special privileges when interacting with the system, enabling them to adapt and improvise the current learning situation to their ever-changing orchestration plan.

(a)   (b)

Fig. 3. (a) The "Allow Simulation" TinkerKey to allow/block simulation for a group, (b) The "Pause Class" TinkerKey to pause the whole class

The scenario is envisioned as follows. The teacher keeps a set of TinkerKeys in his hand while touring the classroom as usual. When he needs to intervene with a group or the class as a whole, he places a card on the group's table (or any group's table, in case of the class TinkerKey). Each TinkerKey triggers a different functionality in the TinkerLamp, such as changing a state or performing an action, which affects either the group for which it is used, or the whole class, thereby helping the teacher improvise the learning activity.

Five TinkerKeys were implemented and tested in TinkerLamp 2.0. The "Allow Simulation" TinkerKey aims to prevent the students from running too many simulations without much reflection or ignoring teacher instructions. By flipping the card over, the teacher can block the students' capability to run simulations. The groups were not authorized to run a simulation without the teacher's permission, requiring contact with the teacher who could then ask them to predict, explain, and compare the performance of the current layout with that of the previous simulation.

The "Pause Class" TinkerKey is used at the class level. It helps the teacher easily and quickly get full attention from the students in order to give instructions or change from a group to a class-wide activity. This TinkerKey blanks out all of the projected feedback from the TinkerLamps on each table. As soon as it is placed on any group's table, the 'pause' command transfers from that group's lamp to the other lamps.

The other three TinkerKeys allow the teacher to intervene with a group and ask questions more effectively than before. These cards hide or show the statistics of the warehouse layouts the group has built. This enables the teacher to ask the students to predict and reflect during the building and simulation session.

The design of the TinkerKeys is lightweight and unobtrusive, making it possible for the teacher to maintain his usual class behaviors. On the other hand, the TinkerKey cards supplement the teacher's abilities by giving him simple but powerful privileges to better orchestrate the class.

3.3 TinkerBoard: Orchestration Awareness Display

There was a problem of class awareness with the TinkerLamp 1.0. Each group moved at their own pace of exploration when working with their own Tinker-Lamp. The teacher had difficulty keeping track of each group's progress, so the time he spent with each group was not optimized. We observed that the pattern of the teacher's movements, and hence the classroom dynamics, was fairly spontaneous and subject to frequent changes. For example, on several occasions, two or more groups made simultaneous requests, and the teacher could not decide which group to help first.

Moreover, class-wide activities were not adequately supported with Tinker-Lamp 1.0. The teachers had no means of displaying the built layouts or intermediate steps taken by a group to the whole class, so class debriefings were difficult to perform. Instead, the teachers asked the apprentices to trace their solutions on the TinkerSheet and reproduce them on the blackboard, which limited the debriefing to only those layouts which were traced. While one could argue that this manual transfer of layouts could be useful in terms of reflection, there was a discontinuity of media and learning during the long process of transferring the layouts.

Fig. 4. Teacher using TinkerBoard for a spontaneous debriefing with his class

The TinkerBoard of TinkerLamp 2.0 (Fig.4) tackles this problem of class awareness and facilitates the conducting of class-wide debriefings.

Class Awareness Display. First, this tool can be used as an awareness display. It displays the whole class history on a big projection board but in a very minimalist manner and requires little intervention and interaction from the teacher. The information provided by this awareness display can facilitate the teacher's orchestration, giving him a mechanism to quickly assess the class progress as a whole and plan his next action.

TinkerBoard includes a) an *event bar* showing what activity each group is doing (building models/doing quizzes/running simulations/etc.) and how intensively, and b) a *layout history* displaying all of the layouts each group has saved during the activity.

TinkerBoard can be beneficial to mediate simultaneous help requests in that the teacher can determine who needs help the most based on the number of saved layouts (how advanced they are in the activity), as opposed to being spontaneous like in TinkerLamp 1.0. By looking at the display, he can also tell if a group is doing too many manipulations. He can then intervene to encourage more thinking and less manipulating.

This information is also designed to support student's reflection and social learning. By looking at the event bar, the students can be more aware of the activity structure of their group and other groups, and hopefully regulate their actions. By looking at the layout history, they can compare the different layouts they have built over time or other layouts from other groups.

Continuity of Activities, Class-Wide Debriefing and Inter-Group Activities. This is the second aspect of orchestration supported by Tinker-Board. TinkerBoard enables the access of any intermediate layouts built by the students, not just the final 'best' ones transferred to the blackboard with Tinker-Lamp 1.0, thereby maintaining the continuity of activity and facilitate class-wide debriefing. There is a dedicated area on the TinkerBoard, called the Comparison Zone, which allows the teacher to explicitly compare different layouts and statistics from different groups during debriefings, explaining their advantages and disadvantages. He does this by choosing interesting layouts from specific groups and displays their statistics on the Comparison Zone for side by side comparisons. Moreover, TinkerBoard provides support for the teacher to conduct a class-wide TinkerQuiz (described below). He can send selected layouts from TinkerBoard to all of the TinkerLamp groups through the network and engage them in a playful class-wide competition.

3.4   TinkerQuiz

TinkerQuiz was designed to introduce a new way of moving from group- to class-level activity using the TinkerLamp, encouraging students to be more reflective but in a fun and engaging way (Figure 5). The TinkerLamp 2.0 supports four TinkerQuiz cards. Each card has a different question, involving the comparison

Fig. 5. (a) A group choosing a response with a TinkerQuiz (b) A group cheering after winning the class quiz

of two warehouse layouts according to a specific criterion. The TinkerQuiz card is small with different colors and icons on it to give it the feel of a game. When a quiz is placed under the lamp and started, two graphical layouts appear above the quiz. A countdown timer also appears, showing how much time remains to finish the quiz. This is intended to deliver a sense of pressure to the students. Students interact with the TinkerQuiz like the TinkerSheet, with a small token. Depending on whether the token is placed on the correct or incorrect answer, it will submit an answer or show the solution, respectively.

The layouts used for the TinkerQuiz are chosen at run-time either by the teacher from the TinkerBoard for a between-groups quiz or randomly by the system among a "museum" of saved layouts for a within-group quiz. This capability allows the teacher to seamlessly move from a group activity (i.e. each group doing quizzes locally) to a class activity (i.e. the whole class doing a class-wide quiz together) just by issuing a command from the TinkerBoard.

4 Evaluation of TinkerLamp 2.0

4.1 Participants and Setup

We conducted an ecologically valid comparison between a baseline *paper/pencil* condition, the *TinkerLamp 1.0* condition, and two alternative variations of *TinkerLamp 2.0*: with and without the TinkerBoard component. A total of 2 teachers and 6 classes were involved in the study: 31 students (2 classes) in the paper/pencil, 30 students (2 classes) in the TinkerLamp 1.0, and 32 students (2 classes) in both TinkerLamp 2.0 conditions.

4.2 Learning Task

We used an authentic learning task that is typically used in the school to teach different types of surfaces involved in the warehouse design process, e.g. raw surface, net storage surface, etc. Each group was asked to collaboratively build models, and then compare and reflect on what they had built to understand the different types of surfaces. In the paper/pencil condition, they drew warehouse models on paper using pens, erasers, and rulers. In the TinkerLamp conditions, the group built the warehouse layouts using the tangible model.

4.3 Task Structure

In total, each classroom trial lasted approximately three hours. The teachers began class by introducing definitions on the blackboard. Then, the class was divided into four groups to perform the learning task. In both conditions, the teacher toured around the room to respond to help requests. At the end of the learning session, the teacher organized a debriefing session where the conclusions of each group were discussed. A post-test, consisting of 12 multiple-choice and 1 open-ended question, was used at the end of the class to evaluate the learning outcomes, in terms of understanding and problem-solving performance.

5   Results

5.1   Learning Outcomes

Statistical tests showed that the TinkerLamp 2.0 system (namely the With-TinkerBoard condition) resulted in higher results in both understanding score and problem-solving score than the TinkerLamp 1.0 and paper/pencil condition. Table 1 summarizes the learning scores of all of the conditions.

Table 1. The average learning outcome scores (and standard deviation)

	Paper/pen	TinkerLamp 1.0	TinkerLamp 2.0 NoTinkerBoard	TinkerLamp 2.0 WithTinkerBoard
Understanding	7.84(2.85)	7.43(2.82)	9.38(2.03)	10.31(1.70)
Problem-solving	5.16(1.70)	5.15(1.78)	6.44(1.65)	6.59(1.53)

Understanding Score. An ANOVA test on a mixed-effect model using group as random factor ($F(3, 21) = 3.98, p < .05$) and a pair-wise Tukey test showed that the scores in the TinkerLamp 2.0 WithTinkerBoard condition were significantly higher than both the TinkerLamp 1.0 ($z = 3.05, p < .01$) and the paper/pencil ($z = 2.60, p < .05$) conditions. None of the other pair-wise comparisons is significant.

Problem-Solving Score. Similar tests found a significant difference between the four conditions in terms of problem-solving ($F(3, 21) = 4.42, p < .01$). The Tukey contrast showed that the WithTinkerBoard condition was significantly higher than both the TinkerLamp 1.0 ($z = 2.72, p < .05$) and the paper/pencil ($z = 2.71, p < .05$) conditions; the NoTinkerBoard condition was marginally higher than both the TinkerLamp 1.0 ($z = -2.42, p = .07$) and the paper/pencil ($z = -2.41, p = .07$) conditions. No other significant difference was found.

5.2   Class Atmosphere and Satisfaction

We distributed a questionnaire to students, asking them to rate the group and class atmosphere and their satisfaction of the class in general. The students felt that the presence of TinkerBoard (which implied the presence of class-wide TinkerQuizz) significantly influenced their perception, when compared with class without the TinkerBoard, in three aspects (confirmed by Wilcoxon-test) by: 1) encouraging more collaboration within their group ($W = 60.5, p < .05$), 2) making the class more fun ($W = 72, p < .05$), and 3) encouraging more comparison of their group's layouts with those of other groups ($W = 65, p < .05$).

This proved that the TinkerBoard fulfilled its goal to bridge the different activities and facilitate the continuity of learning. It enabled class-wide activities for a more playful and collaborative classroom by seamlessly transitioning from the building phase to the debriefing phase and transitioning from group activity to class activity. This board is more than just a monitoring or awareness tool. It is a classroom orchestration tool, supporting both teachers and students at the same time.

5.3   TinkerKey for Empowering Teachers

We observed the teachers using all TinkerKeys throughout the activity in order to pose questions, encourage the students to reflect, and pause the groups to call attention to the class (Table 2). The 'Pause Group' and 'Pause Class' card was used extensively before every debriefing session or class-wide instruction. It clearly helped the teachers gain full class attention compared to previous studies.

Two specific TinkerKeys ('Hide Current Stats' and 'Hide Saved Stats') were used by the teachers throughout the activity to hide statistics for individual groups in order to pose a question. After hiding the stats, the teachers encouraged the students to reflect and discuss the layout before showing them the solution with a TinkerKey card. They also used the 'Allow Simulation' card extensively. We noted students predicting and discussing about the simulation with the teacher because the groups were not authorized to run a simulation without him.

Table 2. The number of use of each TinkerKey with TinkerLamp 2.0

TinkerKey	Number of uses	
	NoTinkerBoard	WithTinkerBoard
1. Hide Current Stats	10	26
2. Hide Saved Stats	12	11
3. Allow Simulation	27	22
4. Pause Group	10	25
5. Pause Class	4	6

In the interview, we were able to confirm our observations that both teachers used all TinkerKeys for the purpose that they were designed. Teacher comments included *"I can use the card to request the students to answer questions and confirm if they're correct. This allows me to vary the activity according to the group expertise and the time available"* and *"Instead of losing time telling the students to be quiet, (with the Pause cards) they have to turn to me and wait for my instructions."*

5.4   TinkerBoard for Class Awareness and Debriefing

The teachers were observed looking at the TinkerBoard often, usually when they finished discussing with a group. Both teachers confirmed this observation, and added that the TinkerBoard was not distracting. They said they used the TinkerBoard to see how much time each group spent building models, running simulations, and saving layouts, and to balance the pace between groups.

We observed that the TinkerBoard enables discussion at all social levels at anytime without having to do any extra interactions with the system: the teacher can discuss with the students just by walking up to the TinkerBoard and referring to the layouts or events permanently and publicly shown on it. Having the TinkerBoard in the class led to 5 more spontaneous debriefings during the activity compared to the other classes.

The class debriefing at the end of the activity was prepared much faster than the traditional blackboard usage. The teachers simply dragged each group's chosen layout into the comparison zone on the TinkerBoard and started the debriefing right after. In addition, the layout history was available during the debriefing, making references to the intermediate solutions and statistics possible.

5.5 TinkerQuiz for Class-Wide Comparison

Although taking place at the end of the activity for a limited amount of time (about 10 minutes), the enthusiasm for these quizzes was notable, as the winning groups always cheered. The students were very excited and the whole classroom turned into a "field" for playful competition.

Both teachers reported that the use of the TinkerQuiz to move from group-level to class-level activity was easy. Consistent with our statistics of class atmosphere, both teachers said that the class TinkerQuiz clearly increased the students' reflection and motivation, and maintained the flow and continuity of activities compared to the previous version of TinkerLamp. They said that the students really enjoyed it and were still talking about it in their next class.

6 Discussion and Conclusion

Our evaluations of the TinkerLamp 2.0 system showed that it fulfilled its design goals. The findings showed that the system provided more options for classroom orchestration by empowering the teachers (TinkerKey), supporting class awareness and facilitating both group and class-wide debriefing (TinkerBoard), as well as encouraging inter-group competition (TinkerQuiz). This orchestration support offered many opportunities for reflection and discussion. The continual transition between group- and class-wide activities supported by the TinkerBoard component seemed to bring a more playful and collaborative atmosphere into the classroom. Although these results still need to be confirmed with a larger sample, it is likely that TinkerLamp 2.0 improved student's learning outcomes (compared to the other conditions) for these reasons.

Overall, the three orchestration tools presented in this paper are diverse in terms of technology use and their orchestration goals. However, learning can be improved if the activities are integrated and exploited at different levels [8]. The combination of our three tools supported the continuity of learning workflow in the classroom by giving the same resource (i.e warehouse layouts) different representations and circulating them in the classroom. We hence recommend future work to consider supporting orchestration by *developing for the whole learning workflow with an ecology of resources*, rather than a stand-alone application.

We showed that orchestration and reflection are *related*. Supporting the teacher with his classroom orchestration led to an improvement in reflection and learning in the classroom. Providing the teacher with appropriate tools that enable him to interact with the group and the class more effectively and efficiently is a way to encourage high-level discussion, at both the group and class level, which is important for learning.

We support the idea that orchestration technologies need to be *flexible* and *minimal*, among other features. These two principles allow teachers to improvise their actions to the unfolding events without adding more workload. Our

TinkerKey cards allow for flexible management of the classroom: the teacher can use any TinkerKey at anytime just by picking out the card he needs from his hand. Due to their minimalist design and light weight, the teacher can easily carry them while touring the class. Similarly, TinkerBoard is flexible and minimal in that it enables reflection at anytime and provides basic but critical awareness about the group's progress in the class. Using this public display, the teacher can spontaneously debrief with the class without having to do any extra interactions.

This paper presents our effort in developing TinkerLamp 2.0, a learning environment that explicitly supports orchestration and can be used in real classrooms. The evaluation results in a promising confirmation of our approach. Supporting classroom orchestration not only facilitated the teacher in dealing with multiple TinkerLamps in the classroom, but also seemed to improve students' learning outcomes. We hope that our experience gives an early example of how orchestration technologies can be developed and how they can impact learning and classroom atmostphere in authentic settings.

Acknowledgments. This research is part of the Dual-T project funded by the Swiss Federal Office for Professional Education and Technology. We would like to thank Olivier Guedat, the teachers and students for their support.

References

1. AlAgha, I., Hatch, A., Ma, L., Burd, L.: Towards a teacher-centric approach for multi-touch surfaces in classrooms. In: ACM ITS 2010, pp. 187–196 (2010)
2. Alavi, H., Dillenbourg, P., Kaplan, F.: Distributed Awareness for Class Orchestration. In: Cress, U., Dimitrova, V., Specht, M. (eds.) EC-TEL 2009. LNCS, vol. 5794, pp. 211–225. Springer, Heidelberg (2009)
3. Alcoholado, C., Nussbaum, M., Tagle, A., Gomez, F., Denardin, F., Susaeta, H., Villalta, M., Toyama, K.: One mouse per child: interpersonal computer for individual arithmetic practice. Journal of Computer Assisted Learning (2011)
4. Bruner, J.S.: Toward a Theory of Instruction. Belknap Press, Cambridge (1966)
5. de Jong, T.: The design of effective simulation-based inquiry learning environments. In: Proc of Conf. on Learning by Effective Utilization of Technologies: Facilitating Intercultural Understanding, pp. 3–6 (2006)
6. DiGiano, C., Patton, C.: Orchestrating handhelds in the classroom with SRI's ClassSync. In: Proc. of CSCL, pp. 706–707 (2002)
7. Dillenbourg, P., Jarvela, S., Fischer, F.: The evolution of research on computer-supported collaborative learning. In: Technology-Enhanced Learning, pp. 3–19 (2009)
8. Dillenbourg, P., Jermann, P.: Technology for Classroom Orchestration. In: Khine, M.S., Saleh, I.M. (eds.) New Science of Learning, pp. 525–552. Springer Science+Business Media, New York (2010)
9. Dillenbourg, P., Zufferey, G., Alavi, H.S., Jermann, P., Do-Lenh, S., Bonnard, Q., Cuendet, S., Kaplan, F.: Classroom orchestration: The third circle of usability. In: Proc. of CSCL, vol. 1, pp. 510–517 (2011)
10. Do-Lenh, S., Jermann, P., Cuendet, S., Zufferey, G., Dillenbourg, P.: Task Performance vs. Learning Outcomes: A Study of a Tangible User Interface in the Classroom. In: Wolpers, M., Kirschner, P.A., Scheffel, M., Lindstaedt, S., Dimitrova, V. (eds.) EC-TEL 2010. LNCS, vol. 6383, pp. 78–92. Springer, Heidelberg (2010)

11. Horn, M.S., Solovey, E.T., Crouser, R.J., Jacob, R.J.: Comparing the use of tangible and graphical programming languages for informal science education. In: CHI 2009, pp. 975–984. ACM, New York (2009)

12. Hornecker, E., Buur, J.: Getting a grip on tangible interaction: a framework on physical space and social interaction. In: CHI 2006, pp. 437–446 (2006)

13. Hornecker, E., Marshall, P., Dalton, N.S., Rogers, Y.: Collaboration, interference: awareness with mice or touch input. In: CSCW 2008: Proc. of the ACM Conf. on Computer Supported Cooperative Work, pp. 167–176 (2008)

14. Jermann, P., Zufferey, G., Dillenbourg, P.: Tinkering or Sketching: Apprentices' Use of Tangibles and Drawings to Solve Design Problems. In: Dillenbourg, P., Specht, M. (eds.) EC-TEL 2008. LNCS, vol. 5192, pp. 167–178. Springer, Heidelberg (2008)

15. Kollar, I., Wecker, C., Langer, S., Fischer, F.: Orchestrating web-based collaborative inquiry learning with small group and classroom scripts. In: TEI 2009: Proc. of the 3rd Int. Conf. on Tangible and Embedded Interaction, pp. 77–84 (2011)

16. Moraveji, N., Kim, T., Ge, J., Pawar, U.S., Mulcahy, K., Inkpen, K.: Mischief: supporting remote teaching in developing regions. In: CHI, pp. 353–362. ACM (2008)

17. Moraveji, N., Morris, M., Morris, D., Czerwinski, M., Henry Riche, N.: Classsearch: facilitating the development of web search skills through social learning. In: CHI, pp. 1797–1806. ACM, New York (2011)

18. Mulholland, P., Anastopoulou, S., Collins, T., Feisst, M., Gaved, M., Kerawalla, L., Paxton, M., Scanlon, E., Sharples, M., Wright, M.: nquire: Technological support for personal inquiry learning. IEEE Transactions on Learning Technologies (2011)

19. Piaget, J.: The future of developmental child psychology. Journal of Youth and Adolescence 3, 87–93 (1974).

20. Price, S., Falcao, T.P., Sheridan, J.G., Roussos, G.: The effect of representation location on interaction in a tangible learning environment. In: Proc. of TEI, pp. 85–92. ACM, New York (2009)

21. Price, S., Rogers, Y.: Let's get physical: the learning benefits of interacting in digitally augmented physical spaces. Comput. Educ. 43(1-2), 137–151 (2004)

22. Prieto, L.P., Villagrá-Sobrino, S., Jorrín-Abellán, I.M., Martínez-Monés, A., Dimitriadis, Y.: Recurrent routines: Analyzing and supporting orchestration in technology-enhanced primary classrooms. Comput. Educ. 57, 1214–1227 (2011)

23. Roschelle, J., Rafanan, K., Estrella, G., Nussbaum, M., Claro, S.: From handheld collaborative tool to effective classroom module: Embedding cscl in a broader design framework. Computers & Education 55(3), 1018–1026 (2010)

24. Schneider, B., Jermann, P., Zufferey, G., Dillenbourg, P.: Benefits of a tangible interface for collaborative learning and interaction. IEEE Transactions on Learning Technologies 4(3), 222–232 (2011)

25. Stanton, D., Neale, H., Bayon, V.: Interfaces to support children's co-present collaboration: multiple mice and tangible technologies. In: CSCL 2002: Conf. on Computer Support Collaborative Learning, pp. 342–351 (2002)

26. Tomlinson, C.: The differentiated classroom: responding to the needs of all learners. Association for Supervision and Curriculum Development (1999)

27. Zuckerman, O., Arida, S., Resnick, M.: Extending tangible interfaces for education: digital montessori-inspired manipulatives. In: CHI 2005, pp. 859–868 (2005)

28. Zufferey, G., Jermann, P., Do-Lenh, S., Dillenbourg, P.: Using augmentations as bridges from concrete to abstract representations. In: BCS HCI 2009, pp. 130–139 (2009)

Understanding Digital Competence in the 21st Century: An Analysis of Current Frameworks

Anusca Ferrari[*], Yves Punie, and Christine Redecker

Institute for Prospective Technological Studies (IPTS),
European Commission, Joint Research Centre,
Edificio Expo, C/Inca Garcilaso 3,
41092 Seville, Spain
{Anusca.Ferrari,Yves.Punie,Christine.Redecker}@ec.europa.eu

Abstract. This paper discusses the notion of digital competence and its components. It reports on the identification, selection, and analyses of fifteen frameworks for the development of digital competence. Its objective is to understand how digital competence is currently understood and implemented. It develops an overview of the different sub-competences that are currently taken into account and builds a proposal for a common understanding of digital competence.

Keywords: Digital Competence, 21st century skills, Frameworks, Key Competences.

1 Pinning Down Digital Competence

The rapid diffusion and domestication of technology [1] is transforming a core competence such as literacy into a 'deictic' concept [2]: rapidly changing in meaning as new technologies appear and new practices evolve. Today, it is argued, we read, write, listen, and communicate differently than we did 500 years ago [3]. It is thus not unreasonable, in our e-permeated society [4], to think of digital competence as a basic need if we are to function in society [5], as an essential requirement for life [6], or even as a survival skill [7]. The concept of digital competence is a multi-faceted moving target. It is interpreted in various ways in policy documents, academic literature, and teaching/learning and certification practices. Just within the European Commission, initiatives and Communications refer to Digital Literacy, Digital Competence, eLiteracy, e-Skills, eCompetence, use of IST underpinned by basic skills in ICT, basic ICT skills, ICT user skills [8]. Academic papers add to this already long list of terms with 'technology literacy' [9], 'new literacies' [3], or 'multimodality' [10]. They also underline how digital literacy is intertwined with media and information literacy [11-14] and is at the core of the 21 century skills [15].

This paper explores how the concept of digital competence is approached in fifteen selected frameworks. The aim of this collection is to identify and analyse examples where digital competence is fostered, developed, taught, learnt, assessed or certified

---

[*] The views expressed in this article are purely those of the authors and may not in any circumstances be regarded as stating an official position of the European Commission.

A. Ravenscroft et al. (Eds.): EC-TEL 2012, LNCS 7563, pp. 79–92, 2012.

to understand which competences are taken into account. The paper is structured as follows. After this first introductory chapter, Chapter 2 reports on the main current academic discourses around digital competence. Chapter 3 summarises the methodology for the collection of the cases and lists the frameworks that have been considered. Chapter 4 compares how the different cases define digital competence; and Chapter 5 maps competence components. Chapter 6 offers some conclusions.

2   Digital Competence Rhetorics

According to the National Council for Curriculum and Assessment (NCCA) [16], there are three frequently cited arguments for promoting ICT in education. The first relates to the potential benefits of ICT for teaching and learning, including gains in students' achievement and motivation. The second acknowledges the pervasiveness of technologies in our everyday lives. As a consequence, the third argument warns against low levels of Digital Competence that need to be tackled to allow all citizens to be functional in our knowledge society [7]. These arguments fuel a series of digital rhetorics [term elaborated from 17], i.e. received discourses built on an elaborated and distinctive theoretical or ideological stand. Among the most notable digital rhetorics, the following inter-twined discourses can be pulled out: the 'digital divide' rhetoric, the 'digital native' rhetoric, the 'digital competence for economic recovery' rhetoric. The term 'digital divide' came into use in the 90s and alludes to the differences in access to ICT and the Internet [18]. As argued by Molnar [19], new types of digital divide have emerged that go beyond access. In this line, Livingstone & Helsper built a taxonomy of uses defining gradations of digital inclusion as a ladder of participation [20]. Instead of delimiting a new binary divide – as was the case in the "Falling through the Net" report [21], which splits haves and have-nots – Livingstone & Helsper propose a continuum of use, which spreads from non-use of the internet to low and more frequent use. A third perspective of the digital divide comes from Erstad, who argues that digital inclusion depends more on knowledge and skills than on access and use [22]. The second digital rhetoric strand builds on the notion of 'digital natives' introduced by Prensky [23] to bring forward the idea that today's generation of young people have grown up surrounded by technologies rather than books and should be taught through technological means rather than traditional ones. The notion has not gone without criticisms: from the fact that these assertions are based on no, or anecdotal, empirical evidence [24]; to the fact that the metaphor has been understood as a claim for the higher digital competence of younger people, who in fact display a high variety of skills and knowledge regardless of the time spent online [25]. The third rhetoric discourse highlighted here argues that to fully participate in life people must be digitally competent [26], and that there is a need to invest in digital skills enhancement for economic growth and competitiveness [27, 28]. Computer-related proficiency is claimed to be the key to employability and improved life chances [26]. In the last decade, competences related to technologies have started to be understood as "life skills", comparable to literacy and numeracy, therefore becoming "both a requirement and a right" [29].

2.1 Digital Competence at the Convergence of Multiple Literacies

ICT usage is becoming more extensive across society: more people are using technologies for more time and for different purposes. The extensiveness of use is moreover derived from the digitalisation of society in general, as many of the activities we undertake have a digital component. As society is becoming digitalized, the competences needed are becoming manifold. For this reason, Digital Competence is currently being defined as closely related to several types of literacy [7, 11, 26], namely: ICT literacy, Internet Literacy, Media Literacy, and Information Literacy. Analysing the repertoire of competences related to the digital domain requires an understanding of these underlying aspects, which will be briefly explained here.

ICT literacy is generally understood as computer literacy and refers to the ability to effectively use computers (hardware and software) and related technologies. Simonson, Maurer, Montag-Torardi & Whitaker [30] define computer literacy as "an understanding of computer characteristics, capabilities and applications, as well as an ability to implement this knowledge in the skilful and productive use of computer applications". The different definitions of ICT literacy developed in the 80s are all along the same lines and have survived unaltered for over twenty years [31].

Internet literacy refers to the proficient use of the Internet. Van Deursen [32] points out that, regardless of the fact that the expression 'Internet literacy' refers to a specific tool or medium, it underlies a basic understanding of computer functioning, and the ability to understand information, media, and to communicate through the Internet. For Hofstetter & Sine [33], Internet literacy relates to connectivity, security, communication and web page development. It should be noted that Internet literacy is quickly evolving, as nowadays web page development is not as central as the proficient use of web 2.0 tools is.

Media literacy is the ability to analyse media messages and the media environment [34]. It involves the consumption and creation of media products for television, radio, newspapers, films and more recently the Internet. Media education is typically concerned with a critical evaluation of what we read, hear and see through the media, with the analyses of audiences and the understanding of the construction of media messages [13]. It involves communication competences and critical thinking. For Ofcom (the UK communication regulator), media literacy is "the ability to access, understand and create communications in a variety of contexts" [35].

Though information literacy has many similarities with media literacy, and is now extremely relevant for Internet use, it is built on the tradition of librarians and started as the ability to retrieve information and understand it. The American Library Association [36] defines it as 'the ability to recognise when information is needed and the ability to locate, evaluate, and use the needed information effectively'.

2.2 Digital Competence as a New Literacy

The above definitions of the different literacies and the digital 'rhetorics' outlined at the beginning of this chapter highlight how discourses around digital competence range from the "tautological to the idealistic", as Livingstone put it [14], from defining it as the ability to use a specific set of tools (e.g. internet literacy as the ability to use the internet) to the understanding of digital competence as an unavoidable requirement [29] for life-fulfilment. Though the above literacies have converged into digital competence, it is more than the sum of its parts: it is not enough to state that digital competence involves what is required for internet literacy, ICT literacy, information literacy and media literacy as there are other components that come into the picture of digital competence. Livingstone [14] states that digital competence is not user dependent but tools dependent – or, it could be argued, application dependent. Reading a printed newspaper or an online one is not the same experience and requires different skills, such as, for instance, the ability to move through hyperlinked texts. Online text perusal requires a more dynamic approach [37] and offers an augmented reading experience. Moreover, computers or smart-phones are generally used through icon-based commands, hence higher cognitive mediation is required [7], as symbolic utterances refer to a system of signs which may not be familiar to everyone, and are underpinned by the ability to read images as texts. Moreover, as Kress [10] argues, changes in the forms and functions of the text –here including visual and audio texts – make the reader a designer of the reading experience. Hyper and multimodal texts allow readers' engagement, as they choose which threads or links to follow, which modes of reading to select. In addition, the decoding and encoding processes are made at faster speed and texts – blogs, newspapers articles, Wikipedia entries –encourage the reader to become an author. Besides, writing is becoming part of the everyday life of the everyday person [38], as many of us write emails, send SMS, and participate in social networks. In a way, these practices – including the 'hyper-intensity' of text or facebook messaging – can be seen as a triumph of the domestication of technologies and their appropriation by the user [39], who plays an active role, shifting from recipient to producer of information and/or media content. Users are moreover becoming engaged in activities they did not necessarily participate in the offline world (an example: the sharing of news or music through social networks, thus acting as a multiplier of information).

3   The Collection of Digital Competence Frameworks

Due to the different terms and understandings of Digital Competence, a literature search was performed for each literacy type outlined above. The search engine Google and the portal 'Google scholar' were chosen, and search items included the different literacy types linked to Digital Competence and combinations with the word 'frameworks'. In addition, searches were carried out in important educational and academic databases (ERIC; Scope). These were complemented with: browsing through curricula in European countries; a review of reports of international organisations working on ICT and learning (e.g. OECD; UNESCO); a review of EU reports, initiatives or funding schemes; and suggestions from colleagues or collaborators. The searches came up with a body of over a hundred cases, from which all the cases that did not constitute a framework were excluded. Here, a framework is understood to be an instrument for the development or assessment of the Digital Competence of a specific target group, according to a set of descriptors of intertwined competences, thus adapting CEDEFOP's definition of framework to our scope [40]. Criteria were then established to limit the number of frameworks to be analysed, namely: fair distribution of target groups; fair geographical distribution;

representation of a plurality of perspectives on digital competence; representation of a plurality of initiative types (from school curricula, to academic papers, to certification schemes). Fifteen frameworks were finally selected for full reporting and analyses (see Table 1 for an overview). Of course, it is acknowledged that these cases represent a partial and qualitative snapshot of how Digital Competence can be translated into learning outcomes.

Table 1. Overview of Frameworks

Name & Target group	Description
ACTIC Target Group: all citizens above 16	ACTIC (Acreditación de Competencias en Tecnologías de la Información y la Comunicación) certifies ICT competences. http://www20.gencat.cat/portal/site/actic
BECTA's review of Digital Literacy Target group: children up to 16	This review provides a model for learners at primary and secondary schools [41]. http://www.timmuslimited.co.uk/archives/117
CML MediaLit Kit Target group: adults	The CML (Centre for Media Literacy) establishes a framework to construct and deconstruct media messages [42]. http://www.medialit.org/cml-framework
DCA Target group: 15-16 years old	DCA (Digital Competence Assessment) is a framework linked to a series of tests for secondary school students [43]. http://www.digitalcompetence.org/
DigEuLit Target group: general population	A 2005-2006 project lead by the University of Glasgow and funded by the European Commission to develop a conceptual framework for Digital Competence [44].
ECDL Target group: adults	ECDL (European Computer Driving Licence) Foundation delivers worldwide a range of certifications on Computer literacy. http://www.ecdl.org/programmes/index.jsp
eLSe-Academy Target group: senior citizens	The eLSe-Academy - eLearning for Seniors Academy - is an online environment adapted to the digital competence needs of senior citizens. http://www.arzinai.lt/else/
eSafety Kit Target group: 4-12 years old children	This initiative aims to support children, their parents/tutors and teachers in safe internet use. www.esafetykit.net
Eshet-Alkalai's framework Target group: general population	This conceptual framework details the multiple literacies that are needed for people to be functional in a digital era [7, 45]
IC3 Target group: students & job-seekers	The Internet and Computing Core Certification by Certiport enhances the knowledge of computers and the Internet. www.certiport.com/Portal/
iSkills Target group: adults	This test from ETS assesses critical thinking and problem-solving skills in a digital environment [46]. http://www.ets.org/iskills/
NCCA ICT framework – Ireland Target group: students	This framework is a guide to embed ICT as a crosscurricular component in primary and lower secondary education [16]. http://www.ncca.ie/en/Curriculum_and_Assessment/ICT/#1
Pedagogic ICT licence –Denmark Target Group: teachers	The Pedagogical ICT Licence offers Danish teachers the opportunity to upgrade their ICT skills. www.paedagogisk-it-koerekort.dk
The Scottish ILP Target group: students	The Scottish Information Literacy Project promotes the understanding and development of information literacy in all education sectors [47]. http://caledonianblogs.net/nilfs/
UNESCO ICT CFT Target Group: teachers	The ICT Competency Framework for Teachers provides guidelines for courses for teachers to integrate ICT in class [48].

The analysis of the content of the selected frameworks aims to answer the following questions:

- How is Digital Competence defined or understood in the selected frameworks?
- What are the main competences that are developed in the selected frameworks?

4   Digital Competence: An Encompassing Definition

In the Communication on Key Competences for Lifelong Learning, the European Commission proposes the following definition of digital competence: "Digital competence involves the confident and critical use of Information Society Technology (IST) for work, leisure and communication. It is underpinned by basic skills in ICT: the use of computers to retrieve, assess, store, produce, present and exchange information, and to communicate and participate in collaborative networks via the Internet" [49]. As the concept of Digital Competence is much debated and multifaceted, as shown above with the discussion of the literature, it comes as no surprise that two thirds of the selected frameworks provide a definition of digital competence. The ten definitions presented in the frameworks have been compared and their main elements have been merged to produce the following encompassing definition of digital competence:

Digital Competence is the set of knowledge, skills, attitudes, abilities, strategies and awareness that is required when using ICT and digital media to perform tasks; solve problems; communicate; manage information; behave in an ethical and responsible way; collaborate; create and share content and knowledge for work, leisure, participation, learning, socialising, empowerment and consumerism.

This working definition has been produced by taking into account all the perspectives of each framework. It can be noted that this definition bears similarities with the European Commission's definition. Moreover, the structure of all definitions provided in the frameworks was found to be quite similar, i.e. assembled on the same building blocks, namely: learning domains, tools, competence areas and purposes. Thus, several cases define the learning domains [50] that are developed in their framework: some frameworks add awareness and strategies to the more expected knowledge, skills, and attitudes, which are the constituent parts of a competence [51]. Half the frameworks that provide a definition insist on skills, while a third mentions awareness. The tools generally include ICTs, only two frameworks explicitly mention media. Regarding the competence areas that are foreseen in the definitions, certainly "use" or "performing tasks" recur most, followed by communication and information management. Finally, the purposes that emerge from this definition are in line with commonly-agreed ones, see for instance the work on monitoring Digital Competence carried out in the frame of the Digital Agenda Scoreboard.[1] It should be stated, however, that purposes should not to be taken as a proxy for competences or competence areas, but should be considered as the context in which the competence may be applied. Although the different frameworks proposed a quite varied list of

---

[1] http://ec.europa.eu/information_society/digital-/scoreboard/docs/pillar/digitalliteracy.pdf

purposes, we felt that there were two missing elements to the picture: "consuming" and "user empowerment". Online shopping is spreading, with 40% of EU citizens buying goods online.[2] However, it is of paramount importance that consumers are aware of the risks connected with online purchases, for instance those resulting from inadequate security settings. To transact safely [52], there are certain competence requirements, which are recognised as a priority in the Digital Agenda [53, Action 61]. In addition, it has been noted that social computing practices allow for user empowerment [54]. As a consequence, we added these two purposes to the working definition as they were not present in the definitions of the frameworks.

5   Areas of Digital Competence

The **NCCA** [16] report claims that most approaches to Digital Competence see skills as tool-dependent: they focus on the practical abilities to use specific software or hardware. This reinforces common visions of digital literacy or media literacy [14]. Although tool-dependent approaches become outdated in no time, they have the advantage of describing skills that are specific and easily measurable [16]. Indeed, the collection provided here presents some frameworks which are oriented at developing skills more than competences and which are structured around the most-used software or tools. For instance, the European Computer Driving Licence (**ECDL**) core programmes consists of 13 modules which mainly aim to make users able to use a specific application, though they are vendor neutral, i.e. not tied to any one brand of software. These modules develop people's skills in using databases, spreadsheets, word processing tools, image editing and presentation software, to give but a few examples. The certification for the "word processing" module includes tasks like creating a new document, formatting text, creating tables, running the spell-check and printing a document. In the same vein, and although it measures content topics together with technology topics, the iSkills test assesses people's ability to use the web (email, instant messaging, bulletin board postings, browsers, search engines); databases (data searches, file management); and software (word processing, spreadsheet, presentations, graphics). The test is built around the assessment of seven types of task, namely: define, access, evaluate, manage, integrate, create, and communicate. An example of a "create" task, as available from the ETS website, is to create a graph from a series of given data, and then answer questions related to the interpretation of the graph. Even though this includes a cognitive component – the interpretation of a graph – the main task is built around a common application, i.e. the spreadsheet package. **IC3** by Certiport provides another example of a tool-related framework. The exams for this certification are explicitly based on Microsoft Windows 7 and Office 2010. The framework is built around three modules, namely: Computing Fundamentals, Key Applications and Living Online. The first module is based on hardware, software and operating systems, thus reflecting a computer engineering approach. The second module has topics on word processing, spreadsheets and presentation software, plus a section covering features common to

---

[2] http://ec.europa.eu/information_society/digital-agenda/scoreboard/docs/scoreboard.pdf

all applications. The third module is described as addressing "skills for working in an Internet or networked environment"[3] and is based on the use of distinctly recognisable tools: online networks, emailing systems, Internet browsers. The section on "the impact of computing and the Internet on Society" is the only one which goes beyond a tool-related certification process, and mainly relates to risks connected to the use of hardware, software and the internet.

It comes as no surprise that the above examples are taken from certification frameworks, which have to satisfy the need for measurability and assessment. This aspect could also be reinforced by the requirements of employers, who could demand abilities in specific hardware/software packages. Although the need for specific skills for employability could be a possible driver for application-oriented programmes, tool-related operational skills are also central in eInclusion initiatives. An example is the **eLSe Academy,** an eLearning environment aimed at senior citizens interested in acquiring or further developing their competences in ICT. Even this course is typically structured on application-based modules: using the learning platform; writing with a computer (word-processors, including word pads); communicating via a computer (emails); and so on. Like the IC3 certification, this case is based on the use of Microsoft Office packages and Windows. The **UNESCO framework for teachers,** even though it is embedded in a more complex structure, includes parts which are tool-oriented. The framework is not about Digital Competence per se, but rather suggests entrenching ICT in every aspect of educational institutions from policy to pedagogy to administration, thus proposing an innovative approach to using technologies in education. However, when detailing the digital competence level expected of teachers, the implementation guidelines suggest a typical application-oriented approach [48]. Many frameworks build on a consolidated though relatively recent tradition. As pointed out by Erstad [55], Digital Competence moved through three main phases. After a first 'mastery phase' (1960s to the mid 80s) where technologies were accessed by professionals who knew programming languages, interfaces became more user-friendly from the mid 80s to the late 90s and were thus opened up to society. This second 'application phase' gave rise to mass certification schemes. As technologies became simpler, they also became more necessary, hence augmenting the population's needs for specific skills in order to "tame" these new tools – and therefore triggering courses targeted at these specific needs. Many eInclusion/eLearning initiatives and digital literacy discourses are built upon this stance, highlighting access and accessibility and tool-related operational skills as their core. From the late 90s, we entered a third phase – the reflective phase– in which the need for critical and reflective skills in the use of technology was widely recognised [55]. Yet in 2004, the NCCA reported that most definitions and approaches to Digital Competence did not take into account higher order thinking skills [16]. Our framework collection cannot confirm this statement, as several of the cases we have gathered here do in fact recognise the importance of reflective and critical uses. However, the modes in which this is translated into learning objectives or competences vary.

---

[3] http://www.certiport.com/portal/common/documentlibrary/
IC3_Program_Overview.pdf

The **iSkills** framework, although it has a central operational component, is an example of an approach which acknowledges thinking skills for Digital Competence and at the same time is still based on applications: *"ICT literacy cannot be defined primarily as the mastery of technical skills. The panel concludes that the concept of ICT literacy should be broadened to include both critical cognitive skills as well as the application of technical skills and knowledge"* [46]. An example might illustrate how the above-mentioned philosophy is translated into assessment of competences. As explained above, the framework is built around seven competence areas. One of these, "Access", implies the collection and/or retrieval of information in digital environments, and therefore is typically endowed with cognitive and critical needs. The two sample tests provided on the website[4] are based on searches within a database, on accurate search terms and correct search strategies (for instance, using Boolean operators or quotation marks). The cognitive dimension is certainly taken into account, although we are left with the impression that this cognitive and critical component is not far from an application-oriented skill. In other words, critical and thinking skills seem to be seen as a means to a specific end, the end being a more efficient use of computers. A similar competence, i.e. "access to information", can be found in **The Scottish Information Literacy Project**, a complex framework where competences are articulated around levels/target groups. For further and higher education, the equivalent of the iSkills "access" competence are the following two competences: "the ability to construct strategies for locating information" and "the ability to locate and access information". These competences include: the articulation of information needs, the development of a systematic method to answer to information needs, the development of appropriate searching techniques (e.g. use of Boolean searches), the use of appropriate indexing and abstracting services, citations index and databases and the use of current awareness methods to keep up to date. Similarities between the two approaches can be found, for instance, in the development of search techniques to select the appropriate information retrieval services (selecting, for instance, the appropriate database). However, the Scottish Information Literacy Project, probably as a consequence of its focus on information literacy rather than digital competence, involves higher order thinking skills and cognitive approaches at a more advanced level.

The cognitive dimension is often associated with access to information. Another case, the **DCA**, develops a competence which links access to information with cognitive skills. The DCA is a test which was originally developed for high school students aged 15-16 and which is currently under development for younger learners. The cognitive dimension translates into the following learning objectives: being able to read, select, interpret and evaluate data and information taking into account their pertinence and reliability. Frameworks for compulsory schooling seem to show a tendency to raise the cognitive dimension of digital competence. Newmann, in charge of a review of digital literacy for children aged 0 to 16 for **BECTA**, in an attempt to simplify the complex terminology this domain generates, proposes looking at digital competence as applying critical thinking skills to technology use [41]. According to

---

[4] See http://www.ets.org/s/iskills/flash/FindingItem.html and
http://www.ets.org/s/iskills/flash/ComplexSearch.html

this reading, digital competence would require both technical skills and critical thinking skills, which are seen as an attribute of information literacy. In the review, Newmann clarifies that the focus is more on thinking skills than on technical ones. In the **NCCA framework**, "thinking critically and creatively" is one of the four foreseen areas of learning.[5] Access to, and evaluation of, information are two important learning outcomes. The novelty of this curriculum consists of its other two learning outcomes; "express creativity and construct new knowledge and artefacts using ICT" and "explore and develop problem-solving strategies using ICT". The NCCA website proposes sample learning activities that could be used by teachers in different subjects to develop these competences, such as organising a digital storytelling project or recording a field trip using a digital camera.

A recurring competence area is what could be called "Ethics and responsibility" and includes a safe, legal and ethical use of the Internet in particular and technologies in general. The **IC3** framework displays 3 application-oriented modules, the third one being called "Living online". After three sections related to applications (Internet, emails and communication networks), a fourth section is about "The Impact of Computing and the Internet on Society" and aims to identify how computers are used in different areas of work, school, and home; the risks of using computer hardware and software; and how to use the Internet safely, legally, and responsibly. While in the IC3 framework, this area constitutes only a small part of the syllabus, in the **eSafety Kit** this issue holds centre stage. Three of the four envisaged competences are based around ethics and responsibility, as in fact this framework, developed for children between the ages of 4 and 12, has the safe use of the internet as its primary scope. Attention to the emotional aspect of dealing with cyber-bullying is a novelty of this framework. Ethics and responsibility are also accounted for in the **NCCA framework**. As part of the forth competence area ("Understanding the social and personal impact of ICT"), students should demonstrate an awareness of, and comply with, responsible and ethical use of ICT.

Several frameworks include "communication" as a competence area. However, it should be remarked that different frameworks do not necessarily concord in the ways they translate this competence into learning outcomes. As a matter of fact, a huge difference can be seen between application-oriented frameworks and more cognitive approaches, as shown in Figure 1.

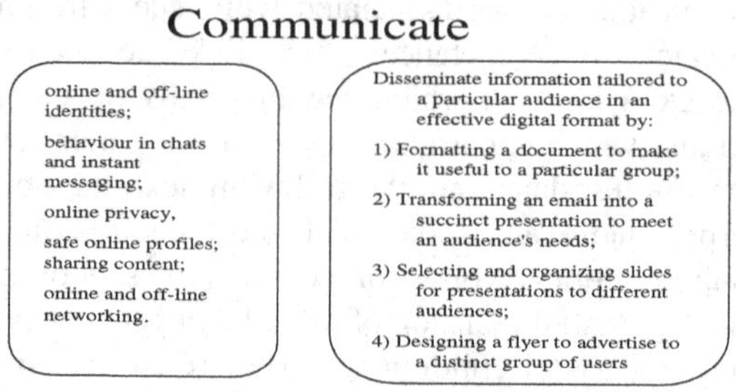

Communicate

online and off-line identities;

behaviour in chats and instant messaging;

online privacy,

safe online profiles; sharing content;

online and off-line networking.

Disseminate information tailored to a particular audience in an effective digital format by:

1) Formatting a document to make it useful to a particular group;

2) Transforming an email into a succinct presentation to meet an audience's needs;

3) Selecting and organizing slides for presentations to different audiences;

4) Designing a flyer to advertise to a distinct group of users

Fig. 1. Two different ways to translate the competence "Communicate"

---

[5] Together with "Creating, communicating and collaborating"; "Developing foundational knowledge, skills and concepts"; and "Understanding the social and personal impact of ICT".

The left hand side of Figure 1 deals with online and off-line identities, privacy, and behaviour. In this framework, the needs for communication in an online environment are interpreted as cognitive needs. At the same time, there is a focus on privacy and security. In addition, there is an interest in comparing the online and off-line worlds, as communicating is a competence that one develops in real as well as virtual contexts. The framework depicted on the right hand side, on the other hand, perceives "communication" as the targeting of information to different audiences through specific software. Therefore, being able to communicate in a digital environment is seen as the ability to format a document, to transform an email into a PowerPoint-like presentation, to organise slides and to design a flyer. It goes without saying that being able to communicate cannot be reduced to the formatting of a text.

6   Conclusions

Several of the frameworks selected for this analysis suggest that technical skills constitute a central component of Digital Competence. In our opinion, having technical skills at the core of a digital competence model obscures the multiple facets of the domain. Digital Competence should be understood, as it is in many frameworks, in its wider sense. The analysis of the 15 selected frameworks underlines several aspects – or areas – of Digital Competence, which can be summarized as follows:

Table 2. Areas of Digital Competence

Area	Description
Information Management	Identify, locate, access, retrieve, store and organize information
Collaboration	Link with others, participate in online networks and communities, interact constructively
Communication and Sharing	Communicate through online tools, taking into account privacy, safety, and correct online behaviour
Creation of Content and Knowledge	Integrate and re-elaborate previous content and knowledge, construct new knowledge
Ethics and Responsibility	Behave in an ethical and responsible way, aware of legal frames
Evaluation and Problem-solving	Identify digital needs, solve problems through digital means, assess the information retrieved
Technical Operations	Use technology and media to perform tasks through digital tools

Each area presented in the table above has been taken from more than one framework. We wish to suggest that technical operations should be considered like any other component of the framework, and not be given the paramount importance they are now. The analysis of the frameworks suggests yet another rhetoric strand: digital competence as mainly based on technical operations. However, many frameworks and initiatives are starting to move away from this perspective and propose a model for the development of digital competence that takes into account higher order thinking skills and that fits in a 21st century skills perspective.

References

1. Silverstone, R., Hirsch, E.: Consuming technologies. Routledge, London/NY (1992)
2. Leu, D.J.: Literacy and technology: Deictic consequences for literacy education in an information age. Handbook of Reading Research 3, 743–770 (2000)
3. Coiro, J., Knobel, M., Lankshear, C., Leu, D.J.: Handbook of research on new literacies. Routledge, New York-London (2008)
4. Martin, A., Grudziecki, J.: DigEuLit: Concepts and Tools for Digital Literacy Development. ITALICS: Innovations in Teaching & Learning in Information & Computer Sciences 5, 246–264 (2006)
5. Gilster, P.: Digital literacy. John Wiley, New York (1997)
6. Bawden, D.: Origins and Concepts Of Digital Literacy. In: Lankshear, C., Knobel, M. (eds.) Digital Literacies: Concepts, Policies & Practices, pp. 17–32 (2008)
7. Eshet-Alkalai, Y.: Digital Literacy. A Conceptual Framework for Survival Skills in the Digital Era. Journal of Educational Multimedia & Hypermedia 13, 93–106 (2004)
8. Ala-Mutka, K.: Mapping Digital Competence: Towards a Conceptual Understanding. In: JRC-IPTS (2011)
9. Amiel, T.: Mistaking computers for technology: Technology literacy and the digital divide (2004)
10. Kress, G.: Multimodality: a social semiotic approach to contemporary communication. Routledge, NY (2010)
11. Bawden, D.: Information and digital literacies: a review of concepts. Journal of Documentation 57, 218–259 (2001)
12. Horton Jr., F.W.: Information literacy vs. computer literacy. Bulletin of the American Society for Information Science 9, 14–16 (1983)
13. Buckingham, D.: Media education: Literacy, learning, and contemporary culture. Polity (2003)
14. Livingstone, S.: The changing nature and uses of media literacy. In: LSE (2003)
15. Rotherham, A.J., Willingham, D.T.: "21st-Century" Skills. American Educator 17 (2010)
16. NCCA: Curriculum Assessment and ICT in the Irish context: a Discussion Paper (2004)
17. Banaji, S., Burn, A., Buckingham, D.: Rhetorics of creativity: a review of the literature (2006)
18. Irving, L., Klegar-Levy, K., Everette, D., Reynolds, T., Lader, W.: Falling through the Net: Defining the digital divide. National Telecommunications and Information Administration, US Deps of Commerce, Washington, DC (1999)
19. Molnár, S.: The explanation frame of the digital divide. In: Proceedings of the Summer School, Risks and Challenges of the Network Society, pp. 4–8 (2003)
20. Livingstone, S., Helsper, E.: Gradations in digital inclusion: children, young people and the digital divide. New Media & Society 9, 671 (2007)
21. McConnaughey, J., Lader, W.: Falling through the net II: new data on the digital divide. National Telecommunications and Information Administration. Department of Commerce, US Government (1998)
22. Erstad, O.: Educating the Digital Generation. Nordic Journal of Digital Literacy 1, 56–70 (2010)
23. Prensky, M.: Digital Natives, Digital Immigrants. On the Horizon 9 (2001)
24. Bennett, S., Maton, K., Kervin, L.: The 'digital natives' debate: A critical review of the evidence. British Journal of Educational Technology 39, 775–786 (2008)
25. Hargittai, E.: Digital Na(t)ives? Variation in Internet Skills and Uses among Members of the "Net Generation". Sociological Inquiry 80, 92–113 (2010)

26. Sefton-Green, J., Nixon, H., Erstad, O.: Reviewing Approaches and Perspectives on "Digital Literacy". Pedagogies: An International Journal 4, 107–125 (2009)

27. Hartley, J., Montgomery, M., Brennan, M.: Communication, cultural and media studies: The key concepts. Psychology Press (2002)

28. European Commission: Europe 2020: A strategy for smart, sustainable and inclusive growth. COM (2010) 2020 (2010)

29. OECD: Learning to change (2001)

30. Simonson, M.R., Maurer, M., Montag-Torardi, M., Whitaker, M.: Development of a standardized test of computer literacy and a computer anxiety index. Journal of Educational Computing Research 3, 231–247 (1987)

31. Oliver, R., Towers, S.: Benchmarking ICT literacy in tertiary learning settings, Citeseer, pp. 381–390

32. Deursen, A.J.A.M.: Internet skills: vital assets in an information society (2010)

33. Hofstetter, F.T., Sine, P.: Internet literacy. Irwin/McGraw-Hill (1998)

34. Christ, W.G., Potter, W.J.: Media literacy, media education, and the academy. Journal of Communication 48, 5–15 (1998)

35. Ofcom: Media Literacy Audit: Report on media literacy amongst children. Ofcom (2006)

36. America Library Association: Presidential Committee on Information Literacy. ALA (1989)

37. OECD: PISA 2009 Results: What Students Know and Can Do. Students performance in reading, mathematics and science. OECD, Paris (2010)

38. Rainie, L., Purcell, K., Smith, A.: The social side of the internet. Pew Research Centre (2011)

39. Silverstone, R.: Domesticating domestication: Reflections on the life of a concept. In: Berker, T., Hartmann, M., Punie, Y., Ward, K.J. (eds.) Domestication of Media and Technology, pp. 229–248. Open University Press, Maidenhead (2006)

40. CEDEFOP: Terminology of European education and training policy. A selection of 100 key terms. Office for Official Publications of the European Communities (2008)

41. Newman, T.: A review of digital literacy in 0 – 16 year olds: evidence, developmental models, and recommendations, Becta (2008)

42. Thoman, E., Jolls, T.: Literacy for the 21st Century. An Overview & Orientation Guide To Media Literacy Education. In: CML (2003)

43. Calvani, A., Cartelli, A., Fini, A., Ranieri, M.: Models and instruments for assessing digital competence at school. Journal of e-Learning and Knowledge Society 4 (2009)

44. Martin, A.: Literacies for the Digital Age. In: Martin, A., Madigan, D. (eds.) Digital Literacies for Learning, Facet, London, pp. 3–25 (2006)

45. Eshet-Alkalai, Y., Chajut, E.: You can teach old dogs new tricks: The factors that affect changes over time in digital literacy. Journal of Information Technology Education 9, 173–181 (2010)

46. International ICT Literacy Panel: Digital Transformation. A Framework for ICT Literacy. ETS (2007)

47. Crawford, J., Irving, C.: The Scottish Information Literacy Project and school libraries. Aslib Proceedings (2010)

48. UNESCO: Unesco ICT Competency Framework for Teachers (2011)

49. European Parliament and the Council: Recommendation of the European Parliament and of the Council of 18 December 2006 on key competences for lifelong learning. Official Journal of the European Union L394/310 (2006)

50. Bloom, B.S.: Taxonomy of Educational Objectives. In: Bloom, B.S., et al. (eds.) The Classification of Educational Goals. Longmans, London (1964) (printed in U.S.A)

51. Westera, W.: Competences in education: a confusion of tongues. Journal of Curriculum Studies 33, 75–88 (2001)
52. Lusoli, W., Bacigalupo, M., Lupiañez, F., Andrade, N., Monteleone, S., Maghiros, I.: Pan-European survey of practices, attitudes & policy preferences as regards personal identity data management. In: JRC-IPTS (2011)
53. European Commission: A Digital Agenda for Europe. COM(2010)245 final (2010)
54. Ala-Mutka, K., Broster, D., Cachia, R., Centeno, C., Feijóo, C., Haché, A., Kluzer, S., Lindmark, S., Lusoli, W., Misuraca, G., Pascu, C., Punie, Y., Valverde, J.A.: The Impact of Social Computing on the EU Information Society and Economy. In: JRC-IPTS (2009)
55. Erstad, O.: Conceptions of Technology Literacy and Fluency. In: Penelope, P., Eva, B., Barry, M. (eds.) International Encyclopedia of Education, pp. 34–41. Elsevier, Oxford (2010)

How CSCL Moderates the Influence of Self-efficacy on Students' Transfer of Learning

Andreas Gegenfurtner[1], Koen Veermans[2], and Marja Vauras[2]

[1] TUM School of Education, Technical University of Munich, Munich, Germany
andreas.gegenfurtner@tum.de
[2] Centre for Learning Research, University of Turku, Turku, Finland
{koen.veermans,marja.vauras}@utu.fi

Abstract. There is an implicit assumption in learning research that students learn more deeply in complex social and technological environments. Deep learning, in turn, is associated with higher degrees of students' self-efficacy and transfer of learning. The present meta-analysis tested this assumption. Based on social cognitive theory, results suggested positive population correlation estimates between post-training self-efficacy and transfer. Results also showed that effect sizes were higher in trainings with rather than without computer support, and higher in trainings without rather than with collaboration. These findings are discussed in terms of their implications for theories of complex social and computer-mediated learning environments and their practical significance for scaffolding technology-enhanced learning and interaction.

Keywords: Computer-supported collaborative learning, self-efficacy, transfer of learning, training, meta-analytic moderator estimation.

1 Introduction

Self-efficacy refers to beliefs in one's capabilities to organize and execute the courses of action required to produce given attainments [1]. Transfer of training is the use of newly acquired knowledge and skills [2,3]. Research indicates that both self-efficacy and reflect students' deep learning [4,5]. There is an implicit assumption in the learning sciences that deep learning is more likely to occur in complex social and technological environments [4]. If it is true that deep learning is associated with higher degrees of self-efficacy [1] and transfer [2,3], then it follows that estimates of the relationship between self-efficacy and transfer of training should be higher in those conditions that afford computer supported collaborative learning (CSCL), because of the positive effects of technology enhancement and social interaction. However, to date, no study has examined the predictive validity of this assumption. As a remedy to this gap, the present meta-analysis sets out to investigate whether higher population correlation estimates between self-efficacy and transfer are found in training conditions that afford computer support and collaboration when compared with other training conditions.

A. Ravenscroft et al. (Eds.): EC-TEL 2012, LNCS 7563, pp. 93–102, 2012.

1.1   Self-efficacy and Transfer of Training

Efficacy beliefs are among the most widely documented predictors of achievement, which has been shown in domains including sports, work, and education [2,6-7]. According to social cognitive theory [1], people with high self-efficacy set high and demanding goals; these goals create negative performance discrepancies to be mastered [1]. Expectations about the perceived efficacy of one's capability to master those discrepancies regulate whether effort is initiated, how much continuous effort is expended, and whether effort is maintained or even increased in face of difficulties during goal attainment. Because the power of self-efficacy to predict task achievement has been so widely documented [1,2,6-10], it seems reasonable to assume that self-efficacy also predicts the initiation, expenditure, and maintenance of efforts toward transfer of training.

If it is true that efficacy beliefs predict sufficient execution of effort to achieve successful outcomes, it follows that efficacy beliefs should also predict successful transfer of training. However, the literature shows mixed evidence. For example, some investigations showed high correlation estimates between self-efficacy and training transfer [11], while other investigations suggested that the magnitude of this relationship is negligible [12]. One possible explanation for the mixed evidence is the influence of sampling error and error of measurement [13] that may have induced biases on the true score population correlation. Therefore, one aim of the present study was to use meta-analytic methods to inquire whether performance self-efficacy, after controlling for sampling error and error of measurement, exhibits a stable influence on transfer and whether this relationship would be higher after training than before training. Another possible explanation for the mixed evidence is that population correlation estimates have been moderated by different study conditions. Identification of boundary conditions has important implications for testing the predictive validity of social cognitive theory [1,2,8,14]. Therefore, a second aim of the study was to identify and estimate the boundary conditions under which self-efficacy and transfer correlate. Inquiring into these characteristics as boundary conditions is significant, because it enables accounting for artifactual variance in the total variance of a correlation, which, in turn, may explain some of the disagreements in the existing literature. Two boundary conditions were analyzed: computer support and collaboration.

1.2   Computer-Supported Collaborative Learning

The rationale for choosing CSCL as boundary conditions was derived from a belief in the learning sciences that "deep learning is more likely in complex social and technological environments" [4]. Deep learning, in turn, is related to higher degrees of transfer [2,3] and self-efficacy [1]. If these assumptions hold, it follows that population correlation estimates of the relationship between self-efficacy and training transfer should be higher in those conditions that afford CSCL. Conditions for CSCL can be examined as (a) computer support and (b) collaboration. For the purpose of this article, computer support was defined as technological material in learning environments intended to promote understanding [15]; collaboration was broadly defined as the working-together of two or more individuals to attain the shared training goals and task at hand [16]. We acknowledge substantial variation in how the term 'collaboration' is defined in the literature [17-18]. We use the term 'collaboration' here for the sake of simplicity and do acknowledge gradual nuances in how key conditions of the nature of joint working (e.g., shared goals, co-construction of knowledge, co-regulation, etc.) are reflected in prior literature to capture different sociopsychological processes of interpersonal coordination and their relation to actualizing motivation [18-20]. An example of training having both computer support and collaboration is [21], who trained participants with a collaborative computer game. An example of a training having computer support but no collaboration is [22], in which they trained participants to use computer software individually and without social interaction. An example of a training including no computer support but collaboration is [23]'s description of nursing team training, which included group discussions, brainstorming, and peer assessment. Finally, an example of a training program including neither computer support nor collaboration was [24], in which participants were trained in a speed-reading skill individually with paper handouts. If it is true that complex social and technological environments promote self-efficacy and transfer [1-4], then it follows that population correlation estimates should be higher in conditions with computer support rather than in conditions without and in conditions with collaboration rather than without. Importantly, population correlation estimates should be highest in conditions affording both computer support and collaboration. Figure 1 illustrates these conditions. The top-left quadrant represents training conditions that neither includes computer support nor collaboration; it is thus assumed to have no particular positive effects. The bottom-left quadrant represents training conditions that include computer support but no collaboration. The top-right quadrant represents training conditions that include collaboration, but no computer support. Finally, the bottom-right quadrant represents training conditions that include both computer support and collaboration.

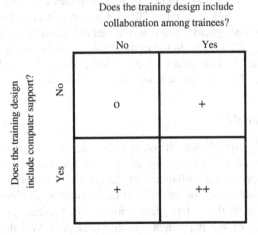

Fig. 1. Hypothesized effects of conditions on the relation between self-efficacy and transfer

1.3   The Present Study—Hypotheses

In summary, the focus of the present study was the relationship between performance self-efficacy and transfer of training. The first aim was to cumulate previous research in order to correct the size of true score population correlations. A second aim of the study was to estimate the moderating effects of computer support and collaboration. Two hypotheses were formulated. Based on social cognitive theory [1], we assumed that transfer of training would be positively related with performance self-efficacy (Hypothesis 1). Based on the assumption that deep learning is more likely to occur in complex social and technological environments [4], we hypothesized that the relationship between self-efficacy and transfer would be more positive in training conditions affording computer support and collaboration (Hypothesis 2).

2   Method

2.1   Literature Searches and Criteria for Inclusion

To test these hypotheses, we used meta-analytic methods [14]. Studies that reported correlations between post-training self-efficacy and transfer of training were located. To be included in the database, a study had to report an effect size r or other effect sizes that could be converted to r (β coefficient; Cohen's d; F, t, or Z statistics). Because the focus of inquiry was on self-efficacy as an individual capacity [1], the database included studies that reported data on individuals. Studies reporting data on group efficacy were omitted. Studies on children as well as animal studies were also excluded, because they represent different premises on training and work performance. Using these inclusion criteria, the literature was searched in three ways. First, the PsycINFO, ERIC, and Web of Science databases were searched using the keywords *self-efficacy*, *behavior change*, *training application*, *training use*, and *transfer of training*. In addition, a manual search of journal issues covering a 25-year period (from January 1986 through December 2010) was conducted. A total of 29 articles, book chapters, conference papers, and dissertations that contributed at least one effect size to the meta-analysis were included in the database. A full list of all included studies is available from the first author. The 29 studies offered a total of $k = 33$ independent data sources. Total sample size was $N = 4,203$ participants.

2.2   Recorded Variables

To answer the study hypotheses, different characteristics were tabulated from the selected research literature. Specifically, each study was coded for effect size estimates, computer support, and collaboration. Effect size estimates included Pearson product-moment correlation r of the self-efficacy–transfer relationship, Cronbach's reliability estimate α of the independent variables (self-efficacy), and Cronbach's reliability estimate α of the dependent variable (transfer). We also coded the first author, publication year, the number of participants, their age (in years), and gender (percentage of females). Computer support for learning afforded during training was coded as 1 = *computer support* and 0 = *no computer support*. Collaboration among participants afforded during training was coded as 1 = *collaboration* and 0 = *no collaboration*. Two independent raters first coded fifteen of the studies. Because intercoder reliability was generally high (Cohen's κ = .91), one rater continued to code the remaining studies. If a study reported more than one effect size, a single composite variable was created to comply with the assumption of independence. As an exception to this rule, linear composites were not created for the theoretically predicted moderator variables, as composite correlations would have obscured moderator effects and prohibited further analysis.

2.3   Meta-analytic Methods Used

Analysis occurred in two stages. A primary meta-analysis aimed to estimate the true score population correlation ρ of the pre- and post-training relationship between performance self-efficacy and transfer of training. A meta-analytic moderator estimation then aimed to identify moderating effects in those relationships.

The primary meta-analysis was done using the methods of artifact distribution meta-analysis of correlations [14]. These methods provide an improvement from earlier statistical formulae when information such as reliability estimates is only sporadically reported in the original studies. First, study information was compiled on three distributions: the distribution of the observed Pearson's r of the transfer–self-efficacy relationship, the distribution of Cronbach's α of the independent variable, and the distribution of Cronbach's α of the dependent variable. Next, the distribution of Pearson's r was corrected for sampling error. Note that the correction was conducted using a weighted average, not Fisher's z transformation, since the latter was shown to produce upwardly biased correlation estimates. The distribution corrected for sampling error was then further corrected for error of measurement using the compiled Cronbach's α reliability estimates. This last step provided the final estimate of the true score population correlations ρ between self-efficacy and transfer. Finally, standard deviations of the corrected observed correlation r_c and of the population correlation ρ were calculated; these were used to derive the percentage of variance attributable to attenuating effects, the 95% confidence interval around r_c, and the 80% credibility interval around ρ.

The meta-analytic moderator estimation followed the primary meta-analysis. Theory-driven nested sub-group analyses were used to estimate the moderating effects of computer support and collaboration. Nested sub-group analysis assumes that the moderator variables are independent and additive in their effects [14]. A criticism of the use of sub-groups is that it reduces the number of data sources per analysis, resulting in second-order sampling error. Although the present study contained a large number of data sources and participants, the possibility of second-order sampling error cannot be completely ruled out. This is therefore indicated when warranted for interpreting the results.

3   Results

3.1   Primary Meta-analysis

Table 1 summarizes the number of studies, participants, and participant characteristics by condition. The mean estimates in Table 1 are age in years and percentage of females. Across all conditions, the uncorrected correlation coefficient r between performance self-efficacy and transfer of training is 0.34 ($k = 33$, $N = 4,158$). The population correlation estimate corrected for sampling error and error of measurement is $\rho = 0.39$ ($SD\rho = 0.23$; 80% $CV = .10$; .68). This estimate is in the positive direction, thus supporting Hypothesis 1. The difference between r and ρ represents a depression of the true score population correlation through sampling error and error of measurement by 14.7%.

Table 1. Number of studies, participants, and participant characteristics by condition[1]

Conditions	k	N	Age M	Age SD	Gender M	Gender SD
Computer support and collaboration	7	730	27.28	04.55	39.60	27.48
Computer support, but no collaboration	7	1,044	26.42	09.90	61.43	17.90
No computer support, but collaboration	17	2,172	31.51	10.11	50.88	15.38
No computer support, no collaboration	2	257	20.70	00.99	21.87	29.89

3.2   Meta-analytic Moderator Estimation

The specific hypothesis was that the relationship between self-efficacy and transfer is moderated by computer support and collaboration. Four conditions were evaluated: trainings with computer support and collaboration (condition 1), trainings with computer support but no collaboration (condition 2), trainings with collaboration but no computer support (condition 3), and trainings with neither computer support nor collaboration (condition 4). There were no systematic age [χ^2 (3,25) = 4.17, ns] or gender [χ^2 (3,21) = 2.77, ns] differences between conditions (computer support, no computer support, collaboration, no collaboration). A nested sub-group analysis of computer support and collaboration as confounding moderator variables signaled two trends. First, computer support and collaboration were highly correlated, with Spearman's $\rho = .44$ (95% $CI = .43$; .46). Second, effect sizes were highest when the training was computer-supported. Third, effect sizes were twice as high in computer-support trainings without collaboration (condition 2) compared to computer-supported trainings with collaboration (condition 1). Table 2 summarizes the results. However, unequal sample sizes and a small cell sizes for condition 4 warrant caution when interpreting these results.

---

[1] k = number of studies, N = sample size, M = mean, SD = standard deviation.

Table 2. Nested moderator effects of computer support and collaboration

Conditions	ρ	SD_ρ	80% CV
Computer support and collaboration	0.31	0.03	0.27; 0.35
Computer support, but no collaboration	0.62	0.07	0.53; 0.71
No computer support, but collaboration	0.30	0.04	0.25; 0.35
No computer support, no collaboration	0.25	0.01	0.24; 0.26

4   Discussion

One aim of this meta-analysis was to cumulate the research of the past 25 years to correct the relationship between performance self-efficacy and transfer of training for sampling error and error of measurement. A second aim was to estimate the moderating effects of computer support and collaboration. The heterogeneity and disagreement in the training literature ultimately led this study to seek a better understanding of whether, to what extent, and under which conditions efficacy beliefs influenced transfer.

The results of the primary meta-analyses suggested positive relationships between post-training self-efficacy and training transfer. These estimates provide support for Hypothesis 1 and are in line with previous literature reviews [2,7,8,24]. These findings empirically support the theoretical assumption that efficacy beliefs influence transfer [1-3], and are consistent with earlier conceptual frameworks in the training literature, such as the integrative model of motivation to transfer training [25].

The results of the meta-analytic moderator estimation suggested systemic effects of computer support and collaboration [2,3,24]. Specifically, computer-supported collaborative learning does not per se promote the relationship between self-efficacy and transfer. The results showed that computer support played a more significant role than collaboration among trainees. Trainings affording CSCL were not generally more effective in promoting efficacy beliefs and transfer than trainings not affording CSCL (see estimates in Table 2). One possible explanation for this unexpected finding may be the form of collaboration [15-16] in the individual study reports. We had no information on how social interaction emerged in the training situations, as the primary studies reported correlation estimates only but did not engage in analyzing interaction with methods currently available [26]. Nor had we information on the degree that the collaborative learning situations were scaffolded or scripted in the original studies. Without sufficient guidance and scaffolding of collaboration activities among training participants, efforts toward collaboration may result in unequal or heterogeneous participation [26], non-reciprocal interpretations of the learning situation [27], and/or lack of co-regulation [18-20]. Future research may take this meta-analytic evidence to test designs for scaffolding collaboration in technology-rich environments intended to promote self-efficacy and transfer. In summary, analysis of the moderating effects of computer support and collaboration illustrate boundary conditions for self-efficacy and transfer in professional training.

Results of this study may have some practical value for the scaffolding of collaborative learning. Specifically, the low confounded moderator effect of computer support and collaboration tends to highlight the danger of ignoring adequate guidance and scaffolding of participatory interactions among trainees in the learning environment. This finding is reported in the literature elsewhere [18,27] and is now reiterated with a special emphasis of how important scaffolded collaboration is for promoting self-efficacy and transfer [2,5]. It can be speculated that the provision of networks can scaffold trainees during training and that post-intervention enhancement of contacts among trainees could facilitate transfer after training. Methodologically, it would be interesting to follow these educational interventions with different methods, including social network analysis, to trace how they influence processes of sharing and co-construction during collaborative team learning and how they promote transfer to typically practice-bound situations at work [7-8,24]. Application to different professional settings could further elucidate the generalizability of the findings in various contexts, including, but not limited to, technology-enhanced medical visualizations [28-32] and computer game-based learning [19].

This study has some limitations that should be noted. One limitation is that the population correlation estimates were corrected for sampling error and error of measurement. This decision was based on the frequent reporting and availability of sample size and reliability information. However, the original research reports may be affected by additional biases, such as extraneous factors introduced by study procedure [14]. Although the estimation of moderators sought to lessen this bias, the true population estimates may be somewhat greater than those reported here. An additional limitation is that some of the relationships in the nested subgroups were based on small sample sizes. However, some authors have noted that correcting for bias at a small scale mitigates sampling error compared to uncorrected estimates in individual studies [14]. Still, although most of the cells contained sample sizes in the thousands, some did contain fewer, which indicates underestimation of sampling error in those few cases and that, therefore, computer-supported collaborative training may show more positive correlation estimates. Finally, the study reports two moderator variables. Although an analysis of the relationship between performance self-efficacy and transfer of training under different boundary conditions clearly goes beyond previous meta-analytic attempts, selection of the boundary conditions was, of course, eclectic and exclusively driven by an interest to better understand technological and social affordances in CSCL environments for motivation and transfer [2,16,17,33,34]. More conditions exist that would warrant inclusion in the meta-analysis and in turn raise concerns of the generalizability of the moderating effects. However, this limitation can be addressed only by additional original research reports that systematically vary different study conditions.

In conclusion, self-efficacy and transfer were assumed to be more positive in complex social and technological environments [4]. The present meta-analytic study sought to test the predictive validity in an examination of the population correlation estimates between self-efficacy and transfer in computer-supported and collaborative training conditions. This examination was done by using meta-analysis to summarize 25 years of research on post-training self-efficacy, by cumulating 33 independent data sources from 4,203 participants, and by examining two confounded moderator variables (computer support and collaboration) on the self-efficacy–training transfer relationship. The findings seem to imply that computer support is more significant for promoting self-efficacy and transfer than is collaboration. Future research is encouraged to extend these first steps reported here to the examination of social and technological conditions moderating self-efficacy and transfer in other educational and learning settings.

References

1. Bandura, A.: Self-Efficacy: The Exercise of Control. Freeman, New York (1997)
2. Gegenfurtner, A.: Motivation and Transfer in Professional Training: A Meta-Analysis of the Moderating Effects of Knowledge Type, Instruction, and Assessment Conditions. Educ. Res. Rev. 6, 153–168 (2011)
3. Gegenfurtner, A.: Dimensions of Motivation to Transfer: A Longitudinal Analysis of Their Influences on Retention, Transfer, and Attitude Change. Vocat. Learn. 6 (in press)
4. Sawyer, R.K.: The New Science of Learning. In: Sawyer, R.K. (ed.) Cambridge Handbook of the Learning Sciences, pp. 1–16. Cambridge University Press, Cambridge (2006)
5. Segers, M., Gegenfurtner, A.: Transfer of Training: New Conceptualizations through Integrated Research Perspectives. Educ. Res. Rev. 8 (in press)
6. Trost, S.G., Owen, N., Bauman, A.E., Sallis, J.F., Brown, W.: Correlates of Adults' Participation in Physical Activity: Review and Update. Med. Sci. Sport. Exec. 34, 1996–2001 (2002)
7. Gegenfurtner, A., Vauras, M.: Age-Related Differences in the Relation between Motivation to Learn and Transfer of Training in Adult Continuing Education. Contemp. Educ. Psychol. 37, 33–46 (2012)
8. Van Dinther, M., Dochy, F., Segers, M.: Factors Affecting Students' Self-Efficacy in Higher Education. Educ. Res. Rev. 6, 95–108 (2011)
9. Gegenfurtner, A., Veermans, K., Vauras, M.: Effects of Computer Support, Collaboration, and Time Lag on Performance Self-Efficacy and Transfer of Training: A Longitudinal Meta-Analysis. Educ. Res. Rev. 8 (in press)
10. Zimmerman, B.J.: Attaining Self-Regulation: A Social Cognitive Perspective. In: Boekaerts, M., Pintrich, P.R., Zeidner, M. (eds.) Handbook of Self-Regulation, pp. 13–35. Academic Press, San Diego (2000)
11. Junttila, N., Vauras, M., Laakkonen, E.: The Role of Parenting Self-Efficacy in Children's Social and Academic Behavior. Eur. J. Psychol. Educ. 22, 41–61 (2007)
12. Yi, M.Y., Davis, F.D.: Developing and Validating an Observational Learning Model of Computer Software Training and Skill Acquisition. Inf. Sci. Res. 14, 126–169 (2003)
13. Brown, T.C., Warren, A.M.: Distal Goal and Proximal Goal Transfer of Training Interventions in an Executive Education Program. Hum. Res. Dev. Q. 20, 265–284 (2009)
14. Hunter, J.E., Schmidt, F.L.: Methods of Meta-Analysis: Correcting Error and Bias in Research Findings. Sage, Thousand Oaks (2004)
15. Kanfer, R.: Self-Regulation Research in Work and I/O Psychology. Appl. Psychol. Int. Rev. 54, 186–191 (2005)
16. Lehtinen, E., Hakkarainen, K., Lipponen, L., Rahikainen, M., Muukkonen, H.: Computer Supported Collaborative Learning. In: Van der Meijden, H., Simons, R.J., De Jong, F. (eds.) Computer Supported Collaborative Learning Networks in Primary and Secondary Education. Project 2017. Final Report. University of Nijmegen (1999)

17. Dillenbourg, P.: What Do You Mean by Collaborative Learning? In: Dillenbourg, P. (ed.) Collaborative Learning: Cognitive and Computational Approaches, pp. 1–19. Elsevier, Oxford (1999)

18. Volet, S., Vauras, M., Salonen, P.: Psychological and Social Nature of Self- and Co-Regulation in Learning Contexts: An Integrative Perspective. Educ. Psychol.-US 44, 1–12

19. Siewiorek, A., Gegenfurtner, A., Lainema, T., Saarinen, E., Lehtinen, E.: The Effects of Computer Simulation Game Training on Participants' Opinions on Leadership Styles

20. Vauras, M., Salonen, P., Kinnunen, R.: Influences of Group Processes and Interpersonal Regulation on Motivation, Affect and Achievement. In: Maehr, M., Karabenick, S., Urdan, T. (eds.) Social Psychological Perspectives, Emerald, New York. Advances in Motivation and Achievement, vol. 15, pp. 275–314 (2008)

21. Day, E.A., Boatman, P.R., Kowollik, V., Espejo, J., McEntire, L.E., Sherwin, R.E.: Collaborative Training with a More Experienced Partner: Remediating Low Pretraining Self-Efficacy in Complex Skill Acquisition. Hum. Factors 49, 1132–1148 (2007)

22. Gibson, C.B.: Me and Us: Differential Relationships Among Goal-Setting Training, Efficacy and Effectiveness at the Individual and Team Level. J. Org. Behav. 22, 789–808 (2001)

23. Karl, K.A., O'Leary-Kelly, A.M., Martocchio, J.J.: The Impact of Feedback and Self-Efficacy on Performance in Training. J. Org. Behav. 14, 379–394 (1993)

24. Gegenfurtner, A., Festner, D., Gallenberger, W., Lehtinen, E., Gruber, H.: Predicting Autonomous and Controlled Motivation to Transfer Training. Int. J. Train. Dev. 13, 124–138 (2009)

25. Gegenfurtner, A., Veermans, K., Festner, D., Gruber, H.: Motivation to Transfer Training: An Integrative Literature Review. Hum. Res. Dev. Rev. 8, 403–423 (2009)

26. Puntambekar, S., Erkens, G., Hmelo-Silver, C.: Analyzing Interactions in CSCL: Methods, Approaches, and Issues. Springer, Berlin (2011)

27. Järvelä, S.: The Cognitive Apprenticeship Model in a Technologically Rich Learning Environment: Interpreting the Learning Interaction. Learn. Instr. 5, 237–259 (1995)

28. Gegenfurtner, A., Lehtinen, E., Säljö, R.: Expertise Differences in the Comprehension of Visualizations: A Meta-Analysis of Eye-Tracking Research in Professional Domains. Educ. Psychol. Rev. 23, 523–552 (2011)

29. Helle, L., Nivala, M., Kronqvist, P., Gegenfurtner, A., Björk, P., Säljö, R.: Traditional Microscopy Instruction versus Process-Oriented Virtual Microscopy Instruction: A Naturalistic Experiment with Control Group. Diagn. Pathol. 6, S8 (2011)

30. Gegenfurtner, A., Jarodzka, H., Seppänen, M.: Promoting the Transfer of Expertise with Eye Movement Modeling Examples

31. Gegenfurtner, A., Seppänen, M.: Transfer of Expertise: An Eye-Tracking and Think-Aloud Experiment Using Dynamic Medical Visualizations

32. Gegenfurtner, A., Siewiorek, A., Lehtinen, E., Säljö, R.: Assessing the Quality of Expertise Differences in the Comprehension of Medical Visualizations. Vocat. Learn. 6 (in press)

33. Veermans, M.: Individual Differences in Computer-Supported Inquiry Learning – Motivational Analyses. Painosalama, Turku (2004)

34. Järvelä, S., Volet, S., Järvenoja, H.: Research on Motivation in Collaborative Learning: Moving beyond the Cognitive-Situative Divide and Combining Individual and Social Processes. Educ. Psychol.-US 45, 15–27 (2010)

Notebook or Facebook? How Students Actually Use Mobile Devices in Large Lectures

Vera Gehlen-Baum and Armin Weinberger

Educational Technology, Saarland University, Saarbrücken, Germany, P.O. Box 151150
{v.gehlen-baum,a.weinberger}@mx.uni-saarland.de

Abstract. In many lectures students use different mobile devices, like notebooks or smartphones. But the lecturers often do not know to what extent students use these devices for lecture-related self-regulated learning strategies, like writing notes or browsing for additional information. Unfortunately mobile devices also bear a potential for distraction. This article shows the results of observational study in five standard lectures in different disciplines and compares it to students' responses on computer use in lectures. The results indicate a substantial divergence between students' subjective stances on how they use mobile devices for learning in lectures and the actual observed, often lecture-unrelated behavior.

Keywords: Lectures, mobile devices, media use.

1 Mobile Devices – Learning Opportunities or Distractions?

More and more students use mobile devices in lectures, either actively, i.e. for writing something down, or passively, i.e. with the mobile device being switched on and stared at, but without any other notable human-machine interaction. To what extent using mobile devices in lectures fosters learning is highly debated. On one hand, mobile devices could support students in their self-directed learning [1] as students get the chance to search for answers or to take notes on the slides. On the other hand, there is a chance of distraction, when students use the mobile devices for lecture-unrelated activities like posting on Facebook or sending Emails to friends [2, 3].

Unfortunately, lecturers do not know what their students use the mobile devices for since their screens are too small to observe or turned away from the teacher. With very little research on this issue there is hardly any understanding on whether notebooks should be allowed, banned or more actively integrated into lectures. To reduce distraction and to make full use of enhancing learning experiences in lectures through notebooks, gathering information on "lecture-related" and "lecture-unrelated" activities with notebooks seems an important first step.

In this article, we discuss general principles of active learning and how technologies can foster those principles in lectures before taking a look at how students actually use mobile devices in lectures and what students think or say they do with mobile devices in large lectures.

A. Ravenscroft et al. (Eds.): EC-TEL 2012, LNCS 7563, pp. 103–112, 2012.
© Springer-Verlag Berlin Heidelberg 2012

2   Active and Self-directed Learning

The lecture format is often criticized for fostering passive behavior and for being inapt for maintaining student focus [4]. Findings of lecture research show that students are required to listen and make notes most of the time [5]. Ideally, students engage in a series of cognitive and metacognitive activities in a focused and active way [6], such as processing and linking what is being taught to prior knowledge, elaborating the learning material with examples, taking notes and monitoring these learning activities to ward off distractions and to continuously examine one's understanding [7]. But, there are also other, lecture-unrelated activities like talking to the neighbor, doing homework or sleeping which can be observed in lectures. Students sometimes have difficulties to identify and focus on the most important aspects of a lecture [7]. Especially students with little prior knowledge and dysfunctional learning strategies find it hard to continuously focus on the most important aspects of a 90 minute lecture.

There is a chance that mobile devices increase that problem and distract students. Both, passive and active use of mobile devices may consume learners' cognitive resources and draw attention away from what is being taught. Passive use may convey stimuli that "catch the eye"; active, lecture-unrelated use may indicate that learners are pursuing other goals than learning [8]. Even with the intention to use the notebook for learning purposes there is a chance that part of the students' attention is consumed by online activities, e.g. by visual indicators of friends being online. So, in order to ignore or minimalize the effects of these kinds of distractions, it seems important to know how students monitor their own learning [e.g. 9, 10].

3   Mobile Devices to Foster Learning in Large Lectures

Mobile devices could foster self-regulated learning, but advanced technology is often paired with simplistic pedagogical models [11]. There is a risk that students use their mobile devices for lecture-unrelated activities and therefore attempt to multitask during the lecture. Based on the idea that the primary task in lectures is to process new information, multitasking here means to apply some of the cognitive resources to additional tasks. As the working memory is limited, lecture-unrelated multitasking in particular could have a negative impact on learning [12].

Fried [2] tested 137 psychology students over 20 lecture sessions with surveys regarding their use of notebooks in class and distraction in lectures and compared them with the results of the American College Test (ACT) and high-school rank (HSR). Her goal was to show that multitasking distraction by notebooks during lectures would lead to lower learning results in the standardized tests. Almost two thirds of the students (64.3 %) reported to use their computers at least once during the sessions and multitasked an average of 17 min per session (75 minutes). Using notebooks correlated negatively with students' focus and test results. Fried discussed the limitations of the findings, given that only self-reported responses where included, based on the assumption that due to social desirability effects students would underreport the number of minutes students spend on multitasking.

Also Kraushaar and Novak [3] found that distractive multitasking behavior has negative effects on academic performance. Kraushaar and Novak [3] studied the notebook use of 55 students in 30 standard lectures (á 75 minutes) of one course by using a questionnaire and installing spyware on students' computers. They categorized notebook use into productive (course-related) and distractive activities. The distractive activities were further divided into surfing, email, instant messaging (IM), PC-operations and miscellaneous. They confirmed that spending more time with distractive multitasking leads to lower academic performance. But for the subcategories this could only be found for using IM during the lecture. One possible explanation for this result is that the spyware did not register for how long the student actively used the distractive environment. While it is possible that some students just opened a page and started listening to the lecture again, synchronous social tasks like IM lead to more distraction as it requires continuous attention.

4   Research Questions

Even though former research indicated that multitasking with notebooks in lecture has a bad influence on learning performance [2] so far little is known about how frequently and which kind of mobile devices are used in standard lectures.

— *RQ1*: Which kinds of mobile devices are used how often by students during large lectures?

So far, studies were mainly conducted in single courses over a longer time period with the students knowing that their use of notebooks is assessed by questionnaire or spyware [3]. There is need to complement this research with covert observational data, i.e. with students being unaware of the fact that their activities are being observed. There is also need to investigate to what extent mobile devices require all of students' attention or are rather used as a background medium as Kraushaar & Novak [3] suspect.

— *RQ2*: Which kind of activities do students engage in with their mobile devices in large lectures?

As their impression on their time and aim of using mobile devices could give an insight of how well learners manage to self-regulate learning activities when bringing mobile devices to the lecture students self-reported data could show differences between what they think they do with mobile devices and what they actually do. If students have metacognitive deficits regarding self-assessing and monitoring their learning processes, there should be differences regarding their self-report on their intention and time spending on mobile devices to what will be observed during lecture.

— *RQ3*: What reasons do students self-report for bringing mobile devices to the classrooms and how do they actually use mobile devices?

5   Method

5.1   Participants

We conducted the study in five standard lectures of education (two lectures), comput-er science (two lectures) and economics (one lecture) collecting data by questionnaire and observation. We gathered 664 student questionnaires of which 331 students re-ported using technology in the lecture. Some of them used their laptop as well as their smartphone.

Table 1. Observed and self-reported use of mobile devices in lectures

	Education Observed (n = 26) / self-reported technology usage (n = 62)	Computer science Observed (n = 38) / self-reported technology usage (n = 136)	Economics Observed (n = 27) / self-reported technology usage (n = 171)
Notebook	25 / 20	31 / 60	25 / 53
Smartphone	1 / 42	7 / 76	2 / 118

While all questionnaires were used to analyze if student used mobile devices, for further analysis we just report data of those students that stated using mobile devices. We also covertly observed a total of 81 students with notebooks and 10 with smart-phones. Table 1 shows the distribution of mobile devices as observed across lectures in education, computer science and economics.

5.2   Procedure

Before the lecture started, the five to seven investigators chose their seats, so they could observe at least one, but most of the time two to four different notebooks or smartphones users. They sat next or behind the students they observed, so that they saw just the screen but did not get further information about the observed student. Also the investigators tried not to be seen during the observation in order to obtain actual student practices. When the lecture started, the lecturers told their audience that an investigation about lecture activities is taking place and that at the end of this lec-ture a questionnaire will be handed to them. The fact that an observation took place as well was only mentioned after the lecture. The investigators started making notes every 30 seconds on the prepared sheets when the lecturer started talking to the class.

5.3   Instruments

Observation. A lecture of 90 minutes was divided in 180 segments, so that every 30 seconds the observer took a look at the observed mobile device and marked the ob-served activity. The activities were classified into lecture-related activities, like mak-ing notes or seeing lecture slides and lecture-unrelated activities like using social networks, seeing online web-sites with non-course materials and watching videos (see Table 2). Also when students downloaded something or the screensaver was activated, it was noted down on an observer sheet, but these kinds of ambivalent activities are not further discussed in this paper.

The activities were further divided into "active" and "passive". When students typed something, obviously read an online article or used their mouse on a web site the activity was marked active as the focus was on the mobile device at that time. Passive use was coded whenever the focus was on the lecturer and his presentation and there were no activities on the mobile device, i.e. the device was switched on, but not interacted with. So, the distinction between active and passive use of mobile devices does not concern whether a device is switched on or off, but whether the student is interacting with and focusing on the device (active) or the lecturer, someone else or something other than the mobile device (passive).

Table 2. Categories of observed activities

Lecture-related activities	Lecture-unrelated activities	ambivalent activities (not reported)
Slides	Lecture-unrelated websites	Browsing the internet for unidentified information
Taking notes	Lecture-unrelated documents	Downloading something
Lecture-related websites	Social networks	Doing some exercises
Lecture-related documents	Email	Browsing the University website
	Chat	Desktop/screensaver
	Games	
	Newspaper	

As informing the students about the observation beforehand could influence their behavior, we told them after the study and used the DGPS recommended practice on ethics as a guideline. Also, we made sure that no connection between the questionnaires and the observations can be established, as the observers did not note down personal aspects like names, numbers etc.

Questionnaire. The questionnaire was designed to get a more accurate impression on students' use of mobile devices as well as their own impression what they used them for. The students should indicate which mobile devices they used and which lecture-related or –unrelated activities they engaged in, like searching further information on the lecture, taking notes, playing computer games or surfing social networking sites like Facebook. This was indicated by a five point scale with values from "not at all" to "very much". In addition, students were asked to indicate for what purpose they used a specific tool. The answers to these open items were coded and categorized into the same categories as the observational data and then divided into the subcategories of "lecture-related" and "lecture-unrelated" activities. New categories were defined if the answers did not fit into one of the predefined categories, e.g. for yet uncharted forms of distractions or communication. After categories had been established, inter-rater reliability was being assessed.

6   Results

With regard to *RQ1* on which kind of mobile devices student use in large lectures, the questionnaire data indicate that half (49.85%) of the audience is using a mobile device at least once in a lecture. The number varied during the lectures as sometimes students came later or left early so not all students who attended the lecture filled out a questionnaire (n = 664).

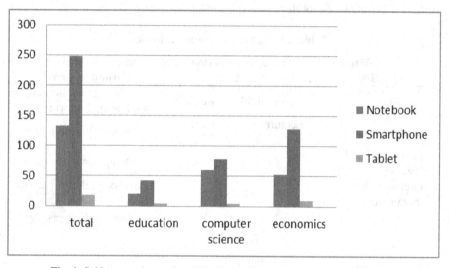

Fig. 1. Self-reported use of mobile devices in lectures (questionnaire data)

The usage of smaller devices couldn't be counted as it is hard to see smartphones when you are not close to them. But the observed use of notebooks indicates that the frequency of using devices as participants indicated (n = 133) (see Figure 1) is consistent with what was observed by the researchers (n = 112). Also here the number varies between different measurement points, as some students store away their device for some time during the lecture. This result was also found during the observation of the 81 students with notebooks.

With regard to *RQ2*, the observations indicate that students were engaged two times more often (51.70%) in surfing lecture-unrelated web-sites and documents than in lecture-related activities (see Figure 2). Whereas active use of mobile devices was stronger associated with lecture-unrelated activities (n = 4015), like communicating through Facebook, lecture-related activities was mostly passive (n = 1460), like looking at the slides of the lecture.

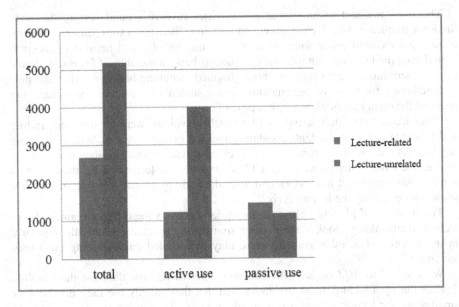

Fig. 2. Observed frequency of active and passive use of lecture-related and -unrelated activities (30 sec intervals)

We differentiated between several categories of active use of mobile devices. The most frequent lecture-related activities on mobile devices are taking a look at the presented slides and taking notes. Students rarely browse for lecture-related websites, however (see Table 3).

Table 3. Frequency of lecture-related and -unrelated activities during the lecture (30 second intervals)

lecture-related activities	
slides	351
notes	218
lecture-related websites	81

lecture-unrelated activities	
lecture-unrelated websites	1445
social networks	679
email	153
chat	76
games	839
videos	170

The most frequent lecture-unrelated activities revolve around surfing the web. Students frequently visit lecture-unrelated websites, like sports sites, different forums or search for downloading some content. Also the use of social networks was common during the lectures with most of the students being a member of Facebook. Some of these activities, e.g. visiting Facebook, implied switching between active and passive activities for some of the students. These students focused on the lecture, but checked the open Facebook page from time to time.

Other lecture-unrelated activities, like watching videos, were far more enthralling and time consuming as students constantly focused on the notebook screen. Wearing headphones during the lecture seemed to clearly indicate that phenomenon. For instance, one student started watching a TV show when the lecture began, then finished some lecture-unrelated homework and started watching another episode of said TV show before leaving the lecture early.

The duration of playing online games varied between the different games. There were students taking a look at their online simulation games, e.g. Farmville, regularly in the lecture, while other students were playing installed games during the whole lecture.

With regard to *RQ3* on how students self-reported on how they use their mobile devices during a lecture, most students stated that they mainly use their notebook or smartphone for lecture-related activities like taking a look at lecture slides or taking notes (see Table 4).

Table 4. Observed vs. self-reported use of mobile devices

	Number of *observed* students actively using mobile devices during the lecture	Number of students *self-reporting* using mobile devices during the lecture
Lecture-related activities		
lecture slides	18 (19.6%)	46 (37.1%)
taking notes	38 (41.3%)	51 (41.1%)
searching	36 (39.1%)	47 (37.9%)
Lecture-unrelated activities		
social networks	44 (47.8%)	19 (15.3%)
chat	10 (10.9%)	4 (3.2%)
emails	22 (23.9%)	15 (12.1%)
communication		28 (22.6%)
unrelated web-sites	56 (60.9%)	
games	14 (15.2%)	7 (5.6%)
video	7 (7.6%)	0 (0%)
distraction		51 (41.1%)

The percentage of students observed taking notes is corresponding to the self-reported one. Also many students mentioned to search for additional information on the lecture to attain deeper understanding. The overall observed number of students doing further research on the internet is rather low (252) compared to other activities like taking notes (805) or social networks (960). But the number of students which report using mobile devices for looking at lecture slides differs from the observed number. Although 37.1% of the students report to display lecture slides on their screens, only 19.6% were actually observed doing so.

In general, the number of students observed doing lecture-unrelated activities is always higher than the self-reported one, although 51 students indicated that they use mobile devices for some sort of distraction in general.

7   Discussion

The study shows that half of the University students use their mobile devices in lectures. While most of the students use smartphones or notebooks, other mobile devices like tablets are not very common in lectures today. The observational data show that most of the students use their mobile devices for lecture-unrelated activities, mostly for surfing on lecture-unrelated websites; this is consistent with prior findings [3]. Other, highly frequent lecture-unrelated activities observed are communicating through social networks and emails. These kinds of lecture-unrelated activities pose a risk of distraction which could hamper learning activities and therefore impoverish learning results [2]. Analyzing active and passive use shows that most of the lecture-related materials, like online slides, do not foster active behavior, e.g. taking notes. In fact, lecture-related use of mobile devices is mostly passive. In contrast, lecture-unrelated activities, like using games or social networks, are typically active. Still, not all of the lecture-unrelated behavior was active. Obviously, students seem to manage some degree of multitasking with passive lecture-unrelated activities, which may not have adverse effects on learning [3]. Because of collecting the data in real classroom scenarios this study aimed to describe how students use their mobile devices. We are currently analyzing differences between students with mobile devices and those without. Some of this data will be presented at the conference. Also, future research may need to inquire how learners are actually dealing with these kinds of distractions successfully with regard to cognitive load and learning outcomes.

There are interesting divergences between observational and questionnaire data. Students may have no good explicit explanation for bringing computers to lectures or may hide their true, lecture-unrelated intentions. Chances are that due to weak self-monitoring strategies students do not entirely realize how much time they spend on lecture-unrelated activities. Perhaps they sometimes do not realize their shift of attention at all.

A lot of Universities install wireless Lan in their lecture halls to give students the possibility to use mobile devices for learning and research. Our results indicate that nearly half of the students accept that offer during lectures – most of them with their smartphone. Even though a lot of students use them not in the intended way, it may be problematic to banish these small mobile devices from lecture halls. Instructional approaches are necessary to help students use mobile devices in lectures intentionally for lecture-related activities [2]. Our future research addresses this issue by suggesting to not ban, but design for involving the devices students bring to the lecture [13]. In such a scenario using an audience response system called Backstage, lecturers would ask students to answer questions as with proprietary clicker systems and allow for students to post lecture-related questions, comments and answers. In this way, students might be facilitated to more actively engage in and better monitor lecture-related activities.

References

1. Greene, J.A., Azevedo, R.: The Measurement of Learners' Self-Regulated Cognitive and Metacognitive Processes While Using Computer-Based Learning Environments. Educational Psychologist 45(4), 203–209 (2010)
2. Fried, C.B.: In-class laptop use and its effects on student learning. Computers & Education 50(3), 906–914 (2008)
3. Kraushaar, J.M., Novak, D.C.: Examining the Effects of Student Multitasking with Laptops during the Lecture. Journal of Information Systems Education 21(2), 11 (2010)
4. Tippelt, R.: Vom projektorientierten zum problembasierten und situierten Lernen - Neues von der Hochschuldidaktik? In: Reiber, K., Richter, R. (eds.) Entwicklungslinien der Hochschuldidaktik. Ein Blick zurück nach vorn, pp. 135–157. Logos, Berlin (2007)
5. Lindroth, T., Bergquist, M.: Laptopers in an educational practice: Promoting the personal learning situation. Computers & Education 54(2), 311–320 (2010)
6. Renkl, A.: Aktives Lernen = gutes Lernen? Reflektion zu einer (zu) einfachen Gleichung. Unterrichtswissenschaft 39, 194–196 (2011)
7. Grabe, M.: Voluntary use of online lecture notes: Correlates of note use and note use as an alternative to class attendance. Computers & Education (2005)
8. Yantis, S.: Stimulus-driven attentional capture. Current Directions in Psychological Science 2(5), 156–161 (1993)
9. Garner, J.K.: Conceptualizing the relations between executive functions and self-regulated learning. The Journal of Psychology 143(4), 405–426 (2009)
10. Hasselhorn, M.: Metacognition und Lernen. In: Nold, G. (ed.) Lernbedingungen und Lernstrategien, pp. 35–64. Gunter Narr Verlag, Tübingen (2000)
11. Roschelle, J.: Keynote paper: Unlocking the learning value of wireless mobile devices. Journal of Computer Assisted Learning 19(3), 12(3), 260–272 (2003)
12. Ericsson, K.A., Kintsch, W.: Long-term working memory. Psychological Review 102(2), 211–245 (1995)
13. Gehlen-Baum, V., Pohl, A., Weinberger, A., Bry, F.: Backstage – Designing a Backchannel for Large Lectures. In: Ravenscroft, A., Lindstaedt, S., Delgado Kloos, C., Hernández-Leo, D. (eds.) EC-TEL 2012. LNCS, vol. 7563, pp. 459–464. Springer, Heidel-berg (2012)

Enhancing Orchestration of Lab Sessions by Means of Awareness Mechanisms

Israel Gutiérrez Rojas[1,2], Raquel M. Crespo García[1], and Carlos Delgado Kloos[1]

[1] Universidad Carlos III de Madrid, Avda. Universidad 30, E-28911 Leganés, Spain
[2] Institute IMDEA Networks, Avda. del Mar Mediterráneo 22, E-28918 Leganés, Spain
{igrojas,rcrespo,cdk}@it.uc3m.es

Abstract. Orchestrating learning is a quite complex task. In fact, it has been identified as one of the grand challenges in Technology Enhanced Learning (TEL) by the Stellar Network of Excellence. The objective of this article is to provide teachers and students with a tool to help them in their effort of orchestrating learning, that makes use of awareness artefacts. Using this powerful mechanism in lab sessions, we propose four different aspects of orchestration as the target for improvement: the **management** of the resources in the learning environment; the **interventions** of the teacher and provision of formative feedback; the collection of evidences for summative **assessment**; and the **re-design** of the activity, adjusting some parameters for future enactments. The proposal has been tested in a real course of Multimedia Applications with junior students (3rd course), measuring the benefits for the orchestration.

Keywords: Orchestration, awareness, lab session, problem-based learning, formative assessment.

1 Introduction

Many Higher Education courses, ranging from engineering to social sciences, have a practical component; that is, the structure of the course is composed by theoretical sessions (lectures) and practical sessions in the computer room, where the students have a computer available to work in the proposed hands-on activity. These face-to-face sessions at the computer lab (henceforth called lab sessions) have a common structure: the teacher proposes a practical task (or set of tasks) to the students; the students work on the proposed task by themselves in their computer; when they encounter a difficulty that cannot overcome by themselves, they raise hand in order to indicate the teacher they have a question; the teacher moves around the room solving questions of the students who raised hand. There are other aspects of the lab session that are specific to the activity in particular: the students work in the computer individually, in pairs or in a group (individual/collaborative activity), there could be one or several teachers (or teaching assistants), the development of the activity could have implications for summative assessment or just have a pure formative component, etc.

In some countries, the education budget has been cut out consequence an increase of the students/teacher ratio in lab sessions. When the ratio is higher than 20, several

A. Ravenscroft et al. (Eds.): EC-TEL 2012, LNCS 7563, pp. 113–125, 2012.
© Springer-Verlag Berlin Heidelberg 2012

problems emerge, intrinsically related to the orchestration of the lab session: the teacher is not able to provide feedback to the students at the same rate that new questions appear; due to the scarcity of the teacher resource, the students compete for the attention of the teacher (e.g., they stand up and wait near to the teacher while she is attending other students, and when she finishes the explanation they grab her attention to help them); the order in which the teacher provide feedback to the students is unfair, regarding parameters like waiting time of the students or progress in the assignment; the teacher has not enough time to check the progress of all the students during the session, mainly the ones who did not ask for help.

In order to mitigate these problems, a tool has been designed that provides teachers with awareness mechanisms in lab sessions. Using the awareness information, teachers gain knowledge about the state of the class: progress of all the students, students who asked for help and when, etc. And they are able to enhance the orchestration of the lab session in several ways: the management of the resources in the learning environment (e.g., feedback time); the interventions and provision of formative feedback; the collection of evidences for summative assessment; and the re-design of the activity, adjusting some parameters for future enactments.

The rest of this article is structured as follows: the next section introduces relevant research about orchestrating learning and awareness; section 3 will be devoted to defining the proposed tool that provides awareness mechanisms to the teacher; in section 4 a validation of such a tool will be presented, based on an experiment in a real setting; finally, in section 5 the conclusions of this work are described as well as some lines of future work related to them.

2   Relevant Literature

Orchestrating learning is a quite complex task. In fact, it has been identified as one of the grand challenges in Technology Enhanced Learning (TEL) by the Stellar Network of Excellence [1]. Moreover, the concept of "orchestrating learning" has different definitions in the TEL community and therefore a different meaning depending of the authors of a publication. In [2], a literature review of orchestrating learning in TEL is carried out; emerging from the review, a conceptual framework is defined consisting of 5+3 aspects of orchestration: 5 aspects about what orchestration is and 3 aspects about how orchestration has to be implemented. Regarding the orchestration definition, the aspects described were (1) design/planning of the learning activities, (2) regulation/management of these activities, (3) adaptation/flexibility/intervention (adaptation of the learning flow to emergent events), (4) awareness/assessment of what happens in the learning process and (5) the different roles of the teacher and other actors. Regarding how the orchestration should be done, (a) pragmatism/practice as opposed to TEL-expert, (b) alignment/synergy to the intended learning outcomes and (c) models/theories that guide the learning orchestration, were the identified aspects. In this work, we have made use of the 5+3 aspects framework in order to structure the contributions to enhance the orchestration by means of awareness mechanisms.

The concept of awareness in the field of Computer Supported Collaborative Work (CSCW) refers to exchange of information among several workers that work on a collaborative activity, regarding status, activity and availability. In this work, we apply these same principles to the context of teaching and learning, considering the awareness as a mechanism that could be used to deal with a complex learning scenario built over different orchestration aspects because it permits teachers and students to get to know better these aspects.

In [3], Alavi et at. use the concept of awareness in the same way in the context of technology-enhanced learning. They analyse the interactions between teacher assistants and learners in recitations sections (sessions of problem-based activities with teaching assistants), and make use of lamps as distributed awareness mechanisms. The lamps are used by students in order to indicate progress (lamp colour) and to request feedback from the teachers (lamp blinking). While the interactions in a recitation section are quite similar to those in lab sessions, both works take a very different approach to overcome the same orchestrations problems: in Alavi's work they focus on using Human Computer Interaction (HCI) aspects (e.g., ambient displays for distributed awareness); instead, our work stresses the importance of recording students' traces and process them to create information useful for the teacher. Therefore, both approaches have advantages: in Alavi's work, the groups of students are aware of their colleagues progress and problems; in our work, the information is processed to offer the teacher a personalised view of the interaction data, and the information could be used after the class to review the session (note: in the context of our work, it is assumed that the assignments are delivered to the students in the form of a web page, which the students interact with). Finally, it is very relevant the stress of both works regarding the importance of the space in the orchestration of face-to-face activities.

In [4], Dong and Hwang introduce the PLITAZ (Pause Lecture, Instant Tutor-Tutee Match, and Attention Zone) system that minimizes learning progress differences. This work is contextualised by the use of software teaching classes that alternate lecture and practice phases. During the practice phase, two strategies are used to attend students' problems: tutor-tutee match (when a student finishes the practice, she is asked to be a reviewer, and after acceptance, she is commanded to help a peer with problems in this practice) and attention zone (students with problems in the practice surrounded by others with the same problem, should be attended first by the teacher in order to prevent isolation). Therefore, in Dong and Hwang's work the space is also a very important factor. The awareness strategy followed by them is very similar to ours, but their main objective is different because they try to minimize the difference of progress among students and our objective is to enhance orchestration.

Regarding pedagogical concepts relevant to this research, we are going to focus on problem-based learning since it is the methodology used in the lab sessions. Collaborative problem-based learning (PBL), usually considered an active learning methodology [5], is an instructional method commonly used for teaching engineering courses: students are organized in small groups and presented with a challenging problem to solve. In this article, Prince concludes that students do not get better assessment results with PBL but they are more motivated and could develop higher-level skills like information retention, problem solving and critical thinking. Barrows [6] presents a taxonomy for problem-based learning and a set of four educational objectives addressed by PBL, much related to this work: (a) the knowledge about the context; (b) the practice and feedback; (c) self-directed skills; and (d) motivation and challenge.

Finally, the concept of formative assessment is also relevant to our research, since the objective of the interactions in the lab sessions is to provide formative feedback to the students. It is stressed the importance of such feedback for awareness of students [7] and teachers [8].

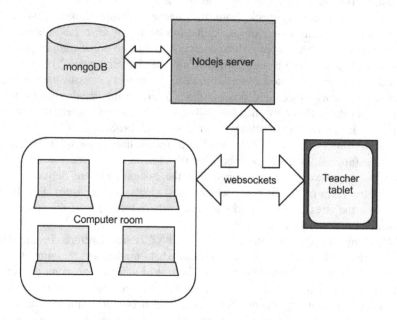

Fig. 1. Awareness system technical architecture

3   Using Awareness Mechanisms for Orchestration

As stated before, we are going to make use of the 5+3 aspects framework described in [2] to classify the different orchestration aspects that we are going to enhance with awareness mechanisms. Four aspects have been identified as part of the orchestration including: the **management** of the resources in the learning environment; the **interventions** of the teacher and provision of formative feedback; the collection of evidences for summative **assessment**; and the **re-design** of the activity, adjusting some parameters for future enactments.

3.1   Awareness System Technical Architecture

The awareness system we have built assumes the context of problem-based learning in lab sessions, being the assignment of the session delivered to the students as a web page. The system is composed of two parts: one embedded in the web page of the assignment and the other one a tablet web interfaces for the teacher. Regarding technologies (as shown in Figure 1), we have used websockets [9] in order to implement real-time communication of events among students and teachers, as the clients are web browsers. For an easy websocket implementation we used nodejs [10] in the back-end (a server-side JavaScript solution) that is able to manage multiple connections opened with the browsers without performance problems. For the data, we have used mongodb[11], a No-SQL database that uses JavaScript as the script language and JSON as data format. In this way, JavaScript and JSON are used in all the stacks (client, server, database) facilitating the integration of the developed components.

3.2   Awareness System Interfaces

In a previous work [12], a exploratory research was presented, consisting of the usage of websocket notifications as a communication backchannel in lab sessions. It also introduced a preliminary version of the students and teacher interfaces. Nevertheless, the design of both interfaces has been changed, taking into account the feedback provided by teachers, students and fellow researchers.

The assignment of the session is a problem-based assignment, composed of several parts or sections. The developed client component for the students (henceforth, student component) analyses the web page of the assignment detecting the sections in the document, and constructs a table of contents. At the beginning of the session, it presents a very simple interface to the students, composed of the following two parts: the main and the aside components.

In the main part of the page (right side of Figure 2), the first section of the assignment is presented (as the "current section" for the students to start with), as well as the references section if existing (list of theoretical references that could be useful for the students during the session); when the students indicate progress (i.e., finished the current section), the next section of the assignment is presented.

On the left side of the screen (aside) there is a fixed part divided in three main regions: (1) in the top: a table of content for the assignment, containing all the sections in the assignment and little circle indicating their status (green: completed, amber: in progress, grey: not initiated); the status circle is clickable for indicating progress (on the current section when it is finished, and on the last finished to undo the progress); at the bottom of this region there is a progress bar indicating the progress of the students in the assignment; (2) in the middle: a red button used to ask a question to the teacher (equivalent to raise hand); when help is requested, the students is prompted to describe the question and the button turns blue; now, the button can be used to indicate that the doubt has been solved (by the teacher or the students themselves) and the student is prompted to describe the answer to her solved question; the position in the queue is also shown to the student because this information is known when raising hand (since you can observe the other students that raised hand too); (3) in the bottom: the name of the students working in the PC; when accessing to the assignment, the first thing they have to do is entering their student id; in this part there is also a button to change the students in case it was necessary.

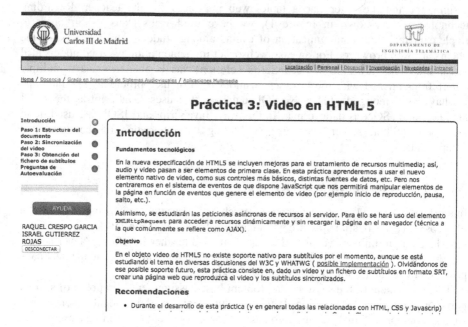

Fig. 2. Students assignment interface

Regarding the tablet web interface for the teacher, it is composed of two parts. The first one (general view, shown in Figure 3) is a representation of the physical classroom where the lab session is carried out. It shows a set of icons representing the PCs in the classroom, which are used as the context for the information of the students working on the PC. In the icons, several types of awareness information are shown. Firstly the background colour of the icon indicates: grey, PC not in use; blue, students working in the PC; or orange to red, the students in the PC asked for help (the colour starts being orange and turns gradually into red in 10 minutes). Secondly, the number in the middle of the icon indicates the current section of the assignment for the students in that PC. Finally, a square in red around the icon of the PC, indicates that these are the students that have been waiting for longer time.

There are also two indicators on the top of the screen: the one on the top-left corner indicates informs about the state of the connection to the server (red: not connected, green: connected); the one on the top-right corner indicates that the teacher has some students waiting for help (red-BUSY: indicates that some students asked for help, green-FREE: indicates that there are no students waiting for help).

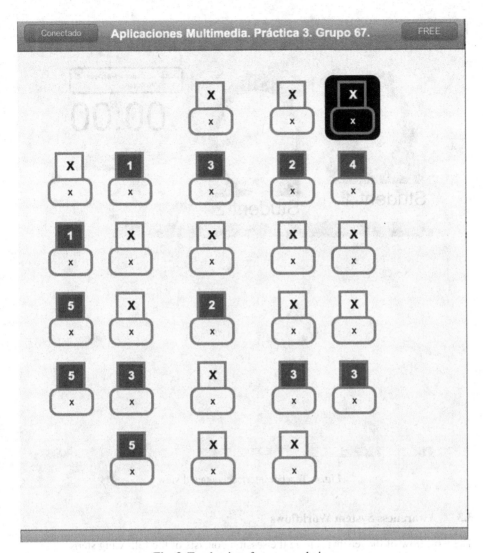

Fig. 3. Teacher interface: general view

The detailed view (shown in Figure 4) appears in the tablet interface when the teacher touches the icon of a PC that is being used by students. In this view, the pictures and names of the students are shown as well as the description of the last question that they asked. Besides that, there is a timer and a button for the teacher to indicate that she is providing feedback to these students. The timer is used for the teacher to be aware of the time devoted in this feedback interaction. When the teacher finishes her intervention, she pushes the button again to stop the timer (or simply push the back button to return to the general view).

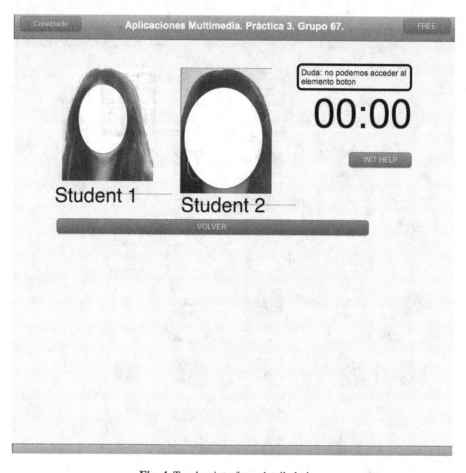

Fig. 4. Teacher interface: detailed view

3.3   Awareness System Workflows

The workflow of the teacher during the session consists of the following steps:

1. connect to the awareness system, indicating the course and session
2. while there is no students waiting for help, she can observe the progress of the students in the general view, and provide feedback to the students proactively
3. when one or more students ask for help, the BUSY/FREE indicator turns red, and a red square indicates the next step (i.e., the students that has been waiting longer for help)
4. when the teacher decides to provide feedback to a group of students, she enters in the detailed view to check the name and pictures of the students and the question they have asked
5. when the feedback starts she presses the button to activate the timer

6. once the feedback has been provided, she presses the button again and checks the general screen for the next group to be attended

The workflow of the students during the session consists of the following steps:

1. connect to the assignment and enter the student id for the system to identify them
2. start working on the first section of the assignment
3. when they finish a section, indicate progress using the progress indicators (colored circles)
4. when they need feedback of the teacher, push the *HELP* button and introduce the description of the question
5. when they solve a question (by the teacher or by themselves), push the *SOLVED?* button and describe the solution to their question
6. if they indicated progress incorrectly, they can use the progress indicator of the last finished section to undo the progress and go back to the previous section

3.4   Awareness System Benefits

The benefits of the introduced awareness system for enhancing the orchestration of the lab sessions are presented following the 5+3 aspects framework as defined above. Moreover, the aspects are presented grouped in two categories: aspects used for orchestration during the session (live) and after the session.

- adaptation/flexibility/intervention: during the session, the awareness information about students progress and help is used by the teacher to plan and execute the intervention for feedback provision
- regulation/management: during the session, the awareness information about the time devoted to a group is used by the teacher to manage the timing of the session
- awareness/assessment: after the session, the teacher makes use of the students' progress at the end of the session to determine which ones worked as expected during the session and reward them with a positive grade
- design/planning: after the session, the teacher reviews the questions and answer of the students and plan the future enactment of the activity

4   Validation of the Awareness System

The system has been validated in a real setting, a course of Multimedia Applications, in 5 sessions with 4 different teachers (2 authors of this work and 2 outsiders). The students in each session ranged from 20 to 30 (10 to 15 groups of 2 students), working in pairs (2 students per PC) in the assignment of the lab session. About 800 events were captured, of the following types:

- connection: students connect to the assignment web page
- finishSection: students indicate a section as completed
- undoFinishSection: students undo the completion of a section
- help: students asked for help
- solved: students indicated that a question have been solved
- initHelp: the teacher starts helping a group of students
- endHelp: the teacher finished the attention to a group of students

In a previous work [13], a set of metrics was presented in order to evaluate an awareness system. One of these metrics is the waiting factor of the lab sessions (ratio between the waiting and the tutoring times). Applying this metric to the collected events, the results obtained were not the expected (see Figure 5).

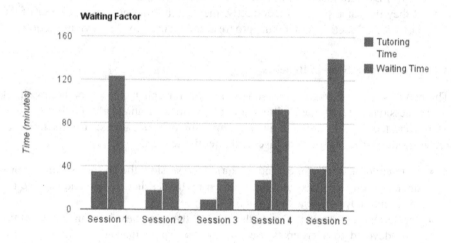

Fig. 5. Tutoring and waiting times per session

The waiting times are always higher than the tutoring times and, therefore, the waiting factor is always greater that 1. Analysing the data of the sessions (shown in Table 1), we found out that the sessions could be categorised in two: sessions 1 to 4 correspond to a very simple assignments and therefore very few questions arose among the students; instead, session 5 consisted of a complicated assignment and the teacher was solving doubts all along the session (34). The data shows that the waiting factor per se cannot be used to determine the time efficiency of the sessions since the session 5 was very efficient but its waiting factor is the worst. Therefore, a new metric should be defined that combines the measures of time and interventions in order to characterise the efficiency of the session. The definition of such a metric is a future work to this contribution.

Table 1. Quantitative data collected in Multimedia Applications

Session number	Tutoring time (min.)	Waiting time (min.)	Waiting factor	Help requests	Interventions
1	34.94	123.59	3.54	17	17
2	18.13	29.11	1.61	13	13
3	9.77	28.76	2.94	14	11
4	28.03	93.33	3.33	12	12
5	38.34	140.46	3.66	37	34

All in all, there could be other factors that conditioned the values obtained. For example, the usage of the system may require a learning curve and thus the parameters could be better in future experiments. Another issue could be that in order to measure the tutoring time, the teachers should press the "INIT HELP" button and, in some occasions, the teachers recognise to have forgotten about doing it.

Nevertheless, besides the quantitative data collected students were asked to fill out a survey about the sessions in which they used the awareness system. The teachers (only the two outsiders) were also interviewed regarding the dynamics of usage of the tool, good/bad features and general feeling.

The main highlight of the interviews are summarised in two good and two bad comments about the systems.

The interviewed teachers identified as the best features of the system that..

- they could find out in a glance which students in the class are working in the session and their progress
- they could be more fair in the distribution of feedback, attending those students who waited longer

They also identified as the worst problems of the system that...

- they did not read the questions of the students in the tool beforehand, but directly asked the students to tell them
- the UI of the teacher interface was improvable, making it adapted to a portable touch device (size of elements, buttons, etc.)

From the students surveys (points in a 1-5 likert scale), it can be stated that the using the tool...

... the tutoring time of the teacher is more fairly distributed (mean 4.38)

... the order of time distribution is more fair (mean 4.46)

... make students concentrate more in solving the problem that in searching for the teacher attention (mean 4.07)

... the students trusts that the teacher is going to help them although they do not raise their hand (mean 4.23)

... the student interface is clear and easy to use (mean 4.00)
... students do not like to write the questions because they prefer to ask her the questions directly

5 Conclusions and Future Work

In this work, a set of enhancements for learning orchestration in lab sessions based on awareness mechanisms have been presented. The enhancements have been organised following the 5+3 framework for orchestration, being the following:

- during the lab session, the teacher is informed with the students progress and help requests in order to plan and execute the **intervention** for feedback provision
- during the session, the teacher is informed about the time devoted to a group in order to **manage** the timing of the session
- after the session, the teacher makes use of the students' events during the session that may be used as evidences for summative **assessment**
- after the session, the teacher reviews the questions and answer of the students and **plan** the future enactment of the activity

The aforementioned enhancements have been validated in a real setting, in a course of Multimedia Applications, demonstrating the effectiveness of the proposed system by means of quantitative (likert scale surveys) and qualitative (interviews) data.

As future work, a lot of lines could be addressed, being the most relevant:

- define metrics that characterise a time efficient system
- integration of formative assessment when completing a section in the assignment: this kind of integration would validate the progress of the students; they would be commanded to deliver a piece of work that proves their progress or do some multiple choice questions for self-assessment
- shared questions: it is a widget for the students to share their questions during the session and they could follow a question of a peer, allowing the teacher to be aware of the most followed questions. A widget based on this principle won the 3rd ROLE widget competition and it is being implemented for the ROLE infrastructure.
- new visualizations of the data collected during the sessions for teacher to better review the session after completion
- provide the students with information about the progress of the class (mean of all students progress) and compare it to her individual progress
- implement new strategies for recommending the next step to the teacher: instead of using always the longer waiting students, different algorithms could be designed following the same principles that the CPU uses for allocating time for processes (FCFS, FJS, Round-Robin)
- a gamification strategy could be implemented for the students to engage more in the session, based on the collected data

Acknowledgments. This research has been partially supported by the project "Learn3: Towards Learning of the Third Kind" (TIN2008-05163/TSI) of the Spanish "Plan Nacional de I+D+i", the Madrid regional project "eMadrid: Investigación y Desarrollo de tecnologías para el e-learning en la Comunidad de Madrid" (S2009/TIC-1650) and the "EEE project" (TIN2011-28308-C03-01) of the Spanish "Plan Nacional de I+D+i".

References

[1] Orchestrating learning, Stellar Network of Excellence, http://www.stellarnet. eu/d/1/1/Orchestrating_learning (last accessed April 02, 2012)

[2] Prieto, L.P., Holenko Dlab, M., Gutiérrez, I., Abdulwahed, M., Balid, W.: Orchestrating technology enhanced learning: a literature review and a conceptual framework. International Journal of Technology Enhanced Learning 3(6), 583–598 (2011)

[3] Alavi, H.S., Dillenbourg, P., Kaplan, F.: Distributed Awareness for Class Orchestration. In: Cress, U., Dimitrova, V., Specht, M. (eds.) EC-TEL 2009. LNCS, vol. 5794, pp. 211–225. Springer, Heidelberg (2009)

[4] Dong, J.-J., Hwang, W.-Y.: Study to minimize learning progress differences in software learning class using PLITAZ system. Educational Technology Research and Development (243) (2012), doi:10.1007/s11423-012-9233-x

[5] Prince, M.: Does Active Learning Work? A Review of the Research. Journal of Engineering Education 93, 223–231 (2004)

[6] Barrows, H.S.: A taxonomy of problem-based learning methods. Medical Education 20(6), 481–486 (1986), doi:10.1111/j.1365-2923.1986.tb01386.x

[7] Dillenbourg, P., Järvelä, S., Fischer, F.: The Evolution of Research in Computer-Supported Collaborative Learning: from design to orchestration. In: Balacheff, N., Ludvigsen, S., de Jong, T., Lazonder, A., Barnes, S. (eds.) Technology-Enhanced Learning. Springer (2009)

[8] Watts, M.: The orchestration of learning and teaching methods in science education. Canadian Journal of Science, Mathematics and Technology Education (2003), http://www.informaworld.com/index/918899916.pdf (retrieved)

[9] Websockets, http://dev.w3.org/html5/websockets/ (last accessed April 02, 2012)

[10] NodeJS, http://nodejs.org (last accessed April 02, 2012)

[11] MongoDB, http://mongodb.org (last accessed April 02, 2012)

[12] Gutiérrez Rojas, I., Crespo García, R., Delgado Kloos, C.: Orchestration and Feedback in Lab Sessions: Improvements in Quick Feedback Provision. In: Kloos, C.D., Gillet, D., Crespo García, R.M., Wild, F., Wolpers, M. (eds.) EC-TEL 2011. LNCS, vol. 6964, pp. 424–429. Springer, Heidelberg (2011)

[13] Gutiérrez Rojas, I., Crespo García, R.: Towards efficient provision of feedback in lab sessions. In: Proceeding of the International Conference on Advanced Learning Technologies, ICALT 2012 (accepted, 2012)

Discerning Actuality in Backstage
Comprehensible Contextual Aging

Julia Hadersberger, Alexander Pohl, and François Bry

Institute for Informatics
University of Munich
{hadersberger,pohl}@pms.ifi.lmu.de, bry@lmu.de
http://pms.ifi.lmu.de

Abstract. The digital backchannel Backstage aims at supporting active and socially enriched participation in large class lectures by improving the social awareness of both lecturer and students. For this purpose, Backstage provides microblog-based communication for fast information exchange among students as well as from audience to lecturer. Rating enables students to assess relevance of backchannel messages for the lecture. Upon rating a ranking of messages can be determined and immediately presented to the lecturer. However, relevance is of temporal nature. Thus, the relevance of a message should degrade over time, a process called aging. Several aging approaches can be found in the literature. Many of them, however, rely on the physical time which only plays a minor role in assessing relevance in lecture settings. Rather, the actuality of relevance should depend on the progress of a lecture and on backchannel activity. Besides, many approaches are quite difficult in terms of comprehensibility, interpretation and handling. In this article we propose an approach to aging that is easy to understand and to handle and therefore more appropriate in the setting considered.

Keywords: Enhanced Classroom, Backchannel, Relevance, Aging.

1 Introduction

Lectures with large audiences is a much-noticed appearance of modern education. In large class lectures students seldom actively participate, despite the fact that active participation is vital for learning success. Several circumstances that favor passivity are provoked by large class lectures [1]: students are often inhibited to speak in front of many peers. They are wary about interrupting the lecturer to ask because their question might only be of minor relevance to the others and thus would merely disturb the lecture; they are also afraid of appearing incompetent when asking many questions [2]. Often students have also difficulties in formulating a question or a comment, especially when dealing with a quite unknown topic. When lectures proceed at a high pace students only have little time to think about the topic and only few opportunities to ask or comment. Besides, in the lecture hall only one person can speak at a time. Whenever several group members engage in a joint discussion, moderation is necessary.

A. Ravenscroft et al. (Eds.): EC-TEL 2012, LNCS 7563, pp. 126–139, 2012.
© Springer-Verlag Berlin Heidelberg 2012

To remedy the shortcomings of large class lectures, much effort has been put in investigating the use of CMC[1] and social media for learning (e.g. [3–5]). We argue that the synchronous use of CMC in the form of a digital backchannel carefully designed for the use in lectures may help to improve the social experience in the classroom. For example, a student may assess the relevance of her question and request for social support. Exchanging on a backchannel allows her to gain confidence to raise a hand. But also the lecturer can utilize the communication on the backchannel to lower the barrier and to stay connected with the audience. The system Backstage [6, 7] is a digital backchannel specifically tailored for the use in large class lectures as part of a research project that aims at advances in both e-learning and social media. Backstage provides carefully designed microblog-based communication by which students can rapidly exchange opinions, questions and comments (cf. Section 2).

Communication on a backchannel can quickly become confusing and incomprehensible without further structuring and filtering, even when the number of participants is small. Furthermore, the relevance and quality may vary entailing the need to filter out irrelevant messages. Therefore, students may rate, i.e. approve or reject, messages. Rating plays an important role for the lecturer: because of the outstanding role and the short time spans during which she can pay attention to the backchannel while lecturing it is hardly possible for her to get a meaningful overview of the backchannel communication without the help of the audience. Rating makes possible to provide her with a top-k ranking of the relevant messages. Also, rating is important for the students because it serves as an instrument to collectively direct the lecturer's attention to what they find particularly relevant for their good reception of the lecture.

However, relevance of lecture-related messages sent during the lecture is of temporal nature. Thus ratings and rankings, for that matter, should depend on time. As the lecture proceeds, topics might change and some questions or comments might become obsolete with respect to the progress of the lecture (while staying relevant and available for discussions and exchange after the lecture). Therefore, some kind of aging is needed. That is, the importance of messages should gradually degrade over time. With aging, attention during the lecture is directed to recent and active messages. Though, determining age on the basis of the physical time does not seem to be reasonable for our purposes. Lectures usually vary in progress. For example, introductory slides might be presented at much a higher pace than a difficult mathematical proof. Aging should rather depend on a lecture-specific measure of time like the activity on the backchannel. The approaches found in the literature seem to be too involved for our needs and difficult to handle in the context of a backchannel for large class lectures. In this article we present an approach to aging that is based on the backchannel activity, and that is highly focused on ease in comprehensibility and handling. It should be noted that although our approach is specifically conceived for Backstage it might also be interesting for other microblogging platforms like Twitter[2], which is discussed in Section 5.

---

[1] Computer-Mediated Communication
[2] http://www.twitter.com

2   A Short Overview of Backstage

Backstage is a digital backchannel for the use in large class lectures. The central part of Backstage is CMC on the basis of microblogs akin to Twitter. Microblogs are short messages comprising only a few words. They seem to be apt for the synchronous use during lectures, since they only contain one information item and may be read and written quickly. Unlike common microblogging, Backstage requires messages to be assigned to predefined categories, e.g. Question or Answer. One rationale behind categories is to convey to the students the kind of communication sought on the backchannel. As mentioned above, messages may be rated by the students to express acceptance or rejection of a message in terms of quality and relevance for the lecture.

A major goal of Backstage is to provide communication and promote student-to-student as well as student-to-lecturer interaction conducive for learning. For this reason, Backstage guides the user's interactions [8]. To provide for context on Backstage the presentation slides are integrated into the users' dashboards (cf. Figure 1).

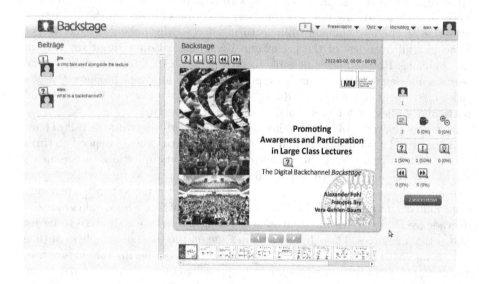

Fig. 1. The lecturer's dashboard on Backstage: the message stream is shown at the left-hand side. The slides are displayed at the center with the categories of messages on top. At the right-hand side the aggregated topic overview is displayed.

To align the backchannel communication with the slides the creation of a message is a well thought process simple and intuitive to perform that, in a manner, is inspired from scripts [9]. It is realized by an iconic drag-and-drop onto the slides to direct users to messages profitable for learning: to write a message the student has to be aware of what she wants to say (both in terms of category and content), and to which part of the slide the message refers to (cf. [8]). That is, on Backstage messages annotate slides, which is also referred to as explicit referencing [10]. As already mentioned, messages on Backstage can also be read, and possibly answered, both by students and by the lecturer at any time after the lecture.

Backstage also provides means to improve the lecturer's awareness. Since due to the script-based user interface every message is necessarily assigned to some predefined category (e.g. Question, Answer, Remark, Too Fast) an aggregated overview showing the distribution of the communication to the categories can be given. For example, such an overview makes possible for the lecturer to quickly become aware during the lecture of many students getting lost, which presumably results in a notable increase in Question- and Too Fast-messages. Besides a topic-related overview, a top-k ranking of messages can be generated showing the k messages that the audience finds particularly relevant. Such a ranking is based on the ratings of students. Thus, rating allows students to direct the lecturer's attention to what that they find relevant. Both kinds of overview, the distribution of messages to the categories and a content-related overview by a top-k ranking supports the lecturer in staying attached to the backchannel.

To support active participation, Backstage allows the conduct of quizzes that are reminiscent to audience response systems (e.g. [11, 12]). Recently, audience response systems have gained much attention. They not only allow to playfully assess students' retention but also help to structure the lecture and activate students at a regular basis. When a quiz is conducted on Backstage, students can only answer the quiz; other functionalities are disabled. After the quiz is finished the results are integrated as ordinary slides that can be annotated and viewed. That is, quizzes can be used for introducing some gamification into the lecture thus providing a kind of break and sustaining the students' attention.

3   Related Work

Prior to presenting our approach to ranking with aging it is reasonable to provide the reader with a short overview of the field. As mentioned above we want to determine a ranking of messages upon the students' ratings. What is understood as rating and ranking is in many cases not so clear, however. Making matters worse, rating and ranking often occur interleaved, since rankings are frequently determined on the basis of ratings. Though, we distinguish between rating and ranking as follows: rating refers to the process of assigning some concrete value to a single message, e.g. "plus" and "minus" or "approve" and "reject". Ranking, in turn, relates two or more messages to each other, thereby specifying a relative (strict) order, for example pairwise comparison of the form "Message A is more relevant than message B".

3.1   Rating

Rating has been applied in various situations. In the Internet it is especially known for its use on commercial websites (rating products or sellers) and in

Web 2.0 applications, to get feedback and find high-quality [13, 14]. Basically rating schemes can be distinguished in two main groups, namely explicit and implicit rating.

The first group – explicit rating – comprises all algorithms that necessitate an intentional vote of a user, which means she is conscious of her evaluation [15, 16]. This kind of obvious rating forces the user to actively think about her judgment, but this can also be seen as an effort so that the user might get discouraged if there is no kind of reward for it [17]. The simplest solution for explicit rating is solely giving users the possibility to "like" an item by voting for it [16], maybe even on a five-star rating scale. Normalization is often used to keep the score within a certain range. The downside of normalization is that the reliability of the average score is not apparent to the users. For example, an item with an average rating of two of five stars voted by only one person does not seem as bad as an item with the same average rating voted by, say, twenty people. However, the opinion of the larger group seems to be more reliable. Another explicit form of rating scheme gives the possibility to not only vote positively for an item, but also negatively or even express neutrality [18–21]. Negative ratings are sometimes desired to give users the possibility to "punish" inappropriate items or behavior. Disadvantageously, calculations with negative values can become complicated, chances are that positive and negative values cancel each other out. As a result, no received votes and a balanced average of votes might be observed as the same overall score.

The second group – implicit rating – extracts rating information from non-rating interactions or data that, however, is interpreted as votes. The user is often unconscious about her influence, since rating happens in the course of using the application [15, 16]. Implicit rating helps to overcome data sparsity, since the user does not need to be motivated to particularly provide for ratings. Different kind of interactions depending on the context can be chosen as a source of rating. Clicks on links or items can be seen as interest and positive feedback, but there could also be "misclicks" which are then misinterpreted [13, 18]. Other interactions might be more reliable, like adding an item to someone's favorites, printing or buying an item or even measuring the time that was spent on an item [17]. Furthermore, answering a question on a discussion board can also be considered as interest in an item and thus, as a positive vote.

Although implicit rating seems to be more complex to handle, since much data has to be analyzed and stored, the retrieval of more reliable data collection in a more timely fashion is possible. On the other hand, explicit rating is the only way to force the user to really consciously form an opinion about an item.

3.2   Ranking

Ranking can be found in various situations, for example online for listing the best game players or ordering search results. For Backstage a ranking is needed that melts the opinions of the users into one single ranking. Basically, there are two different ways to get a collective ranking: aggregating individual ratings or aggregating individual rankings. Furthermore, the collective ranking can be split in two groups, namely non-parametric and parametric solutions.

Non-parametric solutions do not rely on any externally set parameters or weights. The first way, getting a collective ranking by aggregating individual ratings, comprises some simple mathematical solutions that were already mentioned in Section 3.1, like summing up values or calculating the arithmetic mean. As already mentioned, these basic solutions entail different disadvantages, for example positive and negative values cancel each other out. Furthermore, the arithmetic mean makes it easier for new items to get a better overall rating than older ones, since an item can only receive the highest positive overall rating if every rating was that high [14]. More complex ideas entail more complex problems, like finding experts in question-answer-portals. Although it seems to be a good idea to count the number of people a user has already helped, the problem remains that it is not known if she only answered to lay people or other experts [22]. To get individual rankings that can be aggregated to a collective ranking users can be asked to directly order the items according to their opinion. As it is very challenging for users to order many items, comparison based methods are frequently used. Therefore, two ore more items are shown to the user, who has to decide which one she prefers. Repetitive comparison of the winning item against the other alternatives until no items are left result in an individual ranking. This can also be done implicitly, for example while browsing a website with several links on it choosing one link can be interpreted as preference for the clicked link over the other ones [15]. The so-called Hasse method [23] offers the opportunity to create a ranking of items by directly comparing their two or more properties. Disadvantageously items could be incomparable if they are not better or worse in all properties. Hence, the Hasse method might result only in a partial order. Afterwards it can be ranked according to the average positions. The so-called Copeland Score [23] combines the idea of the Hasse method with the direct comparison of items. Like the Hasse method, items with several properties are compared against each other. The Copeland Score for each item denotes the number of wins minus the number of defeats (incomparability is equivalent to zero) while it is compared to all alternatives. Afterwards the items are ranked according to their descending Copeland Score. It has to be noticed that the Copeland Score results in a total, but not necessarily in a strict order, which means there can be two items with the same score.

Each of the above mentioned non-parametric solutions can be combined with, and influenced by, parameters and hence become a parametric solution. Setting the parameters is crucial and can influence the overall computation significantly [23]. Therefore, many experiments are required to find the right configuration. Two interesting projects, using individual ratings to get a collective ranking, shall be mentioned here. The Backchan.nl project [20] is similar to Backstage and includes a formula that combines the so-called voteFactor with the ageFactor. The voteFactor is based on the proportion of positive votes for a message and the number of votes the message received compared to all other messages. This solution is already designed for a very specific context, as it does not only reward positive items but also highly discussed ones. Another algorithm is the Real-Life-Rating [14], an extension of the so-called Bayesian Rating, which is shortly explained in Section 4. This algorithm involves the expertise of users for certain domains and the friendship between users additionally to the rating itself. The Real-Life-Rating algorithm is very elaborate, but also very specific. It seems to be adequate to rather make use of the Bayesian Rating to keep it simple. The last example shows the combination of individual ratings and rankings aggregated at the same time to get a collective ranking. The ranking algorithm for microblog search [24] is based on three different properties. First, the FollowerRank which denotes the number of followers of one user normalized by the total number of her followers and the users she follows. Second, the LengthRank which is the comparison by percentage of this message to the longest message within the search results. Finally, the URLRank is set to a positive constant if the message contains a URL, otherwise it is set to zero. Although this solution can be criticized, as containing a link or being very long does not necessarily constitute a good message, it is a very good example for the smooth transition between rating and ranking. Although all three properties seem to be a ranking due to their name, in fact the two URLRank and FollowerRank are independently set or calculated values without any comparison to other items.

3.3   Aging

In most projects aging is a negative process of losing influence as time goes. Therefore, aging is naturally expressed as some kind of weight decreasing over time and expressing a remaining relevance. The older an item is, the lower its influence on the overall score. As we will see in Section 4 the notion of the term "age" is important.

One solution is based on the half-life parameter as known from the modeling of nuclear decay processes. Therefore, a time-dependent monotonic decreasing function $f(t)$ is included in the algorithm [25], for example the exponential or logistic function. The authors define the time function as $f(t) = e^{-\lambda t}$, where λ is the decay rate $\frac{1}{T_0}$. This algorithm depends on the setting of T_0, which specifies how long it takes to reduce the weight by half. The lower T_0, the faster the decay of the weight and the lower the influence. Another algorithm concerning the freshness of items on social tagging sites [26] divides the timeline in discrete and equi-distant time intervals. The time function a^{m-s} is included into the formula, where a denotes a decay factor between zero and one. While m counts the number of all time slices up to now and s is a indexed variable from one to m, $m = s$ is the current time slice. The fresher a tagging the smaller is the exponent, and the bigger the whole factor. Fresher items have a bigger influence.

In contrast to the above mentioned algorithms, the ageFactor of the system Backchan.nl [20] is not so clear. As the ageFactor is combined with the voteFactor by multiplication it seems obvious at first sight that the aging here is once again some kind of weight. Examples show that voteFactor and ageFactor are inconsistent with one another. Therefore, we solely focus on the ageFactor formula here. The age of a message is defined by the average age of the last five votes the message received. The authors use a constant $\tau = 10^4$ by which the average age is divided to reduce the influence of the age factor. This solution seems to be very intuitive, but it has several drawbacks. First of all, the parameter τ has to be set individually according to each context. It could happen that the ageFactor becomes larger than one if enough time goes by. The inconsistency of this algorithm lies in the fact that with increasing age of an item the ageFactor also increases. Using the ageFactor as defined in [20] with increasing age the influence of such a post is also increased instead of reduced.

4   Discerning Actuality in the Ranking of Messages

For the presentation of aging we first assume that the rating procedure is a black box that yields numeric values for the backchannel messages. According to these ratings messages are sorted in order to obtain a ranking. Various rating schemes of different complexities and requirements may be employed. For the big picture, however, we present in a few words the rating currently used in Backstage. Users may rate a message positively (approval) or negatively (rejection) only once. The overall rating $r(m)$ of a message m is calculated by the following weighted average (e.g. cf. [14]):

$$ r(m) = \frac{1}{\langle \text{NR} \rangle + \text{nr}(m)} \left(\langle \text{NR} \rangle \cdot \langle \text{R} \rangle + \text{nr}(m) \cdot \frac{\text{pos}(m)}{\text{nr}(m)} \right) $$

In the formula above $\langle \text{NR} \rangle$ denotes the average number of ratings of all messages, $\text{nr}(m) = \max(1, \text{pos}(m) + \text{neg}(m))$ denotes the total number of ratings for the message m, $\text{pos}(m)$ is the number of positive ratings the message m received, $\text{neg}(m)$ the negative ratings for m, and $\langle \text{R} \rangle$ denotes the average rating of all messages. As can be seen, positive and negative ratings do not cancel each other out, but the negative ratings weaken the influence of the positive ratings. If the total number of ratings for a message $\text{nr}(m)$ is much smaller than the average number of ratings $\langle \text{NR} \rangle$ the message's rating is dominated by the average rating $\langle \text{R} \rangle$, meaning that not much credit is given to the users who rated the message m. Conversely, if the number of ratings for m is greater than the average number of ratings for a message the rating for m is dominated by the users who rated it. Thus, the rating scheme is biased in as much as it favors the appraisal of the collective over that of the few. However, whether this rating scheme is appropriate for Backstage needs to be investigated in an experiment in the near future.

4.1   Measuring Time and Age in Backstage

To better reflect the progress of a lecture we propose to use the backchannel activity during the lecture to promote aging. The logical time on the backchannel advances after each n-th interaction on Backstage. Both the number n and the specification of what is considered as activity is defined by the lecturer. Activities may comprise sending of messages of certain categories and rating. Since on Backstage rating is performed by (automatically) sending messages of a special rating category, specifying activity amounts to nothing else than selecting categories. Both the number of interactions after which the time advances and the specification of activity on Backstage are very intuitive parameters that can easily be handled by the lecturer, even during a lecture.

A first idea for measuring the age of messages would be to calculate the difference between the current time and the time of creation. However, this solution is inappropriate, since messages that are regularly rated, i.e. active messages, would age at the same pace as messages which are disregarded by the audience. On Backstage, active messages shall age at a lower pace than inactive messages. Thus, it is reasonable to also consider the age of a message's ratings. Hence, aging depends on the attention a message receives: it is promoted when the focus by the audience of a message recedes. A naive approach to determining age might be to calculate the difference of the current time and the time of the most recent rating a message received. This is problematic, though. For example, imagine that many students have rated a post a long time ago, i.e. the message is actually obsolete, but one student revives the message by rating, the message would suddenly, and inexplicably, rejuvenate.

The arithmetic mean over all ratings would solve this issue. However, it is very sensitive to outliers. Many ratings at the same time would be needed to assure that the age of this message can be considered robust. To overcome these difficulties we favor the use of the median as the average age of a message. The median of a frequency distribution is the sampled value of an (artificial) instance that bisects the distribution. For a sequence $(x_1, x_2, \ldots, x_k)$ of k sampled values the median $\overline{x}$ is computed as follows:

$$\overline{x} = \begin{cases} x_{\frac{k+1}{2}} & \text{if k is odd} \\ \frac{1}{2}\left(x_{\frac{k}{2}} + x_{\frac{k}{2}+1} \right) & \text{otherwise} \end{cases}$$

The median is an interesting representative of the central tendency, since it is quite robust against outliers but likewise sensitive enough to reflect relevant changes in the data (cf. [27]). Thus, to determine the age of a message, we determine the median from the ratings' age and from the creation time of the corresponding message. Also considering the time of creation is necessary in the case that a message has not received any ratings at the beginning. Otherwise, the message would not be considered by aging.

4.2   Aging in Backstage

After each n interactions on Backstage aging is promoted and the ranking is updated. We therefore propose the procedure given as pseudo-code in Listing 1.

As can be seen in the given procedure the rating score is obtained by multiplying the positions of a post in the two rankings built upon age and ratings. Since it is possible that two posts may be assigned the same rating they may share the same position in the respective ranking. The final top-k ranking is

---

Algorithm 1. AgingRank: Ranking with Aging

Require: the number k of messages that constitute the ranking
Require: the interaction counter n
  if clockTick(n) **then**
    candidatePosts := getCandidatePosts()
    {*promote aging*}
    for all post **in** candidatePosts **do**
       updateAge(post, calculateMedianAge(post))
    end for
    rankingByAge := sortDescendingByAge(candidatePosts)
    rankingByRating := sortDescendingByRating(candidatePosts)
    {*we assume lists to be 1-indexed*}
    for indexAge := 1 **to** maxIndex(rankingByAge) **do**
       post := getElement(indexAge, rankingByAge)
       {*get the index of the post in the ranking by rating*}
       indexRating := getIndex(post, rankingByRating)
       updateScore(post, indexAge * indexRating)
    end for
    {*sort the candidates by just updated score values*}
    relevantPosts := sortDescendingByScore(candidatePosts)
    resolved := resolveConflicts(relevantPosts)
    result := firstElements(k, resolved)
    updateRanking(result)
  end if

---

then computed by sorting the list of relevant messages according to the messages' scores. However, the given procedure may result in conflicts. For example, two messages, say, m_1 with indexRating = 2 and indexAge = 3, and m_2 with the positions conversed, that is indexRating = 3 and indexAge = 2 would receive the same score 6. Both messages m_1 and m_2 would be assigned the same position in the final ranking. Thus, conflict resolution is necessary.

We propose a simple but eligible approach to conflict resolution: we let the lecturer decide which of the conflicting messages should get higher priority. Therefore, the lecturer specifies in her profile, whether she favors a conservative ranking, i.e. older messages stay in the ranking, or a progressive ranking in which older messages are replaced by newer ones whenever possible. In case of further remaining conflicts we may eventually establish a strict order by resorting to the physical age, since the conflicting messages can then be considered equal in terms of relevance and logical age.

To determine the follow-up ranking it is not necessary to consider the entire message stream. It rather suffices to determine a set of candidates, the number of which depends on the number of interactions n by which aging is promoted. Certainly, the messages listed in the current ranking are also candidates for the follow-up ranking. However, other messages may be candidates as well. For this purpose, consider the example timeline in Figure 2.

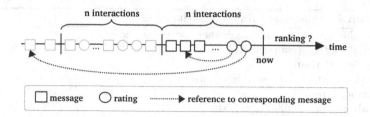

Fig. 2. Example Timeline of Interactions. The rectangles illustrate the points in time at which messages are sent, the circles illustrate the points in time at which messages are rated. The dotted arrows connect the ratings with the rated messages. The dots at the timeline indicate further interactions.

Between two ticks of the logical clock, n interactions are carried out by the users. These interactions may comprise the creation of $x \leq n$ new messages and $y = n - x$ ratings for existing messages. The ratings may refer to up to y messages created during the recent or some earlier time span. All these messages have recently been in the focus of the audience. Thus, besides the currently ranked messages, both the newly created and the newly rated messages are also candidates for the follow-up ranking. Reckoned up, the set of candidate messages comprises not more than $k + n$ messages.

5   Discerning Actuality in Twitter-Based Tools

Although we conceived timely ranking by aging for the digital backchannel Backstage, it most likely might also be of interest for other microblogging platforms based on Twitter. To employ our approach it is sufficient to provide for means to determine relevance ratings, to measure activity, and to set the strategy for updating the rankings. This section illustrates possible applications in both e-learning and non-e-learning fields.

Twitter is the most prominent publicly available generic microblogging service and gained much attention not only by e-learning researchers. Twitter allows to relate microblog messages, so-called tweets, by hashtags. Thus, hashtags make possible to retrieve a coherent line of communication on a topic. Users can follow other users, i.e. become their followers. The tweets of the followed users are displayed at one's own message stream. One may forward messages of followed users to their own followers by a special form of citation, so-called re-tweets: the original message is copied and prefixed with the keyword "RT" followed by the origin user. Thus, a retweet is usually of the form "RT @*originUser* [*original text*]".

Twitter provides a rich API[3] upon which custom microblogging applications can be built. One e-learning backchannel similar to Backstage is Twitterwall[4] [28]. The platform allows the retrieval and display of multiple message streams

---

[3] Application Programming Interface
[4] http://twitterwall.tugraz.at

by specifying hashtags. Furthermore it extends Twitter in that it provides rating of tweets. To extend Twitterwall with aging, the rating scheme that is already integrated can be used. Activity can be measured by the number of messages containing certain hashtags. As for each hashtag Twitterwall displays a separate message stream, it might also be interesting to provide rankings for each of those streams that underly distinct aging.

Also, discerning actuality in tweet rankings directly on Twitter can be accomplished in much the same way as is proposed for Backstage. As mentioned above, Twitter does not provide rating of tweets. However, rating of a tweet can be mimicked by counting the number of users retweeting the tweet. That is, a tweet that is frequently retweeted is heavily focused on by users and may thus be considered relevant. On Twitter, activity can be measured by the number of messages containing certain hashtags and by the number of retweets of those messages. Obtaining a timely ranking of tweets may provide interesting insights into trends in social news broadcast on Twitter.

Another quite interesting field of application might be stock microblogging, e.g. TweetTrader[5] [29]. Among other things, users of TweetTrader estimate in tweets the performance of stock quotations. Using special processable syntax, those tweets are evaluated and aggregated to determine the collective estimation of near-future stock developments. Discerning actuality in a ranking of those estimations might be of great interest for stock microblogging. Ratings in this case might be based on the content of the tweets, i.e. the users' assessments of the stock development. The activity may be specified by the number of tweets sent, for example. A progressive update strategy is likely to be preferred for a ranking in order to always be aware of the most recent estimations. Also, timely ranking of stock quotations might yield interesting outcomes in the analysis of trends.

6   Conclusion and Future Work

This article proposes an intuitive and easy-to-handle approach to discerning actuality in Backstage, a backchannel carefully designed for the use in large class lectures. We show how aging can be used to provide the lecturer with a ranking that considers actuality. The approach in this article favors the activity on the backchannel as the time measure according to which aging of messages is promoted, since the physical time only plays a minor role in determining a lecture's progress. Potential fields of applications are sketched. Since during the development of the presented approach Backstage has undergone several changes, the integration is not yet finished. Furthermore, its usefulness needs to be investigated in an experimental setting. Promoting aging also seems to be valuable for other purposes. For example, further functionalities that aim at supporting the awareness of students and lecturer and making interactions more personal and affectionate are currently under development. Some of these

---

[5] http://tweettrader.net

functionalities also depend on a sort of time and might also require aging. These topics are going to be discussed in a forthcoming paper.

References

1. van de Grift, T., Wolfman, S.A., Yasuhara, K., Anderson, R.J.: Promoting Interaction in Large Classes with a Computer-Mediated Feedback System. In: Proceedings of the International Conference CSCL, pp. 119–123 (2003)
2. Schworm, S., Fischer, F.: Academic Help Seeking. In: Handbuch Lernstrategien [Handbook of Learning Strategies], pp. 282–293 (2006) (in German)
3. Ebner, M., Schiefner, M.: Microblogging – More than Fun? In: Immaculada Arnedillo, S., Isaias, P. (eds.) Proceeding of IADIS Mobile Learning Conference 2008, pp. 155–159 (2008)
4. Ebner, M.: Introducing Live Microblogging: how Single Presentations can be Enhanced by the Mass. Journal of Research in Innovative Teaching 2(1), 108–111 (2009)
5. Holotescu, C., Grosseck, G.: Using Microblogging to Deliver Online Courses. Case-Study: Cirip.ro. Procedia - Social and Behavioral Sciences 1(1), 495–501 (2009); World Conference on Educational Sciences. New Trends and Issues in Educational Sciences, Nicosia, North Cyprus (February 4-7, 2009)
6. Bry, F., Gehlen-Baum, V., Pohl, A.: Promoting Awareness and Participation in Large Class Lectures: The Digital Backchannel Backstage. In: Proceedings of the IADIS International Conference e-society, Spain, Avila, pp. 27–34 (March 2011)
7. Pohl, A., Gehlen-Baum, V., Bry, F.: Introducing Backstage – A Digital Backchannel for Large Class Lectures. Interactive Techology and Smart Education 8(4), 186–200 (2011)
8. Baumgart, D., Pohl, A., Gehlen-Baum, V., Bry, F.: Providing Guidance on Backstage, a Novel Digital Backchannel for Large Class Teaching. In: Education in a Technological World: Communicating Current and Emerging Research and Technological Efforts, Formatex, Spain, pp. 364–371 (2012)
9. King, A.: Scripting Collaborative Learning Processes: A Cognitive Perspective. In: Fischer, F., Kollar, I., Mandl, H., Haake, J.M. (eds.) Scripting Computer-Supported Collaborative Learning, pp. 13–37. Springer US, Boston (2007)
10. Suthers, D., Xu, J.: Kükäkükä: An Online Environment for Artifact-Centered Discourse. In: Proceedings of the 11th World Wide Web Conference (2002)
11. Golub, E.: On Audience Activities During Presentations. Journal of Computing Sciences in Colleges 20(3), 38–46 (2005)
12. Kay, R.H., LeSage, A.: Examining the Benefits and Challenges of using Audience Response Systems: A Review of the Literature. Computers & Education 53(3), 819–827 (2009)
13. Bian, J., Liu, Y., Agichtein, E., Zha, H.: Finding the Right Facts in the Crowd: Factoid Question Answering over Social Media. In: Proceedings of the 17th International Conference on World Wide Web, WWW 2008, pp. 467–476. ACM, New York (2008)
14. Marmolowski, M.: Real-Life Rating Algorithm. Technical Report, DERI (2008)
15. Das Sarma, A., Das Sarma, A., Gollapudi, S., Panigrahy, R.: Ranking Mechanisms in Twitter-Like Forums. In: Proceedings of the 3rd ACM International Conference on Web Search and Data Mining, WSDM 2010, pp. 21–30. ACM, New York (2010)

16. Lerman, K.: Dynamics of a Collaborative Rating System. In: Zhang, H., Spiliopoulou, M., Mobasher, B., Giles, C.L., McCallum, A., Nasraoui, O., Srivastava, J., Yen, J. (eds.) WebKDD 2007. LNCS, vol. 5439, pp. 77–96. Springer, Heidelberg (2009)
17. Nichols, D.M.: Implicit Rating and Filtering. In: Proceedings of the 5th DELOS Workshop on Filtering and Collaborative Filtering, pp. 31–36 (1998)
18. Agichtein, E., Castillo, C., Donato, D., Gionis, A., Mishne, G.: Finding High-Quality Content in Social Media. In: Proceedings of the International Conference on Web Search and Web Data Mining, WSDM 2008, pp. 183–194. ACM, New York (2008)
19. Bian, J., Liu, Y., Agichtein, E., Zha, H.: A Few Bad Votes too Many? Towards Robust Ranking in Social Media. In: Proceedings of the 4th International Workshop on Adversarial Information Retrieval on the Web, AIRWeb 2008, pp. 53–60. ACM, New York (2008)
20. Harry, D., Gutierrez, D., Green, J., Donath, J.: Backchan.nl: Integrating Backchannels with Physical Space. In: Extended Abstracts on Human Factors in Computing Systems, CHI 2008, pp. 2751–2756. ACM, New York (2008)
21. Kamvar, S.D., Schlosser, M.T., Garcia-Molina, H.: The Eigentrust Algorithm for Reputation Management in P2P Networks. In: Proceedings of the 12th International Conference on World Wide Web, WWW 2003, pp. 640–651. ACM, New York (2003)
22. Zhang, J., Ackerman, M.S., Adamic, L.: Expertise Networks in Online Communities: Structure and Algorithms. In: Proceedings of the 16th International Conference on World Wide Web, WWW 2007, pp. 221–230. ACM, New York (2007)
23. Al-Sharrah, G.: Ranking Using the Copeland Score: A Comparison with the Hasse Diagram. Journal of Chemical Information and Modeling 50(5), 785–791 (2010)
24. Nagmoti, R., Teredesai, A., De Cock, M.: Ranking Approaches for Microblog Search. In: Proceedings of the 2010 IEEE/WIC/ACM International Conference on Web Intelligence and Intelligent Agent Technology, WI-IAT 2010, pp. 153–157. IEEE Computer Society, Washington, DC (2010)
25. Ding, Y., Li, X.: Time Weight Collaborative Filtering. In: Proceedings of the 14th ACM International Conference on Information and Knowledge Management, CIKM 2005, pp. 485–492. ACM, New York (2005)
26. Huo, W., Tsotras, V.J.: Temporal Top-k Search in Social Tagging Sites Using Multiple Social Networks. In: Kitagawa, H., Ishikawa, Y., Li, Q., Watanabe, C. (eds.) DASFAA 2010, Part I. LNCS, vol. 5981, pp. 498–504. Springer, Heidelberg (2010)
27. Garcin, F., Faltings, B., Jurca, R.: Aggregating Reputation Feedback. In: Proceedings of the First International Conference on Reputation: Theory and Technology, Gargonza, Italy, vol. 1, pp. 62–67 (2009)
28. Ebner, M.: Is Twitter a Tool for Mass-Education? In: Proccedings of the 4th International Conference on Student Mobility and ICT, Vienna (2011)
29. Sprenger, T.: TweetTrader.net: Leveraging Crowd Wisdom in a Stock Microblogging Forum. In: Proceedings of the 5th International Conference of Weblogs and Social Media, Barcelona, Spain (2011)

Tweets Reveal More Than You Know: A Learning Style Analysis on Twitter

Claudia Hauff[1], Marcel Berthold[2], Geert-Jan Houben[1],
Christina M. Steiner[2], and Dietrich Albert[2]

[1] Delft University of Technology, The Netherlands
{c.hauff,g.j.p.m.houben}@tudelft.nl
[2] Knowledge Management Institute, Graz University of Technology, Austria
{marcel.berthold,christina.steiner}@tugraz.at

Abstract. Adaptation and personalization of e-learning and technology-enhanced learning (TEL) systems in general, have become a tremendous key factor for the learning success with such systems. In order to provide adaptation, the system needs to have access to relevant data about the learner. This paper describes a preliminary study with the goal to infer a learner's learning style from her Twitter stream. We selected the Felder-Silverman Learning Style Model (FSLSM) due to its validity and widespread use and collected ground truth data from 51 study participants based on self-reports on the Index of Learning Style questionnaire and tweets posted on Twitter. We extracted 29 features from each subject's Twitter stream and used them to classify each subject as belonging to one of the two poles for each of the four dimensions of the FSLSM. We found a more than by chance agreement only for a single dimension: active/reflective. Further implications and an outlook are presented.

1 Introduction

Over the last decade, personalization and adaptation in E-learning has become a mainstream component in E-learning systems. Such adaptations provide learners with a personalized learning experience that is either unique to each individual or unique to a particular group of learners. The goals are clear: to keep the learners motivated and engaged, to decrease the learners' frustration, to provide an optimal learning environment and, of course, to increase the learners' expertise in a particular subject.

In order to provide adaptation, the system needs to have access to relevant data about the learner. What is deemed relevant in this context depends on the facilities that are provided by the system. Adaptation can be provided on a number of levels with varying granularity. It can be based on gender [1], on the learners' level of expertise [2, 3], on the learners' culture [4] or on the learners' learning styles [5].

The latter, adaptation according to the learners' learning styles, is also the focus of this paper. We note that there is controversy surrounding the learning style hypothesis [6], which states that enabling a learner to learn with material

A. Ravenscroft et al. (Eds.): EC-TEL 2012, LNCS 7563, pp. 140–152, 2012.

that is tailored to her own learning style will outperform a learner who learns the material tailored to a learning style that is not her own. As of today no studies have conclusively shown that this hypothesis actually holds for a wide range of people. Although learning styles may not yield improved results with respect to objective measures (such as testing the increase in learner expertise), learning styles are of importance for E-learning systems to improve the learners' satisfaction in the material and to keep them engaged by offering them learning that is appropriate for their self-perceived learning style.

At the same time a question raises: Do Twitter users actually provide information about their learning style or how they learn? In paper by [7] the authors investigates why people continue using twitter. Among others it could be shown that users continue using Twitter, because of positive content gratification. Content gratification was comprises by disconfirmation of information sharing and self-documentation (the way users learn, keep track what they are doing, document their life). Therefore it can be argued that tweets are produced to report about users' learning behaviour intentionally. In addition, in this paper data mining is also based on phrases which are derived from exiting questionnaire and should cover some non-intentional phrases in regard to learning behaviour.

Over the years, a number of learning style models have been proposed, among them Kolb's Experiental Learning Theory [8], Fleming's VARK learning styles inventory [9] and Felder-Silverman-Learning-Style-Model (FSLSM) [10, 11]. Independent of the particular model chosen, the procedure to determine a learner's learning style is always the same: the learner fills in a standardized questionnaire (specific to the model) and based on the answers given the different dimensions of the model are determined. One of the problems with this approach is that the learner may be unwilling to spend a lot of effort on this procedure[1]. More importantly though, learners cannot be expected to repeatedly fill in such a questionnaire, which, if a system is used for a long time may become necessary, as there is evidence that learning styles change over time [12]. Thus, an automatic approach to infer the learning style of a learner is likely to be more precise in the long run.

Ideally, we are able to determine the learner's learning style without asking the learner for explicit feedback. One potential solution to this problem lies in the social Web whose rise has made people not merely consumers of the Web, but active contributors of content. Widely adopted social Web services, such as Twitter[2], Facebook[3] and YouTube[4], are frequented by millions of active users who add, comment or vote on content. If a learner is active on the social Web, a considerable amount of information about her is available on the Web and, depending on the particular service used, most of it is publicly accessible. We envision E-learning systems in the future to simply ask the learner about her username(s) on various (publicly accessible) social Web services where the

---

[1] The ILS questionnaire for instance consists of 44 questions.

[2] http://www.twitter.com/

[3] http://www.facebook.com/

[4] http://www.youtube.com/

learner is active on. Then, based on the learner's "online persona", aggregated from the social Web, the system can automatically infer the learner's learning style. We have already shown in previous work [13] that it is possible to derive a basic profile of the learner's knowledge in a particular domain from the learner's activities on the microblogging platform Twitter. In this work now, we are interested to what extent it is possible to derive information about a learner's learning style from the same social Web stream.

In the EU project ImREAL (Immersive Reflective Experience-based Adaptive Learning) intelligent services are developed to augment and improve simulated learning environments among others, to bring real world users data, e.g. content retrieved from tweets, into the simulation to link real world experiences to the simulation. In this paper the following hypothesis is investigated: the information the learner can provide in the learning style questionnaire is already implicitly available in the learner's utterances in the social Web. If this is indeed the case, the research question then becomes of how to extract this implicit information and transform it into the different dimensions of the learning styles models.

We consider the collaborative work of machine learning and psycho-pedagogical approaches presented here as a preliminary study - if we were able to show success in predicting a learner's learning style based on the learner's tweets with a number of simple features, we have evidence that this is a path that is worth investigating further.

The remainder of the paper is organized as follows: in Section 2 related work is presented. Section 3 describes our pilot study and the setup of the experiments. The results are then presented in Section 4 and the paper is concluded with a discussion and an outlook to future work in Section 5.

2   Related Work

We first describe previous work that sheds light on why people use Twitter. Then, we turn to previous works that have attempted what we set out to do too: to infer a learner's learning style from implicit information available about the learner, that is without letting the learner fill in a questionnaire.

2.1   The Use of Twitter in Scientific Research

Two questions that have been investigated by a number of researchers in the past are what is the people's motivation to use Twitter and what do the people actually post about. Java et al. [14] determined four broad categories of tweets: daily chatter (the most common usage of Twitter), conversations, shared information/URLs and reported news. Naaman et al. [15] derived a more detailed categorization with nine different elements: information sharing, self promotion, opinions, statements and random thoughts, questions to followers, presence maintenance, anecdotes about me and me now. Moreover, they also found that the approximately eighty percent of the users on Twitter focus on themselves (they are so-called "Meformers"), while only a minority of users are driven largely by sharing information (the

"Informers"). Westman et al. [16] performed a genre analysis on tweets and identified five common genres: personal updates, direct dialogue (addressed to certain users), real-time sharing (news), business broadcasting and information seeking (questions for mainly personal information). Finally, Zhao et al. [17] conducted interviews and asked people directly about their motivations for using Twitter; several major reasons surfaced: keeping in touch with friends and colleagues, pointing others to interesting items, collecting useful information for one's work and spare time and asking for help and opinions. These studies show that a lot of tweets are concerned with the user herself; we hypothesize that among these user centred tweets, there are also useful ones for the derivation of the learner's knowledge profile.

A number of Twitter studies also attempt to predict user characteristics from tweets. While we are aiming to extract a learner's learning style, Michelson et al. [18] derive topic profiles from Twitter users which are hypothesized to be indicative of the users' interests and expertise. In a number of other works, e.g. [19–21], elementary user characteristics are inferred from Twitter, including gender, age, political orientation, regional origin and ethnicity.

2.2 Learning Style Investigations

A number of previous works exist that infer learners' learning styles based on their behaviour *within* the learning environment. In [22] the outline of such a system is sketched, though no experiments are reported. Garcia et al. [23] investigated to what extent it is possible to infer a learner's learning style (specifically the ILS variant) from the learner's interaction with a Web-based E-learning system and a class of Artificial Intelligence students. They relied on a number of features that model the students' behaviour on the learning system. Some examples of the chosen features are the type of reading material (concrete or abstract), the amount of revision before an exam, the amount of time spent on an exam, the active participation on message boards and chats within the learning environment, the number of work examples accessed and the exam result. The approach was evaluated on 27 students with promising results; the most accurate prediction was possible for the perception dimension (intuitive vs. sensing) with a precision of 77%, followed by the understanding dimension (sequential vs. global) with 63% precision and the processing dimension (active vs. reflective) with 58% precision. The input dimension (visual vs. verbal) was not investigated in this study. In contrast to this work, the features in our experiments are at a lower level - we aim to utilize features that are independent of a particular learning environment and also do not require a specific amount of interaction with the environment first before the learning style can be predicted.

Sanders and Bergasa-Suso [24] also developed a Web-based learning system that monitors user activity to infer the learning styles. Features include the amount of data copied and dragged, the length of the page text, the ratio of text to images, the presence or absence of tables, mouse movements, etc. While initially their predictions did not perform much better than a naive predictor that assigns the majority class to all instances [25], after a number of data post-processing steps, they achieved accuracies well above such a naive predictor for the active/reflective and the visual/verbal dimension[5].

Finally we note that instead of inferring the learning style from the learner's actions within the learning environment, a number of works have also investigated to infer the learning style from other user characteristics such as the Big-Five personality model, e.g. [26].

Our work differs from these previous works in two ways. First of all, our approach is independent of a particular learning environment. We rely on traces the learner left in the past on the social Web. This has the distinct advantage that when a learner starts using a novel E-learning system the learning style can be computed immediately, while in [22–24] a certain amount of interaction is required on part of the learner before the learning style can be inferred. This can also mean that by the time the system has identified the learning style of the learner and is ready to provide material according to the learner's preferences, the learner has already turned away to a better fitting learning system. Secondly, the features we use in our pilot study are very low-level compared to the features in the previous works; we rely on features that can be extracted from any Twitter stream and as such, the results we report here will be the lower boundary of what is possible.

3 Methodology

In line with previous works, in particular [23, 24], we use the following methodology and procedure to investigate our hypothesis: In the period of November 2011 and March 2012, the web-link to a new ILS online version was distributed via different social web network channels such as Twitter, Facebook, LinkedIn and large e-mail lists of different EU-projects and Universities (e.g. University of Graz and Graz University of Technology). In a late stage of this process (end of February), people who tweeted at least once they would be a certain type of learner, e.g. I am an active learner, were directly contacted via Twitter and asked to participate in the survey. Each participant was requested to read the introduction, fill in some personal information such as gender, age, level of education and the degree of which they were familiar with the term learning style. In addition, they were asked to provide their Twitter username and to fill in the ILS items. The instruction included information about the purpose of the study, that the data would be treated anonymously and that each participant had the chance to draw one of three 20 Amazon.com-vouchers. Duration time of filling in all required data was about 15 minutes.

We then evaluate these questionnaires and the found learning styles of each user are our ground truth, that we try to predict in the next stage. We crawl the tweets of the respective Twitter accounts and derive features from them. Then,

---

[5] Please note the the results between different papers are not directly comparable due to differences in the precision formula employed and the number of classes present for each dimension - [23] include a NEUTRAL class for each dimension which is absent in [25] and [24].

we employ a machine learning algorithm to classify each user into the different dimensions based on these features.

Next, we first introduce the learning style model we selected in more detail and then we outline how we derived the features and the machine learning approach.

3.1 The Felder-Silverman Learning Style Model

One of the most popular learning style models is the Felder-Silverman Learning Style Model (FSLSM) [10, 11] which describes the most prominent learning style differences between engineering students on four dimensions:

- **Sensing/intuitive**: Sensing learners are characterized by preferring to learn facts and concentrate on details. They also tend to stick to concrete learning materials, as well as known learning approaches. They like to solve problems by concrete thinking and by applying routine procedures. Intuitive learners on the other hand prefer to learn abstract concepts and theories. Their strengths lie in discovering the underlying meanings and relationships. They are also more creative and innovative compared to sensing learners.
- **Visual/verbal**: This dimension distinguishes learners preferences in memorizing learning material. The visual learner prefers the learning material to be presented as a visual representation, e.g. pictures, diagrams or flow charts. In contrast, verbal learners prefer written and spoken explanations.
- **Active/Reflective**: This dimension covers the way of information processing. Active learners prefer the 'learning by doing' way. They enjoy learning in groups and are more open to discuss ideas and learning material. On the contrary, reflective learners favour to think about ideas rather than work practically. They also prefer to learn alone.
- **Sequential/Global**: On this dimension learners are described according to their way of understanding. Sequential learners learn in small steps and have a linear learning process, focusing on detailed information. Global learners, however, follow a holistic thinking process where learning happens in large leaps. At first, it seems that they learn material almost randomly without finding connections and relations between different areas, but in a later stage, they perceive the whole picture and are able to solve complex problems.

3.2 The Index of Learning Style

The ILS [11] is a self-assessment instrument based on the Learning Style Model [10, 11]. Participants are asked to provide answers to 44 forced-choice questions with two answer options. Each of the four learning style dimensions is covered by 11 items, with an 'a' or b answer option corresponding to one of the poles of the continuum of the corresponding learning style dimension, e.g. active (a) vs. reflective (b). It is suggested to count the frequency of a responses to get a score between 0-11 for one dimension. This method allows a fine gradation of the continuum starting from e.g. 0-1 representing strong preferences for reflective learning till 10-11 strong preference for active learning. Therefore, a preference of a pole of the given dimension may be mild, moderate or strong. Reliability as well as validity analyses revealed acceptable psychometric values. For internal consistency reliability ranging from 0.55 to 0.77 across the four learning style scales of the ILS were found by [27]. Furthermore, factor analysis and direct feedback from students whether the ILS score is representing their learning preferences provided sufficient evidence of construct validity for the ILS.

For the presented study, a new online version of the ILS was created to incorporate a new design, instructions and to add text and check-boxes for required information, such as the Twitter username and some demographic data. We distributed the call for participation on various channels, including university mailing lists and Twitter. In total, 136 people responded and filled in the questionnaire. In a post-processing step we removed subjects: (i) whose Twitter account is protected[6], (ii) whose Twitter account listed less than 20 public tweets, (iii) who provided an invalid or no Twitter ID, and (iv) who did not complete the ILS questionnaire. After this data cleaning process, a total of 51 subjects remained whose learning styles are predicted across all experiments reported in this paper.

3.3   Twitter-Based Features

We derived a set of 29 features from the Twitter stream of each subject. They are listed in Table 1 and can be ordered into four broad classes: features derived from the account information (e.g. number of followers and total number of tweets), features derived from individual tweets whose scores are aggregated (e.g. the percentage of tweets with URLs, the percentage of tweets directed at another user, the average number of nouns or adjectives used by a user), features based on tweet semantics (e.g. the percentage of tweets containing terms indicating anger or joy) and features derived from the external pages that were linked to by the users in their tweets (e.g. the fraction of content words vs. non content words in those pages).

We relied on a number of existing toolkits and resources to derive those features. The tweet processing pipeline is shown in Figure 1. The following steps are executed:

- A Language Detection library[7] is relied upon to determine the language a tweet is written in.
- If the tweet is not in English, the Bing Translation web service[8] is used to translate the text into English.
- The Stanford Part-of-Speech Tagger[9], a library that tags English text with the respective parts of speech (noun, adjective, etc), is relied upon to determine the tweeting style.

---

[6] Tweets of users with a protected user account are not publicly accessible.
[7] http://code.google.com/p/language-detection/
[8] http://api.microsofttranslator.com
[9] http://nlp.stanford.edu/software/tagger.shtml

Table 1. Overview of the 29 features used as input for the classifiers

	Features
Twitter-account based	#tweets, #favourites, #listings, #friends, #followers, $\frac{\#friends}{\#followers}$
Tweet style & behavior	%tweets with URLs, #languages used, %directed tweets, %retweets, %tweets with hashtags, average (av.) and standard deviation (std.) of #terms per tweet, av. and std. of #tagged terms per tweet, av. #nouns per tweet, av. #proper nouns per tweet, av. #adjectives per tweet
Tweet semantics	av. #anger terms, av. #surprise terms, av. #joy terms, av. #disgust terms, av. #fear terms, av. #sadness terms, %emotional tweets
External URLs	av. #images in external URLs, av. $\frac{\#content\ words}{\#non\text{-}content\ words}$ in external URLs

- Boilerpipe[10] is a library that parses web pages that the subjects referred to in their tweets. The output of running Boilerpipe distinguishes between content parts of a web page and non-content parts (copyright notices, menus, etc.). We rely on it to determine the number of actual amount of text (versus images) on a web page.
- Finally, we determine the sentiment of the user by relying on WordNet Affect [28]: it is a set of affective English terms that indicate a particular emotion; there are 127 anger terms (e.g. *mad, irritated*), 19 disgust terms (e.g. detestably), 82 fear terms (e.g. dread, fright), 227 joy terms (e.g. triumphantly, appreciated), 123 sadness terms (e.g. oppression, remorseful) and 28 surprise terms (e.g. fantastic, amazed). Each tweet is matched against this dictionary and the number of emotional tweet for each dimension are recorded.

3.4   Classification Approaches

Since our goal is an initial study on the feasibility of determining one's learning style from a number of tweets, we use two common machine learning approaches: Naive Bayes and AdaBoost[11]. Due to the small number of users, we rely on k-1 cross-validation for training and testing. Furthermore, as the two classes in each dimension are not distributed equally, we set up a cost-sensitive evaluation where an error for the less likely class per dimension was punished with a factor of 5 (the error is punished with a score of 1 for the majority class). The results are reported in terms of the classification precision, recall, F_1 and Cohen's Kappa [29] (κ).

---

[10] http://code.google.com/p/boilerpipe/
[11] We use the Weka Toolkit for our experiments.

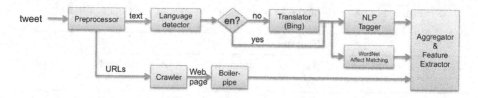

Fig. 1. Tweet processing pipeline

We focus on the last evaluation measure in particular as it measures the inter-annotator agreement, taking into account the element of a chance agreement. Here, the ground truth and the predicted learning style act as the two annotators of the data. A $\kappa \approx 0$ indicates that the annotators agree as often as they would by chance, a value below zero indicates an agreement that is lower than by chance and values above 0 determine different levels of agreement that are better than random agreement. A $\kappa \in (0, 0.2]$ indicates a slight agreement, while $(0.2, 0.4]$ indicate moderate agreement and so on. In general, the larger the value of κ the larger the agreement; when $\kappa = 1$ the agreement is perfect.

4   Results

4.1   Generating the Ground Truth

Due to the odd number of questions in the ILS questionnaire for each dimension, a subject can always be assigned to one of the two opposite ends of the spectrum. In this pilot study, we ignore the strength of the association and we simply assign each subject to the pole with the greater score. The distribution of the subjects across the four dimensions proposed in the ILS approach are presented in Table 2.

Table 2. Distribution of our 51 subjects across the four dimensions of the ILS questionnaire. We report the number of subjects that fall into each category, as well as the mean (μ) and standard deviation (σ) with respect to the score. For comparison, we also report the distribution that were reported in other user studies.

| | | ILS-Twitter study | | | | [25] | [30] |
		#subjects	%	μ	σ	%	μ
Input	**visual**	42	82%	7.31	2.44	76%	8.14
	verbal	9	18%	3.67	2.45	24%	2.86
Processing	**active**	31	61%	6.07	2.35	57%	5.99
	reflective	20	39%	4.91	2.34	43%	5.01
Understanding	**global**	36	71%	6.64	2.41	66%	5.00
	sequential	15	29%	4.34	2.40	34%	6.00
Perception	**intuitive**	35	69%	6.69	2.67	48%	4.32
	sensing	16	31%	4.29	2.68	52%	6.68

It is evident that the split between subjects in the two opposite poles of each dimension is not uniform. To place this distribution in context, we also report the distributions that were found in [25] and [30]. While the visual/verbal and active/reflective dimensions are robust to the subject population, we observe considerable differences among the three studies in the global/sequential and intuitive/sensing dimensions.

Based on the absolute scores, which show the clearest distinction in the visual-verbal dimension as well as the intuitive-sensing dimension, we hypothesize that the classifier will be performing better on those dimensions than the others.

4.2 Results on the Classification Process

In Table 3 we now report the performance our classifiers achieved when classifying the subjects according to the four ILS dimensions. We note, not surprisingly, that classification into the majority class results in high precision and recall values, though if we consider κ we also note that only for a single dimension, namely active/reflective, can we say with relative certainty that the classification approaches perform better than agreement by chance. This holds for both classifiers. The other dimensions show only slightly significant results for one or the other classifier, though not both. Thus, we have to conclude that the simple features we introduced are sufficient for the active/reflective dimension, though they are not indicative for any of the other dimensions in the ILS framework.

Table 3. Results of predicting the different learning style dimensions for our data set

		active	reflective	visual	verbal	global	sequential	intuitive	sensing
Naive	Prec.	0.644	0.667	0.833	0.333	0.668	0.000	0.688	0.333
Bayes	Recall	0.935	0.200	0.952	0.111	0.917	0.000	0.943	0.063
	F_1	0.763	0.308	0.889	0.167	0.786	0.000	0.795	0.105
	κ	0.1547		0.086		-0.109		0.007	
Ada-	Prec.	0.697	0.556	0.814	0.125	0.733	0.364	0.649	0.214
Boost	Recall	0.742	0.500	0.833	0.111	0.725	0.267	0.686	0.188
	F_1	0.719	0.526	0.842	0.118	0.806	0.308	0.667	0.200
	κ	0.2463		-0.058		0.0783		-0.131	

5 Conclusions

Twitter learning style analysis could be used to complete user profiles with respect to learning preferences and as a result they could result in more efficient adaptation and personalization of simulators, e-learning systems or other technology-enhanced learning software. Providing feedback to learners about their learning preferences could be helpful, but it should be relied upon with caution. There have to be explicit explanations that the learning style is a tendency of certain preferences and the assessment does not overrule ones own judgments [11], but rather can be seen as advice or suggestion. Bearing this in mind the Twitter analysis of learning styles could lead to smoother, non-invasive assessment of personal learning preferences.

In this paper, we have performed a first study with the goal to infer a learner's learning style from her Twitter stream. We selected the ILS model due to its validity and widespread use and collected ground truth data from 51 study participants. We extracted 29 features from each subject's Twitter stream and used them to classify each subject as belonging to one of the two poles for each of the four dimensions of the ILS model.

We found a more than by chance agreement only for a single dimension: active/reflective. Here, the agreement was slight to moderate, while for the other three dimensions no agreement between the prediction and the ground truth above agreement by chance was found.

Moreover, there are some limitations inherent in ILS which need to be taken into account. Felder and Spurlin [11] point out the limitation of learning style assessment and the purposes for which it should be used.

We conclude that, while there is some evidence that a Twitter signal contains useful information (as evident in the classification results of the active/reflective dimension), such a classification in general is hard and more complex features need to be derived. Thus, future work will focus on deriving more complex features that are more in agreement with the different learning dimensions, instead of relying on low-level features that can only be somewhat indicative when viewed in isolation.

References

1. Burleson, W., Picard, R.: Gender-specific approaches to developing emotionally intelligent learning companions. IEEE Intelligent Systems 22(4), 62–69 (2007)
2. Kalyuga, S., Sweller, J.: Rapid dynamic assessment of expertise to improve the efficiency of adaptive e-learning. Educational Technology Research and Development 53, 83–93 (2005), doi:10.1007/BF02504800
3. Chen, C.M., Lee, H.M., Chen, Y.H.: Personalized e-learning system using item response theory. Computers & Education 44(3), 237–255 (2005)
4. Blanchard, E., Razaki, R., Frasson, C.: Cross-cultural adaptation of elearning contents: a methodology. In: International Conference on E-Learning (2005)
5. Stash, N., Cristea, A., Bra, P.D.: Adaptation to learning styles in e-learning: Approach evaluation. In: Proceedings of World Conference on E-Learning in Corporate, Government, Healthcare, and Higher Education 2006, pp. 284–291 (2006)
6. Pashler, H., McDaniel, M., Rohrer, D., Bjork, R.: Learning styles. Psychological Science in the Public Interest 9(3), 105–119 (2008)

7. Ivy, L., Cheung, C., Lee, M.: Understanding Twitter Usage: What Drive People Continue to Tweet? In: Proceedings of Pacific-Asia Conference on Information Systems, Taipei, Taiwan (2010)
8. Kolb, D.A.: Experiential learning: Experience as the source of learning and development. Prentice Hall, Englewood Cliffs (1984)
9. Leite, W.L., Svinicki, M., Shi, Y.: Attempted validation of the scores of the vark: Learning styles inventory with multitrait-multimethod confirmatory factor analysis models. Educational and Psychological Measurement 70(2), 323–339 (2009)
10. Felder, R.M., Silverman, L.K.: Learning and teaching styles in engineering education. Journal of Engineering Education 78(7), 674–681 (1988)
11. Felder, R.M., Spurlin, J.: Applications, reliability and validity of the index of learning styles. International Journal of Engineering Education 21(1), 103–112 (2005)
12. Geiger, M.A., Pinto, J.K.: Changes in learning style preference during a three-year longitudinal study. Psychological Reports 69(3), 755–762 (1991)
13. Hauff, C., Houben, G.J.: Deriving Knowledge Profiles from Twitter. In: Kloos, C.D., Gillet, D., Crespo García, R.M., Wild, F., Wolpers, M. (eds.) EC-TEL 2011. LNCS, vol. 6964, pp. 139–152. Springer, Heidelberg (2011)
14. Java, A., Song, X., Finin, T., Tseng, B.: Why we twitter: understanding microblogging usage and communities. In: Proceedings of the 9th WebKDD and 1st SNA-KDD 2007 Workshop on Web Mining and Social Network Analysis, pp. 56–65. ACM (2007)
15. Naaman, M., Boase, J., Lai, C.H.: Is it really about me?: message content in social awareness streams. In: CSCW 2010, pp. 189–192 (2010)
16. Westman, S., Freund, L.: Information interaction in 140 characters or less: genres on twitter. In: IIiX 2010, pp. 323–328 (2010)
17. Zhao, D., Rosson, M.B.: How and why people twitter: the role that micro-blogging plays in informal communication at work. In: GROUP 2009, pp. 243–252 (2009)
18. Michelson, M., Macskassy, S.A.: Discovering users' topics of interest on twitter: a first look. In: AND 2010, pp. 73–80 (2010)
19. Hecht, B., Hong, L., Suh, B., Chi, E.H.: Tweets from justin bieber's heart: the dynamics of the location field in user profiles. In: CHI 2011, pp. 237–246 (2011)
20. Mislove, A., Lehmann, S., Ahn, Y.Y., Onnela, J.P., Rosenquist, J.N.: Understanding the Demographics of Twitter Users. In: ICWSM 2011 (2011)
21. Rao, D., Yarowsky, D., Shreevats, A., Gupta, M.: Classifying latent user attributes in twitter. In: SMUC 2010, pp. 37–44 (2010)
22. Graf, S., Kinshuk: An approach for detecting learning styles in learning management systems. In: Sixth International Conference on Advanced Learning Technologies, pp. 161–163 (July 2006)
23. Garcia, P., Amandi, A., Schiaffino, S., Campo, M.: Evaluating bayesian networks precision for detecting students learning styles. Computers & Education 49(3), 794–808 (2007)
24. Sanders, D., Bergasa-Suso, J.: Inferring learning style from the way students interact with a computer user interface and the www. IEEE Transactions on Education 53(4), 613–620 (2010)
25. Bergasa-Suso, J., Sanders, D., Tewkesbury, G.: Intelligent browser-based systems to assist internet users. IEEE Transactions on Education 48(4), 580–585 (2005)
26. Fang Zhang, L.: Does the big five predict learning approaches? Personality and Individual Differences 34(8), 1431–1446 (2003)

27. Litzinger, T., Lee, S., Wise, J., Felder, R.: Intelligent browser-based systems to assist internet users. Journal of Engineering Education 96(4), 309–319 (2007)
28. Strapparava, C., Valitutti, R.: Wordnet-affect: an affective extension of wordnet. In: Proceedings of the 4th International Conference on Language Resources and Evaluation, pp. 1083–1086 (2004)
29. Landis, J.R., Koch, G.G.: The measurement of observer agreement for categorical data. Biometrics 33(1), 159–174 (1977)
30. Zywno, M.: A contribution to validation of score meaning for felder-solomans index of learning styles. In: Proceedings of the 2003 American Society for Engineering Annual Conference and Exposition (2003)

Motivational Social Visualizations
for Personalized E-Learning

I.-Han Hsiao and Peter Brusilovsky

School of Information Sciences, University of Pittsburgh, USA
{ihh4,peterb}@pitt.edu

Abstract. A large number of educational resources is now available on the Web
to support both regular classroom learning and online learning. However, the
abundance of available content produces at least two problems: how to help
students find the most appropriate resources, and how to engage them into using
these resources and benefiting from them. Personalized and social learning have
been suggested as potential methods for addressing these problems. Our work
presented in this paper attempts to combine the ideas of personalized and social
learning. We introduce Progressor$^+$, an innovative Web-based interface that
helps students find the most relevant resources in a large collection of self-
assessment questions and programming examples. We also present the results
of a classroom study of the Progressor$^+$ in an undergraduate class. The data
revealed the motivational impact of the personalized social guidance provided
by the system in the target context. The interface encouraged students to
explore more educational resources and motivated them to do some work ahead
of the course schedule. The increase in diversity of explored content resulted in
improving students' problem solving success. A deeper analysis of the social
guidance mechanism revealed that it is based on the leading behavior of the
strong students, who discovered the most relevant resources and created trails
for weaker students to follow. The study results also demonstrate that students
were more engaged with the system: they spent more time in working with self-
assessment questions and annotated examples, attempted more questions, and
achieved higher success rates in answering them.

Keywords: social visualization, open student modeling, visualization,
personalized e-learning.

1 Introduction

A large number of educational resources is now available on the Web to support both
regular classroom learning and online learning. However, the abundance of available
content produces at least two problems: how to help students find the most
appropriate resources, and how to engage them into using these resources and
benefiting from them. To address these problems a number of projects have explored
personalized and social technologies. Personalized learning has been suggested as an
approach to help every learner find the most relevant and useful content given the
learner's current state of knowledge and interests [1]. Social learning was explored as

A. Ravenscroft et al. (Eds.): EC-TEL 2012, LNCS 7563, pp. 153–165, 2012.
© Springer-Verlag Berlin Heidelberg 2012

a potential solution to a range of problems, including student motivation to learn [2-5]. In our group's earlier work, these approaches were explored in two systems, QuizGuide [6] and Knowledge Sea II [7]. QuizGuide provides topic-based adaptive navigation support for personalized guidance for programming problems. Knowledge Sea II uses social navigation support to help students navigate weekly reading assignments. These and similar systems demonstrated the value and effectiveness of personalized learning and social learning in E-Learning. However, the combination of these powerful approaches has not been seriously investigated. The work presented in this paper attempts to explore the value of a specific combination of personalized learning and social learning to guide students to the most relevant resources in a course-sized volume of educational content.

2   Related Work

2.1   Open Student Modeling

The research on open student modeling explores the value of making students models visible to, and even editable by, the students themselves. There are two main streams of work on open student modeling. One stream focuses on visualizing the models supporting students' self-reflection and planning; the other one encourages students to participate in the modeling process, such as engaging students through the negotiation or collaboration on construction of the model [8]. Representations of the student models vary from displaying high-level summaries (such as skill meters) to complex concept maps or Bayesian networks. A range of benefits have been reported on opening the student models to the learners, such as increasing the learner's awareness of knowledge development, difficulties and the learning process, and students' engagement, motivation, and knowledge reflection [8-10]. Dimitrova et al. [11] explored interactive open learner modeling by engaging learners to negotiate with the system during the modeling process. Chen et al. [12] investigated active open learner models in order to motivate learners to improve their academic performance. Both individual and group open learner models were studied and demonstrated increased reflection and helpful interactions among teammates. Bull & Kay [13] developed a framework to apply open user models in adaptive learning environments and provided many in-depth examples. Studies also show that students have a range of preferences for how open student modeling systems should present their own knowledge. Students highly value having multiple viewing options and being able to select the one with which they are most comfortable. Such results are promising for potentially increasing the quality of reflection on their own knowledge [14]. In our own work on the QuizGuide system [6] we combined open learning models with adaptive link annotation and demonstrated that this arrangement can remarkably increase student motivation to work with non-mandatory educational content.

2.2   Social Navigation and Visualization for E-Learning

According to Vygotsky's Social Development Theory [15], social interactions affect the process of cognitive development. The Zone of Proximal Development, where learning occurs, is the distance between a student's ability to perform a task under adult guidance and/or with peer collaboration and the student's ability to solve the problem independently. Research on social learning has confirmed that it enhances the learning outcomes across a wide spectrum, including: better performance, better motivation, higher test scores and level of achievement, development of high level thinking skills, higher student satisfaction, self-esteem, attitude and retention in academic programs [16-18].

To support social learning, a visual approach is a common technique used to represent or organize multiple students' data in an informative way. For instance, social navigation, which is a set of methods for organizing users' explicit and implicit feedback for supporting information navigation [19]. Such a technique attempts to support a known social phenomenon where people tend to follow the "footprints" of other people [7, 20, 21]. The educational value has been confirmed in several studies [22-24]. The *group performance visualization* has been used to support the collaboration between learners among the same group, and to foster competition in groups of learners [25]. Vassileva and Sun [25] investigated the community visualization in online communities. They found that social visualization allows peer-recognition and provides students the opportunity to build trust in others and in the group. CourseVis [26] pioneered extensive *graphical performance visualization* for teachers and learners. This helps instructors to identify problems early on, and to prevent some of the common problems in distance learning. A promising, but rarely explored approach is social visualization of open student and group models. Bull and Britland [27] used OLMlets to research the problem of facilitating group collaboration and competition. The results demonstrated that selectively showing the models to their peers increases the discussion among students and encourages them to start working sooner. Our work presented below attempts to further advance this approach.

2.3   Social Comparison

According to *social comparison* theory [28], people tend to compare their achievements and performance with people who they think are similar to them in some way. There are three motives that drive one to compare him/herself to others, namely, self-evaluation, self-enhancement, and self-improvement. The occurrence of these three motives depends on the comparison targets, they are respectively lateral comparison, downward comparison and upward comparison. Earlier social comparison studies [29] demonstrated that students were inclined to select challenging tasks among easy, challenging and hard tasks by being exposed to the proper social comparison conditions. Feldman and Ruble (1977) [30] argued that age differences resulted in different competence and skills in terms of social comparison. As young children grow older, they become more assured of the general competence of their social comparing skills [30]. Later studies showed that social comparison, prompted by the graphical feedback tool, decreases *social loafing* and increases productivity [31]. A synthesis review of years social comparison studies summarized that upward comparisons in the classroom often lead to better performances [32]. Among fifty years of social comparison theory literature, most of the work has been done with qualitative studies by interviews, questionnaires and observation. In this

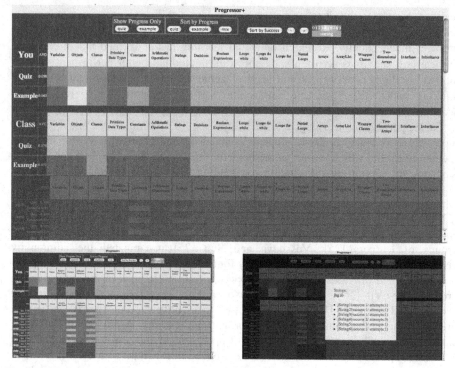

Fig. 1. Progressor+: the tabular open social student modeling visualization interfaces. The open social student model visualization allows collapsing the visualization parts that are out of focus (bottom left) and also provides direct content access (bottom right).

research, we develop a set of quantitative measures for investigating social comparison theory in our target context.

3 Progressor⁺ - An Open Social Student Modeling Interface

In past studies, we explored two open social student modeling interfaces, QuizMap [33] and Progressor [34], to examine the feasibility and the impact of a combined social visualization and open student modeling approach. Both systems use open social student modeling to provide personalized access to one specific kind of learning content – parameterized programming questions for Java. The use of a single kind of context allowed us to ignore the potential complexity of diverse learning content and focus on exploring critical aspects of open social student modeling. At the same time, this meant were unable to explore the scalability of the approach, i.e., its ability to work in a more typical e-learning context where many kinds of learning content may be used in parallel. The goal of Progressor⁺ was to bring our earlier findings up to scale and explore the feasibility of open social student modeling in the context of more diverse learning content. To achieve this goal, we piloted a new scalable tabular interface to accommodate diverse content. The Progressor⁺ system interface is presented in Fig. 1. Each student's model is represented as several rows of a large table with each row corresponding to one kind of learning content and each column corresponding to a course topic. The study presented in this paper has been performed with two kinds of learning content – Java programming questions and Java code examples (thus Figure 1 shows two rows for each student - quiz progress row and example progress row), however, the tabular nature of the proposed interface allows adding more kinds of content when necessary. Each cell is colored coded showing student's progress of the topic. We used a ten-color scheme to represent percentile of the progress. The use of color-coding allows collapsing table rows that are out of focus thus making it possible to present a progress picture of a large class in a relatively small space. This feature was inspired by the TableLens visualization, which is known as highly expressive and scalable [35]. While the interface of Progressor$^+$ was fully redesigned, it implemented most critical successful features discovered in our past studies that we review below.

Sequence: The sequence of the topics provides direction for the students to progress through the course. It also provides flexibility to explore further topics or redo already covered topics. In the QuizMap study [33], the topic arrangement in the treemap visualization was non-sequential. A key issue that emerged was that students had difficulty connecting the course structure and the treemap layout. We improved the design by providing a clear sequence in progressing through the topics in Parallel IntrospectiveViews [36] and Progressor [34] studies. We discovered that students benefited from the guidance offered by the course structure and explored more diverse topics that were appropriate for them at the moment. From these studies we also learned that topic-based personalization in open social student modeling worked more effectively when a sequence feature was implemented. In addition, we have also found that strong students tended to explore ahead of the class and weak students tended to follow them, even for the topics that were beyond the current scope. Therefore, we decided to maintain the "sequence" as one of the important features in Progressor$^+$.

Identity: Identity captures all the information belonging to the student. It is the representation of the student's unique model as well as one of the main entrances to interaction with the domain content. From the QuizMap study [33], we learned that distinguishing aspects of student's own model from the rest of the student models is not enough. This addressed the differences between the student herself and the rest of the class, but it did not carve out a clear model unit that belonged to the student. As we discovered, it is also important to offer a holistic view of individual student progress. In the Parallel IntrospectiveViews [36] study, we utilized the concept of *unity*, which proposed that perception of identity is higher if the model represents unity. This concept makes the students identify themselves with the model and allows them to easily compare themselves each other [12, 13]. In Progressor$^+$, we believe that the simple rows & columns table representation is cohesive and can be easily shown in fragments and recognized as units. Such characteristics could promote the notion of students' identity when interacting with the system.

Interactivity: Interactivity in the visualization of the user model can be implemented in several forms. Based on past studies, we knew that students benefited a lot from accessing content by directly clicking on the student's own model. The idea is simple but effective; the visualization of the user model is not a secondary widget but the main entrance allowing the students to access content directly. Moreover, students are also enabled to interact with content through their peers' models, or interact with their peers by comparing and sorting their performances. In Progressor⁺, the core interactivity is to allow the students to access the content resources directly by clicking on the students' models - the table cells. Meanwhile, other interactivity features are, for example, a collapse-and-expand function allowing the user model visualization to deal with the complexity and the large topic domains [37], or a manipulation function allowing the user to feel in control over his/her model [38].

Comparison: Letting students compare themselves with each other is the key for encouraging more work and better performance [32]. In [33, 34, 39], we found evidence that students interacted through their peers' models. Moreover, the same principle stems from the underlying supporting theory of *Social Comparison*. We believe that socially exposing models implicitly forces the students to perform comparison cognitively. We also learned that lowering the cognitive loads for comparisons could encourage more interactions. Thus, we capitalize our past successful experiences and implement different levels of comparisons: macro- and micro-comparisons. Macro-level comparison allows students to view their own models while at the same time seeing thumbnails of their peers' models. It provides a high level of comparisons, allowing fast mental overlapping of the colored areas between models. Micro-level comparisons occur at the moment a student clicks on any peer models. Progressor⁺ enters in the comparison mode by collapsing the rest of the table rows and displaying the selected peer model with all its details. Both levels of comparison allow students to perform social comparisons at their own free will.

4   Evaluation and Results

To assess the impact of our technology, we have conducted the evaluation in a semester-long classroom study. The study was performed in an undergraduate Object-Oriented Programming course offered by the School of Information Sciences, University of Pittsburgh in the Spring semester of 2012. The system was introduced to the class at the third week of the course and served as a non-mandatory course tool over the entire semester period. Out of 56 students enrolled in the course, 3 withdrew early and 38 out of the remaining 53 were actively using the system. All student activity with the system was recorded. For every student attempt to answer a question or explore an example, the system stored a timestamp, the user's name, the session ids, and content reference (question id and result for questions, example id and explored line number for examples). We also recorded the frequency and the timing of student model access and the peer comparisons. Pre-test and post-test were administered at the beginning and the end of the semester to measure students' initial knowledge and knowledge gain.

Following our prior experience with open student modeling in JavaGuide [40] and Progressor [34], we hypothesized that the ability to view students' models would motivate the students to have more interactions with the system. In particular, we expected that the motivation to work learning content would extend to both kinds of educational content, as in its earlier observed increase in the context of single-kind content collection. To evaluate these hypotheses, we compared the student content usage in three semester long classes that used three kinds of interfaces to access the same collection of annotated examples and self-assessment questions: (1) a combination of a traditional course portal for example access with an adaptive hypermedia system JavaGuide for question access (Column 1 in Table 1); [41] a combination of a traditional course portal for example access and social visualization (Progressor) for question access (Column 2 in Table 1); and (3) an open social student modeling visualization to access both examples and questions through Progressor[+] (Column 3 in Table 1). To discuss the impact on students' motivation and problem solving success, we measure the quantity of work (the amount of examples, lines and questions), *Course Coverage* (the distinct numbers of topics, example, lines and questions) and *Success Rate* (the percentage of correctly answered questions). Table 1 summarizes the system usage for the same set of examples and quizzes in three different conditions.

Table 1. Summary of system usage for three different technologies

		JavaGuide	Progressor	Progressor[+]
Example	N	20	7	35
	Example	19.75	28.71	27.37
Quantity	Line	116.6	219.71	184.18
	Session	5.35	5.50	4.94
	Distinct Topic	9.15	12.28	12.20
Coverage	Distinct Examples	17.3	25.13	27.37
	Distinct Lines	67.1	115.22	141.5
Quiz	N	22	30	38
	Attempt	125.50	205.73	190.42
Quantity	Success	58.31%	68.39%	71.20%
	Session	4.14	8.4	5.18
Coverage	Distinct Topic	11.77	11.47	12.92
	Distinct Questions	46.18	52.7	61.84

4.1   Effects on System Usage

Among 53 registered students, 35 students explored the annotated examples and 38 students worked with self-assessment questions through Progressor[+]. On average, students explored 27.37 examples; accessed 184.18 annotated lines and answered

190.42 questions. We found that there was 38.58%, 57.95% and 51.73% more examples, lines explored and questions answered correspondingly in Progressor⁺ compared to JavaGuide. Although we did not register a significant increase on the usage in Progressor⁺, this still shows that the access through open social student modeling visualization is at least as good as knowledge-based adaptive navigation support, which is considered as a golden standard of personalized information access. As we anticipated, we did not find significant differences in the amount of work done between Progressor and Progressor⁺. This demonstrates that Progressor⁺ was as engaging as Progressor. i.e., the registered increase in the usage of annotated examples did not caused a decrease the self-assessment quizzes usage. Instead, the overall volume of work increased. The quantity results show that open social student modeling that integrates several kinds of content is a valid approach to providing navigational support for multiple kinds of educational content.

In order to demonstrate that our approach is not only valid but also capable of delivering added value, we used other parameters to measure students' learning quality. First, we calculated the number of distinct topics, examples, lines and questions attempted by the student to measure the *Course Coverage*. We found that students were able to explore more topics, examples, lines and questions by using Progressor⁺ than the other two systems. In fact, students explored significantly more distinct lines in Progressor⁺ than with JavaGuide condition, $F(1, 53) = 9.72$, $p < .01$. It suggests that the inclusion of the additional content (examples) into the open social student modeling visualization generated an expected increase of motivation to work with examples while maintaining the motivation to work with questions. However, was it necessary for students to get exposed to more educational content? Was the new technology able to guide students to the right content at the right time? To answer these questions, we have to examine the impact of this technology on students' learning.

4.2   Impacts on Students' Learning and Problem Solving Success

To evaluate students' learning activities, we measured students' pre- and post- tests scores for knowledge gain and used the *Success Rate* to gauge students' problem solving success. Progressor⁺ was provided as a non-mandatory tool for the course, and students were able to learn from other factors, such as assignments, lab exercises etc. Thus, in our target content, it is important to use another parameter to infer students' learning. We chose to measure students' problem solving success. Note that problem solving is an important skill acquired by learning. It has been demonstrated that it could enhance the transfer of concepts to new problems, yield better learning results, make acquired knowledge more readily available and applicable (especially in new contexts), etc. [42, 43].

We found that the students who used Progressor⁺ achieved significantly higher post-test scores ($M=15.0$, $SD=0.6$) than their pre-test scores ($M=3.2$, $SD=0.5$), $t(37)=17.276$, $p < .01$. In addition, we also found that the more example lines the students explored, the higher level of knowledge they gained ($r=0.492$, $p < .01$). With open social student modeling visualization, students also achieved better *Success Rate*. The

Pearson correlation coefficient indicated that the more diverse questions the students tried, the higher success rate they obtained ($r=0.707$, $p<.01$). Similarly, the more diverse examples the students explored, the higher success rate they obtained ($r=0.538$, $p<.01$). We also looked at the value of repeated access to questions, examples and lines. We discovered that the more often the students repeated the same questions and the more often the students repeated studying the same lines the higher success rate they obtained ($r=0.654$, $p<.01$; $r=0.528$, $p<.01$).

4.3   Evidence of Social Guidance

To obtain a deeper understanding of the open social student modeling as a navigation support mechanism, we plot all the students' interactions with Progressor[+] (Figure 2). We categorized the students into two groups based on their pre-test scores (ranging from a minimum 0 to a maximum 20). Due to the pre-test scores being positively skewed, we split the two groups by setting the threshold at score 7. Strong students scored 7 points or higher (7~13) and weak students scored less than 7 (0~6). We color-coded the activities into two colors, orange and blue. Orange dots represent the activities generated by strong students and blue ones are the weak ones. The time of the action is marked on the X-axis and the question complexity on the Y-axis from easy to complex. We found 4 interesting zones within this plot. Zone "A" contains the current activity that students performed along the lecture stream of the course. Students had been working with the system very consistently throughout the first ten weeks. Zone "B" represents the region of after the tenth week. Zone "C" contains all of the attempts to explore earlier content, which the system motivated students to do to achieve mastery of the subject. Zone "D" contains the attempts which students

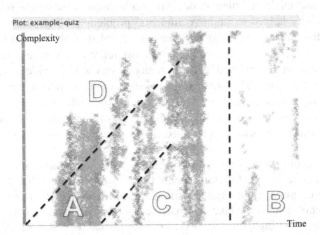

Fig. 2. Time distribution of all examples and questions attempts performed by the students through Progressor[+]. X axis is the Time; Y axis is the complexity of the course. Blue dots represent strong students' actions; orange ones are the weaker ones' actions. Zone "A" – *lecture stream*, zone "B" – *final exam cut (after week 10)*, zone "C" –*work with material from earlier lectures*, zone "D" –*navigating ahead*.

performed ahead of the course schedule. It is not surprising that a lot of the student interactions with Progressor[+] occurred in Zone A. More interesting are Zones C & D. A substantial proportion of the interactions occurred in Zone C. This indicates that the students were self-motivated to go back to achieve better mastery on already introduced topics. Moreover, based on Zone D in the figure, we found that the strong students who already achieved mastery on the current topics were able to use the visual interface to explore topics ahead of the course schedule. In addition, the plot shows that strong students generally explored the content ahead of the weak ones. Such phenomena provided evidence that strong students worked on new topics in Progressor[+] first and left the implicit traces for weak students that were visualized by the interface and provided proper guidance for weaker students. It also demonstrated that the system was actually inviting students to challenge themselves to move a little bit ahead of the course pace instead of passively progressing.

5   Summary

This paper described an innovative tabular interface, Progressor[+], which was designed to help students to find the most relevant resources in a large collection of diverse educational content. The interface provides progress visualization and content access through open social student modeling paradigm. Students were able to navigate through all their peers' models and to perform comparisons from one to another. An exploratory study was conducted. We found that students used Progressor[+] heavily, despite of the non-mandatory nature of the system. We also confirmed the motivational value of the social guidance provided by Progressor[+]. The results showed that the interface encouraged students to explore more topics, examples, lines and questions and motivated them to do some work ahead of the course schedule. The increased diversity helped to improve students' problem solving success. A deeper analysis of the social guidance mechanism revealed that the strong students successfully led the way in discovering the most relevant resources, and provided implicit trails that were harvested by the system and served to provide social guidance for the rest of the class. The study results also demonstrated that the social open student modeling increased student engagement to work with learning content. The students working with Progressor[+] spent more time working with annotated examples and self-assessment questions, attempted more questions, and achieving higher success rate.

While the results in this study were encouraging, we believe that the current approach has not yet reached its full potential. For example, given that students were able to discover more topics and questions by following implicit trails from the stronger students, could we take a proactive role and recommend trails to weak students instead of letting them follow the trails by themselves? According to our past work, providing adaptive navigation support significantly increases the quality of student learning and student motivation to work with non-mandatory learning content. We plan to have a richer integration of open social student modeling with adaptive navigation support. Furthermore, we are motivated to investigate deeper the issues of data sharing and model comparisons in open social student modeling interfaces.

References

1. Kay, J.: Lifelong Learner Modeling for Lifelong Personalized Pervasive Learning. IEEE Transaction on Learning Technologies 1(4), 215–228 (2008)
2. Vassileva, J.: Toward Social Learning Enviroments. IEEE Transaction on Learning Technologies 1(4), 199–214 (2008)
3. Barolli, L., et al.: A web-based e-learning system for increasing study efficiency by stimulating learner's motivation. Information Systems Frontiers 8(4), 297–306 (2006)
4. Méndez, J.A., et al.: A web-based tool for control engineering teaching. Computer Applications in Engineering Education 14(3), 178–187 (2006)
5. Vassileva, J., Sun, L.: Evolving a Social Visualization Design Aimed at Increasing Participation in a Class-Based Online Community. International Journal of Cooperative Information Systems (IJCIS) 17(4), 443–466 (2008)
6. Brusilovsky, P., Sosnovsky, S., Shcherbinina, O.: QuizGuide: Increasing the Educational Value of Individualized Self-Assessment Quizzes with Adaptive Navigation Support. In: World Conference on E-Learning, E-Learn 2004. AACE, Washington, DC (2004)
7. Brusilovsky, P., Chavan, G., Farzan, R.: Social Adaptive Navigation Support for Open Corpus Electronic Textbooks. In: De Bra, P.M.E., Nejdl, W. (eds.) AH 2004. LNCS, vol. 3137, pp. 24–33. Springer, Heidelberg (2004)
8. Mitrovic, A., Martin, B.: Evaluating the Effect of Open Student Models on Self-Assessment. International Journal of Artificial Intelligence in Education 17(2), 121–144 (2007)
9. Bull, S.: Supporting learning with open learner models. In: 4th Hellenic Conference on Information and Communication Technologies in Education, Athens, Greece (2004)
10. Zapata-Rivera, J.-D., Greer, J.E.: Inspecting and Visualizing Distributed Bayesian Student Models. In: 5th International Conference Intelligent Tutoring Systems (2000)
11. Dimitrova, V., Self, J.A., Brna, P.: Applying Interactive Open Learner Models to Learning Technical Terminology. In: Bauer, M., Gmytrasiewicz, P.J., Vassileva, J. (eds.) UM 2001. LNCS (LNAI), vol. 2109, p. 148. Springer, Heidelberg (2001)
12. Chen, Z.-H., et al.: Active Open Learner Models as Animal Companions: Motivating Children to Learn through Interacting with My-Pet and Our-Pet. International Journal of Artificial Intelligence in Education 17(2), 145–167 (2007)
13. Bull, S., Kay, J.: Student Models that Invite the Learner. The SMILI() Open Learner Modelling Framework. International Journal of Artificial Intelligence in Education 17(2), 89–120 (2007)
14. Mabbott, A., Bull, S.: Alternative Views on Knowledge: Presentation of Open Learner Models. In: Lester, J.C., Vicari, R.M., Paraguaçu, F. (eds.) ITS 2004. LNCS, vol. 3220, pp. 689–698. Springer, Heidelberg (2004)
15. Vygotsky, L.S.: Mind and society: The development of higher mental processes. Harvard University Press, Cambridge (1978)
16. Cecez-Kecmanovic, D., Webb, C.: Towards a communicative model of collaborative Web-mediated learning. Australian Journal of Educational Technology 16(1), 73–85 (2000)
17. Johnson, D.W., Johnson, R.T., Smith, K.A.: Cooperative Learning Returns to College: What Evidence is There That it Works? Change: The Magazine of Higher Learning 30(4), 26–35 (1998)
18. Koedinger, K.R., Corbett, A.: Cognitive Tutors: Technology bringing learning science to the classroom. In: Sawyer, K. (ed.) The Cambridge Handbook of the Learning Sciences. Cambridge University Press, New York (2006)

19. Dieberger, A., et al.: Social navigation: Techniques for building more usable systems. Interactions 7(6), 36–45 (2000)
20. Dieberger, A.: Supporting social navigation on the World Wide Web. International Journal of Human-Computer Interaction 46, 805–825 (1997)
21. Wexelblat, A., Maes, P.: Footprints: history-rich tools for information foraging. In: Proceedings of the SIGCHI Conference on Human Factors in Computing Systems: the CHI is the Limit, pp. 270–277. ACM, Pittsburgh (1999)
22. Brusilovsky, P., Sosnovsky, S., Yudelson, M.: Addictive links: The motivational value of adaptive link annotation. New Review of Hypermedia and Multimedia 15(1), 97–118 (2009)
23. Farzan, R., Brusilovsky, P.: AnnotatEd: A social navigation and annotation service for web-based educational resources. New Review in Hypermedia and Multimedia 14(1), 3–32 (2008)
24. Kurhila, J., Miettinen, M., Nokelainen, P., Tirri, H.: EDUCO - A Collaborative Learning Environment Based on Social Navigation. In: De Bra, P., Brusilovsky, P., Conejo, R., et al. (eds.) AH 2002. LNCS, vol. 2347, pp. 242–252. Springer, Heidelberg (2002)
25. Vassileva, J., Sun, L.: Using Community Visualization to Stimulate Participation in Online Communities. e-Service Journal. Special Issue on Groupware 6(1), 3–40 (2007)
26. Mazza, R., Dimitrova, V.: CourseVis: A graphical student monitoring tool for supporting instructors in web-based distance courses. International Journal of Human-Computer Studies 65(2), 125–139 (2007)
27. Bull, S., Britland, M.: Group Interaction Prompted by a Simple Assessed Open Learner Model that can be Optionally Released to Peers. In: Conati, C., McCoy, K., Paliouras, G. (eds.) UM 2007. LNCS (LNAI), vol. 4511, Springer, Heidelberg (2007)
28. Festinger, L.: A theory of social comparison processes. Human Relations 7, 117–140 (1954)
29. Veroff, J.: Social comparison and the development of achievement motivation. In: Smith, C.P. (ed.) Achievement Related Motives in Children. Sage, New York (1969)
30. Feldman, N.S., Ruble, D.N.: Awareness of social comparison interest and motivations: A developmental study. Journal of Educational Psychology 69(5), 579–585 (1977)
31. Shepherd, M.M., et al.: Invoking social comparison to improve electronic brainstorming: beyond anonymity. J. Manage. Inf. Syst. 12(3), 155–170 (1995)
32. Dijkstra, P., et al.: Social Comparison in the Classroom: A Review. Review of Educational Research 78(4) (2008)
33. Brusilovsky, P., Hsiao, I.H., Folajimi, Y.: QuizMap: Open Social Student Modeling and Adaptive Navigation Support with TreeMaps. In: Kloos, C.D., Gillet, D., Crespo García, R.M., Wild, F., Wolpers, M. (eds.) EC-TEL 2011. LNCS, vol. 6964, pp. 71–82. Springer, Heidelberg (2011)
34. Bakalov, F., et al.: Progressor: Personalized visual access to programming problems. In: 2011 IEEE Symposium on Visual Languages and Human-Centric Computing (VL/HCC), Pittsburgh, PA (2011)
35. Rao, R., Card, S.K.: The table lens: merging graphical and symbolic representations in an interactive focus + context visualization for tabular information. In: Proceedings of the SIGCHI Conference on Human Factors in Computing Systems: Celebrating Interdependence, pp. 318–322. ACM, Boston (1994)
36. Hsiao, I.-H., Bakalov, F., Brusilovsky, P., König-Ries, B.: Open Social Student Modeling: Visualizing Student Models with Parallel IntrospectiveViews. In: Konstan, J.A., Conejo, R., Marzo, J.L., Oliver, N. (eds.) UMAP 2011. LNCS, vol. 6787, pp. 171–182. Springer, Heidelberg (2011)

37. Shneiderman, B.: The eyes have it: A task by data type taxonomy for information visualizations. In: Symposium on Visual Languages. IEEE Computer Society, Washington, DC (1996)
38. Kay, J.: Learner know thyself: Student models to give learner control and responsibility. In: International Conference on Computers in Education, ICCE 1997, Malasia, Kuching, Sarawak (1997)
39. Bakalov, F., König-Ries, B., Nauerz, A., Welsch, M.: IntrospectiveViews: An Interface for Scrutinizing Semantic User Models. In: De Bra, P., Kobsa, A., Chin, D. (eds.) UMAP 2010. LNCS, vol. 6075, pp. 219–230. Springer, Heidelberg (2010)
40. Hsiao, I.-H., Sosnovsky, S., Brusilovsky, P.: Guiding students to the right questions: adaptive navigation support in an E-Learning system for Java programming. Journal of Computer Assisted Learning 26(4), 270–283 (2010)
41. Lindstaedt, S.N., Beham, G., Kump, B., Ley, T.: Getting to Know Your User – Unobtrusive User Model Maintenance within Work-Integrated Learning Environments. In: Cress, U., Dimitrova, V., Specht, M. (eds.) EC-TEL 2009. LNCS, vol. 5794, pp. 73–87. Springer, Heidelberg (2009)
42. Dolmans, D.H.J.M., et al.: Problem-based learning: future challenges for educational practice and research. Medical Education 39(7), 732–741 (2005)
43. Melis, E., et al.: ActiveMath: A Generic and Adaptive Web-Based Learning Environment. International Journal of Artificial Intelligence in Education 12, 385–407 (2001)

Generator of Adaptive Learning Scenarios: Design and Evaluation in the Project CLES

Aarij Mahmood Hussaan[1] and Karim Sehaba[2]

Université de Lyon, CNRS
[1] Université Lyon 1, LIRIS, UMR5205, F-69622, France
[2] Université Lyon 2, LIRIS, UMR5205, F-69679, France
aarij-mahmood.hussaan@liris.cnrs.fr
Karim.sehaba@liris.cnrs.fr

Abstract. The objective of this work is to propose a system, which generates learning scenarios for serious games keeping into account the learners' profiles, pedagogical objectives and interaction traces. We present the architecture of this system and the scenario generation process. The proposed architecture should be, insofar as possible, independent of an application domain, i.e. the system should be suitable for different domains and different serious games. That is why we identified and separated different types of knowledge (domain concepts, pedagogical resources and serious game resources) in a multi-layer architecture. We also present the evaluation protocol used to validate the system, in particular the method used to generate a learning scenario and the knowledge models associated with the generation process. This protocol is based on comparative method that compares the scenario generated by our system with that of the expert. The results of this evaluation, conducted with a domain expert, are also presented.

Keywords: Scenario generator, serious games, adaptive system, evaluation protocol.

1 Introduction

Our work is situated in the context of adaptive generation of learning scenario. We define a learning scenario as a suite of structured pedagogical activities generated by the system for a learner keeping into account his/her profile in order to achieve one or more educational goals. We are more specifically interested in the learning scenario generation in serious games [1]. In this area, we propose a system capable of generating dynamically learning scenarios keeping into account the following properties:

- The ability to be utilized in any serious game taking into account its specificities.
- The use of interaction traces as knowledge sources in the adaptation process.

Along with the above mentioned properties, we also aim our system to be reusable with different learning domains and different games as well. Therefore, the different kinds of knowledge presented in the system are organized and separated in a

A. Ravenscroft et al. (Eds.): EC-TEL 2012, LNCS 7563, pp. 166–179, 2012.

multi-layer architecture. These layers represent the learning domain in the form of: *domain concepts*, *pedagogical resources* required to teach these concepts and *serious game resources* that are used to present pedagogical resources to the learner. This separation means that the aspects of any particular layer can be modified without necessarily modifying other layers, hence, rendering the system more reusable.

A trace [2] is defined as a history of learner's actions collected in real-time while the learner is using the serious game. It is considered to be the primary source for the updating of a learner's profile and the domain knowledge. It also serves as knowledge sources in the scenario generation process. Formally, a trace is a set of observed elements temporally located [2][3]. Each observed element represents the learner action on computer environment such as interacting with an educational resource, clicking on a hyperlink, etc.

The idea of automatically generating learning/pedagogical is not new and has been investigated previously by many authors [4][5][6]. However, these systems focuses only on the pedagogical aspects of the problem and do not consider serious games as a potential medium of delivering these scenarios to the learner. Furthermore, not every system defines clearly the separation of the conceptual layer and the pedagogical resource layer which makes them difficult to reuse. Likewise, these systems don't exploit, in general, the learner's traces in the generation process.

Our contribution is situated in the context of the Project CLES[1] (Cognitive Linguistic Elements Stimulation). CLES aims to develop a serious game environment, accessible online, which evaluate and train the cognitive ability for children with cognitive disabilities. In the context CLES, we conducted an evaluation aimed at:

1. Validating the working of the system generator of learning scenario, and
2. Validating the knowledge models that are used by the system to represent different kind of knowledge.

The learning scenario generator is evaluated to confirm the algorithm used to select the different resources (concept, pedagogical resources & game resources). Moreover, the knowledge models are evaluated to verify their functionality in the generation.

The rest of the paper is organized as follows: in section 2 we detail the project CLES, in section 3 a literature review on course generators and serious games is presented. Section 4 presents a brief presentation of our architecture system and section 5 presents the scenario generation process. Section 6 details the evaluation protocol of knowledge models and generator working. We will present the results of the evaluation in Section 7. The next section presents the discussions and conclusions.

2   Application Context

The work on project CLES (Cognitive Linguistic Elements Stimulation) was conducted in collaboration with different partner laboratories. These partners are specializing in serious games development for children with cognitive disabilities, ergonomic design and the study of cognitive mechanisms. This project aims to provide

---

[1] http://liris.cnrs.fr/cles

serious game for training and evaluation of cognitive functions. Eight functions are considered in CLES: perception, attention, memory, visual-spatial, logical reasoning, oral language, written language and transversal competencies.

The serious game developed, in the context of CLES, is called "Tom O'Connor and the sacred statue". This is an adventure game. The protagonist of this game is a character named *Tom*, his task is to search for the sacred statue hidden in a mansion. According to the session, Tom is placed in one of the many rooms in the mansion. Each room has many objects (chair, table, screen etc.). Hidden behind these objects are challenges in the form of mini-games. The user has to interact with these objects to start these mini-games. To move from one room to another and progress in the game, the user has to discover all the mini-games in the room.

Thus, for each of the eight cognitive functions, we have about a dozen mini-games and for each mini-game we've nine levels of difficulty. A more detailed description of games developed in this project is presented in [7].

The role of the scenario generator is to select (according to the learner's profile, his/her interaction traces and his /her therapeutic goals for the session) the mini-games with appropriate difficulty levels, and to put these games in relation with the objects of different rooms of the mansion. This generator should therefore keep in to account:

- What the practitioner has prescribed for his patients
- The knowledge base of the available treatments for the pathology
- Histories of the previous exercises of the learner, stored in the form of traces.
- Specificities of the serious game

The module we develop has to be validated on its theoretical properties (meta-models, models and processes) in the context of the Project CLES (see the sections 6 and 7).

3   Literature Review

The purpose of this section is to present the existing approaches regarding the generation of pedagogical scenarios and serious games, and to show what lacks in the theses approaches and where we are contributing. This literature review is done keeping in mind, among other, the following characteristics of our system, namely:

- General architecture independent of the pedagogical domain and application,
- Usable with serious games, and
- The use of interaction traces for the updating of learner profile and adaptation.

This section is organized in two sections. The first section presents the course generators and the second presents the serious games for learning.

3.1   Course Generators

Learning scenario generation can be divided into two broad categories: course sequencing and course generation. The former selects the best possible pedagogical resource at any time given the performance of the learner and the latter generates an structured course in a single go before presenting it to the learner [8]. A course sequencer by the name of DCG (Dynamic Courseware Generator) is presented in [9]. DCG selects the next pedagogical resource (HTML pages) dynamically according to the current performance of the learner. DCG is heavily dependent on web-based resources and are not suitable for other mediums like (serious games). In WINDS [4], the learner has to either manually navigate through the course or choose from the recommendations offered by the system. However, a complete learning path is not generated for a particular learner, which is required in games like CLES. An expert-system type approach is presented in [10], forcing to enter all the rules beforehand, therefore making it difficult to maintain for a large knowledge base. Statistical techniques are employed in [11] in order to generate a course most suitable to the learner, however, in addition to the relations between the concepts relation between different resources are also maintained. The relations between pedagogical resources are necessary for different resources to be included in the same scenario. This requirement is a limitation where different pedagogical resources are not related (like in project CLES). Case based reasoning is used in a web based system [12] called Pixed (Project Integrating eXperience in Distance Learning). PIXED uses the learners' interaction traces gathered as learning episodes to provide contextual help for learners trying to navigate their way through an ontology-based Intelligent Tutoring System (ITS). They rely on the learners to annotate their traces which can be difficult for cognitive handicapped persons.

A system which combines the techniques of course sequencing and generation is presented in a system called « Paigos » [13]. The authors use HTN-Planning and formalized scenarios to deliver adaptive courses. The manner of construction of Paigos makes it difficult for persons unfamiliar with HTN techniques to use it.

In general, course generators focus on the pedagogical aspects and do not target serious games for delivering their courses. Therefore, it is difficult to use them with serious games. Moreover, the interaction traces are, generally, not used for updating the learner profile & domain knowledge.

3.2   Serious Games

Systems have been proposed to use games for planning and management of business simulation games in [14]. The pedagogical scenario is presented as a tree, providing adaptation according to different learner actions. The construction of tree becomes difficult as the scenario becomes complex. An authoring tool for the creation of 2-dimensional adventure games is presented in [15], personalization is done by predefining the decision tree. A pedagogical dungeon to teach fractions in a collaborative manner is presented in [16]. The interaction traces are used here in the adaptation process. The scenarios are static and the tight coupling between the pedagogical scenario and the gaming interface deprives the approach from reusability. C programming language is taught in [17]. The teachers present to the learner a sequence of learning activities in a Bomberman type game. The manual presentation of learning activities sequences is not practical in case of hundreds of learners. A role playing game is also proposed for the purpose of osteopathic diagnosis [18]. This game also relies heavily on manual teacher intervention.

These systems tightly couple the pedagogical aspects with the gaming aspects i.e. we cannot reuse neither the pedagogical nor the gaming aspects with other games or pedagogical domains. Furthermore, a structured pedagogical scenario is not well defined, mostly; therefore, there isn't a generated personalized pedagogical scenario as well. The learners' interaction traces are also not exploited as well, in general.

4   System Architecture

In this section we present the different kinds of knowledge used in our system and how we've organized them in order to increase reusability. Furthermore, the modeling of this knowledge is also presented along with the general working of the system.

4.1   Knowledge Representation

As mentioned earlier, our objective is to develop a generic system capable of generating dynamically adaptive learning scenarios keeping into account the learners' profile (including their interaction traces) and the specificities of serious games. For this, we propose to organize the domain knowledge in a multi-level architecture. We have considered three types of knowledge (as shown in the figure 1): *domain concept, pedagogical resource* and *serious game resource*. The separation of this knowledge on three layers helps in using change the aspects of one layer without forcibly changing the other layers.

As the name indicates, the first layer contains the domain concepts. These concepts are organized in the form of a graph where the nodes represent the concepts and the edges represent the relation between the concepts.

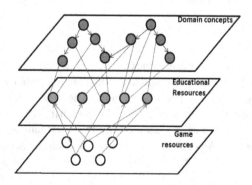

Fig. 1. Knowledge Layer

Formally, the domain knowledge is modeled as $<C, R>$ where, 'C' is the concepts of the domain and 'R' represents the relations between the concepts. Each concept 'C' is defined by $<Id, P>$, where: 'Id' is a unique identifier and 'P' is the set of properties that describe the concept like the author, the date of creation, description of the concept, etc. 'R' is defined by $< C_{From}, T, RC+ >$, where: 'C_{From}' is the origin concept of the relation, 'T' Is the type of the relation and 'RC'(Relation Concepts) = $<C_{To}, F,$ Value $>$ where: 'C_{To}' is the target concept of the relation, the direction of relation is from C_{From} to C_{To}, 'F' is the function that allows propagating the information in the graph in order to update the learner profile. The semantics of the function may differ depending on the type of relation. And 'Value' is the value between the concepts of the relation. This value is used as default in the absence of function 'F'.

We also have created many types of relations [7]. For example, we present here two types of relations:

- Has-Parts $(x, y_1 \ldots y_n)$: indicates that the target concepts $y_1 \ldots y_n$ are the sub-concepts of the super concept x. For example: Has-Parts (Perception, visual perception, auditive perception).
- Required (x, y): indicates that to study concept y it is necessary to have sufficient knowledge of concept x. For example, Required (Perception, Oral Language).

In the context of the project CLES, the domain concept models the eight cognitive functions and relationships that may exist between them.

The second layer contains the pedagogical resources. In general, a pedagogical resource is an entity used in the process of teaching, forming or understanding allowing learning, convey or understand the pedagogical concepts. The pedagogical resources can be of different natures: a definition of a concept, an example, a theorem, an exercise, etc. Formally, each pedagogical resource is defined by a unique identifier, a type, the parameters, an evaluation function, and a set of characteristics (like name, description, name of author etc). As shown in figure 1, each resource can be in relation with one or more domain concepts and vice versa. This relation shows that a resource can be used to understand the concept with which it is related. In the context of project CLES, the pedagogical resource layer contains the mini-games.

The third and final layer contains the game resources. They are static objects that are initialized with dynamic or proactive behavior. In our model, we only consider the game objects that are related to a pedagogical resource. Formally, each game resource is defined by an identifier, the relations with the pedagogical resources with which it is related and a set of characteristics like name, description etc. In the context of project CLES, these resources are the objects of the serious game which are used to hide the pedagogical resources (mini-games).

4.2   System Working

The architecture of our system is shown in the figure 2.

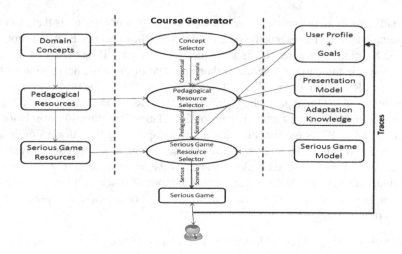

Fig. 2. System Architecture

The process of the system's operation is as follows: (1), the domain's expert(s) feeds the system with the domain's knowledge according to our proposed models and the learners' profile. These models were presented in the previous section. In each learning session, the system is fed with pedagogical goals. These goals are either selected by the learner or are predefined by the system from his/her profile. (2), the system, according to the selected goals and the learner's profile, selects the appropriate concepts from the domain model. This selection is done by the module 'Concept Selector'. The output of this module is the 'Conceptual Scenario'. This conceptual scenario is comprised of concepts along with the competence required to achieve the pedagogical goals.

(3), the conceptual scenario is sent as input to the module 'Pedagogical Resource Selector'. The purpose of this module is to select for each concept, in the conceptual scenario, the appropriate pedagogical resources. These resources are selected according to the 'Presentation Model' and the learner's profile. The latter is represented by a set of properties in the form of <attribute, value> pairs where the *attribute* represents a domain concept and the *value* represents the learner's mastery of that concept. The purpose of the presentation model is to organize the pedagogical resources presented to the learner. The structure of the scenario can be for e.g. starting a scenario by presenting two definitions followed by an example and an exercise. The selection of this model can either be done by the learner or by the teacher (expert) for the learner. The structure of the scenario model can fit the form defined in [13].

Furthermore, the pedagogical resources are then adapted according to the 'Adaptation Knowledge'. The adaptation knowledge is used to set the parameters of pedagogical resources according to the learner's profile and pedagogical goals. The output of this module is a 'Pedagogical Scenario'. This scenario comprises pedagogical resources with their adapted parameters.

(4) The pedagogical scenario is sent as input to the module 'Serious Game Resource Selector'. This module is responsible for associating the pedagogical resources

with the serious game resources. This association is done based on the 'Serious Game Model'. The 'Serious Game Model' is used to associate the type of serious game resource with the types of pedagogical resource. This module produces the 'Serious Scenario' (5).

The learner interacts with the learning scenario via the serious game. As a result of these interactions the learner's interaction traces are generated. These traces are stored in the learner profile and are used to update the profile and consequently modify the learning scenario according to the learner traces, if necessary.

5   Scenario Generator

As mentioned in section 4, the process of learning scenario generation given pedagogical goals and learner's profile is handled by three modules namely 'Concept Selector', 'Pedagogical Resource Selector' and 'Serious Game Resource Selector'. The general functionality of these modules is already defined in section 4. In this section we'll present the textual description of the working of these algorithms.

5.1   Concept Selector

The purpose of this module is to generate a list of domain concepts required to achieve the learning goals. This generation is performed keeping into account the learner's profile. The learning goals are defined as the set of target (domain) concepts along with the competence of each concept required. The generated list of domain concepts is called 'conceptual scenario' in our system. The generation process works as follows; first for each target concept (TC), it is checked (by consulting the learner profile) whether or not this TC is sufficiently known by the learner. If it is sufficiently known by the learner then this TC is ignored and the next TC is looked.

Then the module checks whether or not the TC has some concepts related to it. Some of these concepts, in relation with the concept in question, can be selected to be added in the conceptual scenario. This selection depends on the type of relation between the concepts. In fact, we've identified, for each type of relation, a **strategy for the selection** of concepts. For example, if a learner has chosen a target concept A and A is in a relation of type 'Required' with another concept B (Required (B, A)), then the generator will verify that whether the learner knows sufficiently the concept B. If it's not the case then the generator also includes concept B in the conceptual scenario.

5.2   Pedagogical Resource Selector

The purpose of this module is to select the appropriate resources for every concept in the 'Conceptual Scenario' given a 'Presentation Model (PM)' and learner profile. This selection is outputted in the form of a "Pedagogical Scenario". This contains a list of resources associated with each concept along with their appropriate parameters.

The selection process goes as follows; firstly, for each concept in the 'conceptual scenario' the process searches for the resources of type 'T' as described in the PM. If there is more than one pedagogical resource of type 'T' associated with the concept, then the resource which is not already seen by the learner or not sufficiently known by the learner is added to the list. The process also consults the adaptation knowledge to select the parameters of the resources (in order to adjust the level of difficulty).

5.3   Serious Game Resource Selector

This module associated the pedagogical resources in the 'Pedagogical Scenario' with the serious game resources according to the learner's profile and Serious Game Model (SGM). The result of the execution of this module is a list of game resources called 'Serious Scenario' which contains resulting concepts and the serious game resources initialized with the pedagogical resources and their parameters.

The working of this module is as follows; firstly, for each pedagogical resource in the 'Pedagogical scenario' the serious game resources related to the pedagogical resource are searched. Then for each selected serious game resource, the process consults the learner profile to verify whether the selected resource is appropriate for the learner. If yes then this resource is added to the list.

6   Evaluation Protocol

The first evaluation of our system was conducted in presence of a domain expert. This expert has been a practitioner of cognitive sciences for more than 20 years. The objective of our evaluation, as mentioned earlier, is the validation of:

- The scenario generator's working: more precisely, this means the validation of the concept **selection strategy** which we've defined for each type of relations, and
- The knowledge models: it means to validate the concepts and the relations that we've introduced into the system in the context of project CLES.

For this, the basic strategy that we've adapted is comparative evaluation [19] i.e. it consists in comparing the learning scenarios created manually by the domain expert with the learning scenarios generated automatically by the system for the same input. This input corresponds to the domain knowledge and profile types. Furthermore, during the evaluation process we conduct an Elicitation Interview [20] with the expert. The purpose of this interview is to help the expert in explicating (as much as possible) his/her thinking process, how s/he reasons while creating a learning scenario.

Before conducting the interview we came up with a protocol of evaluation. This protocol is designed to guide us in conducting the evaluation and help us in validating our models and to identify any problems and their source.

This flow of this protocol is depicted in the figure 3. At first, the expert is asked to create a certain number of learner profiles (1). As the expert has a vast experience in his/her respected field, s/he can give us the profiles that are pretty much closer to the.

Ideally, we would like the expert to create a certain number different profiles. The more are the profiles the more it is beneficial for our evaluation. Furthermore, the profiles should also be diverse i.e. different profiles should contain different competencies. This will help us in determining whether our system can handle diverse cases or not. Apart from these profiles we ask the expert to fix some learning objectives for the profiles. Afterwards, we ask the expert to create learning scenarios for each learning objective and each profile.

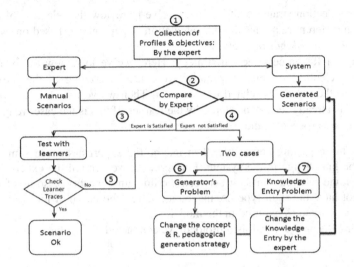

Fig. 3. Evaluation Protocol

Once the expert has identified the profiles and the objectives, we introduce them into the system in order to generate the learning scenarios. Then the two sets of scenarios are compared by the expert (2). This comparison is done by the expert (by an interview of explication where we demand the expert to verbalize his/her thoughts. The expert is filmed during the whole evaluation process.

The result of this comparison will be either the expert will find the scenarios similar or not. If the expert is sufficiently satisfied with the similarity of the scenarios (3), then the scenarios will be presented to real learners. Ideally these learners should've the same profiles as entered in the system. The scenarios will then be presented to the learners. If possible, the learners should be filmed during their interactions with the scenarios. The learners should be asked how difficult are they finding the scenarios. The learners' interaction traces will also help us in answering this question. By analyzing the traces we can determine that a learner is finding the scenario very difficult if s/he is failing constantly in the exercises. Similarly, if the learner is answering the exercises very quickly and correctly then we can conclude that the learner is finding the exercises very easy to solve.

If the learners say that they are finding the scenarios too easy or too difficult (5), then this will imply that either the knowledge entered in the system by the expert can be improved or the system is not generating the scenarios properly. In either case, the protocol to be followed to resolve the problem is defined next.

If as a result of the comparison the expert is not finding the scenarios similar enough (4), then two cases are possible: 1/The system's generator is not working properly (6). 2/ the knowledge entered in the system by the Expert is not correct (7). If the system's generator is not working properly, then we review the following:

1. Concept selection strategy: This means we've to review the selection of concepts based on different relations and the calculation of percentages based on them. Currently we've four kinds of relations.
2. Pedagogical Resource selection strategy: Here, we've to review the pedagogical resource selection strategy. Currently, we select, according to the presentation model, all the resources related to a concept. Then we verify whether a particular resource is already seen & mastered by the learner. If this is the case we ignore that resource and proceed on the next one.

If none of the cases are applicable, then maybe the expert has made some error in entering the knowledge in the system. Furthermore, we can tell the expert that there are either some relations missing between the domain concepts or some of the relations do not have the right type i.e. maybe a relation should be of the type has-parts whereas it is marked as required in the model.

Following the above mentioned protocol we conducted our evaluation.

7   Experiments and Results

We started the evaluation by introducing the domain models in the system. Since the original model of Project CLES is very large, the expert would have found the evaluation of the whole model quite tedious. In fact, there are 8 super concepts and each super concept having at-least 5 sub-concepts and each sub-concept has at-least 5 pedagogical resources. Furthermore, there is also the serious game resources associated with the pedagogical resources. Therefore, we created three mini-models of the original model. All these mini models contain the eight main domain concepts of CLES. The initial arrangements of these concepts are shown in figure 4. All the links between the super concepts are of the type 'Required'. The relation between perception and its sub concepts is of the type Has-Parts.

These super concepts are present in each of the mini-models. In each mini model, in addition to the super concepts, one concept is further detailed. The detailed concepts are: written language, perception and memory. We also prepared six profiles for each model. The profiles are as follows: Profile 1: 8 years, no deficiency in concept x/ Profile 2: 8 years, deficiency in concept x / Profile 3: 14 years, no deficiency in concept x/ Profile 4: 14 years, deficiency in concept x / Profile 5: 18 years, no deficiency in concept x / Profile 6: 18 years, deficiency in concept x.

The concept 'x' is the detailed concept in each model. The choice of these 18 profiles is not arbitrary but they are logically selected, Project CLES targets children between 6 years and 18 years. So the choice covers almost all the age groups. The expert was in agreement with us over the choice of the profiles.

Afterwards, we asked the expert to give sufficient values to the concepts in each profile. The expert defines the values keeping into account the type of the profile for example: lesser values are assigned to the profile with deficiency than those profiles without deficiencies. Afterwards, the objectives for each of the profile are also fixed. These objectives are a bit higher for the profiles without deficiency and vise-versa.

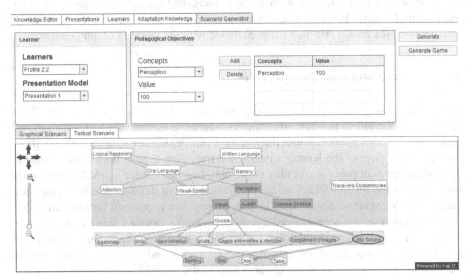

Fig. 4. One of the mini-model on the concept Perception

The whole time the expert was being filmed, with his permission, when he was fixing the values of the profiles. We were also asking the expert question regarding how he was assigning the values and why. Afterwards, we asked the expert to fix the pedagogical objectives for each profile. During this process we also asked questions about how and why he was choosing the pedagogical objectives. As a result of these questioning we discovered many things about the modeling of the domain model and how to select the right pedagogical objectives for a profile.

As soon as the profiles are created and the objectives are set, we introduced them into the system and generated the scenarios via the system. In the meantime, we asked the expert to create the learning scenarios manually. We asked the expert how and why he is selecting the concepts and the pedagogical resources for every profile. Afterwards, we asked the expert to compare the scenarios he created manually with those generated automatically.

The film that was made during the experimentation process is then analyzed by the video analyzing and annotation tool called ADVENE (http://liris.cnrs.fr/advene) [21]. The film made was about two hours long we saw it again and again annotating the important events in the video. These annotations were than analyzed and as a result we discovered some very interesting information. We found out some modifications to be performed in the domain model and some troubles were also detected in the concept selection strategy. In the domain model, we added 5 new relations between concepts, for example the addition of prerequisite relation between Memory and Oral Language. We modified also some concept selection strategy.

Furthermore, we also found that our system only takes into account the learner's profile while setting the pedagogical resources' levels; whereas; the expert was taking into account the *gap* between the profile and the pedagogical objectives. As a result of this evaluation we updated the knowledge models, and corrected the problems with our system. Finally, after the results shown to the expert, the expert seems sufficiently satisfied with the results. He also seems satisfied with the working of our generator.

8   Conclusion and Perspectives

In this paper we presented the working and the architecture of our system. This system is conceived to generate dynamically adaptable learning scenarios for serious games while keeping into account the learner's profile, learner's traces and specificities of serious games. The learner's interaction traces are used in the scenario generation and adaptation process and also while updating the learner's profile. This work took place in the Project CLES where the objective is to develop a serious game for children with cognitive disabilities. In this context, we conducted an evaluation for the verification of the scenario generation process. To conduct this evaluation, we presented an evaluation protocol that we've followed during the evaluation process. Our evaluation was based on comparative strategies and is designed to identify whether the problem exists in the expert's knowledge introduction into the system or in the generation of the scenario, when the expert is not satisfied with the scenario. Moreover, there is also the possibility that the problem exists in both the expert's knowledge introduction and system's generator. However, we pinpointed the problem correctly. However, we can face this problem with future evaluations.

For our future evaluations, we'll like to repeat the process with a number of experts to further verify the system. The tests with real learners will also be conducted to generate real learner traces and then use them to update their profiles. Furthermore, we'll also use them to adapt the scenario if necessary.

Acknowledgements. The authors would like to thank Mr. Philippe Revy, expert speech and language therapist and the director of the society GERIP.

References

[1] Zyda, M.: From visual simulation to virtual reality to games. Computer 38(9), 25–32 (2005)

[2] Clauzel, D., Sehaba, K., Prié, Y.: Enhancing synchronous collaboration by using interactive visualisation of modelled traces. Simulation Modelling Practice and Theory 19(1), 84–97 (2011)

[3] Settouti, L., Prie, Y., Marty, J.-C., Mille, A.: A Trace-Based System for Technol-ogy-Enhanced Learning Systems Personalisation. In: Ninth IEEE International Conference on Advanced Learning Technologies, pp. 93–97 (2009)

[4] Specht, M., Kravcik, M., Pesin, L., Klemke, R.: Authoring adaptive educational hyper-media in WINDS. In: Proceedings of ABIS 2001, Dortmund, Germany, vol. 3(3), pp. 1–8 (2001)

[5] Karampiperis, P., Sampson, D.: Adaptive learning resources sequencing in educational hypermedia systems. Educational Technology & Society 8(4), 128–147 (2005)

[6] Sangineto, E., Capuano, N., Gaeta, M., Micarelli, A.: Adaptive course generation through learning styles representation. Universal Access in the Information Society 7(1-2), 1–23 (2007)

[7] Hussaan, A.M., Sehaba, K., Mille, A.: Tailoring Serious Games with Adaptive Pedagogi-cal Scenarios: A Serious Game for Persons with Cognitive Disabilities. In: 11th IEEE In-ternational Conference on Advanced Learning Technologies, pp. 486–490 (2011)

[8] Brusilovsky, P., Vassileva, J.: Course sequencing techniques for large-scale web-based education. International Journal of Continuing Engineering Education and Life-long Learning 13(1/2), 75–94 (2003)

[9] Vassileva, J.: Dynamic courseware generation: at the cross point of CAL, ITS and au-thoring. In: Proceedings of ICCE, vol. 95, pp. 290–297 (December 1995)

[10] Libbrecht, P., Melis, E., Ullrich, C.: Generating personalized documents using a presen-tation planner. In: ED-MEDIA 2001-World Conference on Educational Multimedia, Hypermedia and Telecommunications (2001)

[11] Karampiperis, P., Sampson, D.: Adaptive learning resources sequencing in educational hypermedia systems. Educational Technology & Society 8(4), 128–147 (2005)

[12] Heraud, J.-M., France, L., Mille, A.: Pixed: An ITS that guides students with the help of learners ' interaction logs. In: 7th International Conference on Intelligent Tutoring Sys-tems, pp. 57–64 (2004)

[13] Ullrich, C., Melis, E.: Complex Course Generation Adapted to Pedagogical Scenarios and its Evaluation. Educational Technology & Society 13(2), 102–115 (2010)

[14] Bikovska, J.: Scenario-Based Planning and Management of Simulation Game: a Review. In: 21st European Conference on Modelling and Simulation, vol. 4 (Cd.) (2007)

[15] Morenoger, P., Sierra, J., Martinezortiz, I., Fernandezmanjon, B.: A documental ap-proach to adventure game development. Science of Computer Programming 67(1), 3–31 (2007)

[16] Carron, J.-M., Thibault, Marty, Jean-Charles, Heraud: Teaching with Game Based Learn-ing Management Systems: Exploring and observing a pedagogical. Simulation & Gam-ing 39(3), 353–378 (2008)

[17] Chang, W.-C., Chou, Y.-M.: Introductory C Programming Language Learning with Game-Based Digital Learning. In: Li, F., Zhao, J., Shih, T.K., Lau, R., Li, Q., McLeod, D. (eds.) ICWL 2008. LNCS, vol. 5145, pp. 221–231. Springer, Heidelberg (2008)

[18] Bénech, P., Emin, V., Trgalova, J., Sanchez, E.: Role-Playing Game for the Osteopathic Diagnosis. In: Kloos, C.D., Gillet, D., Crespo García, R.M., Wild, F., Wolpers, M. (eds.) EC-TEL 2011. LNCS, vol. 6964, pp. 495–500. Springer, Heidelberg (2011)

[19] Vartiainen, P.: On the Principles of Comparative Evaluation. Evaluation 8(3), 371–459 (2002)

[20] Bull, G.G.: The Elicitation Interview. Studies in Intelligence 14(2), 115–122 (1970)

[21] Aubert, O., Prié, Y.: Advene: active reading through hypervideo. In: ACM Hypertext 2005 (2005)

Technological and Organizational Arrangements Sparking Effects on Individual, Community and Organizational Learning

Andreas Kaschig[1], Ronald Maier[1], Alexander Sandow[1], Alan Brown[2], Tobias Ley[3], Johannes Magenheim[4], Athanasios Mazarakis[5], and Paul Seitlinger[6]

[1] University of Innsbruck, Austria
{Andreas.Kaschig,Ronald.Maier,Alexander.Sandow}@uibk.ac.at
[2] University of Warwick, United Kingdom
alan.brown@warwick.ac.uk
[3] Tallinn University, Estonia
tley@tlu.ee
[4] University of Paderborn, Germany
jsm@uni-paderborn.de
[5] FZI Research Center, Germany
mazarakis@fzi.de
[6] Graz University of Technology, Austria
paulchristian.seitlinger@edu.uni-graz.at

Abstract. Organizations increasingly recognize the potentials and needs of supporting and guiding the substantial individual and collaborative learning efforts made in the work place. Many interventions have been made into leveraging resources for organizational learning, ultimately aimed at improving effectiveness, innovation and productivity of knowledge work in organizations. However, information is scarce on the effects of such interventions. This paper presents the results of a multiple-case study consisting of seven cases investigating measures organizations have taken in order to spark effects considered beneficial in leveraging resources for organizational learning. We collected a number of reasons why organizations deem themselves as outperforming others in leveraging individual, collaborative and organizational learning, measures that are perceived as successful as well as richly described relationships between those levers and seven selected effects that these measures have caused.

Keywords: Community, knowledge maturing, knowledge work, multiple-case study, organizational learning.

1 Introduction

Although many concepts, models, methods, tools and systems have been suggested for enhancing learning and the handling of knowledge in organizations [1], there is only scarce information on the effects of these technological and organizational arrangements on the effectiveness of knowledge work [2-7]. While the share of knowledge work [8] has risen continuously during recent decades [9] and knowledge work

A. Ravenscroft et al. (Eds.): EC-TEL 2012, LNCS 7563, pp. 180–193, 2012.

can be found in all occupations, the question remains open whether we can design IT-supported instruments that create positive effects on knowledge work independent of their field of application.

In this paper, we report on how organizations employ IT and organizational instruments to support knowledge work. We analyze these technological and organizational arrangements as levers that are employed and the effects they achieve. In line with dynamic models of organizational learning and knowledge creation, such as the spiral model of knowledge creation [10], the 4I framework [11] or the knowledge maturing (KM) model [12], we put a special focus on how organizations deal with critical knowledge they develop and maintain across the individual, community and organizational level. With a multiple case study that focused on organizations perceiving themselves as successful in sparking positive effects on learning on an individual, community and organizational level, we aim at answering the research question: What successful measures (IT-based and organizational) are applied to evoke positive effects on learning on an individual, community and organizational level?

In pursuing this aim, we relied on qualitative and interpretive methods, based particularly on observation and face-to-face interviews at the work places of the interviewees. The study was conceptualized as a case study with multiple instances the investigation of which relied on a single, coordinated framework of study topics and a common design. We gained multiple perspectives by interviewing several individuals in each case study that together provided rich empirical material on interventions into three levels of learning, traversed when knowledge is passed from individuals' learning and expressing of ideas over informal collectives such as communities to the formal level of organizations.

2   Individual, Community and Organizational Learning

Knowledge is socially constructed and part of workplace practices. Therefore, top down approaches that view knowledge as a decontextualized entity have often met with little success. Thus, it is not surprising that many theories and models start out with learning at the level of individuals. Personal knowledge is defined as the contribution, individuals bring to situations which enables them to think, interact and perform [13]. The "objects" of individual learning include personalized versions of public codified knowledge, everyday knowledge of people and situations, know-how in the form of skills and practices, memories of episodes and events, self-knowledge, attitudes and emotions. The development of practice is reflective, forward-looking and dynamic and seems to work best within a culture that acknowledges the importance of developing practice, expertise and analytical capabilities in an inter-related way so as to be able to support the generation of new forms of knowledge. Those involved in such developments need to have a continuing commitment to explore, reflect upon and improve their practice [14]. At the same time, they play a key role in generating new knowledge and applying it when working in teams with colleagues with different backgrounds and different kinds of expertise [15].

A number of models that connect individual learning at the workplace with (supposed) effects on a community and organizational level have been proposed and discussed in the literature [16]. The 4I framework [11] conceptualizes organizational learning as a dynamic process. It consists of four categories (4Is) of social and psychological processes on different levels: intuiting (individual), interpreting (individual), integrating (group), and institutionalizing (organizational). One premise of the model is that organizational learning includes a tension between exploration (assimilating new learning) and exploitation (using what has already been learned).

The concept of Communities of Practice (COP) has been established as a linking mechanism between individual practice and organizational learning [17]. Individual and collective learning is to a large extent informal based on a continuous negotiation of meaning that takes place within the community [18]. This negotiation captures the way that individuals in the community make sense out of their experiences. Meaning of their experiences is not defined by any external authority, but it is constructed in the COP and constantly negotiated through collaborative processes. Recently, the authors suggested a number of community tools that support these processes [19].

The spiral model on organizational knowledge creation [10], [20] claims that knowledge creation is a social process moving and transforming knowledge from the individual level into communities of interaction that cross organizational boundaries.

The KM model [12] frames a similar stance on this process as goal-oriented learning on a collective level. The model describes knowledge development as a sequence of phases. In its early phases, expressing ideas and appropriating ideas, the model is concerned with learning on an individual level. Similar to [17], the KM model views communities as the main connection between the individual and organizational level in which learning takes place in informal activities, termed distributing phase, yet, might also involve artifacts such as boundary objects [21], created in the formalizing phase, specifically if the boundaries of such communities should be crossed. Communities sometimes also provide the social constellation of choice for ad-hoc training and piloting of new products, processes or practices. Finally, on the organizational level, the model depicts formal training as well as institutionalizing, and ultimately, standardizing. The KM model has been iteratively developed based on evidence gained in an ethnographically-informed study of KM processes and the individual and collaborative activities that happen at the workplace [22] and a survey of a large sample of European companies [23]. The latter study was also used to identify successful examples for KM and companies that were particularly successful.

The present study was conducted in order to gain an in-depth understanding of why and how these cases were successful. This was done mainly by introducing interviewees to the model, guiding them in relating the model with phenomena in their own organizations and then conducting the interview based on these perceived and concrete occurrences of KM. Because the model resonated well with the respondents as the previous studies on the KM model [22], [23] and a pretest had shown, this allowed us to elicit rich stories about concrete cases of KM that were perceived as successfully fostered by deliberately applied technological and organizational arrangements. The results of this analysis should act as a guideline for organizations willing to support KM appropriately.

3 Study Design and Data Collection

To study technological and organizational arrangements, we agreed upon the following topics as the main focus: (1) reasons for better *performing* KM than others; (2) *organizational measures* that are deemed to support KM; (3) ways to overcome *barriers to KM*; (4) *IT-oriented measures* that are deemed to support KM. As the four topics were deemed rather complex and context-specific, we chose the case study approach [24], [25]. For detailed in-depth data collection, multiple sources of information were used, in our cases interviews and observations as well as documents and reports [26]. We followed a holistic multiple-case study approach which is deemed to be more robust than a single-case study design and, furthermore, provided evidence is often seen to be more compelling [25], [27].

We followed a purposeful sampling approach [28] by choosing organizations identified as successful through our previous studies. The unit of analysis is individuals that work and learn in a collective towards a common goal. The plural is important as we did not focus on a single person, but according to the definition of KM on goal-oriented learning on a collective level. This allowed us to triangulate practices within the targeted collective of people and to get a multi-faceted picture of the studied organization. Six European organizations and one network of organizations were investigated. Between two and 15 representatives took part in each case study. The studied cases varied with respect to country, size and sector (see table 1).

Table 1. Studied cases, for classification criteria see OECD and EUROSTAT [29]

Case	Sector	Size	Country	No. of Participants
C1	Service	small	Austria	3
C2	Service	large	Germany	5
C3	Service	large	Poland	7
C4	Service	[network]	United Kingdom	14
C5	Industry	large	Germany	15
C6	Industry	large	Germany	5
C7	Industry	large	Germany	7

Each case study concentrated on collectives of individuals working across departments, subsidiaries or even organizations. To get access to these collectives, interviewees were selected based on a snowball sampling [28]. We defined criteria that interviewees needed to fulfill which helped us to gain valuable data from people who had a broad and informed view about their organization. Interviewees had to have, e.g., a high share of knowledge work; experience in different organizational settings, access to a variety of technical systems; good command of conceptual and management tasks; and strong communication, coordination and cooperation needs.

Data collection was done face-to-face at the workplaces of participants wherever possible. This allowed for direct observation of phenomena in the context of participants' workplaces [28]. We intended (1) to provide cues for participants about important facets surrounding support of KM by technological and organizational arrangements (e.g. by observable artifacts in the participants' work environments), (2) to support the researchers' understanding of the work environments of participants as well as (3) to facilitate joint meaning-making of the technological and organizational arrangements between participant and researcher. To facilitate data collection on the agreed four topics, an interview guideline was developed and adopted by case study teams investigating different organizations. The first page of the interview guideline supported the interviewer in explaining the concept of KM and contained a figure depicting the KM model [12] which was discussed with the interviewees in the context of their organizations. The second page was dedicated to the four topics which were investigated in the sequence described above. The semi-structured interviews were recorded, if allowed, transcribed and analyzed with qualitative content analysis. Besides interviewing, in some cases further methods for data collection, such as focus groups [28], were employed. Between the authors, several face-to-face meetings and teleconferences provided opportunities to exchange lessons learned on case selection, data collection and data analysis.

After conducting the field work, each team analyzed the collected data and created an individual case report structured according to a common template. Once the main findings were summed up and each case study team was aware of results from all case studies, we jointly developed cross-case conclusions, again in a series of teleconferences and face-to-face meetings.

4   Levers, Effects and Their Relationships

We triggered a reflection on certain preconditions that the represented organization meets for performing KM successfully by asking participants about reasons for performing KM better than others in the first topic of the interview. In multiple rounds of joint data analysis [25], we distilled seven effects of interventions for learning on an individual, community and organizational level (see table 2).

Table 2. Effects

Individual	*Increased willingness to share knowledge (C2, C3, C4, C5, C6, C7).* Comprises a communicative environment as well as an attitude of being open-minded towards colleagues' requests and an active provision of knowledge possibly needed by others.
	Openness to change (C1, C3). Describes an organizational culture that prevents the development of permanent consensus. Comprises defrozen thought patterns, overcome rigidity of thinking and sticking in convenient but ineffective action patterns.
	Positive attitude towards knowledge maturing itself (C3, C5, C6, C7). Employees across the organizational hierarchy reflect on potential benefits of putting efforts into KM which is deemed to depend greatly on employees involved in daily work activities and their attitude towards and reflexiveness about it.
Community	*Improved accessibility of knowledge (C3, C4, C6, C7).* Quick accessibility and easy retrieval of knowledge is deemed to positively affect the goal oriented and non-redundant transfer of knowledge within and across communities.
	Strengthened informal relationships (C1, C2, C3, C4, C6). Denote personal ties between colleagues that are usually used to circumvent or shortcut formal procedures in the hierarchical structure. Informal relationships help collaborative reflections upon learning processes and distribution of ideas and information about current activities.
Organization	*Availability of different channels for sharing knowledge (C1, C4, C6, C7).* Availability of different methods or systems used for sharing knowledge that can be related to IT or to organizational measures.
	Improved quality of workflows, tasks or processes (C3). Process improvement instruments, such as best practice process-descriptions, are applied in order to gain improvements with respect to cost, time and quality.

By analyzing the measures that respondents perceived as causing these effects, we surfaced the levers in the sense of technological and organizational arrangements positively impacting the performance of KM. As a result of the cross-case analysis, we structured these levers into five groups (see table 3): (1) soliciting, i.e. levers that trigger employees to provide solutions and ideas for addressing issues present in the organization; (2) guiding, i.e. levers that increase awareness for best practices and standard operating procedures and/or influence the direction or quality of knowledge work; (3) converging, i.e. endorsement of further development of a topic by legitimating to allocate time to an initiative or project where knowledge stemming from different origins can be amalgamated; (4) regular sharing, i.e. the recurring endorsement of sharing knowledge in a defined procedure that could be implemented as a recurring event or as a permanent measure; (5) transferring, i.e. support transmission of knowledge from one group or community to another, for re-use.

Table 3. Levers

Soliciting	*Acting as 'claimant' (C1, C7).* A new idea often needs support from someone creating a demand for it and pulling it towards realization. Ideally, this role is performed by a person who has the capability and authority to stress his/her demand and thus can be a proponent for the new idea. *Fostering competition-based idea management (C3).* Employees present contributions electronically or in an exhibition and the best ideas are awarded. Thus, all members of the organization will learn about other projects and ideas. *Maintaining a best practice database (C3).* A database providing a collection of best practice process-descriptions that have been approved according to a quality assurance concept aimed at improving tasks or processes.
Guiding	*Enabling awareness and orientation (C2, C3).* Continuously documenting (all) business processes and thereby providing transparency for these processes. This is in line with requirements imposed by quality management initiatives. *Offering guidance by supervisors and management (C3, C7).* Raising employees' awareness of knowledge management in general and integrating KM-related topics into the process of management by objectives. *Performing benchmarks (C2, C3, C5, C7).* By performing benchmarks, different units within and across organizations are compared against each other to identify gaps, foster competition foster discussion on possible future measures aiming at an improvement. *Providing organizational guidelines (C2, C3, C7).* Shared sets of rules regarding common ways of performing knowledge work. This includes, e.g., organizing and naming documents and folders on file shares, for approving business process related documents.
Converging	*Allocating competence in projects* (C1, C3, C7). People with different backgrounds working together are perceived as very fruitful because time and legitimation of action is provided which empower project teams to pursue project goals and introduce changes. *Conducting workshops on specific topics (C2, C3, C7).* Topic-oriented meetings where selected employees are brought together to drive a specific topic or to focus on developing a specific skill-set. *Enabling collaborative learning* (C1, C4, C5, C6, C7). Providing tools and services for creating, presenting, discussing, tagging and collecting resources enabling synchronous and asynchronous sharing of information.

Table 3. (*continued*)

Regular sharing	*Conducting regular (team) meetings (C2, C3, C5, C7).* Knowledge transfer is supported by an established procedure of regular team meetings, ensuring fast and target group oriented diffusion of knowledge in both directions along the hierarchy. *Offering formal trainings at regular intervals (C2, C3, C7).* Topics of training courses that are typically selected with respect to identified gaps between employees' competence profiles and needs of organizational units or projects the employee works for. *Providing a flexible working space (C2).* Employees who want or need to work together (e.g. working on a new product or for discussing issues) have the possibility to choose their working place and thereby increase communication effectiveness.
Transferring	*Employing technology-enhanced boundary objects (TEBOs) (C3, C4).* TEBOs (i.e. software-based interactive digital media, which support mediating knowledge sharing across organizational boundaries) are conceived as tools which support situated learning. *Fostering communities of practice (C1, C2, C3, C4, C6).* Regular topic-based meetings for exchanging lessons learnt that were created by employees. These communities of interest are mostly based on informal relationships between members. *Fostering reflection by enabling purpose-oriented task groups (C2, C3, C4, C6).* Groups operating at boundaries between different communities help in extending and deepening the communication. Thus, they enable 'boundary crossing' of knowledge. *Improving access to documented knowledge (C1, C2, C6, C7).* Providing better transparency for finding knowledge contained in documents stored on network drives and by creating a "knowledge library" (e.g., in a wiki) easy to access by knowledge workers. *Installing one supervisor for teams in different subsidiaries (C7).* Leading teams responsible for performing similar tasks in different subsidiaries fosters the transfer of knowledge between them, the development of COPs and facilitates benchmarking.

The relationships between levers and effects that we present in the following were interpreted as causal based on the evidence provided by the perceptions of several interviewees across cases and by a number of stories we obtained supporting them. We provide one selected story of one case study for each effect describing the levers perceived to cause the effect in detail. Moreover, we provide further evidence by one short story reflecting a selected additional case where this effect was also observed. Finally, we discuss the lever-effect relationship in the light of related research.

Increased Willingness to Share Knowledge. In C6, *topics are fostered by community of practice meetings.* Attending a high number of such meetings is accepted by employees of this organization as necessary and helpful to create an efficient working environment and a joint understanding. Open discussions are allowed and supported by different tools like forums and blogs. This technological support is strengthened by giving individuals and teams more responsibility for their projects, for example, negotiating various budget allocations, and offers opportunities to discuss more work- and project-related ideas in the forums within communities. This effect was also observed in case C2. Employees who were more willing to answer colleagues' requests and took part in *community of practice meetings, fostered* by the organization, were supposed to perform "better" with respect to KM. These observations are in line with experiences from other big companies, where knowledge-sharing became a part of organizational culture and thus lead to more efficiency [30]. Also the less hierarchical and formal organizational structure leads to the absence of punishment for not following organizational rules and therefore is also beneficial [31].

Openness to Change. In the software company investigated in C1, an internal Wiki acts as a mindtool that reveals and relate thoughts of different people. Originally implemented to *improve access to documented knowledge*, the wiki additionally provides software support that *enables collaborative learning* processes in early phases of software development. When the wiki has been introduced, the management has *acted as a claimant* influencing the employees to externalize their ideas and problem solutions in form of wiki entries. *Workshops on specific topics* have been proven to defreeze organizational thought patterns as they connect employees with different perspectives and opinions. The wiki-based distribution of ideas among organization's members uncovers different perspectives, fostering diverging thinking during work and preparing employees for a constructive discussion of project meetings. In C3, a Wiki and a *competition-based idea management* scheme is used to collect innovative ideas and put these via discussion and reflection into practice. As an effect, the employees' attitudes towards continuous improvement and open mindedness for organizational changes are fostered. De-freezing thought patterns by means of the Wiki is driven by the complementary processes of accommodation and assimilation [32]. Revealing different perspectives in form of Wiki entries positively affects interpersonal conflicts at a cognitive level. If different perspectives of individuals come into conflict, accommodative processes become operative: an existing conception of a particular problem gets extended and differentiated.

Positive Attitude towards Knowledge Maturing Itself. In C7, senior management actively communicated interest in the prospects of KM. Through *guidance by supervisors* the number of KM-related ideas and projects arose. Also the attitude of supervisors and (middle) management of the organization was positively affected, resulting in evolving projects related to the knowledge management. In this respect, senior management enabled middle and lower level managers to *act as claimants* for further development of selected ideas. In C6, a positive attitude towards KM was evident. This has been expressed by staff and senior management. Especially, an effort was made by fostering topics by *conducting community of practice meetings* and by *fostering reflection by enabling purpose-oriented task groups*. An individual's attitude affecting its intention to act is also discussed in literature. Gee-Woo, Zmud [33] show that attitudes affect individuals' intention to share knowledge. They relied on the theory of reasoned action [34] stating that an individual's decision to engage in a specific behavior is determined by its intention to perform that behavior, which is determined also by its attitude towards it.

Improved Accessibility of Knowledge. For C3, it was possible to improve access to documented knowledge by a company-wide Wiki which is available for employees and contains unrestricted information about all business activities of the company. This wiki allows a quick access to information for all employees. A positive effect arose through *allocating competence in projects* appropriately, through fostering topics by *conducting communities of practice meetings* and *fostering reflection by enabling purpose-oriented task groups*. In C6, knowledge bases and Web 2.0 tools were perceived as essential for *enabling collaborative learning* by fostering the exchange of information about project and company-related aspects. The organization *improved access to documented knowledge* by making project-related information accessible to other departments. The accessibility of knowledge is subject to many individual, organizational and technical obstacles [35]. Despite general cultural and hierarchical issues [31], the quality of social networks is an important key factor for the ability to interact with others and therefore enable access to knowledge from others [36].

Strengthened Informal Relationships. The organization of case C2 *provides office spaces for flexible use* for its employees. Employees are encouraged to choose office spaces close to colleagues they need to communicate with often, from whom they want to learn something. Hence, employees got to know more colleagues which lead to an improvement in their social networks and helped building (informal) relationships. Improved communication channels meant quicker and less bureaucratic answers so that employees are able to ask directly for comments on issues. This is also supported by the organization *fostering topics by* allowing employees to *conduct community of practice meetings* that took place between a number of employees working on similar topics and exchanging lessons learnt and best practices. In C7, *workshops on specific topics* are conducted to foster the creation of informal relationships. Supervisors of different departments or subsidiaries meet regularly to identify employees who might have similar interests/roles/tasks. During one-day workshops, a topic is further developed, participants get to know each other and build informal relationships. Informal relationships are generally considered to be important for informal learning in organizations. The success of the activities participation in group activities, working alongside others, tackling challenging tasks and working with clients is mainly responsible for informal learning dependent on the quality of relationships in the workplace [13]. These informal relationships can be seen as individual social capital, which is considered to be the basis of the social capital of the organization [37].

Availability of Different Channels for Sharing Knowledge. In C7, IT-related and organizational measures are performed to provide different channels for sharing knowledge. From an IT perspective, all subsidiaries are connected via a network and employees are equipped with laptops for *improving the access to documented knowledge*, as well as cell phones and software and hardware for conducting voice and video calls via the Intranet and Internet, hence provide *software support for collaborative learning*. From an organizational perspective, knowledge transfer between different levels of hierarchy is supported by an established procedure of *regular team meetings*. Furthermore, *workshops on specific topics* are used as another medium for. In C4, IT-related and cross-organizational measures were used to provide a range of different channels for sharing knowledge. Use of collaborative software and providing spaces for cross-organizational meetings represented the provision of tools and services for creating, presenting, discussing, tagging and collecting resources enabling synchronous and asynchronous sharing of information. Setting up measures aiming at this effect goes well along with the implications drawn by [38] who emphasizes that different channels are needed for sharing different types of knowledge.

Improved Quality of Workflows, Tasks or Processes. In C3, the only case study which provided us with evidence on this effect, there exist a variety of activities to improve the client's business processes according to a Business Process Model and a best practice model. The *best practice database* maintained by the organization provides a collection of process-descriptions that were approved according to quality assurance procedures. The transition (i.e. the outsourcing) of business processes of the organization's clients is organized according to a highly formalized procedure that is based on experiences of former engagements with other clients. For quality analysis of the revised business processes, key performance indicators are provided. These enable *performing benchmarks* across similar business processes at different clients and were used to identify differences in the performance of different projects with regard to efficiency, effectiveness, value, control etc. Using best practices as an instrument for transferring knowledge between individuals in an organization in order to improve organizational processes is also named for example in [1].

5   Discussion and Limitations

An aggregated view of levers and sparked effects is provided in figure 1. The outer columns depict the levers, grouped according to the five dimensions. The middle column presents the seven effects, mapped to individual, community and organizational level. The arrows represent selected relationships between levers and effects and reflect the stories described in section 4. The levers and effects we are suggesting here may be misunderstood as simple cause and effect relationships. The case descriptions in section 4 show, though, that levers form an intricate network of cause and effect relationships, each of which dependent on the other measures that have been taken. In this sense, all measures need to be carefully designed in cooperation with other levers.

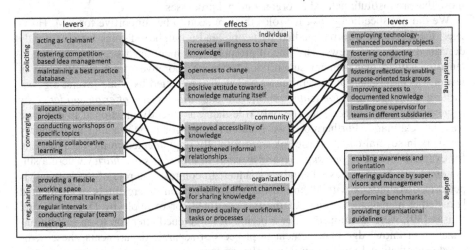

Fig. 1. Levers, effects and relationships

Organizations perceiving themselves as successful support persistent collective learning across individual, community and organizational levels. The levers and arrangements that establish persistent learning across these levels can also be viewed to move a collective such as a community or an entire organization between different poles, e.g., (a) participation and reification, (b) togetherness and separation, (c) individual and group [19], (d) grassroots developments and organizational guidance as well as (e) opening up and filtering. The studied organizations seem to be successful in bridging these typical polarities. For example, cases C1, C3 and C6 illustrate bridging participation and reification (a) when they combine improvements of the accessibility of knowledge through databases or wikis (reification) with formal and informal COP meetings and topical workshops (participation). C7 illustrates bridging togetherness and separation (b) by offering a broad range of synchronous and asynchronous means of communication. And the focus on flexible working and seating arrangements in C2 nicely illustrates flexibly balancing individual and group focus (c).

Concerning the poles *organizational guidance* and *grassroots developments (d)*, various measures of organizational *guidance* had an important role in aligning and structuring organizational practices and processes, and can be a result of formal organizational arrangements, as well as consequence of informal leadership, e.g. through installing best practice guidelines in C3. At the same time, these same organizations strike a balance with measures that allow grassroots developments and the emergence of new ideas, e.g. with idea competition also in C3. Levers and effects on *opening up* and *filtering* (e) also resonate with the polarity between diverging and converging ideas, e.g., in C1 and C3. Knowledge seems to mature along a meandering process between these poles, starting out with opening up for new ideas, filtering those that are handed on to a community, opening up in the community for developing them evolutionarily, filtering those that are formalized into boundary objects, opening up for a competition of good practices identified in several communities and filtering those that are institutionalized as organizational processes.

We did not specifically ask for roles that are seen to be supportive for KM. However, the levers need to be handled by people and a number of roles were explicitly described in the interviews. The role of promoter was mentioned stressing the importance of having management support for levers or, moreover, their involvement in levers was highlighted in several cases and is also reflected by some levers, e.g., *installing one supervisor for teams in different subsidiaries* and *offering guidance by supervisors and management*. The only role that was directly mentioned in one case was the 'claimant'. Furthermore, we found evidence for people acting as boundary spanners in several cases. These roles are formally implemented in the organization, for example in case of *one supervisor for teams in different subsidiaries* where a single employee functions as a boundary spanner. In contrast, the role of a 'claimant' is performed voluntarily without any formal implementation. Interestingly, no dedicated knowledge management roles, of the type outlined for example by Davenport and Prusak [39], were mentioned. After a period of heightened attention to an institutionalization of knowledge management in projects or separate departments, these dedicated organizational units seem to not play an important role in leveraging resources for individual, community and organizational learning. Instead, every employee was seen to be responsible for handling knowledge efficiently and differences between this egalitarian take on knowledge management can be attributed to the primary roles that employees play with respect to the business processes and work practices performed in the organisations.

Although we relied on a sound method and compared the results in a comprehensive cross-case analysis, a few limitations need to be acknowledged. Generally, the limitations are in line with those of comparable empirical studies using purposeful and convenient sampling [28], interviews and observations for data collection and a qualitative methods for data analysis. As the number of seven cases is low, the results of the study are not representative. However, the topics of this study are developed based upon results of a previous study, which involved 139 organizations throughout Europe [23]. Each case study aimed at (parts of) organizations and only a limited number of participants could take part in the study. In this respect, the participants' personal scopes (e.g., responsibilities, interests) may have influenced their perceptions. However, even selecting one person representing a whole organization is a common practice in business and management studies [40]. We relied on at least two interviews per case and selected only interviewees who had a good command of knowledge and learning management in their organization and had gained experience through work being based on offering and applying expertise in different organizational settings. By following these selection criteria, we ensured to gain multiple perspectives on the state-of-play of performed or planned levers to positively affect KM.

6   Conclusions and Future Work

This paper presents the results of seven case studies from four countries using a common set of instruments in order to explore potentials of deliberately applicable technological and organizational arrangements and perceived effects of these levers on individual and organizational learning. The validity of these claims rests on selecting cases that were previously identified as perceiving themselves as particularly successful. The strength of rich stories gathered in interviews conducted directly on the work places of carefully selected multiple individuals per setting is considered a vital aspect for understanding KM processes. Yet the focus on seven cases means that the ways the processes operate in the different contexts are necessarily underplayed. This provides an avenue for future research testing on the one hand the validity of the levers and effects across organizational settings and on the other hand investigating what contextual factors explain differences in the effectiveness of technological and organizational arrangements between organizational settings. The stories report on levers and the effects they spark on learning on an individual, community and organizational level and thus help organizations to select concrete measures to improve individual to organizational learning that are postulated as beneficial if not necessary in a number of theories and models [10-12]. The identification of a temporal order of how to introduce such arrangements of levers that fit well together and ideally intensify their positive effects as well as more in-depth knowledge about how to navigate communities and organizational knowledge bases between the identified poles are further encouraging aspects to be covered in future work.

Acknowledgement. This work was co-funded by the European Commission under the Information and Communication Technologies (ICT) theme of the 7th Framework Programme (FP7) within the Integrating Project MATURE (contract no. 216356).

References

1. Alavi, M., Leidner, D.E.: Review: Knowledge Management and Knowledge Management Systems: Conceptual Foundations and Research Issues. MIS Quarterly 25(1), 107–136 (2001)
2. Drucker, P.F.: Landmarks of Tomorrow. Harper, New York (1959)
3. Kelloway, E.K., Barling, J.: Knowledge Work as Organizational Behavior. International Journal of Management Reviews 2(3), 287–304 (2000)
4. Davis, G.B.: Anytime/Anyplace Computing and the Future of Knowledge Work. Communications of the ACM 45(12), 67–73 (2002)
5. Schultze, U.: On Knowledge Work. In: Holsapple, C.W. (ed.) Handbook on Knowledge Management 1 - Knowledge Matters, pp. 43–58. Springer, Berlin (2003)
6. Thomas, D.M., Bostrom, R.P., Gouge, M.: Making Knowledge Work in Virtual Teams. Communications of the ACM 50(11), 85–90 (2007)
7. Arthur, M.B., DeFillippi, R.J., Lindsay, V.J.: On Being a Knowledge Worker. Organizational Dynamics 37(4), 365–377 (2008)
8. Blackler, F.: Knowledge, Knowledge Work and Organizations: An Overview and Interpretation. Organization Studies 16(6), 1021–1046 (1995)
9. Wolff, E.: The Growth of Information Workers. Communications of the ACM 48(10), 37–42 (2005)
10. Nonaka, I., Takeuchi, H.: The Knowledge Creating Company. In: How Japanese Companies Create the Dynamics of Innovation. Oxford University Press, Oxford (1995)
11. Crossan, M.M., Lane, H.W., White, R.E.: An Organizational Learning Framework: From Intuition to Institution. Academy of Management Review 24(3), 522–537 (1999)
12. Maier, R., Schmidt, A.: Characterizing Knowledge Maturing: A Conceptual Process Model for Integrating E-Learning and Knowledge Management. In: 4th Conference Professional Knowledge Management (WM 2007). GITO, Berlin (2007)
13. Eraut, M.: Informal Learning in the Workplace. Studies in Continuing Education 26(2), 247–273 (2004)
14. Schön, D.A.: Educating the Reflective Practitioner. In: Toward a New Design for Teaching and Learning in the Professions. The Jossey-Bass Higher Education Series. Jossey-Bass, San Francisco (1987)
15. Engeström, Y.: Training for Change. In: New Approach to Instruction and Learning in Working Life, vol. XI, p. 148 S. International Labour Office, Geneva (1994)
16. Crossan, M.M., Maurer, C.C., White, R.E.: Reflections on the 2009 AMR Decade Award: Do We have a Theory of Organizational Learning? Academy of Management Review 36(3), 446–460 (2011)
17. Lave, J., Wenger, E.: Situated Learning: Legitimate Peripheral Participation. Learning in doing. Cambridge University Press (1991)
18. Wenger, E.: Communities of Practice: Learning, Meaning, and Identity. Learning in doing. Cambridge Univ. Press, Cambridge (1998)
19. Wenger, E., White, N., Smith, J.D.: Digital Habitats. In: Stewarding Technology for Communities. Cpsquare, Portland (2009)
20. Nonaka, I., Konno, N.: The Concept of "Ba": Building a Foundation for Knowledge Creation. California Management Review 40(3), 40–54 (1998)

21. Star, S.L., Griesemer, J.R.: Institutional Ecology, 'Translations' and Boundary Objects: Amateurs and Professionals in Berkeley's Museum of Vertebrate Zoology, 1907-39. Social Studies of Science 19(3), 387–420 (1989)
22. Barnes, S.-A., et al.: Knowledge Maturing at Workplaces of Knowledge Workers: Results of an Ethnographically Informed Study. In: 9th International Conference on Knowledge Management (I-KNOW 2009), Graz, Austria (2009)
23. Kaschig, A., Maier, R., Sandow, A., Lazoi, M., Barnes, S.-A., Bimrose, J., Bradley, C., Brown, A., Kunzmann, C., Mazarakis, A., Schmidt, A.: Knowledge Maturing Activities and Practices Fostering Organisational Learning: Results of an Empirical Study. In: Wolpers, M., Kirschner, P.A., Scheffel, M., Lindstaedt, S., Dimitrova, V. (eds.) EC-TEL 2010. LNCS, vol. 6383, pp. 151–166. Springer, Heidelberg (2010)
24. Sake, R.E.: Qualitative Case Studies. In: Denzin, N.K., Lincoln, Y.S. (eds.) The Sage Handbook of Qualitative Research. Sage, Thousand Oaks (2005)
25. Yin, R.K.: Case Study Research - Design and Methods, 4th edn. Applied Social Research Methods Series, vol. 5. Sage, Thousand Oaks (2009)
26. Creswell, J.W.: Qualitative Inquiry and Research Design, 2nd edn. Choosing Among Five Approaches, vol. XVII, p. 395 S. Sage, Thousand Oaks (2007)
27. Herriott, R.E., Firestone, W.A.: Multisite Qualitative Policy Research: Optimizing Description and Generalizability. Educational Researcher 12(2), 14–19 (1983)
28. Patton, M.Q.: Qualitative Research & Evaluation Methods, 3rd edn. Sage, Thousand Oaks (2002)
29. OECD, EUROSTAT: Oslo Manual - Guidelines for Collecting and Interpreting Innovation Data, 3rd edn., p. 162. OECD Publishing (2005)
30. O'Dell, C., Grayson, C.J.: If Only We Knew What We Know: Identification and Transfer of Internal Best Practices. California Mgmt. Review 40(3), 154–174 (1998)
31. Michailova, S., Husted, K.: Knowledge-Sharing Hostility in Russian Firms. California Management Review 45(3), 59 (2003)
32. Cress, U., Kimmerle, J.: A Systemic and Cognitive View On Collaborative Knowledge Building With Wikis. International Journal of Computer-Supported Collaborative Learning 3(2), 105–122 (2008)
33. Gee-Woo, B., et al.: Behavioral Intention Formation in Knowledge Sharing: Examining the Roles of Extrinsic Motivators, Social-Psychological Forces, and Organizational Climate. MIS Quarterly 29(1), 87–111 (2005)
34. Fishbein, M., Ajzen, I.: Belief, Attitude, Intention, and Behavior: An Introduction to Theory and Research, p. 578. Addison-Wesley, Reading (1975)
35. Riege, A.: Three-Dozen Knowledge-Sharing Barriers Managers Must Consider. Journal of Knowledge Management 9(3), 18–35 (2005)
36. Baron, R.A., Gideon, D.M.: Beyond Social Capital: How Social Skills Can Enhance Entrepreneurs' Success. The Academy of Management Executive (1993-2005) 14(1), 106–116 (2000)
37. Inkpen, A.C., Tsang, E.W.K.: Social Capital, Networks, and Knowledge Transfer. The Academy of Management Review 30(1), 146–165 (2005)
38. Christensen, P.H.: Knowledge sharing: moving away from the obsession with best practices. Journal of Knowledge Management 11(1), 36–47 (2007)
39. Davenport, T.H., Prusak, L.: Working Knowledge. In: How Organizations Manage What They Know. Harvard Business School Press, Boston (1998)
40. Bryman, A., Bell, E.: Business Research Methods, 2nd edn. Oxford Univ. Press, Oxford (2007)

The Social Requirements Engineering (SRE) Approach to Developing a Large-Scale Personal Learning Environment Infrastructure

Effie Lai-Chong Law[1], Arunangsu Chatterjee[1], Dominik Renzel[2], and Ralf Klamma[2]

[1] Department of Computer Science, University of Leicester, UK
{elaw,A.Chatterjee}@mcs.le.ac.uk
[2] Chair of Computer Science 5 - Information Systems, RWTH Aachen University, Germany
{renzel,klamma}@dbis.rwth-aachen.de

Abstract. In this paper we reflect on the limitations of applying traditional requirements engineering approaches to the development of a large-scale PLE infrastructure, which is precisely the aim of a technology-enhanced learning project called ROLE. The Social Requirements Engineering (SRE) approach has been proposed as an appropriate alternative. The SRE process is grounded in an agent- and goal-oriented conceptual model. The implementation of SRE prototypes was structured with a five-staged requirement lifecycle: elicitation, negotiation, selection, development and feedback. We report results of the preliminary evaluation of the prototypes and lessons learnt. Several relevant issues have been identified, including the lack of a consensual understanding of key concepts, lurking within Community of Practices (CoP), and cultural differences. Possible solutions are proposed to address the issues, including templates, mandatory voting and prioritisation model.

Keywords: Social requirements engineering, Personal learning environments, Communities of practice, Web 2.0, Prioritization model, Long tail, Voting.

1   Introduction

In charting the roadmap for the field of Requirement Engineering (RE) a decade ago, Nuseibeh and Easterbrook [1] identified three major ideas upheld in RE in the 1990s: understanding the organizational and social context, modelling stakeholders' goals, and resolving conflicting requirements. Built upon these trends, the authors defined six challenges that would face RE in the first decade of the new millennium. The key notions then addressed were: modelling properties of the context of use; integration of formal (e.g. Z notation) and informal requirements elicitation techniques (e.g. contextual inquiry [2]); factors influencing requirements prioritisation and evolution; modelling non-functional requirements; reusability of requirements models; training for requirements practitioners. Despite the lapse of two decades, these trends and challenges remain relevant to today's work in RE, though each of them has progressed to a different extent (e.g. [3][4]). Amongst these somewhat interdependent issues, the one that has become more and more challenging is modelling and analysing context of use, given the ever increasingly heterogeneous stakeholders and

A. Ravenscroft et al. (Eds.): EC-TEL 2012, LNCS 7563, pp. 194–207, 2012.

the associated diverse contexts in which they interact with products and services of interest. This issue is well exemplified by an emerging research topic in the field of technology-enhanced learning, namely Personal Learning Environments (PLE).

Coincidentally, around the time when the RE roadmap was published the notion of PLE was conceived (e.g. [5]). Rooted in the idea of web-based learning and teaching, PLE can be seen as an advance beyond the traditional Learning Management Systems (LMS). In comparison, the former can provide users with higher flexibility and stronger personalisation than the latter, thanks to the mash-up technology that facilitates the integration of different web-based contents, services and applications based on personal needs and preferences [6]. Notwithstanding years of research efforts, a consensual definition of PLE is yet to be reached. Some argue that PLEs are simply a new approach to learning; they are not technical but rather philosophical and pedagogic [8][9]. Based on interactions with the technology-enhanced learning (TEL) community through workshops, surveys and literature reviews[1], we propose to define PLE as *a pedagogy-driven infrastructure that facilitates learners to integrate distributed contents, services, tools and contacts based on personal goals and preferences, thereby enabling them to control their own learning and to connect different learning contexts with the support of communities.* This broad definition implies that the RE process in developing and sustaining a PLE infrastructure is inherently challenging, given the tremendous scope of learners, contexts and artefacts that can be included in use scenarios. Multifaceted modelling and analysis (i.e. organizational, behavioural, domain-specific, quality-based) is deemed necessary and extremely resource demanding.

In exploring alternative cost-efficient approaches to eliciting and analysing requirements from a diversity of stakeholders, traditional RE techniques such as questionnaire, interview and focus group could be used in the early phase of development, but did not scale well in later phases. Owing to the continuously changing and evolving nature of PLEs and hence the requirements stated by multiple highly diverse learner communities, there was a need of a new approach to managing these requirements. The Social Requirements Engineering (SRE) [10] approach with its strength being derived from the Community of Practice [11] perspective is regarded as promising for a community-aware approach that is even suited for very specific requirements from communities in the Long Tail [12].

In the ensuing text, we report the limitations of traditional RE when applied to the emerging PLEs (Section 2). We postulate that the limitations identified can be resolved by the emerging SRE of which the theoretical background is described in Section 3. We then present our first implementation iteration of the SRE process (Section 4). Next we report initial evaluation results (Section 5). Finally, we conclude our paper and describe an outlook to future work (Section 6).

2   Limitations of Traditional Requirements Engineering for PLEs

ROLE (Responsive Open Learning Environment[2]) is a running TEL project that aims to deliver a PLE infrastructure. As the main instrument of requirements elicitation,

---

[1] http://edutechwiki.unige.ch/en/Personal_learning_environment
[2] http://www.role-project.eu/

refinement and evaluation of the project's goals, five international test-beds located in two European countries (Germany, UK) and in China are involved in ROLE. They comprise three academic and two industrial partners. They were deliberately chosen for maximizing diversity in terms of learning domain and motivation, cultural and professional background as well as technical literacy. Furthermore, individual test-beds instantiate a form of transition with learners moving from one institution to another in their professional career. Hence, a general requirement to the intended PLE infrastructure is to render the transition process as seamless as possible. Specifically, ROLE is not about one PLE but rather an interoperable platform, which allows services and service bundles to be mashed-up and re-used by very different learner communities across different institutions. It is required to understand and fulfil their individual and sometimes contradicting needs and priorities in order to develop the infrastructure and to customise tools to support community needs and practices. One phenomenon we observe regularly in this context is the role of community diversity in influencing the expression of requirements for the same technical infrastructure. One prominent example is *privacy,* which can be perceived very differently by different cultural groups. While German institutions have to be guaranteed by law that private data will not be visible to others, students from the test-bed in China give a high priority for being recognised by their teachers, therefore explicitly waiving privacy.

In the early stages of the project a multi-method requirements elicitation approach was used to overcome the difficulties in validating requirements from such heterogeneous learner communities. Some of the key methods included questionnaires, focus groups, workshops, observations, and interviews. While direct communications among the test-bed participants (learners, teachers, administrators), researchers, and developers as well as traceable documentation thereof were a critical success factor for the RE process, it required huge efforts of all the involved parties, especially when there was no appropriate RE software tool supporting the processes. Through the use of these traditional methods a list of preliminary requirements could be negotiated and prioritised by the participants. Some of these methods were also found to be very useful in validating the identified requirements with external stakeholders from the TEL community. However, it was soon realised that managing the RE process with such traditional approaches was impractical and unsustainable for the project.

Although requirements were captured and discussed on an ongoing basis in the project, there was no tool for continuously tracing requirements and their realisation. Requirements were collected and refined for the test-beds in each small development subproject. However, developers only received snapshots of current requirements for all test-beds after the requirements elicitation process was finished. Obviously, those test-beds with the most interesting and clearly communicated requirements and with the highest potential for productive applications could receive strongest attention from the project development team, which consisted of partners from different companies and academic institutions. Consequently, less articulated voices from the other test-beds suffered the risk of being neglected, leading to the *long tail* problem [21].

Furthermore, the traditional RE techniques could involve only a relatively small number of participants but still consume quite a lot of resources. Scaling up to higher orders of magnitude regarding different users and developers in multiple communities with even more requirements will definitely render traditional RE techniques unfeasible. In other words, the traditional RE approaches lack scalability. Another challenge is that the traditional notion of representative users is no longer applicable and needs to be replaced by huge, diverse and distributed communities of users. To capture specific contextual needs of different communities, it is necessary to reach and involve the long tail of the stakeholder communities in the RE process from an early stage of the project. To deal with a situation where the user and the product/service may continuously evolve and adapt depending on the context, a community-oriented approach for early phase RE involving stakeholder communities on a continuous and flexible basis was required. It has been envisioned that with such an approach user requirements can be captured in a community-aware manner. This vision has been driving our research efforts in exploring the emerging RE approach - Social Requirement Engineering (SRE) - of which the theoretical background is described subsequently.

3   Theoretical Background of Social Requirements Engineering

A PLE infrastructure should be designed to enable the combination of learning services by potentially large and diverse learner communities under one roof. A concrete architecture intended to provide community support on this basis requires particular flexibility. A community needs to be able to observe itself, to analyse and maybe even simulate its behaviour in order to evolve its rules of cooperation and to continuously identify and adapt its requirements. The above considerations originate from an operational theory of media - *Transcriptivity Theory* [13], which describes the operational semantics of media artefacts founded on the three basic operations: transcription, localization and addressing. The transcriptivity theory was incorporated into the web-based community software architecture known as ATLAS (Architecture for Transcription, Localization, and Addressing Systems) [14]. In ATLAS, scalable as well as interoperable repositories support networked communities with web service technologies for multi-media content and metadata management. In its reflective conception, ATLAS-based community information systems are tightly interwoven with a set of media-centric self-monitoring tools for the communities. The whole process starts with an initial assessment of community needs. Based on these needs, a socio-technical system is developed, which not only supports the community but also changes the socio-technical context, which in turn generates new needs to be re-assessed and realised. In that sense, RE can be considered as a continuous adaptation process that will definitely require an initial set of impulses in the beginning, but is intended to become self-sustaining eventually.

It is challenging to conceptualise RE for a web-based learner community of which both membership and supporting infrastructure are highly dynamic. It entails a thorough understanding of factors potentially influencing the behaviour of members of such a community. This inquiry is grounded in the notion of Communities of Practice (CoP) [11]. A CoP is a group of people who share a concern or a passion and interact regularly to learn from each other. Individual learning in a CoP is mainly based on "legitimate peripheral participation" [15]. During the participation process, an individual might enter the community as a beginner at the periphery and then gain a more central position over time by the acquisition of cognitive apprenticeship. This acquisition process leads to an intensified inclusion into the social practice of the community. An individual's learning is inherent in the process of social participation in the CoP. Knowledge and learning in a CoP are not abstract models but relations "between a person and the world" [16] or "among people engaged in an activity" [17]. Learning is based on this process of inclusion of outsiders, who gradually become insiders over time in the common practice. Communities of practice themselves can be seen as "shared histories of learning" [11]. In the context of RE, learning refers to the identification of requirements for improving effectiveness and efficiency of community practices, their realisation and adaptation to the resulting new socio-technical context.

According to Wenger [11], three dimensions characterise a CoP: (i) *Mutual engagement* (ME): Community members are required to engage in interactions within their community; (ii) *Joint enterprises* (JE): A common goal of a CoP binding members together is the result of a collective process of negotiation, which reflects the full complexity of mutual engagement; (iii) *Shared repertoire* (SR): Communal re-sources include routines, words, stories, gestures, symbols, genres, actions, tools, ways of doing things, and concepts.

RE should thus be a substantial and recurring part of community practice to constantly adapt to new contexts in terms of optimising practices, for instance, by applying or creating respective technologies and tools. Furthermore, communities are influenced by various external factors called *disturbances*. These disturbances may have negative, positive or neutral influence on the processes within the community. These disturbances keep learning processes alive [18]. Without disturbances, communities are endangered of getting into social or cognitive lock-in situations [19]. As a matter of fact, disturbances come from outside and change the community inside. Thus, the evolvement of communities takes place in a lifelong loop. To map learning communities participating in RE to the CoP concepts, we need a theory that explains socio-psychological aspects, which influence community members through relations between human agents, technologies and resources. Hereby, Actor-Network-Theory [20], according to which no distinction is made between human and non-human actors, can be adopted. Such a non-differentiation between people and technologies intertwines actions, influences, and results of actions.

For the conceptual modelling of our SRE approach, we use the i* framework [21] as a powerful goal and agent oriented framework for modelling community driven processes. i* enables the description of relations between actors in frames of a particular socio-technical system in a clear way [22] and focuses on motives, interests, and options of an actor that play a role in achieving particular goals with the help of the system under scrutiny. A strategic rationale model of the Social Requirements Engineering approach is presented in Fig. 1. The main idea of the SRE approach is to support the constant adaptation process in CoPs with a set of services for expression, tracing, negotiation, prioritization, and realisation of explicit community requirements, leading to a *bazaar-like* SRE environment (or a *requirements bazaar*). Specifically, the approach aims to engage end-users and developers in a negotiation process on the realisation of requirements.

In the realm of TEL, we face an imbalance between many end-users stating valuable requirements and only few developers capable of realising those requirements. Thus, one of the core parts of a requirements bazaar is a service for supporting informed decisions on selecting requirements that are most likely to create

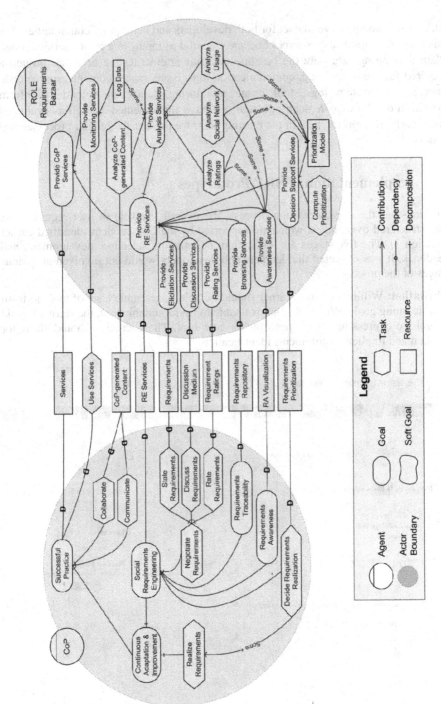

Fig. 1. i* Strategic Rationale Model of the SRE approach [21]. Two main agents: Community of Practice (CoP) and ROLE Requirements Bazaar.

added value and positive impact for both developers and benefitting communities. For this purpose, various data sources become essential as inputs to a prioritisation model. Plain user ratings are only the baseline. Various artefact-testing activities around a required feature or tool (e.g. communication logs, issue tracker contents, source code management system logs, tool usage monitoring data) can potentially inform prioritisation. To explore such potential, we have implemented as well as evaluated some early SRE prototypes. In the ensuing text, we present the related processes and results.

4   Implementation of SRE Prototypes

To kick-start the implementation process, the overall lifecycle of user requirements was structured over stages, which are supported by the CoP services identified earlier (Section 3). The five stages are: elicitation, negotiation, selection, development, and feedback. It was expected that learners and developers would get involved at various stages of the process.

Elicitation: Within a PLE, a learner should be able to assemble a set of tools to fulfil their learning goals. When a learner is unable to find a suitable tool, the learner should be able to express her or his need for the new tool. The elicitation should allow for explicit and implicit requirements identification.

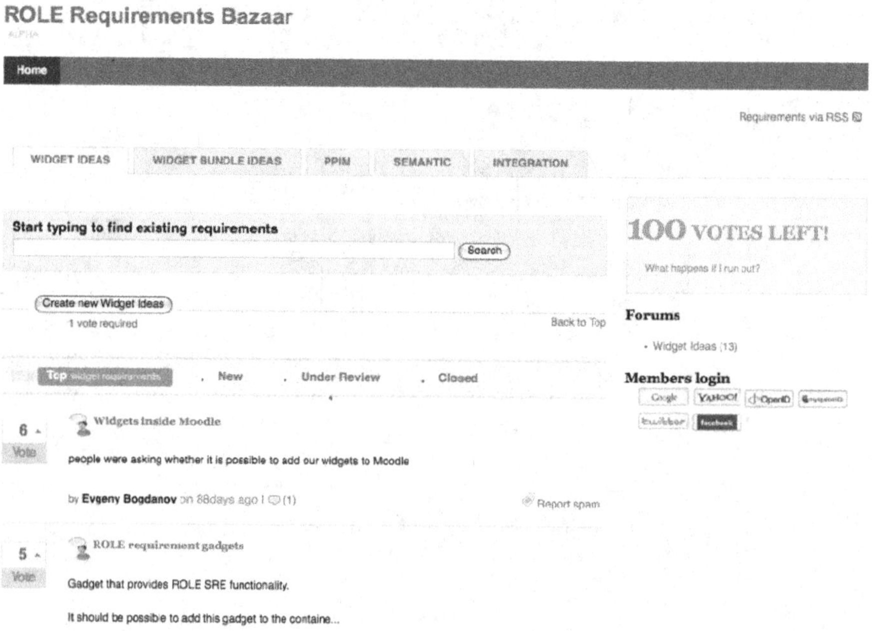

Fig. 2. Web-based requirements elicitation form

In this preliminary implementation phase, elicitation of requirements was limited to 'explicit' method using a simple web-based form. As the community grows, more 'implicit' methods could be used from various Web 2.0-based data streams around the CoP, utilising various Natural Language Processing (NLP) techniques. During the elicitation phase, the learner should be provided with the opportunity to look up similar requirements expressed by other CoP members before creating a new entry to reduce data redundancy (Fig. 2).

Negotiation: Once a requirement has been expressed, it should be visible to the CoPs (learner and developers) to discuss, clarify and refine the stated requirement. This negotiation process should take the form of a threaded discussion similar to a forum. Allocating enough time for this negotiation process should enable the requirement to be thoroughly examined by the community and potentially improve the idea. Due to the qualitative nature of a threaded discussion it may be difficult to analyse longer discussions. Hence, some form of quantitative mechanism such as votes to aid decision-making should be incorporated (the left-hand-panel in Fig. 2).

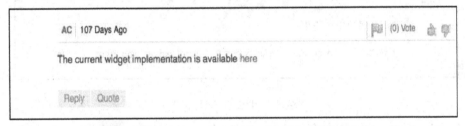

Fig. 3. Quantitative assistance mechanisms during negotiation

A simple like-dislike type mechanism (the upper right hand corner in Fig. 3) as present in popular social media such as Facebook and YouTube would serve the purpose to start with and was implemented around the commenting system. During the negotiation process the learner communities are able to view the requirements and the discussion around them. They are able to voice their opinion, qualitatively via the discussions or quantitatively via a simple ranking mechanism; a requirement with a highest number of votes bubbles up to the top of the list (Fig. 4).

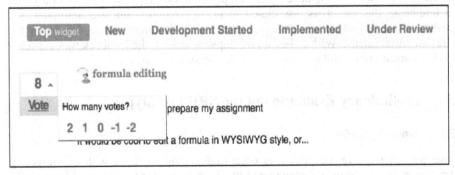

Fig. 4. Ranking mechanism

Besides, community members can be allocated a limited pool of votes that are replenished as soon as their requirement is selected for development (Fig. 4). The rationale is to force judicious use of the votes to rank requirements based on immediate needs.

Selection: This stage, especially its timing, is dependent on the developer community supporting the needs of the associated learner communities. At regular intervals, developers review the list of requirements elicited, refined and ranked by the learner communities. Then they decide on the needs they would attend to in the subsequent development iteration. The decision may depend on a variety of factors like top-ranked community needs, resource availability, and short-term as well as long-term development goals. During this stage, the requirements that are being actively reviewed are clearly marked (e.g. 'Under Review', Fig. 5), enabling the community to further con-tribute to the decision making via comments or votes.

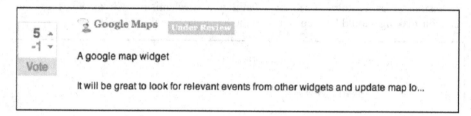

Fig. 5. Requirements review

Development: Once a set of requirements is selected, the developer community needs to add them to the requirements backlog in the project management software for subsequent iterations. This is achieved by integrating the requirement management interface with the project management interface, utilising appropriate APIs. During the pilot, JIRA[3] has been used as the software project management interface, and JIRA APIs have been used for the integration purpose. Once the requirement has been linked to the equivalent JIRA issue, any activity (such as developer comments, status update) around the JIRA issue has been reflected in the equivalent requirements comment stream.

Feedback: As soon as the developers complete the development process of a given requirement, the members associated with the requirement are informed via automated email and status is updated for other community members looking for relevant requirements. Within the current implementation, the feedback was expected via the same commenting system used for the negotiation process.

5   Preliminary Evaluation of the SRE Prototypes

5.1   Issues Identified

Two workshops were independently organized in summer 2011 with one targeting developers and pedagogical experts and the other early stage researchers in the field

---

[3] http://www.atlassian.com/software/jira/overview

of TEL; each involved about twenty participants. During the workshops the SRE approach was first presented to the participants. They were then divided into small groups to design and develop services/widgets for their own learning needs. The participants were demonstrated some example scenarios on how widgets and widget bundles could be used for learning purposes. Within the limited timeframe of the workshops, the participants were able to brainstorm initial ideas and to embark on stage 1 (elicitation) of the SRE process. The expectation was that these activities would then be carried over virtually within the proposed SRE prototype. Qualitative feedback was obtained on SRE during the workshops, and communication and content patterns were observed around the SRE prototype in the subsequent three-month period after the workshops. Additionally, access logs were analysed. Based on the analysis and feedback around the preliminary SRE prototype the following aspects were observed and noted.

Inadequate communication: RE traditionally requires rich and quality communication with stakeholders in order to document current processes and problems for requirements extraction. PLEs are a new way of learning and it will certainly not be a straightforward process to identify requirements for an individual's PLE after a short introduction lasting a couple of hours. Furthermore, a traditional user, for instance, of Learning Management System (LMS) is essentially replaced by a huge and diverse group of learners in the case of PLE. Learners are used to courses and environments designed for them by their teachers and very rarely involved in such an entirely self-regulated process. Asking such learners to enter their "requirements" is a major challenge to them. This eventually results in the problem of communication gap [23] between developers and end-users.

Lack of common understanding about PLEs: There is a lack of consensus within the research community about PLEs. As a result, what is communicated to end-users regarding PLEs varies greatly depending on the speaker's interpretation. This potentially creates a situation where stakeholders (researchers and teachers) might have conflicting views and interests in the development based on their own interpretations, which again is a known problem in RE [24]. This situation is even worse when learners are involved; they may become less confident of expressing their opinions which may be challenged by those who interpret the notion of PLE very differently. Consequently, teachers or researchers would act as "surrogate users" for those learners who choose to remain silent. Such substitution, however, may only reflect users' actual goals, needs and expectations to a limited extent because teachers' or researchers' requirements can be very different from learners'.

Cultural Diversity and language: Difference in stakeholder languages, national and organisational cultures are known RE challenges [25] when dealing with global software development efforts. ROLE test-bed end-users comprise countries with three distinct languages (Chinese, German and English). Such cultural differences may have some undesirable impact on communication and common understanding of requirements.

Lurking: When online communities are involved, lurking seems a norm [26] with only a few members posting regularly [27]. Lurking becomes a threat when too many members choose to lurk rather than contribute. Preece and her colleagues [27] found that the main reasons for lurkers not participating in discussion were several: they felt they did not need to post, they needed to find out more about the group and objectives, they thought they were helpful by being silent (i.e. reducing 'noises'), they could not make the software work. Such issues may lead to low participation rates.

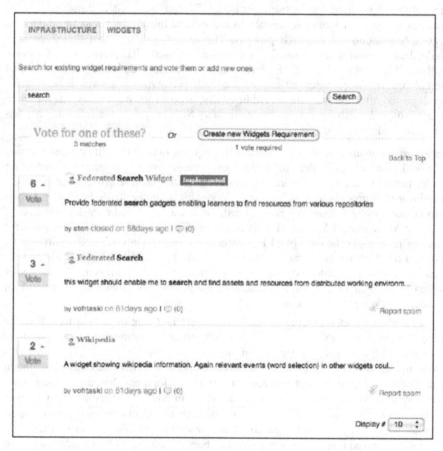

Fig. 6. Two different views – infrastructure and widget – of the SRE prototype

Infrastructure and widgets requirements: A large-scale PLE infrastructure should be able to support end-users to instantiate their personal learning environment. When collecting requirements via the SRE prototype, it has been noticed that quite a number of requirements directly refer to infrastructural aspects in technical jargon (e.g., "exposing semantic data as Linked data"). Such requirement obviously makes sense to developers and researchers, but its presence and visibility to end-users (students, employees, non-technical stakeholders) may deter them from posting as they might feel less confident about the nature of technical information they need to know prior to adding a requirement. However, exposing widget requirements like "I am a math student and would need a widget which supports me in writing or showing mathematical formulae in a widget" probably make more sense to end-users. Hence, it will be better to have two different views: One is where learners are able to view and participate in posting and discussing widget requirements; the view can be presented, for instance, as the requirements dashboard in a widget store where a collection of widgets are accessible. The other is where developers, after reviewing user-generated widget requirements, update regularly the related infrastructural requirements; the view, visible only to developers, can be seen, for instance, in JIRA via SOAP or REST (see Fig. 6 for the proposed change).

5.2   Resolutions Proposed

To address the issues identified some resolutions are proposed:

Understanding of PLE: Links to mini-tutorials presented in the form of videos and blogs can be embedded in the SRE prototypes. Frequently asked questions on PLE such as its major differences from LMS can be included in such blogs to enable learners to come to grips with the new concept of PLE more efficiently.

Templates: As eliciting new requirements can be daunting for end-users, support in the form a structured template with guidelines and samples may ease bootstrapping the process. The end-users then only need to edit such a template, which could be much easier than to use a blank form. Nevertheless, the provision of pre-defined attributes and examples may somewhat restrain the users' views of PLE and thus narrow the scope of requirements.

Mandatory voting to begin with: For most online communities, it is common to observe the manifestation of the "90-9-1 participation rule" [26] with 90% of users acting as lurkers by not actively entering new requirements; it is a well-known 'cold-start' problem [28]. Nonetheless, to mitigate the lurking issue, one feasible resolution is to provide users with the voting mechanism with which they can rate and rank existing requirements to facilitate prioritisation. To encourage contribution as well as to discourage lurking, strategically it may be effective to render voting mandatory and to reward such behaviour by, for instance, a competitive point collection scheme.

Prioritisation model - The current implementation provides a very rough approach to provide weighted priorities for each requirement. This needs to be significantly improved with the possibility of extracting rationales behind their ratings. Additionally, the ability to track roles and affiliations for each of the users may contribute to prioritisation decisions.

6   Concluding Remarks

Designing innovative software for the long tail is challenging, especially when conventional requirement engineering approaches are not applicable, especially when users cannot be identified in a traditional way. PLE is a perfect use case for innovative long tail software. There are potentially millions of users organized in very heterogeneous learning communities with varying needs expressed in very different ways. A major challenge for innovation is to find which requirements are the most promising from the user's point of view. In such a situation, we assume that a combination of popularity of the requirements in the user community, contextual factors and the social status of the users within their communities is a good measure. Based on this elicitation and analysis model, we set up a social requirements engineering process. The process is enabling developers to listen to learners' voices in an efficient manner by making it community-aware. Central to this idea is to create social spaces for learners and developers to meet and exchange requirements within their long tail communities – the Requirements Bazaar. In our first iteration, we have explicitly used Web 2.0 style rating and tagging, and have augmented new elicitation and analysis processes for requirements in large heterogeneous long tail communities. Community-awareness helps developers focusing on the needs of the community. Further steps will be focused on the generation of requirements prioritisations taking into account various requirements-centred data sources such as recorded communication around a requirement, development activity, weighting by importance of people involved using social network analysis (SNA) techniques [29]. All web-based software repositories and development environment platforms add more and more social features to facilitate community-awareness. However, our approach is going beyond that by integrating the software, its distribution platform, and the requirements engineering environment. Consequently, our approach can be integrated in future open source software repositories and web-based development environments.

Acknowledgements. The research leading to these results has received funding from the European Community's Seventh Framework Programme (FP7/2007-2013) under grant agreement no 231396 (ROLE project).

References

1. Nuseibeh, B., Easterbrook, S.: Requirements engineering: A roadmap. In: Proc. ICSE 2000 - Future of Software Engineering Track, pp. 35–46 (2000)
2. Holtzblatt, H., Beyer, H.R.: Requirements gathering: the human factor. Communications of the ACM 38(5), 31–32 (1995)
3. Chung, L., Leite, J.: On Non-Functional Requirements in Software Engineering. In: Borgida, A.T., Chaudhri, V.K., Giorgini, P., Yu, E.S. (eds.) Mylopoulos Festschrift. LNCS, vol. 5600, pp. 363–379. Springer, Heidelberg (2009)
4. Nuseibeh, B., Haley, C.B., Foster, C.: Securing the Skies: In Requirements We Trust. IEEE Computer 42(9), 64–72 (2009)
5. Olivier, B., Liber, O.: Lifelong learning: The need for portable Personal Learning Environments and supporting interoperability standards, http://wiki.cetis.ac.uk/uploads/6/67/Olivierandliber2001.doc
6. Wild, F., Mödritscher, F., Sigurdason, S.: Designing for Change: Mashup Personal Learning Environments. eLearning Papers 9, http://www.elearningpapers.eu/
7. Liber, O., Johnson, M.: Special Issue on Personal Learning Environments. Interactive Learning Environments 16, 1 (2008)
8. Fiedler, S., Väljataga, T.: Personal learning environments: concept or technology? International Journal of Virtual and Personal Learning Environments 2(4), 1–11 (2011)
9. Attwell, G.: The Personal Learning Environments – the future of eLearning? E-Learning Papers 1, 2 (2007)
10. Klamma, R., Jarke, Hannemann, M.A., Renzel, D.: Der Bazar der Anforderungen – Open Innovation in emergenten Communities. Springer, Berlin (2011)

11. Wenger, E.: Communities of Practice: Learning, Meaning, and Identity. Cambridge University Press (1998)
12. Anderson, C.: The Long Tail: Why the Future of Business Is Selling Less of More. Hyperion (2006)
13. Jäger, J.: Transkriptivität - Zur medialen Logik der kulturellen Semantik. In: Jäger, L., Stanitzek, G. (eds.) Transkribieren - Medien/Lektüre, Fink, München, pp. 19–41 (2002)
14. Jarke, M., Klamma, R.: Reflective Community Information Systems. In: ManoIopoulous, Y., Filipe, J., Constantopoulous, P., Cordeiro, J. (eds.) ICEIS 2006. LNBIP, vol. 3, pp. 17–28. Springer, Heidelberg (2006)
15. Lave, J., Wenger, E.: Situated Learning: Legitimate Peripheral Participation. Cambridge University Press (1991)
16. Duguid, P.: The Art of Knowing: Social and Tacit Dimensions of Knowledge and the Limits of the Community of Practice. Information Society 21(2), 109–118 (2005)
17. Østerlund, C., Carlile, P.: How practice matters: A relational view of knowledge sharing. In: Huysman, M., Wenger, E., Wulf, V. (eds.) Communities and Technologies - Proceedings of the First International Conference on Communities and Technologies (C&T 2003), pp. 1–22. Kluwer Academic Publishers, Dordrecht (2003)
18. Klamma, R., Spaniol, M., Denev, D.: PALADIN: A Pattern Based Approach to Knowledge Discovery in Digital Social Networks. In: Tochtermann, K., Maurer, H. (eds.) Proceedings of I-KNOW 2006, 6th International Conference on Knowledge Management, Graz, Austria, pp. 457–464 (2006)
19. Granovetter, M.S.: The strength of weak ties: A network theory revisited. In: Lin, P.M.N. (ed.) Social Structure and Network Analysis, pp. 105–130. Sage, Beverly Hills (1982)
20. Latour, B.: On recalling ANT. In: Law, J., Hassard, J. (eds.) Actor-Network Theory and After, Oxford, pp. 15–25 (1999)
21. Yu, E.: Towards Modelling and Reasoning Support for Early-Phase Requirements Engineering. In: Proceedings of the 3rd IEEE Int. Symp. on Requirements Engineering (RE 1997), Washington D.C., USA, January 6-8, pp. 226–235 (1997)
22. Bryl, V., Giorgini, P., Mylopoulos, J.: Designing socio-technical systems: from stakeholder goals to social networks. Requirements Engineering 14(1), 47–70 (2009)
23. Al-Rawas, A., Easterbrook, S.M.: A Field Study into the Communications Problems in Requirements Engineering. In: Proceedings, Conference on Professional Awareness in Software Engineering (PACE 1996), London (February 1996)
24. Nuseibeh, B.A., Easterbrook, S.M.: Requirements Engineering: A Roadmap. In: Finkelstein, A.C.W. (ed.) The Future of Software Engineering (Companion volume to the Proceedings of the 22nd International Conference on Software Engineering, ICSE 2000). IEEE Computer Society Press (2000)
25. Herbsleb, J.D., Moitra, D.: Global software development. IEEE Software 18, 16–20 (2001)
26. Nielsen. J.: Participatin inequality: Encouraring more users to contribute, http://www.useit.com/alertbox/participation_inequality.html
27. Preece, J., Nonnecke, B., Andrews, D.: The top five reasons for lurking: improving community experiences for everyone. Computers in Human Behavior 20, 201–223 (2004)
28. Schein, A., Popescul, A., Ungar, L., Pennock, D.: Methods and Metrics for Cold-Start Recommendations. In: Proc. of the 25th ACM SIGIR Conference, pp. 253–260 (2002)
29. Wasserman, S., Faust, K.: Social Network Analysis: Methods and Applications. Cambridge University Press, Cambridge (1994)

The Six Facets of Serious Game Design: A Methodology Enhanced by Our Design Pattern Library

Bertrand Marne[1], John Wisdom[2], Benjamin Huynh-Kim-Bang[1], and Jean-Marc Labat[1]

[1] LIP6, University Pierre et Marie Curie, 4 Place Jussieu 75270 Paris, France
{Bertrand.Marne,Benjamin.Huynh-Kim-Bang,Jean-Marc.Labat}@lip6.fr
[2] L'UTES, University Pierre et Marie Curie, 4 Place Jussieu 75270 Paris, France
John.Wisdom@upmc.fr

Abstract. Serious games rely on two main types of competence and expertise: the game designer's and the teacher's. One of the main problems in creating a serious game that is both amusing and educational, and efficiently so, is building a cooperative environment allowing both types of experts to understand each other and communicate with a common language. The aim of this paper is to create such a language using Design Patterns based on our framework: the Six Facets of Serious Game Design. If many design patterns already exist for the game design aspects, they are in short supply on the pedagogical side.

Keywords: Serious Games, Design Patterns, Pedagogy, Game design, TEL, Conceptual Framework, Instructional Design.

1   Introduction

One of the main problems with serious games (SGs) is that if they are designed only by game designers: they may be very entertaining, but knowledge acquisition may not be forthcoming. On the other hand, teachers and trainers may design games that are educationally very efficient, but lacking in the capacity to motivate and engage the player. Our experience of collaborating with design teams and browsing published examples of serious games has led us to the above conclusion and the necessity to create a Design Pattern library to facilitate the cooperation between the different stakeholders in the game design process. We can broadly group the stakeholders into two categories, the pedagogical experts and the game experts. By pedagogical experts or teachers we mean knowledge engineers, teachers, educators, and domain specialists. By game experts we mean game designers, level designers, game producers, sound and graphic designers, and so on. However, defining a serious game is a tall order.

Serious games can be defined as "*(digital) games used for purposes other than mere entertainment*" [1]. This definition is very wide in its scope and to combine fun and learning, we prefer to narrow it down to the notion of the "*intrinsic metaphor*". The latter can be defined as "*a virtual environment and a gaming experience in which the contents that we want to teach can be naturally embedded with some contextual relevance in terms of the game-playing [...]*" [2]

A. Ravenscroft et al. (Eds.): EC-TEL 2012, LNCS 7563, pp. 208–221, 2012.
© Springer-Verlag Berlin Heidelberg 2012

For the moment, a difficulty arises when the teachers and the game experts work together: do they understand the goals of each other? Are they able to communicate efficiently to produce a product that is both educationally efficient and fun to play?

The aim of this paper is to define and describe tools which allow everybody concerned to speak the same language, to be on the same conceptual wave length, and to allow some insight into the design process. We chose to build and review a common solution for these problems: a Design Pattern library to be used within our conceptual framework. We shall therefore focus first on the latter: The Six Facets of Serious Game Design. Then, we shall discuss the previous work on Design Patterns (DPs) and present our library. Finally, we shall present our fieldwork applying the library to it.

2   The Six Facets: A Conceptual Framework for Serious Game Design

Some conceptual frameworks are cited as a method to help designers to blue print serious games. For instance, Yusoff [3] and related work [4], define within his framework the steps to be taken when designing a serious game. The latter do not specify which experts should intervene in each step of the process. On the other hand, Marfisi-Schottman [5] introduces a seven step model, which attributes specific roles and steps to each expert (cognitive and pedagogical experts, storyboard writers, artistic directors, actors, graphic designers, sound managers, etc.). One difficulty, however, with both of these models is that they are sequential and do not easily fit into an iterative design model.

Especially for the serious games based on an intrinsic metaphor, we designed a non sequential and more flexible framework, clearly making explicit the experts needed at each step. We shall present our six facet model and show how it can be used.

Our conceptual framework aims to help evaluate the design process and improve it, either during the design process period or after it (post-mortem) to extract Design Patterns. Each facet is defined by its title, an SG design problem, a general solution and its experts. Previous papers have detailed the facets with numerous examples [6, 7]. Therefore, we shall make only a quick overview of each facet of this framework in order to present how our Design Patterns library will fit into it.

The goal of the first facet, "**Pedagogical Objectives**", is to define the pedagogical content. The general solution is to describe the knowledge model (including misconceptions) of the domain and the educational objectives. The key players here are the pedagogical experts. However, the other participants can gain important information as to how the former work and build a knowledge model.

For Instance, *Donjons & Radon*[1] is an SG meant to help junior high school students to study the transitions of the states of matter. Its Pedagogical Objectives are compiled in a graphed model of a physics course. This model was made by teachers and pedagogical experts with a graphical knowledge and pedagogical modeling tool (*MOT* [8]) and was intensively used by all the stakeholders during the design process.

---

[1] http://www.ad-invaders.com/project.php?id=19

"**Domain Simulation**" (second facet) raises the problem of how to respond consistently and coherently to the correct or erroneous actions of the game players within a specific unambiguous context. The solution consists in defining a simulation based on a formal model of the (educational) discipline. The specialists of this facet are the pedagogical experts.

For instance, as further detailed below in our section about fieldwork (fifth section), the *Donjons & Radon* Domain Simulator was finally based on the water phase diagram model, to ensure the relevancy of the interactions in the game.

The third facet, "**Interactions with the Simulation**" specifies how to engage the players by allowing them to interact with the simulator. The solution is to define the interactions with the formal model through the intrinsic metaphor chosen for the specific SG. The specialists of this facet are the game experts (mainly game designers, level designers, and game producers).

For example, we are working on *Défense Immunitaire*, an SG project meant to teach immunology to junior high students. Interactions with the immunology Simulator are based on the metaphor of the "*Tower Defense*". It is a particular kind of Real Time Strategy (RTS) game. The students must defend a territory (the metaphor of the body) by adjusting the defenses (metaphor of the immune system).

"**Problems and Progression**" (fourth facet) concerns which problems to give the players to solve and in which order. The solution is to design the progression taking into account both required knowledge acquisition (pedagogy) and the progress of the player (fun) from one level to the next. The progression in the game can be viewed as a sequence of challenges (obstacles/problems) that have been overcome. One important point is how to gain feedback concerning the progress made by the player and to transfer it to both the player himself and the trainer. The specialists of this facet are both the pedagogical team and the game experts. Here both groups must be able to communicate clearly and understand each other unambiguously.

For instance in the famous SG *Americas-Army 3*[2] progress in both game and domain competencies are *reified* with rewards badges.

"**Decorum**" (fifth facet) specifies which type of multimedia or fun elements, unrelated to the domain simulation, will foster the motivation of the players. This can be the shape of the avatar, a game within a game, a museum, a hall of fame etc. The specialists of this facet are the game experts. The main objective here is to increase the fun element and consolidate engagement.

For example, the SG *Prévenir la grippe H1N1*[3], is about the flu in a virtual world where cowboys are fighting using soap against different kinds of aliens. The representations for cowboys and aliens create a Decorum with a comical atmosphere based on an absurd situation. These representations, and associated interactions, are made to enhance students' engagement and are not in any way related to the domain simulation.

"**Conditions of Use**" (sixth facet) specify how, where, when, and with whom the game is played. Games can be played by one or several players, in class or online,

---

[2] http://www.americasarmy.com/

[3] http://prevenirh1n1.qoveo.com/

with or without an instructor etc. The specialists of this facet are both the pedagogical team and the game experts: the former to ensure the efficiency of the learning process, the latter to maintain motivation and engagement.

For instance, early in the design of *Donjons & Radon*, the stakeholders decided that the SG should be played during 30-40min sessions, to fit into the French secondary school schedules.

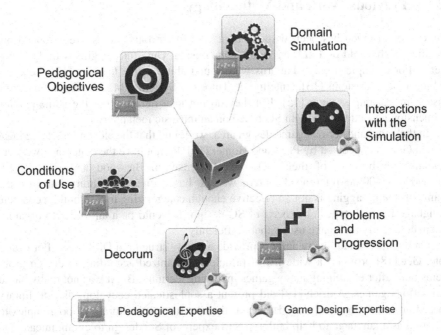

Fig. 1. Graphical representation of The Six Facets of Serious Game Design. For each facet, each type of expertise is shown by an icon.

One benefit of the Six Facets model is to designate the right expert(s) for each design area. But there still remains the problem of how to share the expert knowledge with all those involved within each facet. How can we do this with a view to helping everyone to find their place in the design process, with the goal of improving the combination of fun and pedagogy in serious games based on intrinsic metaphors?

Hopefully the knowledge of the experts is sometimes extracted and set out in the form of "Best Practices". We choose to use the latter approach as Alexander [9] did, by building a Design Pattern library. DPs constitute a set of good practices, focusing on one specific domain (architecture [9], software design [10], SG design [11], etc.), classified so as to be easily retrieved. They can be organized typically in terms of **Pattern Name, Context, Problem, Forces, Solution**, and also **Examples** and **Related Patterns** (if available) facilitating the building of a common pattern language that fosters communication.

Indeed, when Design Patterns are organized by referring to one another, they form what Alexander called a Pattern Language [9]. As far as we are concerned, both Design Patterns and Pattern Languages aim at facilitating the re-use of the best solutions or favoring discussion, brainstorming, and exchange of ideas between game designers and the pedagogical team.

3   Previous Work and Methodology

Since not so much has been written about DPs for serious games, we extended our reading to the field of TEL and video game design. Our study begins with TEL systems. For example, Design Patterns were found about active learning [12], Learning Management Systems [13], Intelligent Tutoring Systems [14] or about analyzing usage in learning systems [15]. But they do not take into account the game-playing dimension needed to design an SG based on an intrinsic metaphor.

In the game and serious game design area, we found that the eleven DPs for Educational Games conceived by Plass and Homer [16] did not have the coherence we were looking for because of their lack of categorization. Barwood and Falstein [17] provided a 400 tag-referenced pattern website based on a DP library but with the same problem: tagging is not an effective enough categorizing tool to build a concrete language for both types of experts of SG design. It would be a tall order to organize them coherently and make use of them efficiently.

On the other hand, many authors provide a highly structured DP library. For example, Gee [18] provides a wide list of principles organized according to design problems built after examining many games involving learning. But it was not really based on serious games. Aldrich [19] did present a sophisticated encyclopædic DP library based on simulations and SGs. However, the structure of the library is too complex to be used as a language to help both types of experts of SG design to communicate. On the other hand, Schell [20] presents one hundred "lenses" in a very understandable visual structure. Unfortunately, the purpose of Shell's lenses is to help designers to build good games and not good serious games.

In the end, we preferred to keep those of Kiili [21], and Björk and Holopaienen [22] especially for their ability to be used for SG design and their compliance with Alexander's [9] and Meszaros' [11] DP library structure. The library of Björk and Holopaienen [22] is both coherent and functional. They created their 200 DPs after interviewing seven game designers. Their aim was to build a catalogue allowing discussion and collaboration. The latter were not intended for the pedagogical aspect of serious games. Nevertheless, Kelle [23] designed a pedagogical meta-structure for Björk and Holopaienen's DPs. Kelle linked the key pedagogical functions with the game design patterns. By mapping the latter they foster the discovery or the adaptation of new DPs specifically designed to mix fun and pedagogy. We have also adapted some of Björk and Holopaienen's DPs such as "Serious Boss" adapted in "Boss Monster (GD)" (DPs from Björk and Holopaienen will appear in Table 1 with the letters GD).

The work of Kiili [21], however, concentrates on serious game design. The weak point is that Kiili [21] conceived a very small library (only eight DPs in six categories), having designed only one game: *AnimalClass*. We have kept some of them and they will appear followed by the letter K.

To collect, adapt patterns and add to the library, we gathered a work team composed of researchers, one game designer and two teachers. Then, in this team, we used an empirical method (bottom-up research) described below. We began with four serious games created by our private partner KTM-Advance[4] (*StarBank, Blossom Flowers, Hairz' Island, Ludiville*), an e-learning company, located in Paris, which has been developing serious games for several years. Unlike many SGs, the latter are not based on quizzes, but use an intrinsic metaphor, thus deploying quite advanced interaction to enhance learning. For example, a builder game (like *Sim-City*) is used to teach the ins and outs of banking. We also used one more serious game design with an intrinsic metaphor, *Donjons & Radon,* developed by a private consortium to which we belong.

We chose to study these five SGs for two reasons: the games were based on intrinsic metaphors, and we had full access to all the design documents. Moreover we made an in-depth analysis of twenty games selected from the Serious Game Classification Library[5] [24]. We selected the games on the same criteria: intrinsic metaphor and access to the greatest quantity of information we could gather. We also conducted interviews with researchers and game designers and the detailed study of two particular design cases.

For each facet of our Six Facets Framework described above, we compiled all our collected data covering different types of design experience, knowledge, and methods. For each facet, we looked for the common problems the designers faced. And we compiled the most interesting answers we had collected in order to build a pattern language as described in the Design Patterns for DP Design provided by Meszaros [11].

4   Our Design Patterns Collaborative Library

Our DP collaborative library is made up of 42 DPs within our Six Facet framework [25]. Table 1 presents the library thus organized. Within each facet, the DPs can be useful for those involved, highlighting the methods used, and form a knowledge base favoring discussion. The ultimate goal of the DPs is to enhance communication between the experts so that the game is both appealing and efficient as a learning process.

We shall first present the list of DPs in the synoptic table, And second, we shall present two examples to illustrate how DPs can best be used. The first one is *"Time for Play / Time for Thought"* (Facet #3: *Interactions with the Simulation*), the second is *"Reified Knowledge"* (Facet #4: *Problems and Progression*). Design patterns are typically written in *italics*.

---

[4] http://www.ktm-advance.com
[5] http://serious.gameclassification.com/EN/

4.1 Synoptic Table of Our Design Pattern Library

Table 1. List of serious game Design Patterns organized in our Six Facets Framework. DPs followed by "(K)" are from Kiili's work [21], and the DPs followed by "(GD)" are from Björk and Holopainen's work [22].

Facet	Design Pattern List	
⊙ **Facet #1:** Pedagogical Objectives	— *Categorizing Skills*	— *Price Gameplay vs. Educational Goals*
⚙ **Facet #2:** Domain Simulation	— *Simulate Specific Cases* — *Build a Model for Misconceptions* — *An Early Simulator*	— *Elements that Cannot be Simulated* — *Do not Simulate Everything*
◉ **Facet #3:** Interactions with the Simulation	— *Museum* — *Social Pedagogical Interaction* — *Serious Boss* — *Protege Effect (K)* — *Advanced Indicators* — *Validate External Competencies* — *Questions – Answers* — *New Perspectives*	— *Pedagogical Gameplay* — *Microworld Interaction* — *Time for Play / Time for Thought* — *Quick Feedbacks* — *Teachable Agent (K)* — *In Situ Interaction* — *Pavlovian Interaction* — *Debriefing*
⚐ **Facet #4:** Problems and Progression	— *Measurement Achievements* — *Surprise* — *Smooth Learning Curve (GD)* — *Fun Rewards*	— *Game Mastery* — *Freedom of Pace* — *Reified Knowledge*
✿ **Facet #5:** Decorum	— *Object Collection* — *Local Competition* — *Loquacious People* — *Graduation Ceremony* — *Fun Context* — *Wonderful World*	— *Narrative Structures (GD)* — *Serious Varied Gameplay* — *Informative Loading Screens* — *Hollywoodian Introduction* — *Comical World*
☁ **Facet #6:** Conditions of Use	— *Two Learners Side by Side*	

Our Collaborative Design Pattern library can be viewed on the internet[6]. We shall detail two examples of our DPs below.

4.2 Pattern: *Time for Play / Time for Thought*[7]

Context: Suppose one starts the *Game-Based Learning Blend* with a list of educational objectives, including high-level knowledge.

Problem: How can one teach high-level knowledge while the player is engaged in the game?

Forces: It is difficult for learners/players to concentrate on the interactions of the game and be engrossed in high-level thinking at one and the same time because of

---

[6] http://seriousgames.lip6.fr/DesignPatterns
[7] http://seriousgames.lip6.fr/site/?Time-for-Play-Time-for-Thought

cognitive overload. We must point out here that video games are often based on in-stantaneous interaction while some knowledge acquisition requires standing back (distance with respect to the problem) and taking time to ponder over what is to be learnt (the reflective phase).

Solution: It is a good idea to use intensive action phases for practice and training; and create less intensive phases for thought and reflection.

Frequent comments compare and contrast playing and learning; whereas, the real antithesis may well be between action (doing something) and reflection (thinking about what one is doing or evaluating what one has done).

In *"Foundation for problem-based gaming"* [21] analyzing problem-based gaming, Kiili highlights the need for reflective phases. The latter are for *"personal synthesis of knowledge, validation of hypothesis laid or a new playing strategy to be tested"*. During action phases, users are engaged emotionally, or focused on a goal, thus they are unlikely to be able to revise or re-structure knowledge acquired during the game. It must be pointed out that those two phases should be part and parcel of the fabric of the game. Video games, like thriller scenarios in the cinema, often provide less-intensive phases for (comic) relief purposes.

Examples: *Warcraft III* and *Plants vs. Zombies* are well-known examples of the *"Tower Defense"* type of video game. In this kind, the transition between phases of action and thought provides the core of the gameplay. There are some serious games of the *Tower Defense* variety, with the time for play separated from the time for thought, for instance *Le Jardinier Ecolo*[8], or *Defense immunitaire* on which we have been working (similar projects already exist [26]).

Uncharted and *L.A. Noire*, are famous video games for their scenarios in which the switch between phases of action and thought is a central element of the story. This is another way to include the flip over between these phases.

Related Patterns: *Instructional Gameplay*: during action phases to allow the player to discover, experience emotionally, or experiment with new knowledge.

Debriefing: during reflective phases, to explain or return to what has been happening during the action.

Reified Knowledge and *Advanced Indicators*: (useful supplementary information providing food for thought) incorporated into the action phases can give the player a bird's eye view of the action.

This DP belongs to the third facet (Interactions with the Simulation) and mainly concerns the game experts but can be extremely useful for the educational team.

4.3   Pattern: *Reified Knowledge*[9]

Context: The particular game that the team is designing involves a variety of compe-tence and knowledge problems.

---

[8] http://www.ludoscience.com/EN/realisation/580-Le-Jardinier-Ecolo.html

[9] http://seriousgames.lip6.fr/site/?Reified-Knowledge

Problem: How can one help users become more aware of their acquired knowledge?

Forces: Several problems arise. How can we make the player aware of the progress he has made for each skill or activity without taking him out of the Flow [27]? How can we use this type of information to enhance his/her motivation and enjoyment of the game?

Solution: Represent items of knowledge or competencies (skills) with virtual objects to be collected. If players have acquired the requisite skill or piece of knowledge, they will be given an object symbolizing this or that knowledge acquisition.

For instance, the users can see their acquisitions either in knowledge or skills embodied in medals, stars or other objects awarded. Every award is placed in a showcase, and thus is exhibited as a means of recapitulating what has been acquired.

Example: In *America's Army 3*, medals can be won when special deeds are accomplished. For example, users win a *"distinguished auto-rifleman"* medal when they have won 50 games as riflemen in combat. Medals, however, do not further player progress in the game; and are more a way of reifying the playing style by rendering it concrete.

In *Ludiville* (a KTM-Advance game for a bank), knowledge about home loans is reified by beautiful trading cards (as in a game called *Magic the Gathering*). Once having learnt a new piece of knowledge, players obtain the related card, which they can use later in the game to meet new challenges.

Related Patterns: *Object Collection*: also used to motivate players who like to collect things.

This DP belongs to the fourth facet (Problems and Progression) and concerns both game experts and teachers, who have to cooperate here.

5   Fieldwork and Discussion

Initially our Design Patterns were tested with a group of twenty students specialized in video game design[10]. They were interviewed and given a questionnaire to fill in after studying the DPs. At the time, our DPs were mainly game design centered, and as the students were knowledgeable in game design they seemingly did not have any use for our DPs. Nevertheless, they showed much interest in some more DPs focusing on pedagogy.

Subsequently, we tested the patterns with two teachers, one working in high school on the body's immune system, the other in college working on a course to help French students understand the US educational system. Both found the DPs useful.

The first project, called *Graduate Admission*, is hypothetical. The English teacher started the project from scratch and used our first DPs to explore game design possibilities. He began by using the Design Pattern *Game-Based Learning Blend*, thus following the procedure used by KTM-Advance game designers [4].

---

[10] They are students of the ENJMIN: *"Ecole Nationale du Jeu et des Medias Interactifs Numériques"* a video game school at Angoulême, France.

He first clearly formulated the educational objectives of the game before designing the storyboard: acquiring the skills and knowledge for admission to an American graduate course; understanding the American higher education system, and the attitudes that Americans have about study and college life; pitfalls that must be avoided (main, most commonly made mistakes.)

Secondly, he used the pattern *Narrative Structure (GD)* to invent a game scenario: A French student in his/her last year at a French university (Bachelor's degree), has met an American visiting Paris. They fall in love and decide to live together. However, the American has been admitted to a graduate school in the US. The French person has decided to apply to the same university. The game consists in acquiring the necessary skills and knowledge to be selected for admission.

In this game project, the thought or reflective phase (*Time for Play / Time for Thought*) could come after the failure to write an acceptable letter or CV. The player should be guided towards understanding the cultural differences, the usage gap between France and the US. Subsequently, the statement of purpose (SOP), which does not exist in France, would probably be a major drawback and a terrible pitfall for a French student. The DP *Debriefing* could be implemented by showing the learner examples of bad SOPs, or by showing his SOP and getting advice from American friends. In other words, *Debriefing* consists in making the player/learner aware of his/her errors and presenting him with the required knowledge necessary for accomplishing the specific task, and especially understanding a higher level cultural trait in depth.

The DPs were useful in helping the teacher to organize his project, outlining the main pedagogical content, creating a simple storyboard. Several students took part in a workshop where they could try out the different phases of the game and acquire symbolic objects. The game prototype was extremely simple and used interconnected web pages, video, and text to show the players how to apply for a university, write a statement of purpose, a résumé, fill in an application form, and prepare an interview.

The human immune system motivated the second game project for junior high students. After consulting the library, the teacher chose the DP *Time for Play /Time for Thought* as it corresponded to one of the main issues when it comes to teaching immunology. Indeed, students find it difficult to focus on the matching mechanisms related to the body defense system and microbes while endeavoring to do the exercises. The DP helped the teacher to choose a specific game play: *Tower Defense*. This kind of game enables players to select their strategies, test them in action, and if they are valid, move on to a reflexive phase during which the initial strategy can be modified if need be.

To conclude, both teachers found the DPs useful and stimulating because the library allowed each of them to find game play solutions for pedagogical problems. This fieldwork with both game design students and teachers demonstrate that DP users are not very interested in DPs focused on their expertise, but more in DPs exploring different or new knowledge. As we had built our first DPs focusing on game design solutions to pedagogical problems, they drew the teachers' attention. On the contrary, game design experts were looking for more educational aspects in our DPs.

For these initial tests of DPs, we were in the context of a single SG designer (teachers or game designers alone). However, we also had the opportunity to try our DPs with some multidisciplinary design teams, closer to the real conditions for which we made the library. On these occasions we also tried to enhance the library. At first, we used facets to identify areas of work, and then we used the Design Patterns to help the team of designers when they were stuck with some problems. For instance, we had relevance problems in the game design of the project *Donjons & Radon*. It is an SG meant to teach water transition phases. The game designer used only schoolbooks to build the core of the puzzles in the game. Doing so, and because he had not a wide knowledge of physics, he made several mistakes concerning the laws of physics. Fortunately, these mistakes were spotted by a physicist during a design meeting. But as the experts were very unlikely to be present during these meetings we were concerned about new mistakes being made.

Thanks to the facets and the DP library, we rapidly established that the design methodology was erroneous: the water behavior was not ruled by a simulation, but had been built by the game designer with simple rules based on the gameplay. And, for that reason, these rules were wrong in many cases. The DP *An Early Simulator* (Inside the second facet: Domain Simulation) was used to convince the game expert to build interactions based on a simulator designed by physics experts. The DPs *Do not Simulate Everything* and *Elements that Cannot be Simulated* helped the physics experts to exclude the kinetic aspect of the changes of the states of matter, and to build a proper simulation using a simple diagram of the water "triple point" (three axes phase diagram of the state of water).

On other occasions, these types of design problems had occurred and we had to refer to several Design Patterns of one of the facets. In some other situations, good examples of design processes led us to build new DPs for our library.

Even if it is very difficult to assess DPs, we are currently working on further evaluation tools for our Design Pattern library. The first aspect of these new evaluation tools is community evaluation. We made a collaborative library, giving everyone the ability to consult, assess, comment on, and even translate, modify and create DPs. When our DP library is well known enough, we hope that the community will give us some qualitative evaluation feedback through comments and modification suggestions. The tracks of the website visitors are also fully recorded in our logs, and we hope that full analysis of the users' navigation will help us to evaluate our library quantitatively.

In order to assess the library, the comparison of two serious games designed with and without DPs is difficult. It is necessary to use indicators to show that games designed with our DPs are better designed than others. We are working on tools to construct a typology of serious games based on the facets. One particular goal of this typology is to serve as an indicator to assess if the game achieved matches the original objectives. Therefore, for a game project, we must determine the *"initial type"* as early as possible and then compare it with the *"final type"* of the game made.

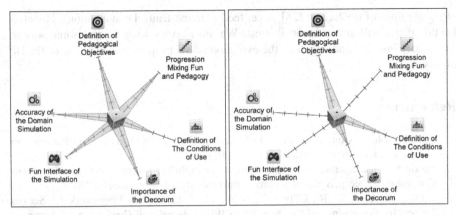

Fig. 2. Each scale represents the valence for each facet. The triangles give the measure of the valence. On the left, there is a stereotype of a serious game with an intrinsic metaphor (it could be the *"initial type"*). On the right there are the valence scales for a typical TEL system (it could be the *"final type"* that designers may not want to obtain).

This typology is based on a valence scale for each of the Six Facets. For instance, for the fourth facet (Condition of Use) there is a tendency to describe very early in the design process the exact conditions of use (e.g. *Donjons & Radon* described above in section 2). However, some designers prefer not to describe the condition of use, and let the users determine it as needed. We are designing some questionnaires to measure these valences for each facet and to set the type of a game project at every step of its design.

We hope to show that the DPs help enhance the compliance with specifications regarding the mix of fun and pedagogy by measuring the gap between the valences measured at the beginning and the end of the serious game projects built with or without them.

6   Conclusion and Future Avenues of Research

By using the Six Facet approach, we have tried to relate the different phases of game design for educational purposes. A team of game designers and teachers should be able to work together and communicate their ideas, brainstorm when necessary, arrive at some kind of holistic coherence.

Thanks to our fieldwork, we have established that when we use our DP library within our Six Facets Framework, it helps the teams to solve some design problems and fosters the communication between stakeholders. Moreover, we noted that our DPs were well suited to the needs of teachers, allowing them to understand the aims, means, and methods of the game experts. Nevertheless there is still much work to be done to help game experts to embrace the pedagogical aspects.

To help in this undertaking, we have created a collaborative web site where those interested can make suggestions, or give us feedback on their experience with DPs, and even create new DPs. We hope to manage this emerging community successfully along the lines of the Bazaar [28] to get feedback and fruitful contributions. Hopefully, this paper will attract some interest. We are also working to find some way to benchmark our DPs and to follow the evolution of SG projects and their use of the DP library.

References

1. Susi, T., Johannesson, M., Backlund, P.: Serious Games: An Overview. Institutionen för kommunikation och information. Skövde (2007)
2. Fabricatore, C.: Learning and Videogames: an Unexploited Synergy. 2000 AECT National Convention - a recap. Springer Science + Business Media, Secaucus, NJ (2000)
3. Yusoff, A., Crowder, R., Gilbert, L., Wills, G.: A Conceptual Framework for Serious Games. In: Proceedings of the 2009 Ninth IEEE International Conference on Advanced Learning Technologies, pp. 21–23. IEEE Computer Society, Riga (2009)
4. Capdevila Ibáñez, B., Boudier, V., Labat, J.-M.: Knowledge Management Approach to Support a Serious Game Development. In: Proceedings of the 2009 Ninth IEEE International Conference on Advanced Learning Technologies, pp. 420–422. IEEE Computer Society, Riga (2009)
5. Marfisi-Schottman, I., George, S., Frank, T.-B.: Tools and Methods fo Effi-ciently Designing Serious Games. In: Proceedings of ECGBL 2010 The 4th European Conference on Games Based Learning, pp. 226–234. Danish School of Education Aarhus University, Copenhagen (2010)
6. Marne, B., Huynh-Kim-Bang, B., Labat, J.-M.: Articuler motivation et apprentissage grâce aux facettes du jeu sérieux. In: Actes de la Conférence EIAH 2011, pp. 69–80. Université de Mons, Mons (2011)
7. Capdevila Ibáñez, B., Marne, B., Labat, J.M.: Conceptual and Technical Frameworks for Serious Games. In: Proceedings of the 5th European Conference on Games Based Learning, pp. 81–87. Academic Publishing Limited, Reading (2011)
8. Paquette, G., Léonard, M., Lundgren-Cayrol, K., Mihaila, S., Gareau, D.: Learning Design based on Graphical Knowledge-Modelling. Journal of Educational Technology and Society 9, 97–112 (2006)
9. Alexander, C., Ishikawa, S., Silverstein, M.: A pattern language. Oxford University Press, US (1977)
10. Amory, A.: Game object model version II: a theoretical framework for educational game development. Educational Technology Research and Development 55, 51–77 (2007)
11. Meszaros, G., Doble, J.: A pattern language for pattern writing. In: Pattern Languages of Program Design-3, pp. 529–574. Addison-Wesley Longman Publishing Co., Inc., Boston (1997)
12. Bergin, J.: Active learning and feedback patterns: version 4. In: Proceedings of the 2006 Conference on Pattern Languages of Programs, pp. 1–6. ACM, Portland (2006)
13. Avgeriou, P., Papasalouros, A., Retalis, S., Skordalakis, M.: Towards a Pattern Language for Learning Management Systems (2003), http://ifets.massey.ac.nz/periodical/6-2/2.html
14. Devedzic, V., Harrer, A.: Software Patterns in ITS Architectures. Int. J. Artif. Intell. Ed. 15, 63–94 (2005)
15. Delozanne, E., Le Calvez, F., Merceron, A., Labat, J.: A Structured Set of Design Patterns for Learner's Assessment. Journal of Interactive Learning Research 18, 309–333 (2007)

16. Plass, J.L., Homer, B.D.: Educational Game Design Pattern Candidates. Journal of Research in Science Teaching 44, 133–153 (2009)
17. Barwood, H., Falstein, N.: The 400 Project, http://www.theinspiracy.com/400_project.htm
18. Gee, J.P.: Good video games + good learning: collected essays on video games, learning, and literacy. Peter Lang, New York (2007)
19. Aldrich, C.: The complete guide to simulations and serious games: how the most valuable content will be created in the age beyond Gutenberg to Google. John Wiley and Sons, San Francisco (2009)
20. Schell, J.: The Art of Game Design: A book of lenses. Morgan Kaufmann (2008)
21. Kiili, K.: Foundation for problem-based gaming. British Journal of Educational Technology 38, 394–404 (2007)
22. Björk, S., Holopainen, J.: Patterns in game design. Cengage Learning (2005)
23. Kelle, S., Klemke, R., Specht, M.: Design patterns for learning games. International Journal of Technology Enhanced Learning 3, 555–569 (2011)
24. Djaouti, D., Alvarez, J., Jessel, J.-P., Methel, G., Molinier, P.: A gameplay definition through videogame classification. Int. J. Comput. Games Technol., 4:1–4:7 (2008)
25. Marne, B., Wisdom, J., Huynh-Kim-Bang, B., Labat, J.-M.: A Design Pattern Library for Mutual Understanding and Cooperation in Serious Game Design. In: Cerri, S.A., Clancey, W.J., Papadourakis, G., Panourgia, K. (eds.) ITS 2012. LNCS, vol. 7315, pp. 135–140. Springer, Heidelberg (2012)
26. Clements, P., Pesner, J., Shepherd, J.: The teaching of immunology using educational: gaming paradigms. In: Proceedings of the 47th Annual Southeast Regional Conference, pp. 1–4. ACM, Clemson (2009)
27. Csíkszentmihályi, M.: Flow: the psychology of optimal experience. Harper-Perennial, New York (1991)
28. Raymond, E.: The cathedral and the bazaar. Knowledge, Technology & Policy 12, 23–49 (1999)

To Err Is Human, to Explain and Correct Is Divine: A Study of Interactive Erroneous Examples with Middle School Math Students

Bruce M. McLaren[1,2], Deanne Adams[3], Kelley Durkin[4], George Goguadze[2], Richard E. Mayer[3], Bethany Rittle-Johnson[4], Sergey Sosnovsky[2], Seiji Isotani[5], and Martin van Velsen[1]

[1] Carnegie Mellon University, U.S.A.
[2] The Center for e-Learning Technology (CeLTech), Saarland University, Germany
[3] University of California, Santa Barbara, U.S.A.
[4] Vanderbilt University, U.S.A.
[5] The University of São Paulo, Brazil
bmclaren@cs.cmu.edu

Abstract. Erroneous examples are an instructional technique that hold promise to help children learn. In the study reported in this paper, sixth and seventh grade math students were presented with erroneous examples of decimal problems and were asked to explain and correct those examples. The problems were presented as interactive exercises on the Internet, with feedback provided on correctness of the student explanations and corrections. A second (control) group of students were given problems to solve, also with feedback on correctness. With over 100 students per condition, an erroneous example effect was found: students who worked with the interactive erroneous examples did significantly better than the problem solving students on a delayed posttest. While this finding is highly encouraging, our ultimate research question is this: how can erroneous examples be *adaptively* presented to students, targeted at their most deeply held misconceptions, to best leverage their effectiveness? This paper discusses how the results of the present study will lead us to an adaptive version of the erroneous examples material.

Keywords: erroneous examples, interactive problem solving, adaptation of problems, self-explanation, decimals, mathematics education.

1 Introduction

An instructional technique that has recently drawn attention from learning science researchers is *erroneous examples*. An erroneous example is a step-by-step description of how to solve a problem in which one or more of the steps are incorrect. Students can be challenged to find the error(s), explain the error(s), and/or fix the error(s) in order to more deeply learn the domain content and develop metacognitive skills. However, the use of erroneous examples for learning is controversial. On the one hand, some teachers fear that presenting errors to students will make them more

A. Ravenscroft et al. (Eds.): EC-TEL 2012, LNCS 7563, pp. 222–235, 2012.

inclined to make those errors [1], which is an idea supported by behaviorist theory [2]. On the other hand, some educators have argued that presenting students with errors for review and discussion can be valuable for learning. For instance, Borasi [3] has argued that mathematics education might benefit from students working with errors, encouraging critical thinking about mathematical concepts and motivating reflection and inquiry.

Our view is that erroneous examples are likely to be helpful to students under three basic conditions. First, the errors should be fictitious examples of *other* students' errors, so the student reviewing the errors is freed from embarrassment – and possible demotivation – of having their own errors exposed. Furthermore, no other real student is put on the spot in front of classmates. Second, the erroneous examples should be *interactive* and *engaging*; in particular, they should be computer-based materials that prompt for explanations, ask students to find and correct errors, and provide feedback. Finally, the erroneous examples should be *adaptively* targeted to the particular needs of individual students. That is, the types of problems presented to students should be aimed at their most deeply held misconceptions and misunderstandings about the target domain.

In short, our hypothesis is that the erroneous examples, presented to students in an interactive and adaptive fashion (e.g., presenting examples when a student is ready, withholding when not), can provide the opportunity to find and reflect upon errors in a manner that will lead to deeper, more robust learning. In this paper we present the results of a study that shows that interactive erroneous examples of others *can* provide learning benefits. Furthermore, we present some data and ideas regarding the next step of our research; that is, making the erroneous examples adaptive to student needs.

2   The Potential of Erroneous Examples for Learning

Research on erroneous examples derives from work on *correct* worked examples, which has attracted much attention in learning science empirical research [4, 5, 6, 7, 8]. Much research has also shown the importance of prompted self-explanation of worked examples, particularly in multi-media learning environments [9]. The theory behind the worked examples effect is that human working memory, which has a limited capacity, is taxed by strictly solving problems, which requires focused thinking, such as setting subgoals. Problem solving consumes cognitive resources that could be better used for learning [10]. The rationale is that worked examples free cognitive resources for learning, in particular, for the induction of new knowledge.

Erroneous examples also appear to free working memory for learning, by providing much of what students need to understand and solve problems, but at the same time, may engage students in a different form of active learning. It appears that erroneous examples may help students become better at evaluating and justifying solution procedures, which, in turn, may help them learn material at a deeper level. Learning with erroneous examples may also be related to the notion of "learning by teaching", as students who find, correct, and explain errors assume a role akin to teaching or tutoring [11].

Some researchers have begun to investigate empirically the use of erroneous examples, attempting to better understand whether, how, and when they make a difference to learning. For instance, Siegler [12] investigated whether self-explaining correct and incorrect examples of mathematical equality were more beneficial than self-explaining correct examples only. He found that students who studied and self-explained both correct and incorrect examples led to the best learning outcomes. Grosse and Renkl [13] studied whether explaining both correct and incorrect examples made a difference to university students as they learned statistics. Their studies also showed learning benefits of erroneous examples but unlike the less ambiguous Siegler results, the benefit was only for learners with higher prior knowledge and for far transfer learning only. When errors were highlighted, on the other hand, low prior knowledge individuals did significantly better, while high prior knowledge students did not benefit, presumably because they were already able to identify errors on their own.

Recently, there has been increasing investigation of *interactive* erroneous examples, those that are computer-based, that allow students editing and correction, and for which feedback is provided. Unlike the Siegler and Grosse and Renkl studies, Tsovaltzi *et al* [14] presented erroneous examples of fractions to students using an interactive intelligent tutoring system with feedback. They found that 6th grade students improved their metacognitive skills when presented with erroneous examples with interactive help, as compared to a problem solving condition and an erroneous examples condition with no help. Older students – 9th and 10th graders – did not benefit metacognitively but did improve their problem solving skills and conceptual understanding by working with interactive erroneous examples that included help.

Encouraged that interactive erroneous examples are promising instructional materials, our project team ran a study of decimal learning, in which we compared an interactive erroneous examples condition to a worked examples condition and a supported (i.e., with correctness feedback) problem solving condition [15]. However, the interactive erroneous examples did not lead to better learning results than worked examples or problem solving, nor was there an interaction between high and low prior knowledge and condition. We attributed this finding to two things. First, the prompted self-explanation of erroneous examples in this study was (potentially) too cognitively taxing. Students were asked to complete explanations of incorrect steps by filling in two phrases of a sentence, using pull-down menus. We observed students struggling with this task, possibly undercutting their math learning. Second, while we presented erroneous examples to students for review and comparison to correct examples, we did not prompt them to *find* and *correct* the errors.

The study presented in this paper was focused, first, on correcting the perceived problems with the prior study's materials and, second, on collecting data so we can learn how to adapt the presentation of erroneous examples to lead to the best possible learning outcomes.

3   The Domain: Learning Decimals

The domain we have focused on for this study is decimals. A variety of studies have shown that many students have difficulty mastering decimals and have common and persistent misconceptions [16, 17, 18], as well as problems that extend into adulthood [19, 20]. For instance, students often treat decimals as if they are whole numbers (e.g. they think 0.15 is greater than 0.8, since 15 is greater than 8, i.e., "longer decimals are larger") they think that decimals less than 1.0 (e.g., 0.23, 0.9) are less than zero. Persistent misconceptions in students' decimal knowledge must be overcome so students can handle everyday tasks (e.g., money calculations) and tackle more advanced mathematics.

Our general approach to addressing decimal learning with erroneous examples has been to develop problems that focus on single, key misconceptions. Based on an extensive literature review, we created short names for and developed a taxonomy of misconceptions that represents 17 misconceptions [15]. The present study focuses on four of these misconceptions, the ones that prior research has shown are most common and contributory to other misconceptions: Megz ("longer decimals are larger", e.g., 0.23 > 0.7), Segz ("shorter decimals are larger", e.g., 0.3 > 0.57), Negz ("decimals less than 1.0 are less than zero"), and Pegz ("the numbers on either side of a decimal are separate and independent numbers", e.g., 11.9 + 2.3 = 13.12).

4   The Study

For the current study we revised the materials from Isotani *et al* [15] by, in the interactive erroneous examples condition, simplifying the self-explanation step – asking students to complete sentences with one multiple-choice phrase instead of two – and by prompting the student to find and fix the errors in the erroneous examples. We also removed all problems, both on the tests and the intervention, related to two of the misconceptions explored in the earlier study (i.e., "multiplication always makes bigger" and "division always makes smaller"). This was done so we could focus on the most common misconceptions. Finally, we also simplified the experimental design, comparing only supported problem solving (PS) and interactive erroneous examples (ErrEx), while dropping the worked examples condition. We did this for two reasons. First, we wanted to compare the most common ecological control condition – that of students solving problems – to the much less typical learning experience of working with erroneous examples. Second, erroneous examples and problem solving are more comparable from a cognitive load perspective. As designed, they both require active problem solving – in the case of erroneous examples, the correction step; in the case of problem solving, generating the solution from the given problem – something worked examples do not (typically) require. Besides discovering whether erroneous examples could make a difference to learning, we had a goal of collecting data to help us determine how to implement automated material adaptation in a subsequent experiment.

4.1   Participants

247 students from a suburban middle school started the study, but 39 were eliminated from analysis (due to missing class or not trying on one or more of the tests, i.e., test score of 0, or close to 0). The remaining 208 participants consisted of 101 male and 107 female students. 105 of the students were in the 6th grade and 103 were in the 7th grade. Ages ranged from 11 to 13 ($M= 11.99$, $SD = .722$). All students worked on the materials exclusively during class time, under the supervision of the teachers and at least one experimenter, as further explained below.

4.2   Design and Materials

The materials consisted of six components, all presented to students on the Internet with all interactions and feedback implemented and logged using the CTAT intelligent tutoring authoring software [21]. The activities of the study are shown in Fig. 1: a pretest, a questionnaire on demographic/math experience (Q1), the intervention problems, a questionnaire asking about the student's experience with the materials (Q2), an immediate posttest, and a delayed posttest.

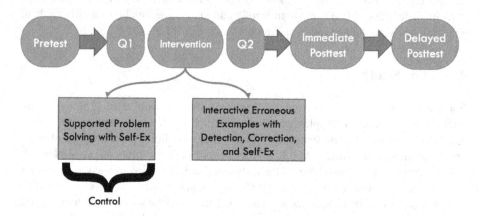

Fig. 1. The study design and sequence of activities

For the pretest and two posttests, three separate but isomorphic tests were constructed (Tests A, B, and C). Every problem on each of the tests was designed to diagnose at least one of the misconceptions of interest – Megz, Segz, Negz, or Pegz. The problems on the tests included placing decimals on a number line, arranging decimal numbers in order by dragging and dropping numbers into a sequence, providing the next two numbers in a given sequence, and defining key decimal concepts by selecting from a multiple-choice list of possible definitions. All tests contained a total of 50 possible answers. For 18 of the problems on each test, distributed relatively equally across the 4 misconception types, students were prompted after answering to indicate how sure they were of their answers on a Likert-scale ranging from 1 ("Not at all") to 5 ("Very Sure"). For the demographic questionnaire (Q1), students were asked their gender, age, and grade level, as well as questions about their experience with decimals and computers. For the post-questionnaire (Q2), students were asked questions such as, "I would like to do more lessons like this", with answers provided on a 5-point Likert scale.

In the intervention, the two groups were presented with isomorphic decimal problems, but with different presentations and ways of interacting with the problems. As shown in Fig. 2, the erroneous examples subjects were (a) presented with an incorrect solution by a fictitious student (upper left panel), (b) prompted to explain what the student had done incorrectly (upper right panel), (c) asked to detect and correct the error (middle left panel) and (d) prompted to explain and reflect on the correct answer (middle right panel and bottom left panel). They received feedback on their responses (i.e., green=correct; red=incorrect; with supportive feedback such as "You've got it. Well done." displayed in the lower right panel).

As shown in Fig. 3, the supported problem solving subjects were (a) asked to solve problems (upper panel) and (b) prompted to explain and reflect on the correct answers. These students also received feedback on their solutions (i.e., green=correct; red=incorrect; with supportive statements, as in the erroneous examples condition).

As shown in Table 1, the intervention comprised a total of 36 problems, 24 of which had interactions such as that illustrated in Figures 2 and 3 (according to condition), and 12 of which were problems to solve (the same across conditions), presented to the students to encourage active processing of the concepts and skills just presented. The problems were arranged in groups of three, each group targeting one of the misconceptions of interest (Megz, Segz, Pegz, and Negz – Highlighting in the table shows the grouping by threes).

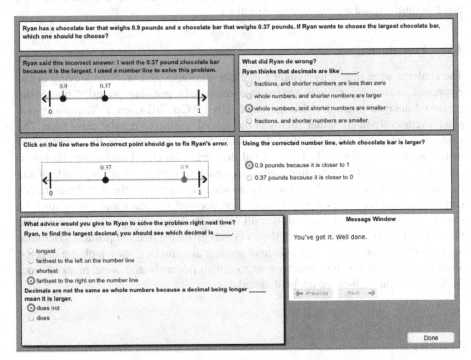

Fig. 2. Sample interactive erroneous example, targeted at the Megz misconception

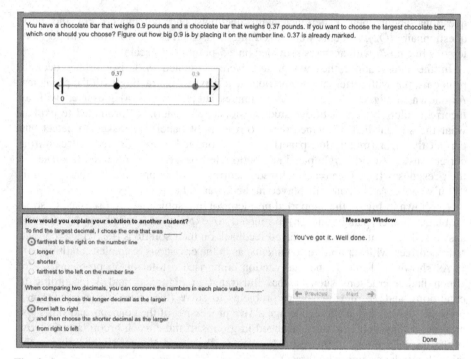

Fig. 3. Sample supported problem to solve, (isomorphic to the erroneous example problem in Fig. 2 and also targeted at the Megz misconception)

4.3   Procedure

The students were randomly assigned to either the supported problem solving or the erroneous example condition and to one of the six possible pretest / immediate posttest / delayed posttest orderings (e.g., ABC, ACB, BCA, etc.). The study took place exclusively in computer rooms in the school, replacing regular class time. The students were given a total of five 43-minute periods to complete the entire set of materials shown in Figure 1. They started the pretest on Day 1 and were allowed to continue immediately to the questionnaires and intervention, using as much of the first three days to work on these materials as needed. If students finished the materials early, they were asked to work on other, non-decimal materials and not disturb the students still working. On the 4th day all of the students were given the immediate posttest and 6 days later, on Day 5, they were all given the delayed posttest. Between the time the students took the immediate posttest and the time they took the delayed posttest, they received no classroom exposure to decimals and were blocked from working with the web-based decimal materials.

Table 1. Design of the Intervention

Supported Problem Solving (PS)	Int. Erroneous Examples (ErrEx)
1. Megz Supported PS 1	1. Megz ErrEx 1
2. Megz Supported PS 2	2. Megz ErrEx 2
3. Megz PS 1	3. Megz PS 1
4. Segz Supported PS 1	4. Segz ErrEx 1
5. Segz Supported PS 2	5. Segz ErrEx 2
6. Segz PS 1	6. Segz PS 1
7. Pegz Supported PS 1	7. Pegz ErrEx 1
8. Pegz Supported PS 2	8. Pegz ErrEx 2
9. Pegz PS 1	9. Pegz PS 1
10. Negz Supported PS 1	10. Negz ErrEx 1
11. Negz Supported PS 2	11. Negz ErrEx 2
12. Negz PS 1	12. Negz PS 1
13. Megz Supported PS 3	13. Megz ErrEx 3
14. Megz Supported PS 4	14. Megz ErrEx 4
15. Megz PS 2	15. Megz PS 2
16. Segz Supported PS 3	16. Segz ErrEx 3
17. Segz Supported PS 4	17. Segz ErrEx 4
18. Segz PS 2	18. Segz PS 2
19. Pegz Supported PS 3	19. Pegz ErrEx 3
20. Pegz Supported PS 4	20. Pegz ErrEx 4
21. Pegz PS 2	21. Pegz PS 2
22. Negz Supported PS 3	22. Negz ErrEx 3
23. Negz Supported PS 4	23. Negz ErrEx 4
24. Negz PS 2	24. Negz PS 2
25. Megz Supported PS 5	25. Megz ErrEx 5
26. Megz Supported PS 6	26. Megz ErrEx 6
27. Megz PS 3	27. Megz PS 3
28. Segz Supported PS 5	28. Segz ErrEx 5
29. Segz Supported PS 6	29. Segz ErrEx 6
30. Segz PS 3	30. Segz PS 3
31. Pegz Supported PS 5	31. Pegz ErrEx 5
32. Pegz Supported PS 6	32. Pegz ErrEx 6
33. Pegz PS 3	33. Pegz PS 3
34. Negz Supported PS 5	34. Negz ErrEx 5
35. Negz Supported PS 6	35. Negz ErrEx 6
36. Negz PS 3	36. Negz PS 3

5   Results

Due to bugs in four of the problems on all three of the tests the data for those problems was removed from all students, leaving a total possible score of 46 for each test for all students. The mean score for the pretest, immediate posttest, and delayed posttest, per

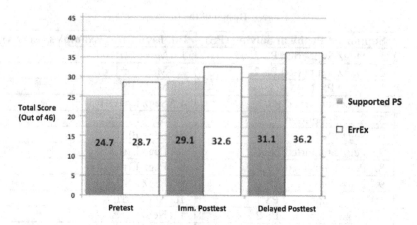

Fig. 4. Mean scores of pretest, immediate posttest, and delayed posttest

condition, is shown in Figure 4. Standard deviations were 9.4, 9.6, 9.5, 9.0, 9.2, and 7.5 for the left-to-right scores of Figure 4.

All students significantly improved their test performance from pretest to immediate posttest ($t(207)$ = -8.058, p < .001, mean increase of 9%, effect size d = .44) and from pretest to delayed posttest ($t(207)$ = 14.496, p.< .001, mean increase of 15%, effect size d = .75). An independent sample t-test revealed that the ErrEx group performed significantly better on the pretest than the Supported PS group, t (206) = 3.045, p = .003. An ANOVA revealed that there was no significant difference between the order of the three tests (A, B, C), F (5,202) = 1.293, MSE = .057, p = .268; thus, we assume the three tests were truly isomorphic to one another.

To determine whether the conditions differed significantly on the immediate and delayed posttest, an ANCOVA with pretest as a covariate was used. Results showed:

1. There were no significant differences in performance on the immediate posttest between the two groups, $F(1,205)$ = .768, MSE = 34.97, p = .382, d = .38.
2. On the delayed posttest there was a significant difference, in favor of the ErrEx condition, $F(1,205)$ = 9.896, MSE = 349.08, p = .002, d = .62.

In other words, although students in the ErrEx condition did not do significantly better than the students in the Supported PS condition on the immediate posttest, they *did* do significantly better on the delayed posttest, after a six-day delay. This occurred even though none of the students received further decimal training or practice between the immediate and delayed posttests.

To see whether higher prior knowledge students benefitted more from erroneous examples than low prior knowledge students, as they had in prior studies (e.g., [13]), we divided the participants using a median split into a high prior knowledge group (score of 26-46, total of 101 students) and a low prior knowledge group (score of 8-25, total of 107 students). ANCOVAs on posttest scores, with pretest covariates, showed the same outcome as 1-2 above, regardless of prior knowledge. That is, for *both* low and high prior knowledge learners there were no significant differences between conditions on the immediate-posttest, while the differences on the delayed-posttest were significantly better for the erroneous examples condition.

6   Discussion

Our study provides clear evidence that erroneous examples can help students learn. The delayed posttest results show that the students who were presented with interactive erroneous examples learned better than those who were presented with supported problems to solve. This suggests that erroneous examples may provide a deeper learning experience, one that can help students build upon their initial understanding of decimals, leading to a deeper understanding over time.   More specifically, it appears that erroneous examples might encourage generative processing (i.e., deeper cognitive processing that organizes the material and relates it to prior knowledge). This generative processing may be due to an effect referred to as "desirable difficulty" [22], in which problems of a more challenging form have been shown to lead to delayed learning benefits. The interactive erroneous examples of this study were very likely the harder of the two intervention types – and almost certainly less familiar to students – so they may have helped long-term retention.

On the other hand, we expected that higher prior knowledge students would get more benefit than lower prior knowledge students from erroneous examples, similar to the results of [13], but we did not find that higher prior knowledge students benefited more. Perhaps our materials, unlike those of the Grosse and Renkl study, were designed straightforwardly enough so that even lower prior knowledge students could easily follow, interact with, and learn from the examples without incurring excessive cognitive load. Indeed, one of our goals in this study was to simplify and streamline the prior year's study materials, in which no erroneous examples effect was found [15]. We did the streamlining by, for instance, making the self-explanation statements pure multiple choice, rather than sentence construction items with multiple components. The Grosse and Renkl work was also different in that it focused on errors related to confusing problem types instead of deeply entrenched misconceptions, which is what the current study focused on.   In other words, erroneous examples may be more helpful for students with low prior knowledge when they involve common misconceptions.

One caveat to our results is that more than a single variable differs between the erroneous examples and the supported problem solving materials. For instance, while students in both conditions had to self-explain their work, the erroneous examples condition had the extra self-explanation step, prompting students to explain why the fictitious student may have made the given error. Yet, our goal in this study, which we view as one step in an exploration of how erroneous examples might benefit learning, was not to demonstrate the full generality of erroneous examples. Rather, our objective was to see if we could find an advantage to erroneous examples – which our results have clearly shown – and, in future studies, isolate the instructional features of the materials that might account for the benefits. It is also worth repeating that our erroneous examples intervention, while differing in multiple ways from supported problem solving, has still demonstrated advantages to the most obvious and common control condition, that of students solving problems in conventional fashion.

7   Future Work

Our ultimate goal is to determine *when* and *with what students* we should present erroneous examples, and also to determine *what types of erroneous examples* to present to students. We have been pursuing this objective by modeling students with a Bayes Net of decimal misconceptions [23], which is updated when students take the tests described above. The Bayes Net represents the misconceptions that a student might have – Megz, Segz, Negz, and Pegz – and is updated based on carefully crafted test questions that probe for each of the misconceptions. Our isomorphic tests, A, B, and C, contain 9 Megz problems, 10 Segz problems, 10 Pegz problems, and 9 Negz problems, after the 4 buggy problems are eliminated (there are also 8 problems that are targeted at a more general misconception called Regz, which contributes to all of the other misconceptions). Students can either get these problems correct, in which case the probability of the targeted misconception drops, they can get them incorrect in an unexpected way, in which case the misconception is only partially increased, or they can get them incorrect in a way that provides direct evidence for the misconception in the Bayes Net. The tests were designed so that the misconception problems are relatively evenly distributed across the tests. Some of the misconception problems have possible answers that can indicate more than one misconception. The details of the Bayes Net are discussed in [23]. Our approach was inspired by the similar implementation of Stacey *et al* [24].

Given how the Bayes Net of each of the 208 students in the present study were updated, we calculated mean probabilities over all misconceptions: Segz=0.37; Megz=0.31; Pegz=0.15; Negz=0.15. Furthermore, we created *misconception profiles* for all of the students, based on the order of probability of each of the misconceptions for each student. For instance, a student with a Megz probability of 0.92, Segz probability of 0.75, Pegz probability of 0.32 and Negz probability of 0.2 would have a misconception profile of Megz>Segz>Pegz>Negz. Table 2 summarizes the misconception profiles of all the students by most prominent misconception, i.e., the misconception that has the highest probability.

Table 2. Summary of the misconception profiles of all 208 students

Description	#	Pre	Megz	Segz	Pegz	Negz	General Misconception Profile
Megz is the most prominent misc.	42	18.1	**0.97**	0.55	0.28	0.19	Megz>Segz>Pegz>Negz
Segz is the most prominent misc.	60	19.3	0.34	**0.89**	0.10	0.21	Segz>Megz>Negz>Pegz
Pegz is the most prominent misc.	58	33.9	0.03	0.02	**0.22**	0.03	Pegz>Megz>Negz>Segz
Negz is the most prominent misc.	48	34.3	0.00	0.00	0.00	**0.19**	Negz>Pegz>Megz>Segz

Key: The "#" column is the number of students with this misconception profile. The "Pre" column is the number of items, on average out of 46, that students in this row got correct on the pretest. The values under the "Megz", "Segz", "Pegz", and "Negz" columns are the average probabilities, according to the Bayes Net, that students in this row have each of these misconceptions.

As can be seen, the students were reasonably well distributed across the most prominent misconception categories, but there are stark differences in the mean values. Note that students who displayed the Megz ("longer decimals are larger") and Segz ("shorter decimals are larger") misconceptions as their most likely misconception, show a very high probability for actually having those misconceptions (see bold items in rows 1 and 2), while the students who displayed the Pegz ("each side of the decimal is separate and independent") and Negz ("decimals less than 1.0 are less than zero") misconceptions as most likely, show a much lower probability for actually having those misconceptions (see bold items in rows 3 and 4). Furthermore, the pretest scores of the Megz and Segz students are dramatically lower than the Pegz and Negz students. Finally, the *other* possible misconceptions of the Megz and Segz students have a much high probability than those of the Pegz and Negz students.

What does this tell us? First, these results are in line with the math education literature, which clearly indicates that Megz and Segz are the most likely decimal misconceptions of middle school math students. Having recently learned whole numbers and fractions, middle school students are very susceptible to the mistake of thinking longer decimals are larger (as is so with whole numbers) or that shorter decimals are larger (as is so with shorter denominators in fractions). Students who struggle with either (or both) of these misconceptions are much more likely to do poorly on decimal tests. Second, these results give us some clues about how to adapt our materials to particular students. Clearly, our system is more likely to be successful in helping students by emphasizing the Megz and Segz problems. Our initial plan is, not surprisingly, to provide more intervention problems aimed at the misconceptions for which students have shown they may have, according to the Bayes Net and the resulting misconception profiles. We will retrieve the misconception profile for each student and then provide that student with an intervention curriculum catered to that profile. The curriculum associated with each misconception profile will be weighted toward providing more problems aimed at that student's highest-probability misconception, less problems at the next highest probability misconception, and so forth. Considering the three-problem "problem groups" of Table 1, given a student with misconceptions in the order A, B, C, D, we might present 4 problem groups aimed at misconception A; 3 problem groups aimed at misconception B; 2 problem groups aimed at misconception C; and 1 problem group aimed at misconception D. Curricula will be variable, though, dependent on how different the probabilities are within a profile, e.g., given the clear need to ameliorate the Negz misconception in the misconception profile of the last row of Table 2, we might present a student that has such a profile with many more Negz problem groups than any other problem groups. We are well positioned to identify the most likely curricula needed; we have mined data for all of the 208 students in the study from the Bayes Net, created their misconception profiles, and have quantitative data to guide our approach. For instance, we have discovered that, of the 24 possible profiles (all of the permutations of the 4 misconceptions), 6 profiles never occur. Within specific profiles we have also discovered that some students have high probability values, very close to 1, while others have very low probability values, very close to 0, suggesting that even within each misconception profile, we will want to adjust curricula per student.

Besides this relatively straightforward adaptation proposal, we will investigate more sophisticated strategies. For instance, it is likely that misconceptions are causally interrelated, to a certain extent, so we will investigate ways to identify causality and use it to make adaptation decisions. The misconception profiles could also be adjusted based on either (or both) the Likert or questionnaire data that we collect as part of our study. For instance, a student who says he or she is "very sure" of an incorrect (and misconception) answer would lead to a more weighted update of the Bayes Net than a student who says he or she is "unsure" of an incorrect answer.

8   Conclusion

This paper has presented a study that provides evidence that interactive erroneous examples may be helpful to learning, especially over time, when a student has had an opportunity to reflect. Our next step is to investigate how we can adapt our erroneous examples material according to user models represented as Bayes Nets of decimal misconceptions. We will investigate a relatively straightforward adaptation strategy to see if it can be helpful to learning and then explore more complex strategies.

Acknowledgements. The U.S. Department of Education (IES), Award No: R305A090460, provided support for this research. We also thank the Pittsburgh Science of Learning Center, NSF Grant # 0354420, for technical support of our work.

References

1. Tsamir, P., Tirosh, D.: In-service mathematics teachers' views of errors in the classroom. In: International Symposium: Elementary Mathematics Teaching, Prague (2003)
2. Skinner, B.F.: The behavior of organisms: An experimental analysis. Appleton-Century, New York (1938)
3. Borasi, R.: Reconceiving Mathematics Instruction: A Focus on Errors. Ablex Publishing Corporation (1996)
4. Catrambone, R.: The subgoal learning model: Creating better examples so that students can solve novel problems. Journal of Experimental Psychology: General 1998 127(4), 355–376 (1998)
5. McLaren, B.M., Lim, S., Koedinger, K.R.: When and how often should worked examples be given to students? New results and a summary of the current state of research. In: Proceedings of the 30th Annual Conference of the Cognitive Science Society, pp. 2176–2181. Cog. Sci. Society, Austin (2008)
6. Renkl, A., Atkinson, R.K.: Learning from worked-out examples and problem solving. In: Plass, J.L., Moreno, R., Brünken, R. (eds.) Cognitive Load Theory. Cambridge University Press, Cambridge (2010)
7. Sweller, J., Cooper, G.A.: The use of worked examples as a substitute for problem solving in learning algebra. Cognition and Instruction 2, 59–89 (1985)
8. Zhu, X., Simon, H.A.: Learning mathematics from examples and by doing. Cognition and Instruction 4(3), 137–166 (1987)

9. Roy, M., Chi, M.T.H.: The self-explanation principle in multimedia learning. In: Mayer, R.E. (ed.) The Cambridge Handbook of Multimedia Learning, pp. 271–286. Cambridge University Press, New York (2005)

10. Sweller, J., Van Merriënboer, J.J.G., Paas, F.G.W.C.: Cognitive architecture and instructional design. Educational Psychology Review 10, 251–296 (1998)

11. Leelawong, K., Biswas, G.: Designing learning by teaching agents: The Betty's Brain system. Int'l Journal of Artificial Intelligence in Education 18(3), 181–208 (2008)

12. Siegler, R.S.: How does change occur: A microgenetic study of number conservation. Cognitive Psychology 28, 225–273 (1995)

13. Grosse, C.S., Renkl, A.: Finding and fixing errors in worked examples: Can this foster learning outcomes? Learning and Instruction 17(6), 612–634 (2007)

14. Tsovaltzi, D., Melis, E., McLaren, B.M., Meyer, A.-K., Dietrich, M., Goguadze, G.: Learning from Erroneous Examples: When and How Do Students Benefit from Them? In: Wolpers, M., Kirschner, P.A., Scheffel, M., Lindstaedt, S., Dimitrova, V. (eds.) EC-TEL 2010. LNCS, vol. 6383, pp. 357–373. Springer, Heidelberg (2010)

15. Isotani, S., Adams, D., Mayer, R.E., Durkin, K., Rittle-Johnson, B., McLaren, B.M.: Can Erroneous Examples Help Middle-School Students Learn Decimals? In: Kloos, C.D., Gillet, D., Crespo García, R.M., Wild, F., Wolpers, M. (eds.) EC-TEL 2011. LNCS, vol. 6964, pp. 181–195. Springer, Heidelberg (2011)

16. Irwin, K.C.: Using everyday knowledge of decimals to enhance understanding. Journal for Research in Mathematics Education 32(4), 399–420 (2001)

17. Resnick, L.B., Nesher, P., Leonard, F., Magone, M., Omanson, S., Peled, I.: Conceptual bases of arithmetic errors: The case of decimal fractions. Journal for Research in Mathematics Education 20(1), 8–27 (1989)

18. Sackur-Grisvard, C., Léonard, F.: Intermediate cognitive organizations in the process of learning a mathematical concept: The order of positive decimal numbers. Cognition and Instruction 2, 157–174 (1985)

19. Putt, I.J.: Preservice teachers ordering of decimal numbers: When more is smaller and less is larger! Focus on Learning Problems in Mathematics 17(3), 1–15 (1995)

20. Stacey, K., Helme, S., Steinle, V., Baturo, A., Irwin, K., Bana, J.: Preservice teachers' knowledge of difficulties in decimal numeration. Journal of Mathematics Teacher Education 4, 205–225 (2001)

21. Aleven, V., McLaren, B.M., Sewall, J., Koedinger, K.R.: A new paradigm for intelligent tutoring systems: Example-tracing tutors. International Journal of Artificial Intelligence in Education 19(2), 105–154 (2009)

22. Schmidt, R.A., Bjork, R.A.: New conceptualizations of practice: Common principles in three paradigms suggest new concepts for training. Psych. Sci. 3(4), 207–217 (1992)

23. Goguadze, G., Sosnovsky, S., Isotani, S., McLaren, B.M.: Evaluating a bayesian student model of decimal misconceptions. In: Proceedings of the 4th International Conference on Educational Data Mining (EDM 2011), pp. 301–306 (2011)

24. Stacey, K., Sonenberg, E., Nicholson, A., Boneh, T., Steinle, V.: A Teacher Model Exploiting Cognitive Conflict Driven by a Bayesian Network. In: Brusilovsky, P., Corbett, A.T., de Rosis, F. (eds.) UM 2003. LNCS, vol. 2702, pp. 352–362. Springer, Heidelberg (2003)

An Authoring Tool
for Adaptive Digital Educational Games

Florian Mehm, Johannes Konert, Stefan Göbel, and Ralf Steinmetz

Multimedia Communications Lab (KOM), Technische Universität Darmstadt, Germany
{florian.mehm,johannes.konert,stefan.goebel,
ralf.steinmetz}@kom.tu-darmstadt.de

Abstract. Digital educational games, especially those equipped with adaptive features for reacting to individual characteristics of players, require heterogeneous teams. This increases costs incurred by coordination and communication overhead. Simultaneously, typical educational games have smaller budgets than normal entertainment games. In order to address this challenge, we present an overview of game development processes and map these processes into a concept for an authoring tool that unifies the different workflows and facilitates close collaboration in development teams. Using the tool, authors can create the structure of a game and fill it with content without relying on game programmers. For adding adaptivity to the game, the authoring tool features specific user support measures that assist the authors in the relatively novel field of creating non-linear, adaptive educational experiences. Evaluations with users recruited from actual user groups involved in game development shows the applicability of this process.

Keywords: Digital Educational Game, User Modeling, Player Modeling, Authoring Tool.

1   Introduction

Digital Educational Games promise to combine the strengths of computer games (high acceptance especially among adolescents, high immersion, motivation and inherent learning by design) with the educational value of e-learning systems. It has long been suggested that this mixture can be beneficial to learning [24][25], and educational games have been on the market for a long time. A possible means for increasing the effectiveness and enjoyment of an educational game is the introduction of adaptivity, allowing a game to be customized for a specific player based on assessment of their state of learning or other characteristics. This approach has been used widely in e-learning tools (e.g. [6]) and can lead to a higher effectiveness of the resulting application [7]. However, in the context of games, it has seen only few adopters. Possible explanations for this are the increased efforts and the associated increase in production costs.

As will be detailed in section 3, the creation of an adaptive educational game requires special care during design, writing, and later on in the production because of

A. Ravenscroft et al. (Eds.): EC-TEL 2012, LNCS 7563, pp. 236–249, 2012.
© Springer-Verlag Berlin Heidelberg 2012

the need for adaptable paths through the game which allow the game to be customized for a player. Also, since more content is required, production costs rise.

We propose that the major hurdles in the production of an adaptive educational game can be overcome by optimizing the production process. This is achieved by mapping the traditional roles and workflows to an authoring tool specialized in adaptive educational games, allowing close collaboration and minimizing overhead due to coordination in game production teams. Section 4 describes a concept for such an authoring tool, which is then realized prototypically in the Serious Game authoring tool StoryTec as shown in section 5.

2   State of the Art

As the focus of this paper lies on authoring tools for adaptive educational games, basic details of game-based learning cannot be expanded upon in detail here for the sake of brevity. For such basics, the reader is referred to Prensky's foundational text in [24] and a current account of developments in [10].

2.1   Adaptive Technology in e-Learning

In the area of e-Learning, adaptivity has been used in many commercial and academic projects [14]. Two areas where this approach has long been researched are adaptive hypermedia systems and intelligent tutoring systems (ITS). As specified in several publications, such systems typically feature several models used for adaptivity. Shute and Towle [26] name the following: A *content model*, indicating the learning domain and interdependencies between knowledge; a *learner model* summarizing characteristics of the learner; and an *instruction model* binding the two previously mentioned by assuring the learner is provided with the right information or assessment at the right time. The actual adaption is then handled by a software component referred to as an *adaptation engine*. Usually, the model of the learner is created by assessment using tests involving computer-readable exercise formats (such as clozes, multiple choice questions or drag & drop exercises). This user model is then used to select the content to present to the user.

2.2   Adaptive Digital Educational Games

In the field of adaptive educational games, fewer examples abound. For entertainment games, one of the main fields of work so far has been the work on dynamically adjusting difficulty e.g. [11]. In the field of procedural content generation, current ideas include the application of generation algorithms while the game is running, based on the current state of the game [27]. Lopes and Bidarra [16] provide an overview of challenges and methods in this field.

In addition to the models adaptive e-learning systems use, adaptive (educational) games can also account for play preferences, thereby building a *player model* and using this model for adaptation purposes. The PaSSAGE [28] project uses a model sorting players into one of 5 possible categories, other player modeling approaches use the model presented by Bartle [2].

Maciuszek [18] describes an architecture for educational games utilizing the strengths of the role-playing game genre that combines game-based learning with work from intelligent tutoring systems in order to create adaptive educational games. The 80Days project [13] created a game architecture allowing adaption of an educational game both on the local level (giving hints, changing difficulties, ...) and the global level (different learning paths, ...). Bellotti et al. [3] describe a refined architecture for adaptive serious games which treats adaptivity as an optimization problem and proposes to use genetic programming for solving this problem.

What unifies the cited examples is that assessment in the games is handled to be minimally disruptive of the gameplay. This is captured by the notion of *evidence* being collected from the game whenever a player completes or fails a task in the game [23].

2.3   Authoring Tools for Educational Games

In this section, we provide an overview of authoring tools which have been created specifically for the purpose of educational games. Tools in adaptive e-learning are often based on existing e-learning authoring tools which allow the creation of learning objects and add the possibility to control adaptive features. An example of this can be found in [5].

The major example of tools for educational games is the e-Adventure authoring tool [22], which is conceptualized as an authoring tool for adventure games. Using a simplified authoring language which user can program by selecting from a list of possible actions and conditions, non-programmers are addressed by this tool. By confining the tool to one genre, the realization of the authoring process can rely on a set of assumptions which limit the choice of authors and prohibits the creation of games from other genres than adventures. e-Adventure does not provide an automatic means of adaptation; however, adaptivity is possible by using the means of the authoring toolkit.

3   Adaptive Digital Educational Games

This section provides an overview how adaptation can be introduced into educational games, including the specialized requirements on game content resulting from this.

3.1   Narrative Game-Based Learning Objects

As pointed out in section 2, several possible axes along which adaptation in games can be carried out are available. We propose to choose narration (adaption of the game's story, play (using a player model as described above) and learning. This choice of adaptation axes is consistent with the previously presented concept of

Narrative Game-Based Learning Objects (NGLOB), for in-depth information about this readers are referred to [9].

In order to structure the game for a game engine that can handle adaptation, a minimal unit of gameplay must be chosen. At a later stage, this allows authors to work on the game content in a similarly structured way, see section 4. In our work we propose the concept of a scene, similar to movies or stage plays. Thereby, a scene has a minimal context, involving a fixed set of characters, props, as well as logical objects necessary for the game engine, such as variables. Scenes can be hierarchically organized to allow better structuring of game content. For example, it is common to have the notion of a level in a computer game, in which graphical assets such as the level geometry or other features, such as background sounds, are shared. By hierarchically organizing scenes, these features are inherited by scenes lower in the hierarchy from those above them.

Each scene can then be seen as a Narrative Game-Based Learning Object and accordingly be annotated with relevant information about the scene for the adaptation algorithm. For storytelling purposes, this involves the narrative context of the scene, i.e. the function in the game's narrative this scene has. In order to formalize this, we make use of available narrative structures, such as the Hero's Journey (as mentioned in [9]). In order to adapt for gaming preferences, we utilize the notion of a player model capturing the different interest of gamers. As an example, the player model presented by Bartle [2] is used. However, the concept is flexible in this regard and allows other, similar models to be used or for authors to create their own player models customized for the game genre or content. Finally, for the purpose of learning, a learner model based on Competency-based Knowledge Space Theory (CbKST) [1] is used in which a scene can have different prerequisite competencies that are required to understand the educational content presented in the scene.

3.2   Adaptive Algorithms

In a non-adaptive game, the unfolding of the game's story is controlled directly by the choices made by the player inside the space of options provided to him or her by the game's author. In the concept described here, this is modeled by transitions between scenes. After a transition has been triggered by an action of the player, the game switches from the old to the new scene.

For making the game adaptive, different possible paths through the game have to be created, each allowing adaptation by choosing a different variation based on the current state of the information about the player. In the presented model involving scenes and transitions as links between scenes, several methods for providing such paths are possible. One model is that of transitions which are marked as "free". Using this kind of transition, an author does not connect a given player input directly with a fixed transition, but rather with a set of possible transitions. Based on the possible scenes indicated by the free transitions and the models of the player, an adaptive algorithm chooses the most appropriate in the current context. This variation has the advantage that an author has a direct overview of the possible points the player can get to from a certain action and plan alternatives explicitly.

The second possible method for authors to indicate adaptive choices to the game engine is by providing pools of scenes. In this variation, instead of modeling each possible transition between scenes explicitly, all scenes are placed in a container, thereby implying a net of transitions linking all pairs of scenes. When the game gets to a section of the game modeled in this way, a sequence of free choices from the available scenes can be made before this section is left again. This allows a very modular approach to adaptive game authoring, since a large pool of scenes with different content, gameplay and parts of the narrative can mean that the game is assembled at runtime and can be adapted very specifically to the player in each play session. However, this model is at the same time more abstract for authors (especially those not used to creating non-linear experiences) since no fixed order of the scenes in the pool is apparent, making storytelling, the creation of clearly designed learning paths and a learning curve in the gameplay harder. Here, the use of a rapid prototyping tool as described in section 5.2 becomes paramount for authors to quickly test their choices.

3.3   Assessment

In order to update the models the adaptive engine uses at runtime, assessment has to take place. In the presented concept, the interactivity in the game (inputs to the game and outputs from the game) is modeled as sets of stimuli and responses. Each stimulus is an action carried out by the player, such as clicking a button in the game. Responses are sequences of actions, high-level instructions for changes in the game.

Each action by the player which can be interpreted to yield information about his or her current state is annotated with the corresponding information. For example, finishing a task requiring knowledge of certain facts indicates that the player has gained this knowledge while playing, while a choice between different story continuations with varying levels of action or social interaction can give information about the player's game preferences. In essence, this provides the adaptive engine with an interpretation of game evidence [23] as noted in section 2.2.

Each of the updates to the user models of the player should be balanced with previous information in order to lower the effects of errors of measurement and of concept drift [4], i.e. when a player initially prefers action-laden sequences and later grows more interested in social interaction. Therefore, different update strategies are possible. We propose a simple weighted update function, which takes into account the old value of a certain attribute of the player model with a weight alpha and the updated value with weight *1 – alpha*. The factor *alpha* then determines the importance of older information compared with newer updates.

4   Authoring Processes

The following section describes game development processes that are usually found in the creation of educational games and how these processes are mapped in the authoring concept described in this paper. The analysis of game development processes is based on various accounts, including [12] and a study carried out by the authors with a German educational game studio.

4.1   Educational Game Development Processes

The game development of an educational game in general is similar to a regular entertainment digital game, with the additional challenges of providing educational content. Traditional roles found in game development include *game designers*, who are tasked with setting up the game's story, world, characters and gameplay. *Technicians*, i.e. game programmers and associated roles, are then tasked with creating the technical infrastructure of the game and realizing the gameplay, while *artists* (graphical artists, sound artists, …) create the necessary assets such as 3D models or sounds for the game. Finally, the game's quality is assured by *testers* before being released.

In the creation of an educational game, the above user groups receive new tasks and simultaneously new groups are added. This already indicates the increased complexity compared to entertainment-only games. The development team is augmented with *domain experts*, who introduce specialized knowledge about the target domain, as well as *pedagogues* in order to establish an educational design of the game. Common tasks for these groups include the creation of exercises or exercise pools, whereby in practice commonly general purpose tools are used for the creation and dissemination of the created content to the rest of the team.

The core game development team, as mentioned above, also receives more tasks as compared to the development of an entertainment-only game. Since one major purpose of the educational game is the presentation of educational content, the game design has to be adjusted for this, either by providing possibilities for placing learning content in the game or by adapting the gameplay itself in such a way as to be educational. An example for the former could be an adventure game placing educational content in the dialogue with a character, while an example for the latter is a physics game involving actual simulated physics-based puzzles the player has to solve by simulation.

The necessity for close integration of educational content continues from the design to all aspects of the game, including the art production required to produce assets which conform to the educational content and the game programmers realizing educational features or adding mechanisms for adaptivity.

4.2   Challenges Related to Adaptivity

When adding adaptivity to a Digital Educational Game, another layer of complexity is introduced in the processes. The design (especially concerning the narrative and the gameplay) has to be adjusted towards allowing adaptivity by providing several different ways to play the game or by different variations for the paths the players can take through the game.

Another challenge for designers and storywriters lies in the dynamic and algorithmic nature of adaptivity. The effect of the narrative paradox captures this to a certain degree [17]: while in a classical, linear medium such as a movie the consumer of the entertainment product has no freedom in choosing how the experience continues, a player in a game can influence the continuation of the game. This leads to the author of the game not being completely in control of the narrative, but having to foresee the possible actions of the player and providing the gameplay and story for each action. On top of this, adaptivity adds an element of uncertainty for the author since the actual continuation of the game depends on the current state of the user models, thereby potentially differing in each play session. Authors, especially those trained mainly with classical media, can find it difficult to retain an overview of the whole game and the flow of events in the game with the player and the adaptive algorithm influencing it continually. This effect can similarly be observed in e-Learning [8].

4.3   Mapping Authoring Processes into Authoring Tools

The challenges addressed in this section have been mapped into the authoring tool concept described in this paper and realized in the authoring tool StoryTec. In the following, we provide an overview how processes have been mapped into a unified authoring tool. Foss and Cristea [8] present a set of imperatives how e-learning authoring tools should support authors of adaptive content. These imperatives are in line with our concept.

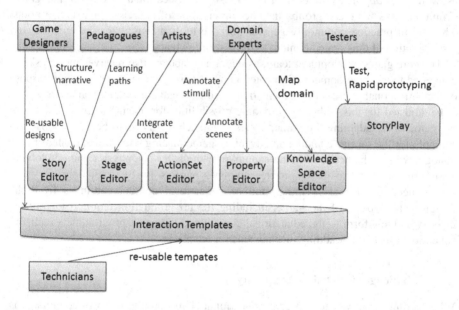

Fig. 1. Some of the possible mappings between user groups and the components of the described authoring tool. Note that not all combinations are shown, for example, game designers could also use StoryTec and StoryPlay in conjunction for storyboarding and rapid prototyping.

Figure 1 shows some of the mappings between users in the educational game development process and the components of the authoring tool. In general, user groups are separated in the way they interface with the authoring tool in order to contribute work towards a game. On a technical level, game programmers are tasked with building a basic framework on which the game can be run. This includes an environment in which the game can be executed (a game engine) as well as templates encapsulating various types of gameplay found in the game. Of course, the selection and specifics of these templates are governed by the initial game design carried out by game designers. These templates are programmed and integrated into StoryTec to be used in the creation of a specific game [21].

On the other hand, most users found in the game development process collaborate by using the authoring tool directly via its normal interface. Game designers set up the structure of the game and the individual levels/rooms by the visual interface. Artists add to these structures the finished assets, for example background graphics for the rooms the designers created. By providing a visual programming approach, game designers can directly manage the high-level flow of the game, while lower-level details are then handled by the interaction templates provided by the programming team.

For domain experts and pedagogues, special components are provided. For example, the creation of a knowledge structure for the learner model on which adaptation is based can be carried out in a graphical editor which visualizes the structure of the game's learning domain. Apart from this, they work in the same environment as game designers and therefore both groups are able to see the results of each other's work and collaborate. This helps in creating a common basis for communication about the tasks at hand.

Adaptivity is a central part of the authoring tool and therefore visible to all user groups. By means of the Story Editor providing a visual overview of the whole story and all paths through the game as well as the adaptive parts, users are supported in retaining an overview of the adaptive features of the game. The added effort for the creation of different, adaptable paths through the game is mitigated by variations being quickly creatable using the interaction templates and the possibility of copying and varying existing structures. Finally, a rapid prototyping tool with specialized visualization for adaptive algorithms and user models assists authors in understanding how the game will typically react during a game session by quickly testing out variations, with a prototype version of the game.

5   StoryTec

5.1   Authoring Tool

In this section, the authoring tool StoryTec[1] (cf. [19] among others) is described as a realization of the concept shown above, incorporating the workflows and processes as detailed in the last section. Special focus will be laid on the support of the creation of adaptive educational games.

The main user interface of StoryTec as seen in figure 1 is the principal interface for all user groups collaborating directly in the authoring tool, including game designers, artists and domain experts. Therefore, all important information is provided visually in the interface, and all functions for editing the game rely on simple concepts instead of programming languages or other more technical systems.

---

[1] Available at http://www.storytec.de

The main overview of the whole project is the Story Editor, in which scene hierarchies, objects and transitions are created and visualized. Authors create the structure of the game in this editor or re-use a provided structural template. Furthermore, the adaptive systems of free transitions and scene pools are available directly in the Story Editor, allowing all collaborating users to see and manipulate them.

The process of defining scene contents is carried out in the Stage Editor and relies on the interaction templates included in StoryTec or added by game programmers for a specific game (genre). Therefore, this allows quick editing of the content of the game, since each scene is equipped with an interaction template handling the details of gameplay implementation in the game, only requiring the input of the necessary content and settings. Since this approach of encapsulating gameplay allows the creation of several games in the same genre, it can lower the costs of production by fostering re-use and more rapid development cycles.

Fig. 2. The main components of the user interface of StoryTec (from top left in clockwise direction): Stage Editor, Objects Browser, Property Editor, Story Editor

The two other user interface elements in the standard configuration of StoryTec are the Objects Browser and Property Editor. The former is used for providing an overview of all available content objects and for adding them to any scene via drag & drop, while the latter is used to change parameters for objects and scenes.

Interactivity on a low level (graphics rendering, sound playback, camera control in 3D games) is intended to be handled by the game engine and the interaction templates due to their inherent complexity. Authors are empowered to configure high-level rules by using the ActionSet Editor (cf. figure 3), which connects each Stimulus (cf. section 3) with a set of Actions that should be applied in the game at runtime. Boolean conditions allow branching, thereby reacting to the current state of the game. Additional Actions are provided for assessment purposes, i.e. to update the user models for adaptivity. These actions are to be used whenever a Stimulus can be interpreted to indicate a change in a user model (e.g. a player solving a task that requires understanding a certain piece of knowledge). Since this system again is available to all collaborating users and does not require previous knowledge in game programming, it can increase collaboration and support rapid prototyping by allowing designers to quickly test game prototypes without waiting for a programmed prototype.

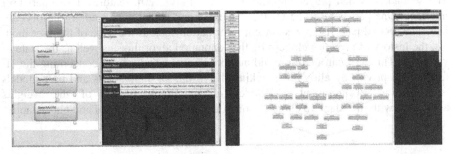

Fig. 3. Left: The ActionSet Editor of StoryTec, enabling non-programmers to structure the interactive flow of the application. **Right:** The Knowledge Space Editor. Boxes indicate facts or competencies, arrows indicate dependencies between them.

Specific support for the creation of educational games is offered in the Knowledge Space Editor (cf. figure 3). This editor assists in modeling the knowledge domain the game is based on by visualizing competencies and facts and the dependencies between them as a graph of boxes and arrows. A dependency here indicates that a certain competency A has to be understood before competency B can be addressed, a notion found in Competency-Based Knowledge Space Theory.

5.2   Runtime Environment

The complete StoryTec prototype also includes a full runtime environment for playing games created with the authoring tool. This includes, on the one hand, a player application intended to be used for evaluation purposes and as a rapid prototyping tool, and a multi-platform player application on the other hand.

As a basis for all provided player applications, several components are important to mention. The projects which are created by StoryTec are interpreted by a component referred to as Story Engine, which is linked to the game engine. The Story Engine acts as a high-level command instance, dispatching commands to the game engine and other components based on the parameters and actions the authors have set up in StoryTec. Concretely, it relays all gameplay commands to the game engine and the interaction template implementations included in the game engine and receives stimuli back from the game engine. For adaptivity purposes it includes the user models of the players whose updates it carries out based on information from the game as well as the algorithms for choosing how to continue in the game. Whenever an update in the game calls for an adaptive choice, all possible variations are considered (all free transitions or all scenes in a scene pool) and assigned a numerical value indicating their appropriateness when seen from a narrative, learning or play perspective. Depending on the overall goal of the play session, these values are then weighted in order to result in a choice of next scene. The chosen scene is that which yields the highest weighted sum of all values and conforms to all further constraints (e.g. not visiting the same scene again in a scene pool). For details of this process, see [9].

As described above, this basic architecture is realized in two player applications. The "StoryPublish" player is intended for cross-platform publishing a finished game. The second provided player application, "StoryPlay" [20], features a two-part user interface. One part is reserved for the gameplay and is therefore similar to StoryPublish. The second part visualizes current information such as the state of the user models, the history of previous choices by the adaptation algorithms as well as the state of variables. This tool can therefore aid authors in evaluating games concerning the effects of adaptivity by allowing checking the results of annotations and user models early during development. A slider allows quick tuning of the weights associated with the adaptive choices along the narrative, educational and play adaptivity axes.

6   Evaluation

Several evaluations of the presented approach have been undertaken (see also [19]). In the following, results of the evaluation studies will be highlighted. The subjects of the first evaluation were students involved in a course on Serious Games without previous exposure to StoryTec, with the study's focus being the general usability of StoryTec. It was carried out with 26 participants (1 female, 25 male, m = 25.2 years, SD = 3.71 years).

Table 1. Results of the usability questionnaire (Values range from 1 to 7)

Basic principle	Mean value	Standard deviation
Suitability to the task	4.74	0.88
Self-description	3.51	0.93
Controllability	5.48	0.77
Conformity with user expectations	4.55	1.06
Error tolerance	3.42	0.80
Suitability for individualization	4.42	0.72
Suitability for learning	5.14	0.78

The test was carried out with a variation of the "Thinking Aloud" method. The participants were assigned one of three roles: one participant read out the tasks aloud, the executing participant was given control of the computer running StoryTec, and an observer was asked to watch closely and give comments to the other two participants.

In this way, the participants were encouraged to have conversations about the tasks at hand and how they could be solved in StoryTec. Afterwards, the participants were asked to rate StoryTec in a questionnaire based on the usability standard ISO 9241-10.

The results of the questionnaire, aggregated to the seven basic usability principles of the standard, are shown in table 1. The examination of the questionnaire results shows that there is a tendency of the study participants to rate the ergonomics of StoryTec positively.

In the second study, three professional game developers (aged 31, 37, 46; 2 male, 1 female) were first asked to complete a short series of tasks in StoryTec and give feedback by thinking aloud and commenting on their experience. After this first stage, the participants were lead through a guided interview during which they were questioned on the usability of StoryTec in their individual domains (including game design and game programming of educational games) as well as their assessment of the effects of using StoryTec for the creation of an educational adventure game. All participants commented that they were very interested in the approach of StoryTec and that they could imagine a finished version of the software being used in actual development of educational games. In the evaluated state the participants were able to imagine the tool being best suitable for storyboarding and prototyping. For the use in game production, they would require more detailed parameters than the prototype version offered.

Apart from tests concerning only the usability of StoryTec when seen in isolation and a focus group evaluation, a larger comparison study with the goal of comparing StoryTec with e-Adventure was carried out. A set of N = 47 test subjects were recruited from a university course on serious games (8 male, 39 female, age range from 21 to 32 years (m= 24.79; SD= 2.62;)). The experiment set-up consisted of a task that was phrased to be equally accomplishable in both tools which consisted of three tasks moving from simple to more complex interaction with the respective authoring tool. During each evaluation session, a group of up to 8 participants (each with an individual PC) was instructed to work for 25 minutes on the tasks in the first authoring tool and then for 25 minutes in the other authoring tool. The order in which the tools were evaluated was randomized per group of participants, with n(1) = 25 participants starting with StoryTec and n(2) = 22 starting with e-Adventure. After this, the participants were asked to fill out a questionnaire with individual sections for each authoring tool, again based on the areas of the ISO 9241-10 standard. Additionally, some background information, an assessment of the perceived level of mastery of the respective authoring tool and a comparative question between the authoring tools and demographical data was asked for.

Initial results indicate that StoryTec (m= 4.58; SD=1.17;) was preferred compared to e-Adventure (m= 4.21; SD=0.78;) by the participants (p= .084). Male participants were observed to rate StoryTec higher (p=.023) while female participants did not see a significant difference between the two tools (p > .20), which could be due to the low ratio of female participants. This interaction between gender (male, female) and tool (StoryTec, e-Adventure) borders significant (p=.072).

This evaluation also included performance data, including the time participants required for solving each task and the resulting project files, which are reviewed based on an objective set of rules for completeness and correctness. The result of this analysis has not yet been fully compiled and is therefore excluded here.

7   Conclusion

In this paper, we have presented an approach that maps the processes commonly found in the development of adaptive educational games into a unified authoring tool which allows structured and transparent collaboration between the involved user groups. It addresses the major problems found in the development of educational games, namely the higher costs of production due to differing tools and operation methods of different groups and the increased need for communication in order to collaborate effectively. Furthermore, the challenges faced when creating an adaptive game, including the need for authors to retain an overview of the game in its adaptive form and to create additional content for adaptive variations was included in the concept.

The concept has been realized as the authoring tool StoryTec and the associated player applications: StoryPublish for cross-platform publishing and for actual players; StoryPlay as a rapid prototyping and evaluation tool for authors. Evaluations of StoryTec have shown that users from the actual involved user groups (game developers, domain experts) have assessed the usability and usefulness of StoryTec in educational settings positively.

References

1. Albert, D., Lukas, J.: Knowledge spaces: theories, empirical research, and applications. Routledge (1999)
2. Bartle, R.: Hearts, clubs, diamonds, spades: Players who suit MUDs. Journal of MUD Research 1(1), 19 (1996)
3. Bellotti, F., et al.: Adaptive Experience Engine for Serious Games. IEEE Transactions on Computational Intelligence and AI in Games 1(4), 264–280 (2009)
4. Black, M., Hickey, R.J.: Maintaining the performance of a learned classifier under concept drift. Intelligent Data Analysis 3(6), 453–474 (1999)
5. Bontchev, B., Vassileva, D.: Courseware Authoring for Adaptive E-learning. In: 2009 International Conference on Education Technology and Computer, pp. 176–180 (2009)
6. Brusilovsky, P.: Developing adaptive educational hypermedia systems: From design models to authoring tools. Information Sciences, 377–409 (2003)
7. Conlan, O., Wade, V.: Evaluation of APeLS – An Adaptive eLearning Service Based on the Multi-model, Metadata-Driven Approach. In: De Bra, P.M.E., Nejdl, W. (eds.) AH 2004. LNCS, vol. 3137, pp. 291–295. Springer, Heidelberg (2004)
8. Foss, J.G.K., Cristea, A.I.: The Next Generation Authoring Adaptive Hypermedia: Using and Evaluating the MOT3. 0 and PEAL Tools. Complexity, 83–92 (2010)

9. Göbel, S., Wendel, V., Ritter, C., Steinmetz, R.: Personalized, Adaptive Digital Educational Games Using Narrative Game-Based Learning Objects. In: Zhang, X., Zhong, S., Pan,
   Z., Wong, K., Yun, R., et al. (eds.) Edutainment 2010. LNCS, vol. 6249, pp. 438–445.
   Springer, Heidelberg (2010)
10. Harteveld, C.: Triadic Game Design. Springer (2011)
11. Hunicke, R., Chapman, V.: AI for Dynamic Difficulty Adjustment in Games. Assessment,
    91–96 (2003)
12. Kelly, H., et al.: How to build serious games. Communications of the ACM (2007)
13. Kickmeier-Rust, M.D., Albert, D. (eds.): An Alien's Guide to Multi-Adaptive Educational
    Computer Games. Informing Science Press, Santa Rosa (2012)
14. Knutov, E., et al.: AH 12 years later: a comprehensive survey of adaptive hypermedia
    methods and techniques. New Review of Hypermedia and Multimedia 15(1), 5–38 (2009)
15. Kunin, T.: The construction of a new type of attitude measure. Personnel Psychology
    51(4), 823–824 (1955)
16. Lopes, R., Bidarra, R.: Adaptivity Challenges in Games and Simulations: A Survey. IEEE
    Transactions on Computational Intelligence and AI in Games 3(2), 85–99 (2011)
17. Louchart, S., Aylett, R.S.: Solving the Narrative Paradox in VEs – Lessons from RPGs. In:
    Rist, T., Aylett, R.S., Ballin, D., Rickel, J. (eds.) IVA 2003. LNCS (LNAI), vol. 2792, pp.
    244–248. Springer, Heidelberg (2003)
18. Maciuszek, D., Martens, A.: A Reference Architecture for Game-based Intelligent Tutor-
    ing. In: Felicia, P. (ed.) Handbook of Research on Improving Learning and Motivation
    through Educational Games Multidisciplinary Approaches. IGI Global (2011)
19. Mehm, F., et al.: Authoring Environment for Story-based Digital Educational Games. In:
    Kickmeier-Rust, M.D. (ed.) Proceedings of the 1st International Open Workshop on Intel-
    ligent Personalization and Adaptation in Digital Educational Games, pp. 113–124 (2009)
20. Mehm, F., et al.: Bat Cave: A Testing and Evaluation Platform for Digital Educational
    Games. In: Proceedings of the 3rd European Conference on Games Based Learning. Aca-
    demic Conferences International, Reading (2010)
21. Mehm, F., et al.: Introducing Component-Based Templates into a Game Authoring Tool.
    In: Dimitris Gouscos, M.M. (ed.) 5th European Conference on Games Based Learning, pp.
    395–403. Academic Conferences Limited, Reading (2011)
22. Moreno-Ger, P., et al.: Adaptive Units of Learning and Educational Videogames. Journal
    of Interactive Media in Education 05, 1–15 (2007)
23. Peirce, N., et al.: Adaptive Educational Games: Providing Non-invasive Personalised
    Learning Experiences. In: 2008 Second IEEE International Conference on Digital Game
    and Intelligent Toy Enhanced Learning, pp. 28–35 (2008)
24. Prensky, M.: Digital Game-Based Learning. Paragon House (2007)
25. Rieber, L.P.: Seriously considering play: Designing interactive learning environments
    based on the blending of microworlds, simulations, and games. Educational Technology
    Research & Development 44(2), 43–58 (1996)
26. Shute, V., Towle, B.: Adaptive E-Learning. Educational Psychologist 38(2), 105–114
    (2003)
27. Smith, G., et al.: PCG-based game design: enabling new play experiences through proce-
    dural content generation. In: Proceedings of the 2nd International Workshop on Procedural
    Content Generation in Games PCGames 2011 (2011)
28. Thue, D., et al.: Learning Player Preferences to Inform Delayed Authoring. Psychology,
    158–161 (2007)

A Dashboard to Regulate Project-Based Learning

Christine Michel[1,2], Elise Lavoué[1,3], and Laurent Pietrac[1,4]

[1] Université de Lyon, CNRS
[2] INSA-Lyon, LIRIS, UMR5205, F-69621
[3] Université Jean Moulin Lyon 3, MAGELLAN, LIRIS, UMR5205
[4] INSA-Lyon, AMPERE, UMR5005, F-69621
{Christine.Michel,Laurent.Pietrac}@insa-lyon.fr,
Elise.Lavoue@univ-lyon3.fr

Abstract. In this paper, we propose the dashboards of the Pco-Vision platform to support and enhance Project-Based Learning (PBL). Based on the assumption that Self-Regulated Learning (SRL) is a major component of PBL, we have focused our attention in the design of a dashboard to enhance SRL in PBL. We describe the characteristics of PBL and show why a dashboard can help involved SRL processes, more particularly self-monitoring and self-judgment. We provide a categorization of the information to be presented on dashboards to help students involved in a PBL situation; by taking into account both the project and the learning goals. Finally we have conducted an experiment using the Pco-Vision platform with 64 students involved in a 6-months PBL course; results show that, whereas students rather use direct communication for tasks related to the self-monitoring process, the dashboard appears to be of great importance to enhance the self-judgment process, especially by presenting the information about the way of carrying out the activities.

Keywords: Self-Regulated Learning, Project-Based Learning, Dashboard.

1 Introduction

In this paper, we study how to support Project-Based Learning (PBL), which is a teaching and learning model that organizes learning around projects. PBL combines the project goals (the aim to achieve) and the learning goals (the knowledge to learn in the course). Actually, we observe that the implementation of PBL in engineering schools, universities or professional training do not benefit from all its capacities, because it is often action (according to the Kolb's learning cycle) which is favored to the detriment of reflection and personal experience [1]. Action involves students in the PBL situation, but is not sufficient to help them to acquire new knowledge and skills, like learning to collaborate or learning to manage a project. Our approach considers Self-Regulated Learning (SRL) as a major component of PBL to bring learners to self-reflect on their experience and to apply metacognitive skills.

Our research aims at designing a dashboard to support the SRL processes in project-based activities, by providing useful information to students. In the first part

A. Ravenscroft et al. (Eds.): EC-TEL 2012, LNCS 7563, pp. 250–263, 2012.

of the paper, we study the SRL processes involved in PBL and justify the use of a dashboard to enhance self-monitoring and self-judgment processes. We base on this study to provide a categorization of the information useful for learners to regulate themselves in PBL. We then describe the software prototype Pco-Vision, which offers dashboards that present this information on the shape of indicators. In the second part, we detail the results of an exploratory study conducted in real conditions, with 64 students involved in a 6-month PBL course. We were interested in studying the utility of dashboards for students, with regard to the activities to carry out. We more particularly focused on the utility of the indicators presented on the dashboards of the Pco-Vision platform.

2   State of the Art

2.1   Project-Based Learning

PBL is often applied in the case of complex learning, which aims to help learners acquire various linked skills or develop their behaviors [2]. Collaborative learning through project-based work promotes abstraction from experience, explanation of results, and understanding of conditions of knowledge applicability in real world situations; it also provides the experience of working in teams [3]. The main characteristic that makes PBL different from other instructional methods is its problem-centered content structure. It affects the learning and reasoning process: the teachers do not organize and assign the tasks and the learning does not consist in a simple fact-collection [4]. Instead, PBL learners have to engage in a more or less inquiry process: the organization of the learning activity is only defined on a macro-schedule and students have to define actions to do to solve the problem. Moreover, the PBL situations are often carried out on a long-term, usually several months. The characteristics of PBL arouse the complexity of the learning, which requires learners to deal with the management of the actions to establish by taking into account the time and team constraints. So in a long time and collective project, learners have to regulate themselves individually and collectively.

2.2   Self-regulated Learning and Group Awareness

Self-regulation can be defined as *"self-generated thoughts, feelings and actions that are planned and cyclically adapted to the attainment of personal goals"* [5]. Zimmermann's loop of self-regulated learning consists of three aspects: forethought, performance and self-reflection. In the social Cognitive Theory of Self-Regulation they are called self-monitoring, self-judgment, and self-reaction [6]: *"Self-monitoring is the mechanism by which individuals gain information on their own performance by setting realistic goals and evaluate their progress toward them. Self-judgment involves the processes by which individuals compare their performance or actions to the personal standards they developed in a particular domain. [...] Self-reaction represents the activities undertaken to regulate actions."* These processes are conventionally viewed as being predominantly individualistic [7]. According to the knowledge building model elaborated by Pata and Laanpere [8], SRL in an organizational context should also consider that in order to perform intrinsically motivated learning, learners have to align their learning activities to their organizational learning goals, the learning activities of other members of the organization and their own learning goals.

According to Carmen and Torres [9], the characteristics of self-regulated learners are self-motivation, employment of learning strategies and active participation in learning on a behavioral, motivational and metacognitive level. But it is very difficult for students to regulate their learning without help, since it requires complex skills. Unless provided with the appropriate tools, most people are not proactive enough to initiate a learning process or simply do not know how to learn [7]. The usual way to help learners to regulate their learning in a collective context is to give them group awareness tools.

The group awareness as been well defined by Buder and Bodemer [10] as knowledge about the social and collaborative environment the person is working in (e.g., knowledge about the activities, presence or participation of group members). Group awareness tools supply information to students to facilitate coordination and regulation of activities in the content space (i.e., efforts aimed at problem-solving, such as exchange of information or discussion of answers and alternatives) or the relational space (i.e., efforts to establish a positive group climate and to ensure effective and efficient collaboration [11]). They were developed in the CSCL area to foster the acquisition of group awareness, which is helpful for efficient group performance by presenting social comparison and guide for activities [12]. However, awareness tools are not meant for supporting long-term activities, by linking the activities with the goals to reach. Furthermore, according to [12], awareness approaches must be supplemented by with knowledge-related ones in order to describe an individual's state of being informed and having perceived information about others' knowledge.

2.3   A Dashboard to Regulate Project-Based Learning

In the context of project management, dashboards are used to present the project goals and the organization's goals and to support the collaborative work of the teams in a long-term perspective. This type of dashboard provides information resources that support distributed cognition. They are intended to provide information at a glance and to allow easy navigation to more complete information on analysis views [2]. In order to well manage a project, a dashboard must present three types of information: the state of the tasks carried out; the values of some specific characters (e.g. Coordination of resources, Scope, Time, Cost) listed by the Project Management Institute (PMI); and a performance analysis based on the relevance, effectiveness and efficiency (REE) of the resources used as compared to the results obtained [13]. When the dashboard is linked with shared workspace tools, the resources used for carrying out the tasks are directly exploited to increase awareness and cooperation in the team [14] in order to help workers to coordinate their on-going activities within the shared resources.

In the context of Project-Based Learning (PBL) and according to an organizational approach [7], the learners need personalized information about the organization's objectives and expectations; the learning activities and achievements of co-workers; and learners' own progress with regard to their current learning goal(s). For example, Siadaty et al. [15] have developed a Learning Pal tool, which offers functionalities divided into three main sets: a) harmonization with organizational goals, b) aligning to organizational members' learning goals, and c) aligning to individual learning goals. We also think that the behavior awareness is as important as knowledge awareness in PBL, since students have to acquire various linked skills [2] and to carry out complex tasks and activities [3]. According to Scheffel et al. [16], the key to SRL is self-monitoring and self-reflecting one's own behavior. Monitoring their learning activities helps learners to become aware of their actions and that could then lead to an adjustment of their behavior.

In PBL, the monitoring process is done through the specification of goals and strategy planning. It is complex because students have to consider both the project goals (and their associated strategy) and the learning goals (and their associated strategy). The first ones are defined to support the activities to carry out to achieve the project outcomes. The second ones are defined to support the cognitive and metacognitive processes involved in the learning of project management (i.e. target academic and social skills). The formalization of these goals is necessary due to the time and team constraints. The results of the actions have to be presented to the team in order to help the project management and to help each member to do the self-monitoring and self-judgment on the academic and social skills applied. Our work relies on the assumption that the use of a dashboard can help these processes.

3   A Dashboard for Students in PBL Situations

3.1   Information to Present on a Dashboard in PBL Situations

In this part, we first provide a categorization of the information useful for learners to regulate themselves in PBL. It is based on both project management systems and group awareness issues. We then describe the Pco-Vision platform we have developed to support the SRL processes in a PBL course. We distinguish two types of information: (1) the information about the individuals' and groups' goals, which direct the activities and (2) the information about the activities: the way of carrying out them and their results.

Information about Goals

- *The project goals.* The project group defines the project goals into the master plan. With regard to the PMI recommendations, most useful information is about project integration, human resource, time and scope. It can be represented by planned tasks, to which human resources are allocated. Information about pre-identified risks is also useful. So the project goals information is really closed to information managed with classical project management tools.

- *The learning goals.* The learning goals are the knowledge to acquire. In the context of PBL, there are two types of learning goals. The course learning goals are the knowledge required to achieve the course. They are described into the curriculum of the course and correspond to pre-defined evaluation rules. The project learning goals are the knowledge necessary to achieve the project. When the goal is different for each project, each group has its own knowledge to acquire (e.g. about the subject of the project or the programming language to use). The members of a group could also have different learning goals, since they have different tasks to achieve.

Information about the Activities. In order to help the self-monitoring process and to reinforce the learners' motivation, we advise to present both information about the way the activities are carried out and their results.

- *The way of carrying out the activities.* The way of carrying out the activities relates to the learners' and groups' behaviors and state of mind during the project. On an objective way, the behavior and the level of motivation can be represented for instance by the time spent working on the project (individually and collectively), the social organization (the members who work together), or the modalities of work (presence or distance, individual or collective). On a more subjective way, the way of carrying out the project can be described by the level of motivation and the state of mind expressed by students during their work: satisfaction of themselves, feeling of efficacy and efficiency or their situation within their group.
- *The results of the activities.* With regards to the goals, the results of the activities relate, on the one hand, to the progress in the project and, on the other hand, to the level of knowledge acquired. These results have to be presented taking into account the target level of achievement defined for each goal. Furthermore, the results concern different actors: some are defined for each member individually and others for the whole project group.

In this categorization, we make the hypothesis that the awareness of the project goals compared to the results of the activities will help the self-monitoring of the skills applied to achieve the project. On the same way, the awareness of the learning goals compared to the results of the activities will help the self-monitoring of the skills applied to achieve the course. The self-judgment process will be supported by information about the way of carrying out the activities. The motivation and behavior of the team can be viewed as an indirect judgment of the pairs. The modality of work or time spent can be viewed as a standard of comparison. Finally, the whole information gives to student the possibility to make causality links between goals, actions and results. It so can help the self-reaction process. This is especially helpful to build complex skills that require an evolution of behavior.

3.2   The Dashboards of the Pco-Vision Platform

General Presentation. We adopt an iterative and participatory design approach to develop a platform to support PBL. The software prototype Pco-Vision is the result of a second development cycle and is based on a paper prototype named MESHAT [2]. It has been designed mainly to test the utility of the information presented on dashboards in a PBL course. Pco-Vision is a web-based platform, which offers several functionalities thanks to five drop-down menus (see Fig. 1). (a) A home menu provides a video demonstration to help students to use the platform. (b) A data capture menu offers several data entry forms to students. (c) A dashboards menu gives access to an individual and a group awareness project tools (dashboards). (d) A collaboration menu gives access to a blog and an agenda for the group. (e) A documentation menu gives access to resources useful for the project (e.g. models of document) and for the course achievement (e.g. a learning contract).

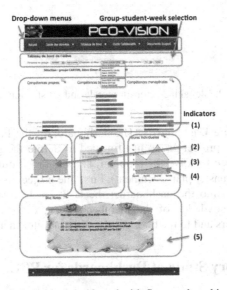

Fig. 1. Individual dashboard with five numbered indicators

Dashboards. Pco-Vision offers two dashboards: a collective dashboard for the group and an individual dashboard for each student. Students can click on indicators to access an analytical view.

- *The individual dashboard.* The individual dashboard (see Fig. 1) offers to students an overview of their activities thanks to five indicators: (1) their level for the knowledge they have to acquire, in comparison with the target level defined in the master plan, (2) their state of mind during the last four weeks, on the shape of two curves (morale and satisfaction), (3) the tasks to do, on the shape of a post-it note, (4) the individual working time in comparison with the collective working time, on the last four weeks (two different curves), (5) the key events that students note (like in a blog).
- *The project group dashboard.* The group dashboard presents six indicators: (1) the workload (the working time of the group, with regards to the planned tasks), (2) the

ratio of individual to collective work, (3) the level of knowledge acquired with regards to the defined target level, (4) the problems that occurred and the number of times, (5) the state of mind (mean, minimum, maximum and deviation of the level of satisfaction of the members), (6) the progress of the tasks.

In order to support the analytical process of reflexive behavior understanding, students can explore information about the way they are doing activities [17]. For example on Fig. 2, students can view the global evolution of their morale and satisfaction during the first months of the project, and navigate dynamically on this view.

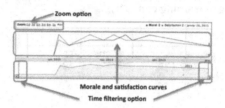

Fig. 2. Global indicator for morale and satisfaction

The Intended use of the Dashboards. The dashboards of the Pco-Vision platform could be used when the project group has finished to define the goal of the project (the solution they propose to solve the problem) and the broad outline of the project plan (formalized in the master plan document). First, the project manager could report the planed organization into the dashboards: the tasks to do (list, affectations and dates) and the target knowledge of the members. Then, the dashboards could be opened to all the members and to the tutors assigned to monitor them.

4   An Exploratory Study of Dashboards for PBL

4.1   Context and Participants

The participants on this study included a convenience sample of 64 students from an engineer school in France, registered on a 6-month PBL course. The students worked on a self-regulated way by group of 8-9 students, on a specific real industrial problem. The learning goals of the course were to learn to manage a project and to learn to collaborate. This course is conducted since 12 years and each year the students are asked to develop their own dashboard in order to coordinate the tasks and the team. As the use of a dashboard is already integrated into the framework of this course, we chose this context to study the use of a dashboard for self-regulated learning.

The students of the experimental group (n=24) used the dashboards of Pco-Vision during the project. In the paper, this group is named "Pco-Vision group". The students of the control group (n=40) used a dashboard that they have developed by their own, mainly on spreadsheets. Concerning the PBL course, all the students followed the same rules and had the same evaluation criteria. The groups had their own tutor that they met every week in order to take advices. They also each had a dedicated project room during the 6 months in order to work by group.

4.2   Data Collection and Analysis

The results are based on a quantitative statistical analysis of questionnaires. At the end of the course, all the students (n=64) answered a questionnaire composed of 28 questions. First questions concern the utility of the dashboard in the PBL context and is measured by yes/no questions. Other questions concern the relevance of the use of dashboards in order to manage some specific tasks and the other means the students used to manage their project. These questions were measured according to a Lickert scale (1 to 4). The students from the Pco-Vision group (n=24) answered an additional questionnaire about the general design of Pco-Vision and the relevance of the indicators presented on Pco-Vision. This questionnaire is composed of 24 questions measured according to a yes/no scale (11 questions) and a Lickert scale (1 to 4) (13 questions). This study aims to answer the following questions:

- Is a dashboard useful in the context of a PBL course?
- How often the dashboard and other external means are used to support individual and/or group activity regulation? We more particularly studied two types of regulation processes: self-monitoring and self-judgment.
- Amongst the indicators presented on the dashboards of Pco-Vision, which were the more frequently used by the students?

The questions related to the use of a dashboard (and external tools) for some tasks are intended to measure their relevance for these tasks. As recommended in [18], they are also intended to measure the individual perception of cognitive or metacognitive processes involved and so give information about the student's SRL aptitudes. The two distinct groups (Pco-Vision group and control group) allow us to study the impact of the use of a given dashboard (in Pco-Vision) on the students' SRL aptitudes, in comparison with the students who develop their own dashboard.

5   Results

5.1   General Utility of a Dashboard in a PBL Course

Utility of a Dashboard. The students rather agree with using a dashboard in the context of the course (see Table 1). For the majority of them, the context of the course (in university and in presence) and the size of the groups (only 8 or 9 students) do not limit the relevance of the use of dashboards. We make the assumption that the large number of "no answers" involves that the students that had Pco-Vision did not feel concerned by these questions.

Table 1. Utility and limits of the use of a dashboard in the study (Pco-vision *vs* control group)

Utility and limit of dashboard		No answer	No	Yes
Utility of a dashboard	pco	1.00%	8.33%	**87.50%**
for the course	cont	10.00%	10.00%	**65.00%**
Limits related to the	pco	91.67%	0.00%	**8.33%**
size of the team	cont	0.00%	92.50%	**7.50%**
Limits related to the	pco	91.67%	0.00%	**8.33%**
academic context	cont	0.00%	95.00%	**5.00%**

As a major result of the analysis work, Tables 2 and 3 show the tasks carried out by both groups of students with their group and individual dashboards. We observe a significant difference between the responses of the two groups. On the one hand, as the responses of the control group are almost identically distributed, we cannot identify a strong tendency in their use of their dashboard. Furthemore, we observe a large part of "no answer" responses in the control group, which means that the students did not understood or did not felt concerned. On the other hand, we can identify a tendency in the use of Pco-Vision. The students mostly used punctually Pco-Vision for the self-monitoring tasks we asked them in the questionnaire. We also observe that they more often used Pco-Vision in order to support their self-judgment individually (by checking their working time) and within their group (by checking the others' tasks, working time, moral and satisfaction). We deduce that having Pco-Vision has encouraged the students to have self-regulation processes, especially self-judgment.

Table 2. Frequency of use of the dashboard for the group work (Pco-Vision *vs* control group)

Tasks related to group work		No answer	Never Used	Punctually Used	Often Used	Frequently used
Self-monitoring process						
Define de director plan	pco	4.17%	12.50%	**70.83%**	12.50%	0.00%
	cont	**32.50%**	47.50%	12.50%	7.50%	0.00%
Adapt the director plan	pco	4.17%	33.33%	**62.50%**	0.00%	0.00%
	cont	**32.50%**	45.00%	12.50%	7.50%	2.50%
Coordinate the progress	pco	4.17%	33.33%	**62.50%**	0.00%	0.00%
of the tasks	cont	**27.50%**	12.50%	25.00%	15.00%	20.00%
Coordinate the work	pco	4.17%	33.33%	**62.50%**	0.00%	0.00%
of the group	cont	**27.50%**	20.00%	25.00%	10.00%	17.50%
Self-judgment process						
Check the others' tasks	pco	4.17%	33.33%	**45.83%**	**12.50%**	**4.17%**
	cont	**30.00%**	25.00%	25.00%	12.50%	7.50%
Check the others' working	pco	4.17%	20.83%	**41.67%**	**25.00%**	**8.33%**
time	cont	**30.00%**	15.00%	17.50%	22.50%	15.00%
Check the moral and	pco	4.17%	12.50%	**50.00%**	**16.67%**	**16.67%**
satisfaction of the others	cont	**27.50%**	17.50%	15.00%	22.50%	17.50%

Table 3. Frequency of use of the dashboard for individual work (Pco-Vision *vs* control group)

Tasks related to individual work		No answer	Never Used	Punctually Used	Often Used	Frequently used
Self-monitoring process						
Check my tasks to do	pco	4.17%	33.33%	**58.33%**	4.17%	0.00%
	cont	**25.00%**	45.00%	10.00%	7.50%	12.50%
Regulate my way of work	pco	4.17%	29.17%	**58.33%**	8.33%	0.00%
	cont	**27.50%**	32.50%	10.00%	10.00%	20.00%
Analyze the way the	pco	4.17%	29.17%	**58.33%**	4.17%	4.17%
project is realized	cont	**30.00%**	35.00%	22.50%	7.50%	5.00%
Analyze my way of	pco	4.17%	29.17%	**62.50%**	0.00%	4.17%
work in team	cont	**32.50%**	30.00%	20.00%	10.00%	7.50%
Self-judgment process						
Check my working time	pco	4.17%	12.50%	**29.17%**	**41.67%**	12.50%
	cont	25.00%	10.00%	15.00%	37.50%	12.50%

Table 4. External means used to plan and monitor the tasks and to collaborate (Pco-Vision *vs* control group)

External means used	Pco-Vision Group			Control Group		
	No answer	No	Yes	No answer	No	Yes
To plan the work						
Project management tool	12.50%	41.67%	45.83%	0.00%	**70.00%**	30.00%
Electronic Gantt chart	12.50%	41.67%	45.83%	0.00%	**75.00%**	25.00%
Paper Gantt chart	8.33%	20.83%	**70.83%**	0.00%	**72.50%**	27.50%
Online shared calendar	12.50%	41.67%	45.83%	0.00%	**80.00%**	20.00%
Face to face discussions	4.17%	0.00%	**95.83%**	0.00%	40.00%	**60.00%**
Discussions formalized into the dashboard	12.50%	25.00%	**62.50%**	0.00%	55.00%	**45.00%**
To monitor the project progress						
Face-to-face discussions	0.00%	0.00%	**100%**	0.00%	40.00%	**60.00%**
Synthesis of a team member	12.50%	16.67%	70.83%	0.00%	55.00%	45.00%
To organize the collaboration						
Email (to send files)	0.00%	0.00%	**100 %**	0.00%	27.50%	**72.50%**
Other platforms (googledocs)	8.33%	41.67%	50.00%	0.00%	**72.50%**	27.50%
Chat	16.67%	37.50%	45.83%	0.00%	**97.50%**	2.50%
SMS	8.33%	12.50%	**79.17%**	0.00%	50.00%	50.00%

The Association of Tools with Dashboards. We questioned the students about the other means or tools they used, in association with their dashboards. Both groups have regularly used oral discussions to plan the tasks and to monitor the progress of the project. To keep tracks of the discussions, some students (principally in the Pco-Vision group) have entered the results of the discussions into the dashboards and have made synthesis. Both groups used communication tools (email and SMS) to organize the collaboration.

We observe significant differences between the two groups in the use of tools: the students of the Pco-Vision group have more often used planning and collaboration tools than the students of the control group. In order to plan the tasks, most of the students of the Pco-Vision group used a paper Gantt chart, and about half of them also used a project management tool, an electronic Gantt chart and a shared calendar. About half of the students of the Pco-Vision group also used other platforms (like googledoc) and a chat tool to collaborate. Only few students of the control group used these planning and collaboration tools.

5.2   Specific Analysis of the Use of Pco-Vision

General Evaluation. According to the results presented in Table 5, Pco-Vision was seen as a constraint with limited benefit to the students' activity. Approximately half of the students consider that the group dashboard and the individual one (see Fig. 1) are adapted. Most useful indicators are synthesis ones (i.e. presented on the overview of the dashboard). Indeed, only few students have used dynamic indicators to see an analytical view (see Fig. 2). The indicators are globally considered coherent and rather useful but not sufficient. More precisely, the students consider indicators not very relevant and that they do not well reflect the reality. These results could be partly explained by an insufficient general design of Pco-Vision: the information loading process is low and must be improved; the data input process is based on a manual reporting activity (done one time per week) and so the data may not be very accurate, even missing.

Table 5. General evaluation of Pco-Vision

	No answer	No	Yes
General statement about pcovision			
Seen as a constraint	8.3%	37.5%	54.2%
Have given benefice	20.8%	50.0%	29.2%
Group Dashboard was adapted	4.2%	45.8%	**50.0%**
Individual Dashboard was adapted	4.2%	41.7%	**54.2%**
Evaluation of Indicators			
Coherent	4.2%	29.2%	**66.7%**
Relevant	4.2%	**58.3%**	37.5%
Reflect the real activity	4.2%	**58.3%**	37.5%
Useful	4.2%	50.0%	45.8%
Sufficient	12.5%	**66.7%**	20.8%
Use of synthesis indicators	4.2%	37.5%	**58.3%**
Use of dynamic indicators	8.3%	**75.0%**	16.7%

Usefulness of Indicators. The most used indicators are about the way of carrying out the activities. More precisely, the students mainly used Pco-Vision to see the state of mind (morale and satisfaction) of the group, the workload of the group, the ratio of individual to collective work and their individual working time. The less used indicators are those related to the results of the activities: the level of knowledge and skills acquired (at the individual and group level) and the progress in the project. In connection with the hypothesis made in section 3.1, we deduce that the students mainly used the information that helps the self-judgment process than the information supporting the self-monitoring process.

Table 6. Usefulness of the Pco-vision indicators

Pco-Vision Indicators	No answer	Never Used	Punctually Used	Often Used	Frequently used
Group dashboard					
Workload	4.17%	29.17%	29.17%	**25.00%**	**12.50%**
Ratio of individual to collective work	4.17%	37.50%	20.83%	**29.17%**	**8.33%**
Level of knowledge acquired	0.00%	**66.67%**	**20.83%**	12.50%	0.00%
Problems	0.00%	54.17%	29.17%	12.50%	4.17%
State of mind	0.00%	20.83%	33.33%	**33.33%**	**12.50%**
Tasks achievement progress	0.00%	**54.17%**	**33.33%**	12.50%	0.00%
Individual dashboard					
Skills required for project achievement	8.33%	**29.17%**	**41.67%**	20.83%	0.00%
Level of knowledge (technical skills)	8.33%	**29.17%**	**50.00%**	12.50%	0.00%
Level of knowledge (social skills)	8.33%	**37.50%**	**50.00%**	4.17%	0.00%
State of mind	8.33%	20.83%	45.83%	20.83%	4.17%
Tasks to do	8.33%	37.50%	25.00%	25.00%	4.17%
Working time	8.33%	29.17%	8.33%	**41.67%**	**12.50%**
Self-notes, keys events	8.33%	41.67%	25.00%	16.67%	8.33%

6   Discussion and Conclusions

In this paper, we have first provided a categorization of the information to present on a dashboard to enhance Self-Regulation Learning (SRL) processes in a Project-Based Learning (PBL) course. This categorization distinguishes the information about goals (project and learning goals) and the information about the activities (the way of carrying out them and their results). We based on this categorization to design the dashboards of the Pco-Vision software prototype that has been used by 64 students in a PBL course during six months. We conducted a study in the context of this course by the way of two questionnaires and we think that the results of the study have implication for the design and the integration of a dashboard in a PBL course.

The students agree with the importance of using a dashboard in their PBL course. However, the students with Pco-Vision felt the use of the dashboard as a constraint with limited benefice to their activity. But, although Pco-Vision has design problems that explain a rather negative opinion of the students, the results allowed us to identify a positive tendency in the use of the dashboard for some specific tasks that help the self-monitoring and self-judgment processes. We so first deduce that having Pco-Vision has encouraged the students to apply self-regulation processes.

The dashboard has been mainly used to support the self-judgment process, thanks to the information on the way of carrying out the individual and group activities, especially the working time and the state of mind. The students also used the dashboard to compare themselves or to check if their involvement in the project (quantity of work) was visible.

The dashboard has been little used to support tasks related to the self-monitoring process. Indeed, the students prefer direct communication (face-to-face, Email, SMS) or tangible tools (paper support) to plan the tasks, to monitor the progress of the project and to organize the collaboration within the team. However, we observed that the students of the Pco-Vision group are more inclined than those of the control group to use instrumented tools, more particularly planning and collaboration tools,

As a perspective, we will improve the design of the dashboards of Pco-vision, so as to facilitate their use. Indeed, the students have trouble reporting each week the data related to the way they carry out their activities and so consider that the indicators do not well reflect the reality. Moreover, the students did not succeed in using knowledge indicators (the level for the knowledge to acquire) or deep indicators presented in analytical views.

We make some hypothesis to explain these results and to determine our future work. We think that the presentation of the indicators on the dashboards has to be rethought, so as to help their use to support complex tasks (monitor the collaborative work, support metacognitive process) and so enhance self-regulation. For instance, we will use more spatial or temporal information views in order to improve the work with indicators during the monitoring and reacting steps. We thing also useful to offer other interaction functions of flexible display layouts [15] in order to let students manipulate indicators, choose their presentation and modify their view-size and position according to their importance and weight. Finally, we will improve the data input process, for instance by offering contextualized and flexible data input interfaces centered on each unit of task carried out. We will also automate a part of the data input process by linking the dashboards with other tools used to carry out the project (e.g. Gantt chart).

References

1. Thomas, J., Mengel, T.: Preparing project managers to deal with complexity – Advanced project management education. International Journal of Project Management 26, 304–315 (2008)
2. Michel, C., Lavoué, E.: KM and Web 2.0 Methods for Project-Based Learning. MEShaT: a Monitoring and Experience Sharing Tool. In: Ifenthaler, D., Isaias, P., Spector, J.M., Kinshuk, S.D. (eds.) Multiple Perspectives on Problem Solving and Learning in the Digital Age, Heidelberg, pp. 49–66 (2011)
3. Jeremic, Z., Jovanovic, J., Gasevic, D.: Semantically-Enabled Project-Based Collaborative Learning of Software Patterns. In: IEEE International Conference on Advanced Learning Technologies, pp. 569–571. IEEE Computer Society, Los Alamitos (2009)

4. Hung, W.: The 9-step problem design process for problem-based learning: Application of the 3C3R model. Educational Research Review 4, 118–141 (2009)
5. Zimmerman, B.: Attaining self-regulation: A social cognitive perspective. In: Handbook of Self-Regulation, pp. 13–40 (2000)
6. Gravill, J., Compeau, D.: Self-regulated learning strategies and software training. Information & Management 45, 288–296 (2008)
7. Littlejohn, A., Margaryan, A., Milligan, C.: Charting Collective Knowledge: Supporting Self-Regulated Learning in the Workplace. In: Ninth IEEE International Conference on Advanced Learning Technologies, ICALT 2009, pp. 208–212. IEEE (2009)
8. Pata, K., Laanpere, M.: Supporting Cross-Institutional Knowledge-Building with Web 2.0 Enhanced Digital Portfolios. In: Eighth IEEE International Conference on Advanced Learning Technologies, ICALT 2008, pp. 798–800. IEEE (2008)
9. Carmen, M., Torres, G.: Self-Regulated Learning: Current and Future Directions. Educational Psychology 2, 1–34 (2004)
10. Buder, J., Bodemer, D.: Group Awareness Tools for Controversial CSCL Discussions: Dissociating Rating Effects and Visualized Feedback Effects. In: 9th International Conference on Computer Supported Collaborative Learning (CSCL 2011), Hong Kong, pp. 358–365 (2011)
11. Janssen, J., Erkens, G., Kirschner, P.A.: Group awareness tools: It's what you do with it that matters. Computers in Human Behavior 27, 1046–1058 (2011)
12. Engelmann, T., Dehler, J., Bodemer, D., Buder, J.: Knowledge awareness in CSCL: A psychological perspective. Comput. Hum. Behav. 25, 949–960 (2009)
13. Marques, G., Gourc, D., Lauras, M.: Multi-criteria performance analysis for decision making in project management. International Journal of Project Management 29, 1057–1069 (2011)
14. Treude, C., Storey, M.-A.: Awareness 2.0: staying aware of projects, developers and tasks using dashboards and feeds. In: Proceedings of the 32nd ACM/IEEE International Conference on Software Engineering, vol. 1, pp. 365–374. ACM, New York (2010)
15. Siadaty, M., Jovanovic, J., Pata, K., Holocher-Ertl, T., Gasevic, D., Milikic, N.: A Semantic Web-enabled Tool for Self-Regulated Learning in the Workplace. In: 11th IEEE International Conference on Advanced Learning Technologies (ICALT), pp. 66–70. IEEE (2011)
16. Scheffel, M., Wolpers, M., Beer, F.: Analyzing Contextualized Attention Metadata with Rough Set Methodologies to Support Self-regulated Learning. In: IEEE 10th International Conference on Advanced Learning Technologies (ICALT), pp. 125–129. IEEE (2010)
17. Kuo, C.-H., Tsai, M.-H., Kang, S.-C.: A framework of information visualization for multi-system construction. Automation in Construction 20, 247–262 (2011)
18. Moos, D., Azevedo, J.: Monitoring, planning, and self-efficacy during learning with hypermedia: The impact of conceptual scaffolds. Computers in Human Behavior 24, 1686–1706 (2008)

Lost in Translation from Abstract Learning Design to ICT Implementation: A Study Using Moodle for CSCL

Juan Alberto Muñoz-Cristóbal, Luis Pablo Prieto, Juan Ignacio Asensio-Pérez, Iván M. Jorrín-Abellán, and Yannis Dimitriadis

GSIC-EMIC Group, University of Valladolid, Spain
{Juanmunoz,lprisan}@gsic.uva.es, {juaase,yannis}@tel.uva.es,
ivanjo@pdg.uva.es

Abstract. In CSCL, going from teachers´ abstract learning design ideas to their deployment in VLEs through the life-cycle of CSCL scripts, typically implies a loss of information. It is relevant for TEL and learning design fields to assess to what extent this loss affects the pedagogical essence of the original idea. This paper presents a study wherein 37 teachers' collaborative learning designs were deployed in Moodle with the support of a particular set of ICT tools throughout the different phases of CSCL scripts life-cycle. According to the data from the study, teachers considered that the resulting deployment of learning designs in Moodle was still valid to be used in real practice (even though some information is actually lost). This promising result provides initial evidence that may impulse further research efforts aimed at the ICT support of learning design practices in the technological context dominated by mainstream VLEs.

Keywords: CSCL, Moodle, learning design, life-cycle, VLE, translation.

1 Introduction

Effectiveness of collaborative learning depends on multiple factors, including the way interactions among learners are promoted, structured, and regulated [1]. Such learner scaffolding may be achieved through Computer Supported Collaborative Learning (CSCL) scripts, that can take the form of computationally interpretable specifications of a desired collaboration process [2]. CSCL scripting can be considered a specific form of learning design [3], focused on collaborative learning pedagogical principles and techniques.

Different approaches in the literature identify the phases that CSCL scripts go through during their "life-cycle", from initial inception to enactment. For instance, phases for specification, formalization, simulation and deployment are proposed in [2], while design, instantiation and enactment are mentioned in [4,5]. Additionally, operationalization is used instead of instantiation (i.e. design, operationalization and execution) in [6].

As we can see, the phases considered in such life-cycle can change depending on the methodologies and tools used, or on other factors. Moreover, the script life-cycle does not need to be lineal, with perfectly differentiated phases [6]. Nevertheless,

A. Ravenscroft et al. (Eds.): EC-TEL 2012, LNCS 7563, pp. 264–277, 2012.

different approaches have in common that, from the CSCL script's conception in the mind of its author, up to its final form ready to be used in a concrete computer-supported scenario, the script has to traverse different human or computer agents, in which it is completed, particularized or modified. It is also noteworthy that, in many educational institutions around the world, the technological environment in which CSCL scripts are deployed, executed or enacted (depending on the approach followed) often is a Virtual Learning Environment (VLE) [7] such as *Moodle*[1], *Blackboard*[2], *Sakai*[3] or *LAMS*[4].

However, supporting the life-cycle of CSCL scripts using different software tools until its deployment in a widespread VLE may introduce changes in the original idea of the learning designer [8]. Typically, several software agents (e.g. design authoring tools, instantiation tools, VLEs, etc.) and human agents (teachers, instructional designers, etc.) will be involved in this script life-cycle, with different data models and different conceptions/understandings of the design, respectively. Thus, in the end, the result (e.g., a course in *Moodle*) may not reflect the original abstract ideas and the pedagogical intention of the designers (e.g. a teacher), due to the multiple translations performed during the whole process.

To the best of our knowledge, there is a dearth of examples in literature studying these transformations from abstract inception to deployment in a particular VLE. However, we do believe that this transformation is highly relevant for the learning design and technology enhanced learning (TEL) research fields. If the changes are too large, the pedagogical essence of the original design idea may be fatally modified, and the resulting course may no longer be valid to be enacted in the teacher's class (which somewhat decreases the usefulness of making learning design decisions explicit). Finding means of applying learning design tools and methods to existing, widespread ICT learning environments, is an issue that can "make or break" the applicability and impact of learning design on a wider scale.

The objective of this paper is to study the CSCL script life-cycle of a set of 37 CSCL learning designs devised by higher-education teachers from different disciplines, in the context of two professional development workshops. The paper tries to clarify at what points of the scripts life-cycle the information changes, what is the nature of those changes, how much information and what information is lost. Our ultimate goal is to ascertain how these changes affect the fidelity of the result in a VLE, to be enacted in a real situation by a teacher.

The structure of the paper is as follows. Section 2 presents the problem of the translations when going from a learning design idea to a VLE-based infrastructure, following the life-cycle of CSCL scripts. In Section 3, 37 designs from two workshops are analyzed, to evaluate to what extent the final result in a widespread VLE (such as *Moodle*) maintains the pedagogical essence of the original idea. Section 4 discusses the results, and finally, the main conclusions and future research lines are described in Section 5.

---

[1] http://moodle.org (Last access 3/22/2012)
[2] http://www.blackboard.com (Last access 3/22/2012)
[3] http://sakaiproject.org (Last access 3/22/2012)
[4] http://www.lamsfoundation.org (Last access 3/22/2012)

2   The Problem of Translations in the Life-Cycle of CSCL Scripts

Several ICT tools may support one or more phases of the life-cycle of CSCL scripts (e.g., a set of learning design tools can be found in *The Learning Design Grid*[5]). Some tools, as e.g. *Reload*[6], *Collage* [9] or *ReCourse*[7], focus on supporting the design phase and follow a particular specification of learning design language [10,11] like IMS LD [12], while others employ their own proprietary data model (*CompendiumLD*[8], *Pedagogical Pattern Collector*[9]). Other tools focus on the instantiation phase, as e.g. *InstanceCollage* [5] and *CopperCore*[10], or cover both design and instantiation, such as *WebCollage*[11]. Finally, *GLUE!-PS* [13] is a tool dealing with instantiation and deployment that allows deploying learning designs from multiple learning design language/authoring tool to multiple VLEs.

On the other hand, most widespread VLEs like *Moodle*, *Sakai* or *Blackboard* focus only on enactment/execution. LAMS, on the contrary, provides support to the complete life-cycle (including learning design), and it is an example of an easy-to-use integrated approach. However, such an all-in-one approach does not allow taking advantage of affordances provided by other design tools and thus, sharing and reusing design resources outside the *LAMS* VLE becomes difficult for practitioners. Finally, *LAMS* is not as widespread as *Moodle* or *Blackboard*[12], and therefore it may not be available (or practical) for many teachers, due to institutional VLE choices.

All in all, there is a diversity of computer agents (tools) potentially involved in the CSCL script life-cycle. Additionally, it is frequent to have more than one human agent (teachers, instructional designers, etc.) using the aforementioned tools in different moments of the process. It is thus important to know what occurs with the pedagogical essence of a script along this process, from being an idea in the mind of, e.g., a teacher, up to its crystallization as a set of resources ready to be used in a VLE.

In general terms, the information in the script can change each time it traverses machine or human agents: because of human or machine action, or due to a human to machine interaction. For instance, information can be lost when a third party, e.g. an instructional designer, interprets a teacher design. Also, data may be modified to be adapted to the specific data model used by a supporting tool. Information may be lost as well because of a lack of expertise of the user of any of the supporting tools, or due to missing information in the formalization or interpretation of the learning design.

---

[5]   http://www.ld-grid.org/resources/tools (Last access 3/22/2012)
[6]   http://www.reload.ac.uk/ldesign.html (Last access 3/22/2012)
[7]   http://tencompetence-project.bolton.ac.uk/ldauthor/ (Last access 3/22/2012)
[8]   http://compendiumld.open.ac.uk (Last access 3/22/2012)
[9]   http://tinyurl.com/ppcollector3 (Last access 3/22/2012)
[10] http://coppercore.sourceforge.net/ (Last access 3/22/2012)
[11] http://pandora.tel.uva.es/wic2 (Last access 3/22/2012)
[12] Three month Traffic Rank at alexa.com (03/30/2012): blackboard.com: 1,709; moodle.org: 4,285; lamsfoundation.org: 937,821.

Figure 1 shows an example of three generic ICT tools supporting the life-cycle of a CSCL script. Each time the CSCL script traverses a human or a machine agent (an ICT tool in the example), the information in the script can change.

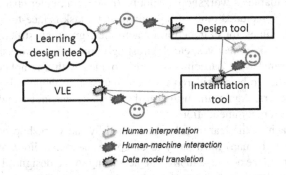

Fig. 1. Potential points of modification/loss in a typical CSCL script life-cycle

So far, research initiatives have not analyzed the information loss that might occur during the complete life-cycle, and therefore evidence should be provided on the fidelity of the final product (e.g. a course or activity ready to be used in a VLE), compared to the original learning design idea. If such a product has lost the pedagogical essence of the original idea, it may not result useful to be enacted by the teacher in a real situation. Thereby, it is necessary to study the degree of alignment of this "reified script" and the pedagogy underlying the original learning design idea. Evidence on the information loss may contribute to the design and development of appropriate supporting tools, and help researchers in understanding the complete life-cycle.

3 A Study: From Learning Designs in a Workshop to Moodle

3.1 Description of the Study

In order to study the loss of information when following the CSCL script life-cycle from abstract design to a widespread VLE, two workshops on professional development were conducted and analyzed at the University of Valladolid, the first one in June and September 2011, and the other in February 2012. The workshops focused on designing CSCL activities and participants were faculty members from multiple fields (e.g. Computer Science, Medicine, Biology, etc), with varying ICT abilities. Both workshops had a blended learning format, with two 4-hour face-to-face sessions and a number of tasks to be accomplished on-line between sessions. The first session was devoted to the creation of a technology-enhanced collaborative learning design by means of a Pyramid collaborative pattern [14]. After this initial session, each participant was asked to particularize such a learning design to one of his/her own courses, using a collaborative pattern. The designs produced by teachers were free-form, natural language descriptions of the design ideas, often with accompanying graphical schemata. Even though participants were free to choose any collaborative pattern for their designs, the Pyramid was recommended because of its relative simplicity. Nevertheless, descriptions of other patterns such as Jigsaw, Think-Pair-Share or Brainstorming, were available as workshop handouts. In addition, other characteristics were recommended to be included in the designs: whether a task is face-to-face, blended or remote, estimated times for completion of activities, grouping structures, ICT tools used to support a task, objectives, etc. Interestingly, the second workshop introduced this in a more formal way: participants were provided with a template identifying a list of characteristics to be considered for their inclusion in the learning designs to be generated by teachers. Again, the usage of the template was not mandatory and was solely intended as a recommendation.

Afterwards, each of the learning design created by the workshop participants was used as input by a (human) third party to complete the remaining CSCL script life-cycle phases, to produce a course in *Moodle* according to the designs. Twelve of these designs were completed in the first workshop, and twenty five more in the second one. The third party role was played by an ICT-expert researcher, who used the *Web-Collage* learning design tool to convert the teachers' designs into computationally interpretable scripts. Then, the scripts were deployed automatically in *Moodle* using *GLUE!-PS*.

Figure 2 shows the particular CSCL script life-cycle employed in this study, with an example of the life-cycle of one of the scripts, as well as the critical points (in green, red and black) where information of the script might have been lost. The context described so far serves to settle the research question driving the whole study:

[QG]: *Does the final result of the designs in Moodle maintain the pedagogical essence well enough to remain usable by their original authors (faculty)?*

In order to answer the research question, we employed a mixed evaluation approach [15], gathering both quantitative and qualitative data.

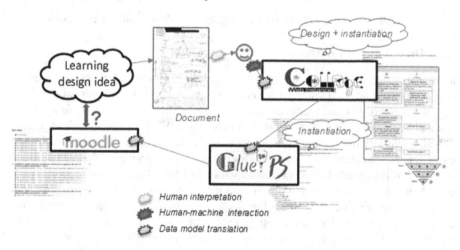

Fig. 2. CSCL script life-cycle and points of change in the use case

3.2   Context and Methodologies of the Study

As mentioned above, the study was carried out in the context of two workshops at the University of Valladolid, on the topic of design and deployment of advanced collaborative activities using ICT. To help with the planning and organization of the evaluation, we followed the *Evaluand*-oriented Responsive Evaluation Model (CSCL-EREM) [16], using a variety of quantitative and qualitative data gathering techniques. The model is deeply focused on the *Evaluands* (the subject under evaluation), and it is framed within the Responsive Evaluation approach [17]. According to this, the model is oriented to the activity, the uniqueness and the plurality of the *Evaluand* to be evaluated, promoting responsiveness to key issues and problems recognized by participants at the site. The model includes three core parts (*Perspective*, *Ground* and *Method*) that could be taken into account while doing an evaluation, a representation diagram to help evaluators in the planning stage, and a set of recommendations to write the report of the evaluation. The emphasis of the *Perspective* has to do with the point of view from which we are conducting the evaluation. *Ground* is the context in which the *Evaluand* takes place or is intended for. *Method* is the sequence of steps that lead the evaluation process [18].

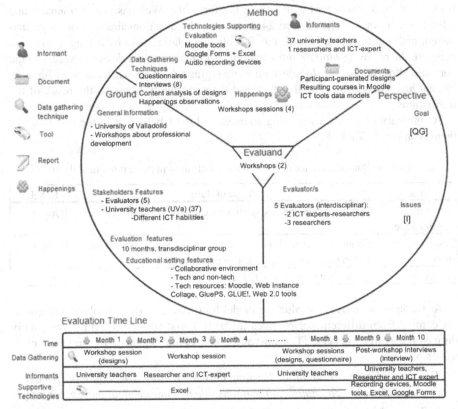

Fig. 3. Planning of the evaluation using the CSCL-EREM model. [QG] refers to the research question (section 3.1) whereas [I] refers to considered evaluation issue (section 3.2).

Figure 3 shows the planning diagram of the evaluation conducted, using the afore-mentioned CSCL-EREM model. The diagram shows that the *Evaluand* corresponds to the two workshops. The *Perspective* is that of a *research* work. The *Ground* is a context of the two workshops already mentioned, wherein the participants were 37 university teachers, and the organizers 5 interdisciplinary researchers (the evaluators). The workshops' environment was collaborative, and in a mixed form of technological and not technological, using both physical materials (e.g. pen/paper) and ICT tools.

The *Data Gathering Techniques* used in the evaluation process were: interviews (8); Web-based questionnaires (24); naturalistic observations of the 4 *Happenings* (workshops' face-to-face sessions); as well as a quantitative content analysis of the designs. Such a content analysis of the designs consisted of: structuring the designs in facets (or characteristics); studying the occurrence of each facet in the 37 designs; and, analyzing where those facets were lost in the CSCL script life-cycle. The content analysis performed in both workshops was confronted in an iterative way, finding that in the second one, more facets were considered. This way, the analysis of the design contents of the first workshop was enriched by incorporating the new facets arisen from the second one.

In the second workshop, additionally to the aforementioned content analysis, feedback from the teachers was gathered, in the form of a Web-based questionnaire and interviews. In total, we processed 24 answers to the questionnaire (out of 25 participants), and eight interviews with the aim of triangulating data by asking teachers to compare the resulting *Moodle* infrastructure with their original designs. Access to the corresponding deployed *Moodle* course was granted to all participants (with both student and teacher roles) so that they had the opportunity to assess the result of the translations.

A summary of the data gathering sources, and the labels used in the text to quote them is shown in Table 1.

Table 1. Data sources for evaluation and labels used in the text to quote them

Data source	Type of data	Labels
Web-based questionnaire	Quantitative ratings and qualitative explanations of the teachers	[Quest]
Designs content analysis	Quantitative data about facets, occurrence of facets, and facets lost in translations (quantitative data analysis)	[Content]
Interviews	Qualitative interview with teachers	[Interview]

As recommended by the evaluation model followed, the study involved 5 researchers coming from different perspectives in the ICT and education fields, who jointly defined the evaluation *Issue* (Tension) as the conceptual organizer of the whole evaluation process:

[I]:   *Does the final result of the learning designs in Moodle maintain the pedagogical essence well enough to remain usable by their original authors (faculty)?*

According to the method followed, the *Issue* is split into a set of more concrete *Topics* with the aim of helping researchers to illuminate it. Following the same rationale, each *Topic* is operationalized in a number of *Information Questions* that give insight on each topic. This way, a set of *Information Questions* helps in the understanding of a particular *Topic*; a set of *Topics* illustrates the *Issue*, that functions as conceptual organizer of the evaluation, helping to better understand our *Evaluand*. Figure 4 shows *Topics* [T] and *Information Questions* [IQ] defined and it illustrates the relation between *Information Questions*, *Topics*, *Issue* and *Evaluand*.

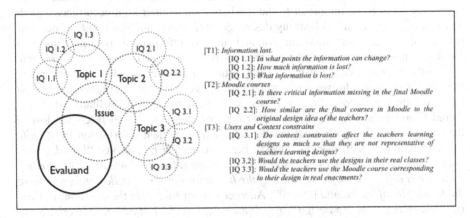

Fig. 4. Topics and Information Questions

3.3   Results and Evidences

Regarding the *Information lost* [T1], teachers' designs were structured in characteristics (or facets). 47 facets were detected in the 37 designs:

<u>Design general facets:</u>
Title, author, teacher, design date, context, description, collaborative pattern, general objective, objectives, competences, number of students, number of sessions, grouping, total duration, previous requirements, routines, backup plan, ICT tools, total temporal extension, contents, face-to-face/remote duration, subject, resources, work method, student estimated attendance.

<u>Facets in the sequence of activities:</u>
Title, session, session duration, task duration, duration, face-to-face/remote, pictures/drawings, actor, task description, number of students, grouping, student tasks, teacher tasks, instrument/artefact/resource, non-ICT tools, ICT tools, time between sessions, routines, objectives, phase/level, deliverable, physical space structure.

We calculated the occurrence of these facets, and we analyzed what information is lost and where, by comparing the facets with the data models and user interfaces of the different ICT tools involved (*WebCollage*, *GLUE!-PS* and *Moodle*).

The analysis of the translations carried out [Content] shows that most information is lost in the tool used to generate a computerized script from the teachers' designs, (*WebCollage*, in our case). Figure 5 shows the facets with occurrence over 40% (i.e. that appear in more than 40% of the teachers' designs), and whether they are supported (green) or not (striped red) by each of the ICT tools used in this concrete instance of the CSCL script life-cycle. In addition, Figure 5 shows that most of the facets *not* supported by *WebCollage*, are not supported either by the rest of tools (*GLUE!-PS* and *Moodle*). Interestingly, we found out that 47,37% of the facets identified in the original designs (with the aforementioned 40% occurrence) would be lost in the resulting courses in *Moodle* (red in Figure 5) [Content], since they were not present in one or more of the tools in the involved life-cycle. One would expect that this loss of facets with high occurrence should have a great effect on the final result, since they seem to be important to the teachers (due to their high occurrence). Such lost facets mostly relate to learning design *general characteristics* (context, description, number of students, total duration and subject), information about *time and sessions*, and information about whether the task is *face-to-face or remote*. On the topic of *Moodle courses* [T2], it would be interesting to uncover whether teachers notice the loss of critical information or not [IQ 2.1]. Triangulating the quantitative data from the aforementioned facet analysis, with quantitative and qualitative data gathered from teachers feedback, we found out that although several facets are lost in the translations [Content], most teachers don't miss critical information in *Moodle* implementation of their designs [Quest]. E.g., one teacher commented *"I think everything is included but when I compare it with the one (Moodle) I use in my course, my structure is different, maybe because of the limitations of the Moodle configuration of the University of Valladolid* [Quest]". Another teacher said: *"[to the question: did you miss something from the design?] No, I don't think so [...] Maybe the description of some of the tasks, or some missing questionnaire [...] the general activity schema is well developed* [Interview]".

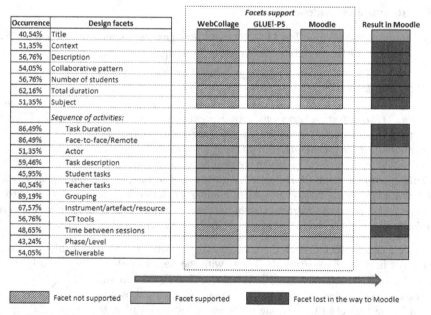

Occurrence	Design facets	WebCollage	GLUE!-PS	Moodle	Result in Moodle
40,54%	Title				
51,35%	Context	▨	▨	▨	■
56,76%	Description				■
54,05%	Collaborative pattern		▨		
56,76%	Number of students	▨	▨	▨	■
62,16%	Total duration	▨	▨	▨	■
51,35%	Subject	▨	▨	▨	■
	Sequence of activities:				
86,49%	Task Duration	▨	▨		■
86,49%	Face-to-face/Remote				
51,35%	Actor				
59,46%	Task description				
45,95%	Student tasks				
40,54%	Teacher tasks				
89,19%	Grouping				
67,57%	Instrument/artefact/resource				
56,76%	ICT tools				
48,65%	Time between sessions	▨	▨	▨	■
43,24%	Phase/Level				
54,05%	Deliverable				

Facets support

▨ Facet not supported   □ Facet supported   ■ Facet lost in the way to Moodle

Fig. 5. Facets with occurrence in more than 40% of the analyzed learning designs, as well as ICT tools that support them (in green) or not (in red ruled), and facets lost in the way to Moodle in one or more ICT tools (red)

Most of the teachers (67% [Quest]) gave positive feedback about the similarity of the final course in *Moodle*, when compared to their initial idea [IQ 2.2]. Such positive feedback was confirmed in the qualitative answers in [Quest] where, for instance, one comment was "*Yes (it is similar), although I think it would be good to have a graphical sketch* of the design/pattern used [Quest]", and in the interviews ("*[Do you think the course represents the design faithfully?] More or less it does. The activity structure was correctly built [...] I think there is a problem with one activity, which should be individual and was in group* [Interview]"). Some teachers reported some misinterpretation in the design: "*[to the question: what did you think about the generated course?] There was a small problem [...] you interpreted that there were 8 documents, but it was the same one for all the class (in the end, it seemed that s/he assumed that the 8 links referred to different documents, not to the same one)* [Interview]".

About *Users and Context constraints* [T3], results show that the context, tasks and indications of the workshop imposed constraints to the designs made by teachers [IQ 3.1]. 76% of the teachers answered in [Quest] that they changed their way on designing learning activities. This was confirmed by the questionnaire qualitative data. For example, a teacher wrote that "*[...] work in groups is something I had considered before, but I had rejected the idea because of the complexity [...]* [Quest]", while another commented "*[...] never limited myself to a collaborative work pattern (I had no idea they existed!) [Quest]*", and another wrote "*[...] Another important change is the introduction of ICTs in the class for the work in groups* [Quest]".

Although it seems that the teachers changed their way of designing (which is expected given that the workshops dealt with learning how to do learning design), it is interesting that most of them would use the workshop learning designs in real practice [IQ 3.2]. 75% answered they could use the design in practice [Quest]. This finding is also confirmed in the qualitative answers of [Quest], where, for example, one teacher comments "*It is something I can do. I see that it is feasible to include this kind of activities progressively [...]* [Quest]". On the other hand, some teachers think that using collaborative designs in real practice is difficult: "*At this moment I cannot apply activities like these because of program limitation and available time* [Quest]".

Also, most of the teachers (67% [Quest]) would use the *Moodle* course corresponding to their design in real enactments [IQ 3.3]. This element was confirmed with the qualitative data in questionnaire and interviews. An example is the comment of a teacher: "*Yes [I would use it], it would save time, although it would require tuning* [Quest]". Other teacher commented "*[to the question: would you use your design in real practice immediately?] This same thing I designed [...] I think I could do it [...] there were a couple of technical problems that I would have to work out [...] I will probably try this* [Interview]". Most teachers confirmed in interviews that students would be able to use the *Moodle* course: "*[to the question: would your students be able to use it?] I think they would, Moodle is not the problem [...] the problem is the tedious work of forming groups, creating documents, reforming groups...* [Interview]". Also, some teachers did not like the appearance of the course in *Moodle*. For instance, a teacher said "*[to the question: would you be able to use it?] I think it has to be simple [...] I don't see it very complex, it is simple, but [...] It is not appealing to the eye [...] it is the presentation [...] I did not expect it to be like this (the list of links activities presentation in Moodle) [Interview]"*). Another commented: *"[to the question: would you use this design in real practice] The Moodle as it is now [...] it limits too much, is not very interactive [Interview]"*.

Table 2 shows a summary of the findings, and the supporting data sources.

Table 2. Main findings in the evaluation

Topic	Finding	Support data
[T 1]	47,37% of facets with an occurrence over 40% are lost in the resulting courses in *Moodle*.	[Content]
[T 1]	42,11% of facets with an occurrence over 40% are not supported by *WebCollage*, same % are not supported in *GLUE!-PS*, and 36,84% are not supported by *Moodle*.	[Content]
[T 2]	Not much critical information lost in the final *Moodle* course	[Content][Quest][Interview]
[T 2]	Most of the teachers gave positive feedback about the similarity of the final course in *Moodle* to their initial idea	[Quest][Interview]
[T 3]	The context, tasks and indications of the workshop imposed constraints to the designs made by teachers	[Content][Quest]
[T 3]	The differences were mainly in the using of collaborative activities and patterns, and ICT tools (the focus of the workshops)	[Quest]
[T 3]	Most of teachers would use the learning designs in real classes	[Quest][Interview]
[T 3]	Most teachers would use the *Moodle* course in real enactments	[Quest][Interview]

4   Discussion

We have found evidences showing that, in the particular situation studied, with its inherent constraints, most of the teachers consider that the final result of their learning designs in *Moodle*, although not exactly like their initial idea, is similar enough to be used. Most of the opinions were positive about the course in *Moodle*, and they did not notice either too much or too critical information loss, even though the quantitative study of the designs content showed that a considerable amount of information was lost in translations from initial designs to *Moodle*.

Our first finding, the constraints imposed on the teachers' designs by the context of the study, was somehow expected. The study was conducted in the context of two professional development workshops, and participants were being trained in designing collaborative activities supported by ICT. Given that the designs generated by participants were not representative of their own designing style so far, it was important to obtain evidences of the feasibility of the generated designs to be used by the teachers in real practice after the workshop. The results in this regard are promising, due to the positive feedback in quantitative ratings in questionnaire and qualitative answers both in questionnaires and interviews. Also, qualitative data in the questionnaire shows evidences that the main changes in the designs were the inclusion of collaboration, and ICT support in the activities. Such indications of produced learning design feasibility show that this research is highly relevant for the TEL/CSCL field. However, the reduced scope of the presented study, with very similar contexts and constraints, is the main limitation of the present work. We have studied only a particular case, with several pedagogical constraints and imposing certain restrictions to the creativity of participant teachers. In any case, this is an unprecedented case of end-to-end life-cycle study, and thus, other studies in other contexts are a clear line of future research.

As we mentioned in Section 2, information lost in the CSCL script life-cycle can take place by the action of human and software agents. The present work is more slanted to the technological side, being more focused in data translations than in the pedagogical side of the designs. Also, human and software agents were the same in both workshops, which is another limitation of the study. Future work including different agents and comparing results with other technological solutions, and other human agents interpreting results is thus another clear path to extend this research.

The chosen technological solution is another interesting feature of the study, since it allows to complete the CSCL script life-cycle in an automated way, going from multiple different design authoring tools (or learning design languages), to multiple different VLEs (by using the GLUE!-PS architecture [13]). This technological solution imposes further data losses that other solutions (which may directly translate from an authoring tool to a VLE [8,19,20]) may not incur. However, the positive results of the study, even in this unfavorable technological setting, are promising for these kinds of solutions trying to apply learning design to mainstream VLE educational scenarios.

Probably the most striking result of this study is that, in the learning designs included in the study, almost 50% of the facets identified (with an occurrence in more of 40% of the designs) were lost in the translation. Despite this fact, teachers didn't seem to notice, in general, a loss of critical information. Most facets could be considered to refer to contextual descriptions, although some of them seemed significant *a priori* (like time or sessions information). Thus, further research regarding the relevance of the design facets for the usability in the real practice using widespread VLEs should be undertaken. More specifically, a deeper study of the final result (i.e. the deployed *Moodle* course) should be performed by the original authors (e.g., using the *Moodle* course in a real class), in order to detect any particular relevant facets that may had gone unnoticed by the teachers in the visual review of the *Moodle* courses.

5   Conclusions

The results discussed above are limited to the context and scenario of the case studied, but this kind of results could be of high interest to researchers working on the support of CSCL scripts, and can motivate further research in this field. ICT tools supporting the CSCL script life-cycle can be improved taking input from similar studies as, for instance, the ICT tools involved in the present research could be enhanced to include some of the lost facets with high occurrence detected in content analysis (e.g. time or sessions information).

Also, further research can be conducted considering the combination of ICT tools that GLUE!-PS is able to support, studying the pedagogical effects of real enactments in different VLEs and Web 2.0 platforms (e.g. Blogs or Wikis), and using different learning design tools. Moreover, the effect of human agents is also an interesting research line that this work could motivate. Studying how the interpretation of a design in the different life-cycle phases affects the pedagogical essence, or the interpretation and formalization processes themselves when using a particular ICT tool, or even how changes in the learning design (due to human or computer agents) affect the reusability of a learning design. All those could be questions for further research that are relevant not only for the learning design field, but for TEL practice as a whole.

Acknowledgments. This research has been partially funded by the Spanish Ministry of Economy and Competitiveness Projects TIN2008-03-23, TIN2011-28308-C03-02 and IPT-430000-2010-054, and the Autonomous Government of Castilla and León Project VA293A11-2. The authors thank the participants in the teacher workshops, the rest of the GSIC/EMIC research team, as well as the Learning Design Theme Team funded by the European Union through the Stellar Network of Excellence for Technology Enhanced Learning (FP7-IST-231913) for their contributions.

References

1. Dillenbourg, P.: Over-scripting CSCL: The risks of blending collaborative learning with instructional design. In: Kirschner, P.A. (ed.) Three Worlds of CSCL. Can we Support CSCL?, pp. 61–91. Open Universiteit Nederland, Heerlen (2002)
2. Weinberger, A., Kollar, I., Dimitriadis, Y., Mäkitalo-Siegl, K., Fischer, F.: Computer-supported collaboration scripts: Theory and practice of scripting CSCL. Perspectives of educational psychology and computer science. In: Balacheff, N., Ludvigsen, S., Jong, T.D., Lazonder, A., Barnes, S., Montandon, L. (eds.) Technology-Enhanced Learning. Principles and Products, pp. 155–174. Springer (2008)
3. Koper, R.: An Introduction to Learning Design. In: Koper, R., Tattersall, C. (eds.) Learning Design Modelling and Implementing Networkbased Education Training, vol. 2003, pp. 3–20. Springer (2005)
4. Hernández-Leo, D., Villasclaras-Fernandez, E.D., Asensio-Pérez, J.I., Dimitriadis, Y., Retalis, S.: CSCL Scripting Patterns: Hierarchical Relationships and Applicability. In: Sixth International Conference on Advanced Learning Technologies, pp. 388–392 (July 2006)
5. Villasclaras-Fernandez, E.D., Hernandez-Gonzalo, J.A., Hernandez-Leo, D., Asensio-Perez, J.I., Dimitriadis, Y., Martinez-Mones, A.: InstanceCollage: A Tool for the Particularization of Collaborative IMS-LD Scripts. Educational Technology & Society 12(4), 56–70 (2009)
6. Vignollet, L., Ferraris, C., Martel, C., Burgos, D.: A Transversal Analysis of Different Learning Design Approaches. Journal of Interactive Media in Education (2) (2008)
7. Dillenbourg, P., Schneider, D.K., Synteta, P.: Virtual Learning Environments. In: Dimitracopoulou, A. (ed.) Proceedings of the 3rd Hellenic Conference "Information & Communication Technologies in Education", Greece 2002, pp. 3–18. Kastaniotis Editions (2002)

8. Bower, M., Craft, B., Laurillard, D., Masterman, L.: Using the Learning Designer to develop a conceptual framework for linking learning design tools and systems. In: 6th International LAMS & Learning Design Conference, Sydney, Australia, LAMS Foundation (2011)

9. Hernández-Leo, D., Villasclaras-Fernández, E.D., Asensio-Pérez, J.I., Dimitriadis, Y., Jorrín-Abellán, I.M., Ruiz-Requies, I., Rubia-Avi, B.: COLLAGE: A collaborative Learning Design editor based on patterns. Educational Technology & Society 9(1), 58–71 (2006)

10. Botturi, L., Derntl, M., Boot, E., Figl, K.: A Classification Framework for Educational Modeling Languages in Instructional Design. In: IEEE International Conference on Advanced Learning Technologies (ICALT 2006), pp. 1216–1220. IEEE Press, Kerkrade (2006)

11. Conole, G.: Designing for Learning in an Open World. Explorations in the Learning Sciences, Instructional Systems and Performance Technologies, vol. 4. Springer (2012)

12. IMS Global Learning Consortium: IMS Learning Design Information Model (2003), http://www.imsglobal.org/learningdesign/ldv1p0/imsld_infov1p 0.html (accessed March 22, 2012)

13. Prieto, L.P., Asensio-Pérez, J.I., Dimitriadis, Y., Gómez-Sánchez, E., Muñoz-Cristóbal, J.A.: GLUE!-PS: A Multi-language Architecture and Data Model to Deploy TEL Designs to Multiple Learning Environments. In: Kloos, C.D., Gillet, D., Crespo García, R.M., Wild, F., Wolpers, M. (eds.) EC-TEL 2011. LNCS, vol. 6964, pp. 285–298. Springer, Heidelberg (2011)

14. Hernández-Leo, D., Asensio-Pérez, J.I., Dimitriadis, Y.: Computational representation of Collaborative Learning Flow Patterns using IMS Learning Design. Educational Technology & Society 8(4), 75–89 (2005)

15. Martínez, A., Dimitriadis, Y., Rubia, B., Gómez, E., de la Fuente, P.: Combining qualitative evaluation and social network analysis for the study of classroom social interactions. Computers & Education 41(4), 353–368 (2003)

16. Jorrín-Abellán, I.M., Stake, R.E., Martínez-Moné, A.: The needlework in evaluating a CSCL system: the evaluand oriented responsive evaluation model. In: Proceedings of the 9th International Conference on Computer Supported Collaborative Learning, CSCL 2009, vol. 1, pp. 68–72. International Society of the Learning Sciences (2009)

17. Stake, R.E.: Standards-Based and Responsive Evaluation. SAGE Publications, Inc., University of Illinois at Urbana-Champaign, USA (2004)

18. Jorrín-Abellán, I.M., Stake, R.E.: Does Ubiquitous Learning Call for Ubiquitous Forms of Formal Evaluation? An Evaluand Oriented Responsive Evaluation Model. Ubiquitous Learning: An International Journal 1(3), 71–82 (2009)

19. Burgos, D., Tattersall, C., Dougiamas, M., Vogten, H., Koper, R.: A First Step Mapping IMS Learning Design and Moodle. Journal of Universal Computer Science (JUCS) 13(7), 924–931 (2007)

20. Perez-Rodriguez, R., Caeiro-Rodriguez, M., Fontenla-Gonzalez, J., Anido-Rifon, L.: Orchestrating groupware in Engineering Education. In: Frontiers in Education Conference (FIE), pp. F3D-1–F3D-6. IEEE (2010)

The Push and Pull of Reflection in Workplace Learning: Designing to Support Transitions between Individual, Collaborative and Organisational Learning[*]

Michael Prilla[1], Viktoria Pammer[2], and Silke Balzert[3]

[1] University of Bochum, Information and Technology Management, Universitaetsstr. 150, 44780 Bochum, Germany
[2] Know-Center, Infeldgasse 21A, Graz, Austria
[3] Institute for Information Systems at German Research Center for Artificial Intelligence (DFKI), Saarland University, Campus, Bld. D3 2, Saarbruecken, Germany

Abstract. In work-integrated learning, individual, collaborative and organisational learning are deeply intertwined and overlapping. In this paper, we examine the role of reflection as a learning mechanism that enable and facilitates transitions between these levels. The paper aims at informing technological support for learning in organisations that focuses on these transitions. Based on a theoretical background covering reflection as a learning mechanism at work as well as the abovementioned transitions, and on observations in two organisations (IT consulting, emergency care hospital unit), we argue that such technological support needs to implement two inherently different, yet complementary mechanisms: push and pull. "Push" subsumes procedures in which reflection outcomes transcend individual and collective ownership towards the organisation through efforts made by the reflection participants. "Pull" subsumes situations in which the effort of managing the uptake of results from reflection is shifted away from the reflection participants to third parties in the organisation. We illustrate each mechanism with an application built to support it.

Keywords: Reflective Learning, Organisational Learning, Software Design.

1 Introduction

One of the challenging aspects of work-integrated learning is that individual, collaborative and organisational learning are deeply intertwined and mutually dependent. Reflection as a mechanism of learning at work has great potential to support transitions between these levels of learning, as it concern both single experiences as well as

---

[*] The project "MIRROR - Reflective learning at work" is funded under the FP7 of the European Commission (project number 257617). The Know-Center is funded within the Austrian COMET Program - Competence Centers for Excellent Technologies - under the auspices of the Austrian Federal Ministry of Transport, Innovation and Technology, the Austrian Federal Ministry of Economy, Family and Youth and by the State of Styria. COMET is managed by the Austrian Research Promotion Agency FFG.

A. Ravenscroft et al. (Eds.): EC-TEL 2012, LNCS 7563, pp. 278–291, 2012.

comprehensive topics. Naturally, this affects the technological support that is required within organisations to support learning. The goal of this paper is to shed light on transitions between individual, collaborative and organisational learning (ICO transitions) based on theory and two empirical studies in order to inform technology design.

2   Theory: Reflection and Levels of (Informal) Learning

2.1   Reflection as a Mechanism of Informal Learning at Work

Reflection as we mean it can be defined as "those intellectual and affective activities in which individuals engage to explore their experiences in order to lead to new understandings and appreciations" [1] and has been recognized as a common process in everyday work – be it done by individuals [2, 3] or by groups reflecting collaboratively [4, 5]. It consists of three elements [1]: Going back to experiences, re-evaluating these experiences in the light of current insights and knowledge, including experiences of others, and deriving knowledge for future activities from this, including the planning and implementation of changes.

Given this understanding of reflection, it is obvious that besides other mechanisms such as problem based learning, reflection is a core mechanism of (informal) learning at work [6]. Its grounding in previous experiences binds it closely to the context of work and its clear focus on outcomes distinguishes it from less fruitful modes of thinking about past work such as rumination. In addition, reflection is not only bound to negative experiences and problems, but also concerns positive experiences, which may result in deriving good practice. Understanding reflection as based on own experiences means that **reflective learning can occur at an individual or collaborative level**, where the critically examined experiences are the experiences of an individual or shared experiences within a group. This does not mean that reflection is limited to contribute to individual and group levels only: On the contrary, examples from the studies presented below show that individual or collaborative learning by reflection is a powerful mechanism to create and refine organisational knowledge.

Recently, **computer support for reflection has been identified as a vital field of technology enhanced learning** (e.g. [7–10]). There are also various theoretical models on reflection these tools are aligned to [1–3], but these tools and models mainly consider reflection as a cognitive process. Only recently [4], reflection groups have been integrated into this discussion and support for collaborative reflection has been worked on specifically (e.g. [11]). Furthermore, [12] have described a model of computer-supported reflective learning support specifically in work settings, in which they describe the various roles that tools can play in the reflection process and generally allow reflection to happen collaboratively. However, there is currently no model considering the transitions between individual, collaborative and organisational learning.

2.2   Levels of Learning and Knowledge: Individual, Collective, Organisational

The description of reflection given above already indicates that its outcome – knowledge on (how to improve) work practice – may have impact on different levels: an individual may learn for herself and take the corresponding knowledge to a group of peers, while reflection in groups may result in individual outcomes (e.g. when peers collaboratively reflect individual experiences and learn for their own work) or be relevant for the groups as a whole (e.g. when a group reflects its rules for cooperation and changes them afterwards). Depending on group size and members as well as on the topics being reflected about, outcomes from group reflection may also add to organisational learning in that they provide knowledge on needs or how to change organisational practice. The influence of reflection and tools to support it on transitions between these levels, however, has not been researched intensively so far.

Learning as understood here denotes a "change in understanding, perspective or behaviour" in the broadest sense and it is the individual, a group of people (in the context of work: a team) or an organisation that learns [13]. For organisations, we understand learning as the improvement of an organisation's task performance over time, as well as the change of target values that measure an organisation's task performance [14]. Thus, we comprehend the learning process at an organisational level as structural changes, affecting individuals and groups, and subsequently individual and collaborative learning processes, within the organisation (for a more comprehensive explanation see [15]). Following this, we distinguish **individual and collaborative learning by the kind of learning process:** while individual learning can be considered a cognitive process, collaborative learning is social and happens in communication, e.g. when a team reflects on their performance. This means that collaborative learning support also needs to take into account support for communication and cooperation. **Organisational learning is based on the outcomes of these processes and characterised by the result of learning** (an organisation's task performance over time changes). This also means that the knowledge learned on an organisational level is a result of individual and collaborative learning processes.

There are many approaches describing the **relationship** between individual, collaborative and organisational levels of knowledge. Among the most popular, Nonaka (1994) [16] describes a spiral model of organisational knowledge creation, which starts at an individual level and brings knowledge to collective and organisational levels, "when all four modes of knowledge creation are ‚organisationally' managed to form a continual cycle" [16]. For this, the model describes a continuous cycle of socialisation (exchanging tacit knowledge), externalization (articulating tacit knowledge), combination (relating different bits of knowledge to each other, thus creating new knowledge) and internalisation (integrating explicit knowledge into one's own context). Learning in this sense takes place when internalisation is done by an individual. Kimmerle et al. (2010) [17] add that in the same way as learning takes place for individuals, groups (and thus organisations) learn by explicating knowledge, e.g. by making rules for their cooperation explicit. These concepts show that reflection tools need to provide correspondent functions such as making tacit reflection outcomes available for others (i.e. communicating it), explicitly sustaining outcomes from group reflection and relate outcomes to individual or group contexts to support the transition between levels of learning.

Other approaches describe **transitions** between levels of knowledge as a communicative and contextualizing process. Stahl, for example, regards collaborative learning, which takes individual knowledge to the collective level, as a continuous interchange of perspective taking and perspective making, meaning that the individual level can only be transcended by interpreting "the world through some else's eyes" [18]. Herrmann and Kienle (2008) [19] add that a "shared context" can only be maintained by "contextual communication", meaning that learning needs to take place with a close relation to what people are learning about. In a similar approach, Beers et al. (2005) [20] describe a process in which knowledge is created by the abovementioned processes of externalization and internalisation complemented by negotiation to create a common ground from different perspectives and integration to relate insights organisational knowledge (see **Fig. 1**). Although [20] do not explicitly refer to organisational knowledge, this processes shows how knowledge is created by group and individual efforts and thus builds a base for ICO transitions.

For reflection tools to bridge between levels of knowledge, the process depicted in **Fig. 1** means that there is a need to **explicitly intertwine perspectives** of reflection participants (external knowledge), to **foster communication** between reflection participants (shared knowledge, common ground) and **to relate outcomes to organisational standards and processes** (constructed knowledge). Regarding the informal and experience-bound nature of reflection, this provides a challenge for tool design.

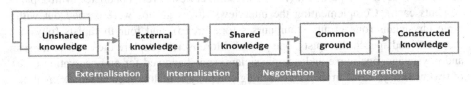

Fig. 1. Transitions between individual and collective knowledge (Beers et al. 2005)

2.3   Research Goals

Based on the gaps identified by the review of existing work on reflective learning and ICO transitions, the work presented here follows two goals related to each other:

1. **Exploring reflection in practice and learning about its influence on work life an learning:** As there is not enough work available on the role of reflection in work and learning – especially not on ICO transitions –, its understanding and support depend on exploring it in practice. This will be tackled in Sect. 3 and 4.
2. **Developing an understanding and a framework enabling reflection and ICO transition support:** Our work is directed towards IT support for reflection in practice and thus, one major goal is to develop a framework for this support from our empirical work. Results on this goal are described in Sect. 5 and 6.

3   A Study on Individual, Collaborative and Organisational Learning by Reflection

3.1   Study Methodology

In order to understand reflection better (see goals above), we conducted a study using a variety of qualitative methods such as interviews, focus groups and observations in the participating organisations. The usage of these methods was targeted towards finding out **what role learning by reflection plays in everyday work life** (see goals given above). Note that consequently the study results mostly point towards "what is" and to a much smaller degree to "what could be" – however, they form a base for the design of tools supporting reflection by e.g. diminishing existing barriers or motivating currently unused opportunities for reflection. For an extensive description of study design, tools such as interview guidelines and results see [21].

3.2   Participants

The emergency care hospital unit observed in our empirical studies specializes in the treatment of neurological diseases such as stroke and epilepsy. Here, interviews were conducted with 3 physicians and 4 nurses. To cover explicitly the organisational perspective on learning, 4 interviews were conducted with representatives of the management board, from quality management, from human resources and from the advanced education department (two of these interviews were conducted with 2 participants each). Complementing the interviews, focus groups were carried out with three physicians (one group), four nurses (one group), and four therapists (physiotherapists and speech therapists, one group). Moreover, to explore the work of nurses and physicians, one nurse and one physician were shadowed for an observation time of two workdays each. In the IT consulting company, which is specialized on creating and adapting customer relationship solutions to small and medium companies, interviews were conducted with 8 sales and business consultants. To cover explicitly the organisational perspective on learning, interviews were also conducted with 2 members of management (from HR department and from the management board). In addition, two sales consultants were observed during two working days each. Table 1 gives an overview of the study participants.

Table 1. Participants in the study

Organisation	Interviews	Observations	Focus Groups
Hospital	2 nurses, 1 physician, 1 therapist 6 representatives of management	Two days each: 1 nurse, 1 physician	3 physicians, 4 nurses, 4 therapists
IT Consulting	8 sales / business consultants 2 members of management	Two days each: 2 sales consultants	-

3.3   Data Analysis

Interviews and focus groups were transcribed, and observations were documented. From this raw material, we extracted stories that describe reflection in practice, including good practice, barriers and shortcomings. For the work described in this paper, we focused on those stories that involve transitions between individual, collaborative and organisational learning.

4   Study Results: Stories about Transitions between Individual, Collaborative and Organisational Learning

4.1   Example 1 (Emergency Care Unit)

During our observation, a patient with an acute stroke and in very bad condition was admitted to the emergency room of the ward. After a short time, the responsible physician realized that this was not a routine case but a very critical one. The standard procedure in this case is to give an internal alarm, which causes the head physician and an emergency team to immediately come to the ward. The physician told the present nurse to use her internal telephone to give the alarm, as there was no alarm button in the room. The nurse vaguely remembered the procedure of giving the alarm and started it immediately. However, the alarm did not go off and the helpers did not arrive in the next minutes. The nurse therefore called the head physician and the emergency team directly; they came to the emergency room and took care of the patient.

After this situation, the nurse started to reflect on his failed attempt to give the alarm (and why it failed) by going through the procedure he had applied in the emergency room again and again. As he did not find a reason, he included other nurses into reflection, but they did not have much experience with the procedure and could not help him. It was only when he started to discuss and analyse the situation with the head nurse that they discovered the head nurse had had a similar experience. They realized that the emergency procedure as described in the hospital's quality manual was too complicated to be performed under the stress of treating a patient in bad shape. After this, the head nurse added the issue to the agenda for the regular ward meeting. In this meeting, the nurse who had experienced the problem explained the case to the others and the head nurse explained the problem behind it. Some of the nurses reported that they had had similar problems before. As a result, they agreed to practice essential procedures more often, that the telephone procedure should be changed and that there should be an emergency button in each patient room. As the latter two changes could not be implemented by ward staff but are subject to hospital-wide quality standards or infrastructural decisions, the head nurse agreed to promote the proposal and to talk to the quality manager in order to change the procedure.

4.2   Example 2 (Emergency Care Unit)

In one ward meeting we observed, which was attended by all nurses and physicians working at the ward, a nurse mentioned that she had been thinking a lot about the way physicians treated patients during the ward round in the morning. She thought that physicians took too little time to talk to patients and that taking more time would make the patients feel more secure and to receive better care. Other nurses remembered that they had witnessed similar situations and supported her. The physicians started to reflect on recent cases in which they did the ward round and agreed that most of the time they could have taken more time for the patients if the ward round would not interfere with a follow up meeting they had to attend. It was then agreed to start the ward round earlier from this time on, and that the physicians would take extra time spent in patients' rooms talking to the patients. This, however, was only implemented in one ward and there was no comparison with the practice of other wards and no sharing of practice with others.

4.3   Example 3 (IT Consulting)

After a time of good success in selling products and services to customers, some sales consultants of the company realized that they were losing more pitches than they were used to. Each consultant had thought about reasons for this, but nobody had a clear idea how to change this. In the monthly meeting of sales consultants, in which they usually iterate through current activities, one consultant mentioned that he had experienced problems in winning pitches over the company's competition recently. The other consultants realized that this was not only their problem and reported similar issues. As a result of this, they focused the meeting on pitches that had been lost recently and started to reflect on potential reasons for these losses by going through the experiences reported by the respective consultant for several pitches. They found that in most pitches the customer had asked the consultant to demo the system. However, the approach of the company was to not use demo systems but to invite potential customers to the site of reference customers in order to show them a fully-fledged system and how well it suited the needs of the respective customer. The consultants realized that in most cases of lost pitches the responsible consultant reported that the customer was dissatisfied with the lacking demo system and that competitors had demo systems with them during pitches. They decided that from now on they would take a demo system with them. The head consultant agreed to talk to the company's IT department in order to set up a demo system with realistic data. He also reported this to the management, who agreed to change the company standards to include demo systems into the process for customer acquisition. However, management complained that they had not known about the problem earlier and that this had caused severe losses of orders.

5   Analysis of Study Results

5.1   Observed Transitions between Individual, Collaborative and Organisational Learning By Reflection

Analysing the examples described above, a common pattern of the transitions between individual, collaborative and organisational learning can be found (see Fig. 2):

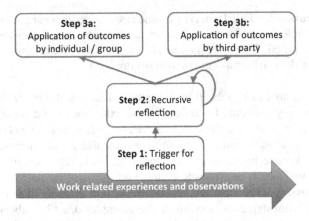

Fig. 2. Transition model. One salient work experience triggers (step 1) a reflection process (2 – either individual or collaborative reflection). The reflection outcomes may lead to consecutive reflection processes (recursion into step 2). The outcomes can then either be applied directly by the reflection participants (step 3a) or by third parties (step 3b).

1. **Trigger for Reflection:** In all examples above, an individual or a group makes a work-related experience that triggers reflection of an individual or group (in all examples above this is an individual). For example, the emergency procedure does not work properly in an emergency situation (example 1) or more sales pitches are lost than it is usually the case (example 3). It should be noted, however, that this trigger does not need to be explicit, nor does the decision to reflect need to be taken explicitly: Reflection might also be triggered during group conversations – then the trigger implicitly leads to reflection, but is still there.

2. **Recursive Reflection:** The trigger leads individuals or a group to **reflect** on their observations and experiences in order to find a resolution. For example, the nurse asks a colleague, tries to find out the emergency procedure and looks for solutions to the problem in example 1; the sales consultants start to reflect on lost pitches as one of them reports on a lost pitch (example 3). In line with the theory described in section 2, the transition from individual knowledge to a collective level depends on communication in all examples. Communication starts the spiral of knowledge and crosses the boundary between individual and collaborative learning. Such reflection may also result in the externalization of rules and their integration into existing knowledge and practice, as illustrated in example 2 when staff agrees on changes in the ward round schedule. This stage is **recursive**, as individual reflection can trigger reflection in groups, which in turn triggers individual or group reflection again and so on – example 1 illustrates this nicely.

3. **Application of Outcomes:** After (potentially recursive) phases of reflection, in all examples the reflection participants achieve a learning outcome (insights into reasons for a problem, partial or full solutions, plans for action etc.). However, there are two alternatives concerning the implementation of outcomes:
   (a) **Application by reflection participants:** Sometimes, outcomes can be implemented by the reflection participants (individuals or groups). In the examples above it was always, finally, a group. Examples for this are the decision to extend training in example 1 or the decision to adapt the ward round schedule in example

(b) **Application by Third Party:** In some cases, outcomes cannot or can only partially be implemented by the reflection participants, and a decision or action of others is required. Examples for this are the changes in emergency handling (example 1) or the need for a demo system (example 3).

These observations can be related back firstly to Boud et al. (1985) [1], who distinguish as three key elements for reflection the experience(s) on which reflection is based (step 1), re-examining these experiences (step 2, recursive) and deriving applicable outcomes (steps 3a and 3b). Secondly, we see that for transferring individual to collective knowledge by reflection, the foremost needs are communication support and enabling individuals or groups to relate to outcomes of earlier reflection sessions. Both are necessary when iterating between reflection sessions in step 2, as well as when moving from step 2 to 3a or 3b. In the examples described above, we also see that (iterative) reflection makes knowledge more generally applicable. The recursion into step 2 thus obviously contains the knowledge creation and transition process of Beers et al. (2005) [20] that lies between externalization and integration (see Fig. 2).

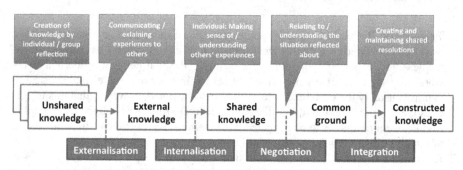

Fig. 3. The knowledge creation and transition process by Beers et al. (2005) [15], complemented with influences of reflection (top level elements)

5.2   Lost Learning Opportunities

As interesting as analysing what we observed, is **what we did not observe**. In the study, we observed how learning by reflection currently happened in the investigated organisations. The examples given above show successful learning in organisations. However, it is also interesting to **speculate where opportunities for learning were "lost" in practice**. Indications for lost learning opportunities stem from the above described stories as well as from within the rest of our collected data. The quality manager at the hospital for instance was very interested in proactive input from the operational levels and reported that she was dependent on this input in order to "verify" that the organisational processes work in practice. This is well argued e.g. by [22], who report that successful companies of knowledge workers ensure that their employees are motivated in "investing" into the company by sharing their insights and applying their knowledge proactively. At another point, we were told by an employee of the IT consulting company that he focused mainly on his core responsibilities because it was too frustrating to make suggestions for improvements outside this sphere and not seeing them followed-up. This obviously is a barrier for employees to invest time in thinking about how organisational processes could be improved.

Relating back to the transition model in Fig. 2, we see that the report of the employee corresponds to step 2 with a lack of following up, neither by initiating subsequent reflection sessions, nor by applying outcomes in the sense of step 3a (because others need to be involved) or relating insights to others who could act on gained insights. This points to a crucial factor that determined and constrained the success of the three examples described in section 4: In all these examples, one individual (a nurse in both examples 1 and 2, and a sales consultant in example 3) was motivated and able to follow up on own observations and to initiate subsequent collaborative reflection sessions. In each reflection session, again at least one person (the head nurse in example 1, the physicians in example 2, the head consultant in example 3) took responsibility for continuing the recursive reflection, for applying outcomes of the reflection session or for communicate further the gained insights (to the quality manager in example 1, to the company management and the IT department in example 3). The examples also contain learning opportunities lost, as both in examples 1 and 3, multiple nurses / sales consultants had already experienced the same problem without changing work practice in the end (before our stories started). This means that often, there is a **barrier to the propagation of knowledge** (gained through reflection in our examples). This barrier lies between what individuals and groups concerned with operative work can achieve and reflect and what third parties such as management or other groups can implement. This is a known challenge for organisational learning by reflection, as per definition the knowledge needs to be created out of the work experiences of an individual or a group (i.e. the operative level). Organisational learning however can often only be implemented at management levels of hierarchy.

6   Synthesis and Outlook: Push and Pull Mechanisms for Transitions between Individual, Collaborative and Organisational Learning

The transition model (see Fig. 2 and Sect. 5.1) shows the transitions between observations rooted in work experiences (trigger experiences), reflection sessions, and the application of outcomes. It expresses that typically multiple, iterative reflection sessions are required to create organisational learning out of individual and collaborative reflection and that in organisations the reflection participants and the people who implement reflection outcomes may be different people (step 3b).

In order to fully understand ICO transitions, the communication mechanisms that underlie the transition model need to be understood. All examples described in Sect. 4 are characterized by a **"push"-mechanism of communication**. By "push"-mechanism of communication we mean, that the reflection participants actively initiate the communication necessary to move between stages and that the chain of communication finally reaches the organisational level: The reflection participants (either all or one out of a group) push information to other people than current reflection participants, and thus initiate iterative reflection sessions or the application of insights and solutions on superior and, finally, organisational levels.

The lost learning opportunities discussed in Sect. 5.2 are lost because those reflection participants who had valuable observations or insights did not or could not push this information to other people. A mechanism of communicating or applying outcomes may have helped in these situations. This mechanism would need to shift the burden of communication or application from reflection participants to other people within the organisation, who are capable of or responsible for the reflection or application of insights on the organisational level. We call this a **"pull" mechanism:** It assumes that there are stakeholders within an organisation who are interested in pulling together valuable observations and insights from knowledge workers within the company.

Clearly, both mechanisms may also work without technology support: For the "push"-mechanism this can be seen in our examples, for the "pull" mechanism, verbal communication or paper based workflows also work. However, technology can provide benefits: For instance, technology can support communication by facilitating documentation of experiences, sense-making by relating knowledge expressed by others to own knowledge, relating insights to the original experiences by allowing rich and hyperlinked documentation, and support maintenance of shared solutions in context of the original rationale. For a fine-grained discussion of computer support for reflection (but not specifically targeted towards ICO transitions), we refer also to [12].

Below we describe two apps that are designed to support ICO transitions, but using different communication mechanisms. The design of both apps was informed by the theoretical considerations discussed in this paper.

6.1   Example ICT Support for the Push-Mechanism: The Talk Reflection App

The Talk Reflection App (see also [11]) is designed for physicians, nurses and carers who need to regularly lead emotionally straining conversations with patients and their relatives – these conversations often include conveying bad news and the like. The app (see **Fig. 1**) supports the documentation, individual (A) and collaborative (B) reflection and sustainment of learning outcomes (C) on conversations between medical staff and patients or their relatives. Such conversations are already being documented on paper as part of staff's work, but this documentation lacks relevant information for reflection such as how straining the conversation was for the physician (F and H: the spider diagram show assessments of emotions for a conversation).

Relating back to Beer's model (Fig. 4), the App supports documenting experiences and, by making comments to documented talks (E), encourages deeper observations of own and shared documentation. It also supports preparing observations for communication with others (by marking cases, G) as well as communication about specific conversations via the sharing functionality (D). Furthermore, it enables collaborative work on shared material via comments and respects the need to relate insights to initial observations by providing a comment function (E). The app thus enables recursive reflection into groups that may then implement solutions found for talking to relatives – in the case of the hospital this would be a group of head physicians. The burden of initiating (recursive) reflection sessions, and communicating further any insights is left to the reflection participants (push-mechanism): physicians can use the app and reflect on conversations with a group on their ward, senior physicians can use and reflect on outcomes from these reflection in the meeting of different wards and finally, the head physicians of the hospital can reflect on these outcomes and decide whether to apply them in the whole hospital The app is currently being evaluated at the hospital that also participated in the empirical work (cf. [11]).

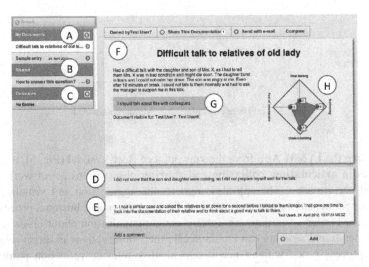

Fig. 4. The Talk Reflection App used to reflect on conversations with patients and relatives (A, F), including functionality to share own experiences with others (B, D), an illustration of own assessments of performance during talks (H), comments (E) and markup for discussion needs (G).

6.2 Example ICT Support for the Pull Mechanism: The Issue Articulation & Management App

The Issue Articulation & Management App (IA&M App, see Fig. 5) is designed to bridge the gap between those who create knowledge relevant for the organisation in daily work (operational workers) and third parties, who are able to implement it (non-operational stakeholders or decision makers) – it thus implements the pull mechanism. In this context, the app supports the articulation of issues (observations, insights, etc.), their relation to tasks and the corresponding business processes as well as the provision of different visualisations according to stakeholder's needs and requirements. These visualisations may take the form of e.g. EPC notation (A), and can be enriched by notes on how many issues, classified according to the type assigned, are related to certain tasks (B). By clicking on such an annotated task, a list of the corresponding issues is derived, which allows for detailed insights in the issues process participants have inserted (C). A tag cloud based on tags provided by the users while articulating an issue gives an overview of topics discussed (D).

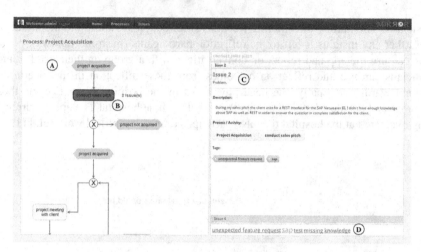

Fig. 5. The Issue Articulation & Management App

Analysing the **IA&M App** from the viewpoint of Beer's model (Fig. 5), it supports employees in **articulating, collecting** and **sharing** observations about work experiences and performance. The app also supports the aggregation and visualisation of these observations and relates these to specific work tasks and business processes. In this way, the burden of initiating point-2-point communication about observations is shifted: Operational workers can communicate their observations in direct relation to business process tasks, without needing to identify relevant reflection participants, engaging them in reflection, etc. Non-operational stakeholders like for instance a quality manager can take over this task by having access to such observations and insights – and if necessary can initiate reflection sessions and/or application of insights (pull mechanism). Furthermore, an aggregated overview of annotations can help to identify topics that are discussed and annotated over all work processes, indicating starting points for reflective learning on an organisational level. If many people indicate a problem with a specific work task in a business process for example, it may be valuable for the corresponding decision maker to reflect about changing the underlying working routines or even the whole business process. The IA&M App is currently being evaluated at the hospital and planned to be evaluated at the IT consulting company that participated in the empirical work.

Acknowledgements. The work presented in this paper is a joint effort and supported by numerous researchers being part of the MIRROR project. We thank them for their contribution. In addition, we thank Andrea, Anne, Dominik, Manuel and Volker for allowing us to gain deep insights into their respective company.

References

1. Boud, D.: Reflection: Turning experience into learning. Kogan Page, London (1985)
2. Kolb, D.A.: Experiential learning: Experience as the source of learning and development. Prentice-Hall, Englewood Cliffs (1984)
3. Schön, D.A.: The reflective practitioner. Basic books, New York (1983)
4. Dyke, M.: The role of the Other'in reflection, knowledge formation and action in a late modernity. International Journal of Lifelong Education 25, 105–123 (2006)
5. Hoyrup, S.: Reflection as a core process in organisational learning. Journal of Workplace Learning 16, 442–454 (2004)
6. Eraut, M.: Informal learning in the workplace. Studies in Continuing Education 26, 247–273 (2004)
7. Fleck, R., Fitzpatrick, G.: Teachers' and tutors' social reflection around SenseCam images. International Journal of Human-Computer Studies 67, 1024–1036 (2009)
8. Lin, X., Hmelo, C., Kinzer, C.K., Secules, T.J.: Designing technology to support reflection. Educational Technology Research and Development 47, 43–62 (1999)
9. Loo, R., Thorpe, K.: Using reflective learning journals to improve individual and team performance. Team Performance Management 8, 134 (2002)
10. Scott, S.G.: Enhancing Reflection Skills Through Learning Portfolios: An Empirical Test. Journal of Management Education 34, 430–457 (2010)
11. Prilla, M., Degeling, M., Herrmann, T.: Collaborative Reflection at Work: Supporting Informal Learning at a Healthcare Workplace. In: Proceedings of the ACM GROUP (2012)
12. Krogstie, B., Prilla, M., Knipfer, K., Wessel, D., Pammer, V.: Computer support for reflective learning in the workplace: A model. In: Proceedings of ICALT (2012)
13. Harri-Augstein, E.S., Thomas, L.F.: Learning Conversations: The Self-organised Learning Way to Personal and Organisational Growth. Taylor & Francis (1991)
14. Argyris, C., Schön, D.A.: Organisational learning II: Theory, method, and practice. Addison-Wesley (1996)
15. Balzert, S., Fettke, P., Loos, P.: A Framework for Reflective Business Process Management. In: Sprague, R.H. (ed.) Proceedings of HICSS (2011)
16. Nonaka, I.: A dynamic theory of organisational knowledge creation. Organisation Science, 14–37 (1994)
17. Kimmerle, J., Cress, U., Held, C.: The interplay between individual and collective knowledge: technologies for organisational learning and knowledge building. Knowledge Management Research & Practice 8, 33–44 (2010)
18. Stahl, G.: Collaborative information environments to support knowledge construction by communities. AI & Society 14, 71–97 (2000)
19. Herrmann, T., Kienle, A.: Context-oriented communication and the design of computer supported discursive learning. International Journal of CSCL 3, 273–299 (2008)
20. Beers, P.J., Boshuizen, H.P.A., Kirschner, P.A., Gijselaers, W.H.: Computer support for knowledge construction in collaborative learning environments. Computers in Human Behavior 21, 623–644 (2005)
21. Wessel, D., Knipfer, K.: Report on User Studies. Deliverable D1.2, MIRROR IP (2011)
22. Kelloway, E.K., Barling, J.: Knowledge Work as Organisational Behaviour. International Journal of Management Reviews 2, 287–304 (2000)

eAssessment for 21st Century Learning and Skills

Christine Redecker*, Yves Punie, and Anusca Ferrari

Institute for Prospective Technological Studies (IPTS),
European Commission, Joint Research Centre,
Edificio Expo, C/Inca Garcilaso 3,
41092 Seville, Spain
{Christine.Redecker,Yves.Punie,Anusca.Ferrari}@ec.europa.eu

Abstract. In the past, eAssessment focused on increasing the efficiency and effectiveness of test administration; improving the validity and reliability of test scores; and making a greater range of test formats susceptible to automatic scoring. Despite the variety of computer-enhanced test formats, eAssessment strategies have been firmly grounded in a traditional paradigm, based on the explicit testing of knowledge. There is a growing awareness that this approach is less suited to capturing "Key Competences" and "21st century skills". Based on a review of the literature, this paper argues that, though there are still technological challenges, the more pressing task is to transcend the testing paradigm and conceptually develop (e)Assessment strategies that foster the development of 21st century skills.

Keywords: eAssessment, Computer-Based Assessment (CBA), competence-based assessment, key competences, 21st century skills.

1 Rethinking 21st Century Assessment

Assessment is an essential component of learning and teaching, as it allows the quality of both teaching and learning to be judged and improved [1]. It often determines the priorities of education [2], it always influences practices and has backwash effects on learning [3]. Changes in curricula and learning objectives are ineffective if assessment practices remain the same [4], as learning and teaching tend to be modelled against the test [2]. Assessment is usually understood to have two purposes: formative and summative. Formative assessment aims to gather evidence about pupils' proficiency to influence teaching methods and priorities, whereas summative assessment is used to judge pupils' achievements at the end of a programme of work [2].

Assessment procedures in formal education and training have traditionally focused on examining knowledge and facts through formal testing [4], and do not easily lend themselves to grasping 'soft skills'. Lately, however, there has been a growing awareness that curricula – and with them assessment strategies – need to be revised to

---

* The views expressed in this article are purely those of the authors and may not in any circumstances be regarded as stating an official position of the European Commission.

A. Ravenscroft et al. (Eds.): EC-TEL 2012, LNCS 7563, pp. 292–305, 2012.

more adequately reflect the skills needed for life in the 21st century. The evolution of information and communication technologies (ICT) has contributed to a sudden change in terms of skills that students need to acquire. Skills such as problem-solving, reflection, creativity, critical thinking, learning to learn, risk-taking, collaboration, and entrepreneurship are becoming increasingly important [5]. The relevance of these "21st Century skills" [6] is moreover recognised in the European Key Competence Recommendation [7] by emphasizing their transversal and over-arching role. To foster and develop these skills, assessment strategies should go beyond testing factual knowledge and aim to capture the less tangible themes underlying all Key Competences.

2   Looking beyond the e-Assessment Era

At the end of the eighties, Bunderson, Inouye and Olsen [8] forecasted four generations of computerized educational measurement, namely: *Generation 1:* Computerized testing (administering conventional tests by computer); *Generation 2:* Computerized adaptive testing (tailoring the difficulty or contents or an aspect of the timing on the basis of examinees' responses); *Generation 3:* Continuous measurement (using calibrated measures to continuously and unobtrusively estimate dynamic changes in the student's achievement trajectory); *Generation 4:* Intelligent measurement (producing intelligent scoring, interpretation of individual profiles, and advice to learners and teachers by means of knowledge bases and inferencing procedures).

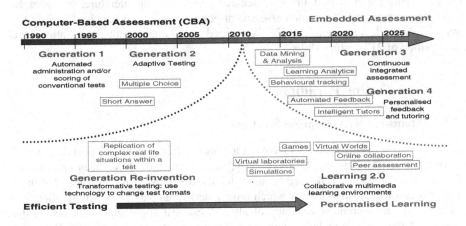

Fig. 1. Current and future e-Assessment strategies. Source: IPTS on the basis of [8-10]

These predictions from 1989 are not far off the mark. The first two generations of eAssessment or Computer-Based Assessment (CBA), which should more precisely be referred to as Computer-Based *Testing*, have now become mainstream. The main challenge currently lies in making the transition to the latter two, the era of Embedded Assessment, which is based on the notion of "Learning Analytics", i.e. the interpretation of a wide range of data about students' proficiency in order to assess academic progress, predict future performance, and tailor education to individual students [11]. Although Learning Analytics are currently still in an experimental and development phase, embedded assessment could become a technological reality within the next five years [11].

However, the transition from computer-based testing to embedded assessment requires technological advances to be complemented with a conceptual shift in assessment paradigms. While the first two generations of CBA centre on the notion of *testing* and on the use of computers to improve the efficiency of testing procedures, generation 3 and 4 seamlessly integrate holistic and personalised assessment into *learning*. Embedded assessment allows learners to be continuously monitored and guided by the electronic environment which they use for their learning activities, thus merging formative and summative assessment within the learning process. Ultimately, with generation 4, learning systems will be able to provide instant and valid feedback and advice to learners and teachers concerning future learning strategies, based on the learners' individual learning needs and preferences. Explicit testing could thus become obsolete.

This conceptual shift in the area of eAssessment is paralleled by the overall pedagogical shift from knowledge-based to competence-based assessment and the recent focus on transversal and generic skills, which are less susceptible to generation 1 and 2 e-Assessment strategies. Generation 3 and 4 assessment formats may offer a viable avenue for capturing the more complex and transversal skills and competences crucial for work and life in the 21st century. However, to seize these opportunities, assessment paradigms need to shift to being enablers of more personalised and targeted learning processes. Hence, the open question is how we can make this shift happen. This paper, based on an extensive review of the literature, provides an overview of current ICT-enabled assessment practices, with a particular focus on the more recent developments of ICT-enhanced assessment tools that allow the recognition of 21st century skills.

3   The Testing Paradigm

3.1   Mainstream eAssessment Strategies

First and second generation tests have led to a more effective and efficient delivery of traditional assessments [9]. More recently, assessment tools have been enriched to include more authentic tasks and to allow for the assessment of constructs that have either been difficult to assess or which have emerged as part of the information age [12]. As the measurement accuracy of all of these test approaches depends on the quality of the items it includes, item selection procedures – such as Item Response Theory or mathematical programming – play a central role in the assessment process [13]. First generation computer-based tests are already being administered widely for a variety of educational purposes, especially in the US [14], but increasingly also in Europe [15]. Second generation adaptive tests, which select test items based on the candidates' previous response, allow for a more efficient administration mode (less items and less testing time), while at the same time keeping measurement precision [9]. Different algorithms have been developed for the selection of test items. The most well-known type of algorithmic testing is CAT, which is a test where the algorithm is designed to provide an accurate point estimation of individual achievement [16]. CAT tests are very widespread, in particular in the US, where they are used for assessment at primary and secondary school level [10, 14, 17]; and for admission to higher education [17]. Adaptive tests are also used in European countries, for instance the Netherlands [18] and Denmark [19].

3.2   Reliability of eAssessment

Advantages of computer-based testing over traditional assessment formats include, among other: paperless test distribution and data collection; efficiency gains; rapid feedback; machine-scorable responses; and standardized tools for examinees as calculators and dictionaries [17]. Furthermore computer-based tests tend to have a positive effect on students' motivation, concentration and performance [20, 21]. More recently, they also provide learners and teachers with detailed reports that describe strengths and weaknesses, thus supporting formative assessment [20]. However, the reliability and validity of scores has been a major concern, particularly in the early phases of eAssessment, given the prevalence of multiple-choice formats in computer-based tests. Recent research indicates that scores are indeed generally higher in multiple choice tests than they are in short answer formats [22]. Some studies found no significant differences between student performance on paper and on screen [20, 23], whereas others indicate that paper-based and computer-based tests do not necessarily measure the same skills [10, 24].

Current research efforts concentrate on improving the reliability and validity of test scores by improving selection procedures for large item banks [13], by increasing measurement efficiency [25], by taking into account multiple basic abilities simultaneously [26], and by developing algorithms for automated language analysis, that allow the electronic scoring of long free text answers.

3.3   Automated Scoring

One of the main drivers of progress in eAssessment has been the improvement of automatic scoring techniques for free text answers [27] and dedicated written text assignments [28]. Automated scoring could dramatically reduce the time and costs associated with the assessment of complex skills such as writing, but its use must be validated against a variety of criteria for it to be accepted by test users and stakeholders [29].

Already, assignments in programming languages or other formal notations can be automatically assessed [30]. For short-answer free-text responses of around a sentence in length, automatic scoring has also been shown to be at least as good as that of human markers [31]. Similarly, automated scoring for highly predictable speech, such as a one sentence answer to a simple question, correlates very highly with human ratings of speech quality, although this is not the case with longer and more open-ended responses [17]. Automated scoring[1] is also used for scoring essay-length

---

[1] Examples include: Intelligent Essay Assessor (Pearson Knowledge Technologies), Intellimetric (Vantage), Project Essay Grade (PEG) (Measurement, Inc.), and e-rater (ETS).

responses [10] where it is found to closely mimic the results of human scoring: the agreement of an electronic score with a human score is typically as high as the agreement between two humans, and sometimes even higher [10, 17, 29]. However, these programmes tend to omit features that cannot be easily computed, such as content, organization and development [32]. Thus, while there is in general a high correlation between human and machine marking, discrepancies are higher for essays which exhibit more abstract qualities [33]. A further research line aims to develop programmes which mark short-answer free-text and give tailored feedback on incorrect and incomplete responses, inviting examinees to repeat the task immediately [34].

4   The Tutoring Paradigm

Integrated assessment is already implemented in some technology-enhanced learning (TEL) environments, where data-mining techniques can be used for formative assessment and individual tutoring. Similarly, virtual worlds, games, simulations and virtual laboratories allow the tracking of individual learners' activity and can make learning behaviour assessable. Many TEL environments, tools and systems recreate learning situations which require complex thinking, problem-solving and collaboration strategies and thus allow for the development of 21st century skills. Some of these environments include the assessment of performance. Assessment packages for Learning Management Systems are currently under development to integrate self-assessment, peer-assessment and summative assessment, based on the automatic analysis of learner data [35]. Furthermore, data on student engagement in these environments can be used for embedded assessment, which refers to students engaging in learning activities while an assessment system draws conclusions based on their tasks [36]. Data-mining techniques are already used to evaluate university students' activity patterns in Virtual Learning Environments for diagnostic purposes. Analytical data mining can, for example, identify students who are at risk of dropping out or underperforming,[2] generate diagnostic and performance reports,[3] assess interaction patterns between students on collaborative tasks,[4] and visualise collaborative knowledge work.[5] It is expected that, in five years' time, advances in data mining will enable the interpretation of data concerning students' engagement, performance, and progress, in order to assess academic progress, predict future performance, and revise curricula and teaching strategies [11].

Although many of these programmes and environments are still experimental in scope and implementation, there are a number of promising technologies and related

---

[2] For example: the Signals system at Purdue University,
`http://www.itap.purdue.edu/tlt/signals/` the Academic Early Alert and Retention System at Northerm Arizona University,
`http://www4.nau.edu/ua/GPS/student/`
[3] `http://www.socrato.com/`
[4] `http://research.uow.edu.au/learningnetworks/seeing/snapp/index.html`
[5] `http://emergingmediainitiative.com/project/ learning-analytics/`

assessment strategies that might soon give rise to integrated assessment formats that comprehensively capture 21st century skills.

4.1   Intelligent Tutoring Systems and Automated Feedback

Research indicates that the closer the feedback is to the actual performance, the more powerful its impact is on subsequent performance and learner motivation [37]. Timing is the obvious advantage of Intelligent Tutoring Systems (ITSs) [38], that adapt the level of difficulty of the tasks administered to the individual learners' progress and needs. Most programmes provide qualitative information on why particular responses are incorrect [37]. Although in some cases fairly generic, some programmes search for patterns in student work to adjust the level of difficulty in subsequent exercises according to needs [39]. Huang et al. [40], for example, developed an intelligent argumentation assessment system for elementary school pupils. The system analyses the structure of students' scientific arguments posted on a moodle discussion board and issues feedback in case of bias. In a first trial, the system was shown to be effective in classifying and improving students' argumentation levels and assisting them in learning the core concepts. Currently, ITSs such as AutoTutor[6] and GnuTutor[7] [41] teach students by holding a conversation in natural language. There are versions of AutoTutor that guide interactive simulation in 3D micro-worlds, that detect and produce emotions, and that are embedded in games [42]. The latest version of AutoTutor has been enabled to detect learners' boredom, confusion, and frustration by monitoring conversational cues, gross body language, and facial features [43]. GnuTutor, a simplified open source variety of AutoTutor, intends to create a freely available, open source ITSs platform that can be used by schools and researchers [41].

ITSs are being widely used in the US, where the most popular system "Cognitive Tutor" provides differentiated instruction in mathematics which encourages problem-solving behaviour to half a million students in US middle and high schools. The programme selects mathematical problems for each student at an adapted level of difficulty. Correct solution strategies are annotated with hints. Students can access instruction that is directly relevant to the problem they are working on and the strategy they are following within that problem.[8] Research indicates that students who used Cognitive Tutor significantly outscored their peers on national exams, an effect that was especially noticeable for students with limited English proficiency or special learning needs [44]. Research on the implementation of a web-based intelligent tutoring system "eFit" for mathematics at lower secondary schools in Germany confirms this finding: children using the tutoring system significantly improved their arithmetic performance over a period of 9 months [45].

ITSs are also used widely to support reading. SuccessMaker's Reader's Workshop[9] and Accelerated Reader[10] are two very popular commercial reading software products

---

[6] http://www.autotutor.org/

[7] http://gnututor.com/

[8] http://carnegielearning.com/static/web_docs/
2010_Cognitive_Tutor_Effectiveness.pdf

[9] http://www.successmaker.com/Courses/c_awc_rw.html

[10] http://www.renlearn.com/ar/

for primary education in the US. They provide ICT-based instruction with animations and game-like scenarios. Assessment is embedded and feedback is automatic and instant. Learning can be customized for three different profiles and each lesson can be adapted to student's strengths and weaknesses. These programmes have been evaluated as having positive impacts on learning [39]. Anotherexample is iSTART,[11] a Web-based tutoring programme to teach reading strategies to young, adolescent and college-aged students. Animated agents are used to develop comprehension strategies such as paraphrasing and predicting. As the learner progresses through the modules, s/he creates self-explanations that are evaluated by the agent [46]. iSTART has been shown to improve the quality of students' self-explanation and this, in turn, was reflected in improved comprehension scores [47]. Similarly, an automated reading tutor for children starting to read in Dutch automatically assesses their reading levels, provides them with oral feedback at the phoneme, syllable or word level, and tracks where they are reading, for automated screen advancement or for direct feedback to them [48].

4.2   Immersive Environments, Virtual Worlds, Games and Simulations

Immersive environments and games are specifically suitable for acquiring 21st century skills such as problem-solving, collaboration and inquiry, because they are based on the fact that what needs to be acquired is not explicit but must be inferred from the situation [49]. In these environments, the learning context is similar to the contexts within which students will apply their learning, thus promoting inquiry skills; making learning activities more motivating; and increasing the likelihood that acquired skills will transfer to real-world situations [46]. It has been recognised that immersive game-based learning environments lead to significantly better learning results than traditional learning approaches [50].

Assessment can be integrated in the learning process within virtual environments and games. A virtual world like Second Life, for example, uses a mixture of multiple choice and environment interaction questions for both delivery and assessment of content, while encouraging an exploratory attitude to learning [51]. However, Virtual worlds, games and simulations are primarily conceived of as learning rather than testing environments. In science education, for instance, computer simulations, scientific games and virtual laboratories provide opportunities for students to develop and apply skills and knowledge in more realistic contexts and provide feedback in real time. Simulations may involve mini-laboratory investigations, or "predict-observe-explain" demonstrations. Dynamic websites, such as Web of Inquiry,[12] allow students to carry out scientific inquiry projects to develop and test their theories; learn scientific language, tools, and practices of investigation; engage in self-assessment; and provide feedback to peers [52]. Simulations provided by Molecular Workbench,[13] for example, emulate phenomena that are too small or too fast to observe, such as

---

[11] iSTART stands for interactive strategy training for active reading and thinking, see: http://129.219.222.66/Publish/projectsiteistart.html
[12] http://www.webofinquiry.org
[13] http://mw.concord.org/modeler/index.html

chemical reactions or gas at the molecular level. These visual, interactive computational experiments for teaching and learning science can be customized and adapted by the teacher. Embedded assessments allow teachers to generate real-time reports and track students' learning progress. Programmes like these usually provide opportunities for students to reflect on their own actions and response patterns [39]. Some science-learning environments have embedded formative assessments that teachers can access immediately in order to gauge the effectiveness of their instruction and modify their plans accordingly [53].

Furthermore, a variety of recent educational games for science education could, in principle, integrate assessment and tutoring functionalities similar to ITSs. ARIES (Acquiring Research Investigative and Evaluative Skills) is a computerized educational tool which incorporates multiple learning principles, such as testing effects, generation effects, and formative feedback [54]. Another example is Quest Atlantis,[14] which promotes causal reasoning skills, subject knowledge in physics and chemistry, and an understanding of how systems work at both macro and micro level.

Some game environments include feedback, tutoring and monitoring of progress. In River City, for example, students use their knowledge of biology and the results of tests conducted online with equipment such as virtual microscopes to investigate the mechanisms through which a disease is spreading in a simulated 18th century city. Prompts gradually fade as students acquire inquiry skills. Data-mining allows teachers to document gains in students' engagement, learning and self-efficacy [46, 55]. Similarly, the Virtual Performance Assessment project[15] relies on simulated, game-like environments to assess students' ability to perform scientific inquiry. The assessment is performed in authentic settings, which allow better observation and measurement of complex cognition and inquiry processes [6].

4.3   Authentic Tasks Employing Digital Tools

Practical tasks using mobile devices or online resources are another promising avenue for developing ICT-enabled assessment formats. A number of national pilots assess tasks that replicate real life contexts and are solved by using common technologies, such as the internet, office and multimedia tools. In Denmark, for example, students at the commercial and technical upper secondary schools have, since 2001, been sitting Danish language, Maths and Business Economics exams based on CD-ROMs with access to multimedia resources. The aim is to evaluate their understanding of the subjects and their ability to search, combine, analyse and synthesise information and work in an inter-disciplinary way.[16] A new pilot was launched in 2009 which aims to develop use of full Internet access during high stakes formal assessments. Similarly, in the eSCAPE project, a 6-hour collaborative design workshop replaced school examinations for 16 year-old students in Design and Technology in eleven schools

---

[14] http://atlantis.crlt.indiana.edu/
[15] http://vpa.gse.harvard.edu/a-case-study-of-the-virtual-performance-assessment-project/
[16] http://www.cisco.com/web/strategy/docs/education/DanishNationalAssessmentSystem.pdf

across England. Students work individually, but within a group context, and record assessment evidence via a handheld device in a short multimedia portfolio. The reliability of the assessment method was reported as very high [6, 56]. In the US, College Work and Readiness Assessment (CWRA) was introduced in St. Andrew's School in Delaware to test students' readiness for college and work, and it quickly spread to other schools across the US. It consists of a single 90-minute task that students must accomplish by using a library of online documents. Students must address real-world dilemmas (like helping a town reduce pollution), making judgments that have economic, social, and environmental implications, and articulate a solution in writing.

This approach has proved particularly fruitful for the assessment of digital competence. The Key Stage 3 ICT tests (UK), for example, require 14 year-old students to use multiple ICT tools in much the same way they are used in real work and academic environments [10]. The project led to the development of authentic tasks to assign to students, who completed tests of ICT skills in a virtual desktop environment [56]. Similarly, the iSkills[17] assessment aims to measure students' critical thinking and problem-solving skills in a digital environment. In a one-hour exam real-time, scenario-based tasks are presented that measure an individual's ability to navigate, critically evaluate and understand the wealth of information available through digital technology. The national ICT skills assessment programme in Australia [57] is designed to be an authentic performance assessment, mirroring students' typical 'real world' use of ICT. In 2005 and 2008, students completed tasks on computers using software that included a seamless combination of simulated and live applications.

5   The Conceptual Leap

As the examples outlined above illustrate, we are currently witnessing a great number of innovative developments, both in the area of Computer-Based Testing and in the area of embedded assessment and intelligent tutoring, which indicate promising avenues for capturing complex key competences and 21st century skills. In the past, research focused on the technological side, with more and more tools, functionalities and algorithms to increase measurement accuracy and to create more complex and engaging learning environments with targeted feedback loops [10, 17, 36]. Given that learning analytics could, in the future, replace explicit testing, (e)Assessment will become far closer interwoven with learning and teaching and will have to respond to and respect the pedagogical concepts on which the learning process is based. To further exploit the potential of technologies, research efforts should therefore not only focus on increasing efficiency, validity and reliability of ICT enhanced assessment formats, but must also consider how the pedagogical and conceptual foundations of different pedagogical approaches translate into different eAssessment strategies. The possibility to use data from the learning process itself for assessment purposes in an objective, valid, reliable and comparable way renders explicit testing obsolete. Thus,

---

[17] http://www.ets.org/iskills/

we need to re-consider the value of tests, examinations, and in particular of high stakes summative assessments, that are based on the evaluation of students' performance displayed in just one instance, on dedicated tasks that are limited in scope [39]. If traditional computer-based formats continue to play an important role in high-stakes tests, there is a risk that learning processes and outcomes will fail to address the key competences and complex transversal skills needed in the 21st century, despite the technological availability of embedded assessment formats and the implementation of competence-based curricula. We have seen in the past that with computer-based testing, formative and summative assessment formats have diverged. As a result, schools are now starting to use computer-based multiple choice tests as a means of training for high stakes examinations [58]. Since assessment practice tends to affect teaching and learning practices, the danger is to focus on "scoring well" rather than on developing complex key competences which will be increasingly important in the future [5]. If new assessment strategies are not deployed, computer-based assessment will further increase the gap between learning and testing and the necessary competences to be acquired will not be given the relevance they deserve. If, on the other hand, formative and summative assessment become an integral part of the learning process, and digital learning environments become the main source for grading and certification, there is a need to better understand how information collected digitally should be used, evaluated and weighted to adequately reflect the performance of each individual learner. If genuinely pedagogical tasks, such as assessing and tutoring, are increasingly delegated to digital environments, these have to be designed in such a way that they become a tool for teachers and learners to communicate effectively with one another.

Embedded assessment should be designed to respect and foster the primacy of pedagogy and the role of the teacher. Judgements on the achievement and performance of students that are based on data collected in digital environments must be based on a transparent and fair process of interpreting and evaluating these data, which is mediated by digital applications and tools, but ultimately lies in the hands of teachers and learners. Given that teachers will be able to base their pedagogical decisions and judgements on a wider range of data than in the past, pedagogical principles for interpreting, evaluating, weighing and reflecting on these different kinds of data are needed. Therefore, to facilitate the development and implementation of environments that will effectively serve learners and teachers in developing 21st century skills, technology developers and educators need to enter into a dialogue for the creation of effective learning pathways. Hence, it is important to ensure that further progress in the technological development of environments, applications and tools for learning and assessment is guided by pedagogical principles that reflect the competence requirements of the 21st century.

6   Conclusions

This paper has argued for a paradigm shift in the use and deployment of Information and Communication Technologies (ICT) in assessment. In the past, eAssessment has mainly focused on increasing efficiency and effectiveness of test administration;

improving the validity and reliability of test scores; and making a greater range of test formats susceptible to automatic scoring, with a view to simultaneously improving efficiency and validity. However, despite the variety of computer-enhanced test formats, eAssessment strategies have been firmly grounded in the traditional assessment paradigm, which has for centuries dominated formal education and training and is based on the explicit testing of knowledge. However, against the background of rapidly changing skill requirements in a knowledge-based society, education and training systems in Europe are becoming increasingly aware that curricula and with them assessment strategies need to refocus on fostering more holistic "Key Competences" and transversal or general skills, such as so-called "21st century skills". ICT offer many opportunities for supporting assessment formats that can capture complex skills and competences that are otherwise difficult to assess. To seize these opportunities, research and development in eAssessment and in assessment in general have to transcend the testing paradigm and develop new concepts of embedded, authentic and holistic assessment. Thus, while there is still a need to technologically advance in the development and deployment of integrated assessment formats, the more pressing task at the moment is to make the conceptual shift between traditional and 21st century testing and develop (e-)Assessment pedagogies, frameworks, formats and approaches that reflect the core competences needed for life in the 21st century.

References

1. Ferrari, A., Cachia, R., Punie, Y.: Innovation and Creativity in Education and Training in the EU Member States: Fostering Creative Learning and Supporting Innovative Teaching. In: JRC-IPTS (2009)
2. NACCCE: All Our Futures: Creativity, Culture and Education (1999)
3. Ellis, S., Barrs, M.: The Assessment of Creative Learning. In: Sefton-Green, J. (ed.) Creative Learning, pp. 73–89. Creative Partnerships, London (2008)
4. Cachia, R., Ferrari, A., Ala-Mutka, K., Punie, Y.: Creative Learning and Innovative Teaching: Final Report on the Study on Creativity and Innovation in Education in EU Member States. In: JRC-IPTS (2010)
5. Redecker, C., Leis, M., Leendertse, M., Punie, Y., Gijsbers, G., Kirschner, P., Stoyanov, S., Hoogveld, B.: The Future of Learning: Preparing for Change. In: JRC-IPTS (2010)
6. Binkley, M., Erstad, O., Herman, J., Raizen, S., Ripley, M., Rumble, M.: Defining 21st Century Skills. In: Griffin, P., McGaw, B., Care, E. (eds.) Assessment and Teaching of 21st Century Skills, pp. 17–66. Springer, Heidelberg (2012)
7. Council of the European Union: Recommendation of the European Parliament and the Council of 18 December 2006 on key competences for lifelong learning. (2006/962/EC). Official Journal of the European Union, L394/10 (2006)
8. Bunderson, V.C., Inouye, D.K., Olsen, J.B.: The four generations of computerized educational measurement. In: Linn, R.L. (ed.) Educational Measurement, pp. 367–407. Macmillan, New York (1989)
9. Martin, R.: New possibilities and challenges for assessment through the use of technology. In: Scheuermann, F., Pereira, A.G. (eds.) Towards a Research Agenda on Computer-Based Assessment. Office for Official Publications of the European Communities, Luxembourg (2008)

10. Bennett, R.E.: Technology for Large-Scale Assessment. In: Peterson, P., Baker, E., McGaw, B. (eds.) International Encyclopedia of Education, vol. 8, pp. 48–55. Elsevier, Oxford (2010)

11. Johnson, L., Smith, R., Willis, H., Levine, A., Haywood, K.: The 2011 Horizon Report. The New Media Consortium (2011)

12. Pellegrino, J.W.: Technology and Learning - Assessment. In: Peterson, P., Baker, E., McGaw, B. (eds.) International Encyclopedia of Education, vol. 8, pp. 42–47. Elsevier, Oxford (2010)

13. El-Alfy, E.S.M., Abdel-Aal, R.E.: Construction and analysis of educational tests using abductive machine learning. Computers and Education 51, 1–16 (2008)

14. Csapó, B., Ainley, J., Bennett, R., Latour, T., Law, N.: Technological Issues for Computer-Based Assessment. The University of Melbourne, Australia (2010)

15. Moe, E.: Introducing Large-scale Computerised Assessment Lessons Learned and Future Challenges. In: Scheuermann, F., Björnsson, J. (eds.) The Transition to Computer-Based Assessment. Office for Official Publications of the European Communities, Luxembourg (2009)

16. Thompson, N.A., Weiss, D.J.: Computerized and Adaptive Testing in Educational Assessment. In: Scheuermann, F., Björnsson, J. (eds.) The Transition to Computer-Based Assessment. Office for Official Publications of the European Communities, Luxembourg (2009)

17. Bridgeman, B.: Experiences from Large-Scale Computer-Based Testing in the USA. In: Scheuermann, F., Björnsson, J. (eds.) The Transition to Computer-Based Assessment. Office for Official Publications of the European Communities, Luxembourg (2009)

18. Eggen, T.J.H.M., Straetmans, G.J.J.M.: Computerized Adaptive Testing of Arithmetic at the Entrance of Primary School Teacher Training College. In: Scheuermann, F., Björnsson, J. (eds.) The Transition to Computer-Based Assessment. Office for Official Publications of the European Communities, Luxembourg (2009)

19. Wandall, J.: National Tests in Denmark – CAT as a Pedagogic Tool. In: Scheuermann, F., Björnsson, J. (eds.) The Transition to Computer-Based Assessment. Office for Official Publications of the European Communities, Luxembourg (2009)

20. Ripley, M.: Transformational Computer-based Testing. In: Scheuermann, F., Björnsson, J. (eds.) The Transition to Computer-Based Assessment. Office for Official Publications of the European Communities, Luxembourg (2009)

21. Garrett, N., Thoms, B., Alrushiedat, N., Ryan, T.: Social ePortfolios as the new course management system. On the Horizon 17, 197–207 (2009)

22. Park, J.: Constructive multiple-choice testing system. British Journal of Educational Technology 41, 1054–1064 (2010)

23. Hardré, P.L., Crowson, H.M., Xie, K., Ly, C.: Testing differential effects of computer-based, web-based and paper-based administration of questionnaire research instruments. British Journal of Educational Technology 38, 5–22 (2007)

24. Horkay, N., Bennett, R.E., Allen, N., Kaplan, B., Yan, F.: Does it Matter if I Take My Writing Test on Computer? An Empirical Study of Mode Effects in NAEP. Journal of Technology, Learning, and Assessment 5 (2006)

25. Frey, A., Seitz, N.N.: Multidimensional adaptive testing in educational and psychological measurement: Current state and future challenges. Studies in Educational Evaluation 35, 89–94 (2009)

26. Hartig, J., Höhler, J.: Multidimensional IRT models for the assessment of competencies. Studies in Educational Evaluation 35, 57–63 (2009)

27. Noorbehbahani, F., Kardan, A.A.: The automatic assessment of free text answers using a modified BLEU algorithm. Computers and Education 56, 337–345 (2011)
28. He, Y., Hui, S.C., Quan, T.T.: Automatic summary assessment for intelligent tutoring systems. Computers and Education 53, 890–899 (2009)
29. Weigle, S.C.: Validation of automated scores of TOEFL iBT tasks against non-test indicators of writing ability. Language Testing 27, 335–353 (2010)
30. Amelung, M., Krieger, K., Rösner, D.: E-assessment as a service. IEEE Transactions on Learning Technologies 4, 162–174 (2011)
31. Butcher, P.G., Jordan, S.E.: A comparison of human and computer marking of short free-text student responses. Computers and Education 55, 489–499 (2010)
32. Ben-Simon, A., Bennett, R.E.: Toward a more substantively meaningful automated essay scoring. Journal of of Technology, Learning and Assessment 6 (2007)
33. Hutchison, D.: An evaluation of computerised essay marking for national curriculum assessment in the UK for 11-year-olds. British Journal of Educational Technology 38, 977–989 (2007)
34. Jordan, S., Mitchell, T.: e-Assessment for learning? The potential of short-answer free-text questions with tailored feedback. British Journal of Educational Technology 40, 371–385 (2009)
35. Florián, B.E., Baldiris, S.M., Fabregat, R., De La Hoz Manotas, A.: A set of software tools to build an author assessment package on Moodle: Implementing the AEEA proposal. In: 10th IEEE International Conference on Advanced Learning Technologies, ICALT, pp. 67–69 (2010)
36. Ridgway, J., McCusker, S.: Challenges for Research in e-Assessment. In: Scheuermann, F., Pereira, A.G. (eds.) Towards a Research Agenda on Computer-Based Assessment. Office for Official Publications of the European Communities, Luxembourg (2008)
37. Nunan, D.: Technology Supports for Second Language Learning. In: International Encyclopedia of Education, vol. 8, pp. 204–209. Elsevier, Oxford (2010)
38. Ljungdahl, L., Prescott, A.: Teachers' use of diagnostic testing to enhance students' literacy and numeracy learning. International Journal of Learning 16, 461–476 (2009)
39. Looney, J.: Making it Happen: Formative Assessment and Educational Technologies. Promethean Thinking Deeper Research Papers 1 (2010)
40. Huang, C.J., Wang, Y.W., Huang, T.H., Chen, Y.C., Chen, H.M., Chang, S.C.: Performance evaluation of an online argumentation learning assistance agent. Computers and Education 57, 1270–1280 (2011)
41. Olney, A.M.: GnuTutor: An open source intelligent tutoring system based on AutoTutor. Cognitive and Metacognitive Educational Systems: Papers from the AAAI Fall Symposium FS-09-02, 70–75 (2009)
42. Graesser, A.: Autotutor and the world of pedagogical agents: Intelligent tutoring systems with natural language dialogue. In: 22nd International Florida Artificial Intelligence Research Society Conference, FLAIRS-22, vol. 3 (2009)
43. D'Mello, S., Craig, S., Fike, K., Graesser, A.: Responding to Learners' Cognitive-Affective States with Supportive and Shakeup Dialogues. In: Jacko, J.A. (ed.) HCI International 2009, Part III. LNCS, vol. 5612, pp. 595–604. Springer, Heidelberg (2009)
44. Ritter, S., Anderson, J.R., Koedinger, K.R., Corbett, A.: Cognitive tutor: Applied research in mathematics education. Psychonomic Bulletin and Review 14, 249–255 (2007)
45. Graff, M., Mayer, P., Lebens, M.: Evaluating a web based intelligent tutoring system for mathematics at German lower secondary schools. Education and Information Technologies 13, 221–230 (2008)

46. Means, B., Rochelle, J.: An Overview of Technology and Learning. In: Peterson, P., Baker, E., McGaw, B. (eds.) International Encyclopedia of Education, vol. 8, pp. 1–10. Elsevier, Oxford (2010)

47. DeFrance, N., Khasnabis, D., Palincsar, A.S.: Reading and Technology. In: Peterson, P., Baker, E., McGaw, B. (eds.) International Encyclopedia of Education, vol. 8, pp. 150–157. Elsevier, Oxford (2010)

48. Duchateau, J., Kong, Y.O., Cleuren, L., Latacz, L., Roelens, J., Samir, A., Demuynck, K., Ghesquière, P., Verhelst, W., Hamme, H.V.: Developing a reading tutor: Design and evaluation of dedicated speech recognition and synthesis modules. Speech Communication 51, 985–994 (2009)

49. de Jong, T.: Technology Supports for Acquiring Inquiry Skills. In: Peterson, P., Baker, E., McGaw, B. (eds.) International Encyclopedia of Education, vol. 8, pp. 167–171. Elsevier, Oxford (2010)

50. Barab, S.A., Scott, B., Siyahhan, S., Goldstone, R., Ingram-Goble, A., Zuiker, S.J., Warren, S.: Transformational play as a curricular scaffold: Using videogames to support science education. Journal of Science Education and Technology 18, 305–320 (2009)

51. Bloomfield, P.R., Livingstone, D.: Multi-modal learning and assessment in Second Life with quizHUD. In: Conference in Games and Virtual Worlds for Serious Applications, pp. 217–218 (2009)

52. Herrenkohl, L.R., Tasker, T., White, B.: Pedagogical practices to support classroom cultures of scientific inquiry. Cognition and Instruction 29, 1–44 (2011)

53. Delgado, C., Krajcik, J.: Technology and Learning - Supports for Subject Matter Learning. In: Peterson, P., Baker, E., McGaw, B. (eds.) International Encyclopedia of Education, vol. 8, pp. 197–203. Elsevier, Oxford (2010)

54. Wallace, P., Graesser, A., Millis, K., Halpern, D., Cai, Z., Britt, M.A., Magliano, J., Wiemer, K.: Operation ARIES!: A computerized game for teaching scientific inquiry. Frontiers in Artificial Intelligence and Applications 200, 602–604 (2009)

55. Dede, C.: Technological Support for Acquiring Twenty-First -Century Skills. In: Peterson, P., Baker, E., McGaw, B. (eds.) International Encyclopedia of Education, vol. 8, pp. 158–166. Elsevier, Oxford (2010)

56. Ripley, M.: Transformational Computer-based Testing. In: Scheuermann, F., Björnsson, J. (eds.) The Transition to Computer-Based Assessment. New Approaches to Skills Assessment and Implications for Large-scale Testing. Office for Official Publications of the European Communities, Luxembourg (2009)

57. MCEECDYA: National Assessment Program. ICT Literacy Years 6 and 10 Report 2008. Ministerial Council for Education, Early Childhood Development and Youth Affairs (MCEECDYA) (2008)

58. Glas, C.A.W., Geerlings, H.: Psychometric aspects of pupil monitoring systems. Studies in Educational Evaluation 35, 83–88 (2009)

Supporting Educators to Discover
and Select ICT Tools with SEEK-AT-WD

Adolfo Ruiz-Calleja[1], Guillermo Vega-Gorgojo[1], Areeb Alowisheq[2],
Juan Ignacio Asensio-Pérez[1], and Thanassis Tiropanis[2]

[1] University of Valladolid. Paseo Belén, 15, Valladolid, Spain
[2] University of Southampton. Web and Internet Science Group, Southampton, UK
adolfo@gsic.uva.es, guiveg@tel.uva.es, aaa08r@ecs.soton.ac.uk,
juaase@tel.uva.es, tt2@ecs.soton.ac.uk
http://www.gsic.uva.es/
http://www.ecs.soton.ac.uk/

Abstract. Several educational organizations provide Information and
Communication Technology (ICT) tool registries to support educators
when selecting ICT tools for their classrooms. A common problem is how
to populate these registries with descriptions of ICT tools that can be
useful for education. This paper proposes to tackle it taking advantage
of the information already published in the Web of Data, following a
Linked-Data approach. For this purpose, SEEK-AT-WD is proposed as
an infrastructure that automatically retrieves ICT tool descriptions from
the Web and publishes them back once they are related to an educational
vocabulary. A working prototype containing 3556 descriptions of ICT
tools is presented. These descriptions can be accessed either from a web
interface or through a search client that has also been developed. The
potential of the proposal is discussed by means of a feature analysis
involving educators, data publishers and registry administrators. Results
show that it is much easier to update the data associated to SEEK-AT-
WD in comparison to other approaches, while educators obtain from a
single registry thousands of tool descriptions collected from the Web.

1 Introduction

The use of technology to support learning activities is nowadays generalized.
Recently, the Web 2.0 movement and the proliferation of Web-based tools have
caused the appearance of thousands of Information and Communication Technology (ICT) tools that are frequently used to support learning activities [1].
In this paper, all these ICT tools that can potentially be used in educational
situations are considered educational ICT tools. Some of them are specific for
education, but others have been successfully employed in the classroom even if
they were not designed for educational purposes [2], such as wikis or chats. Due
to the number of educational ICT tools, it is specially challenging to discover
and select those with the required capabilities for a particular learning situation.

Experience shows that specialized search facilities are very helpful to inform
educators when discovering and selecting ICT tools [3,4]. For this reason, several

A. Ravenscroft et al. (Eds.): EC-TEL 2012, LNCS 7563, pp. 306–319, 2012.

educational organizations maintain their own ICT tool search systems, such as CoolToolsForSchools[1], SchoolForge[2] or Ontoolsearch[3]. All these search facilities rely on registries that require a significant effort to be populated and maintained updated. Some of them, such as Ontoolsearch, depend on an expert that assumes this effort. Other examples, such as CoolToolsForSchools, trust the community of educators not only to consume tool descriptions, but also to publish them following a Web 2.0 approach [5]. In any case, they require human intervention to build their collection of descriptions from scratch. This limitation could be overcome if educational tool descriptions could be automatically obtained from other repositories. Nevertheless, this scenario can not be reached unless educational ICT tool registries are federated and able to share data using compatible formats and schemas [6].

With the aim of federating datasets, Linked Data [7] has been proposed as a methodology to publish data on the Web in a way that avoids heterogeneity of data formats and schemas, envisioning a Web-scale federation of datasets. Several providers are publishing their data according to this methodology, creating the so-called Web of Data [8]. This Web of Data is very promising to enable the federation of educational datasets [9]. In fact, some organizations publish educational-specific data, such as the Government of United Kingdom[4] or the University of Southampton[5]. Despite this, there are no educational-specific ICT tool datasets in the Web of Data, although several data sources provide information that can be useful in education. For example, DBpedia [10] mirrors part of the Wikipedia to the Web of Data, including hundreds of descriptions of ICT tools that are commonly used in the classroom.

In this context, this paper proposes SEEK-AT-WD (Support for Educational External Knowledge About Tools in the Web of Data), a Linked-Data-based registry of educational ICT tools. Its key idea is to take advantage of ICT tool data already available on the Web, automatically provide it an educational value by identifying the potential learning tasks they might support, and publish it again to be consumed by educational applications or imported by other federated educational datasets. This data could be used by educational search systems thus offering a big collection of updated descriptions of ICT tools to educators, while the cost of maintaining this data is dramatically reduced. As a proof of concept, a search client for SEEK-AT-WD has been developed. Then, a feature evaluation is used to compare SEEK-AT-WD with other registries of educational ICT tools.

The rest of the paper is structured as follows: section 2 further motivates the development of SEEK-AT-WD analyzing current approaches, as well as showing a motivating scenario that cannot be supported by them. Section 3 presents the design and development of SEEK-AT-WD, an infrastructure able to support the aforementioned scenario. Then, section 4 presents a feature evaluation,

---

[1] http://cooltoolsforschools.wikispaces.com/
[2] https://schoolforge.net/
[3] http://www.gsic.uva.es/ontoolsearch/
[4] http://data.gov.uk/
[5] http://data.southampton.ac.uk/

comparing SEEK-AT-WD with other educational ICT tool registries. Finally, conclusions and outlook are provided in section 5.

2   Existing Approaches to Support the Discovery and Selection of Educational ICT Tools

This section firstly describes different approaches followed by current registries of educational ICT tools regarding the way they collect and publish their data. Then, it proposes a desired scenario that can not be supported by current approaches since they are not federated to the Web of Data. Several requirements are extracted out of this scenario that should be fulfilled in order to support it.

2.1   Obtaining and Publishing Educational Descriptions of ICT Tools

Several educational organizations maintain their own registries of ICT tools, offering search facilities for educators to retrieve their data. These registries present a wide diversity on how their data is collected and made available for consumption. This subsection shows this diversity by describing several examples of educational registries of ICT tools.

When authoring data, some educational organizations depend on experts that update their registries of ICT tools. Some examples are Ontoolsearch or Software Library of Universidad Complutense de Madrid (Sisoft)[6]. Their main disadvantage is that organizations have to allocate resources to deal with the creation and updateness of their tool datasets. In addition, only some experts selected by the organization are allowed to include data in the registry. These drawbacks can be somehow alleviated following a Web 2.0 approach [5]. As an example, Schoolforge provides an interface where tool providers can add new tool descriptions. In addition, it also allows its users to review existing tools, thus enriching these descriptions with their experience using them. A similar approach is followed by CoolToolsForSchools, but it trusts the community of educators both to add new tools and to enrich existing descriptions.

On the other hand, it is also interesting to discuss how these registries publish their data, thus showing if it is feasible for third parties to reuse it. In this regard, some evaluation criteria has been proposed by Tim Berners-Lee [7], and further explained [8, chap. 2], about how to publish data on the Web to make it usable by third parties. This criteria proposes a five-star rating scheme, where the more stars a dataset gets, the easier it is to retrieve data from it.

The first star is obtained by all the datasets that publish open data on the Web. This star is obtained both by CoolToolsForSchools and Ontoolsearch, but Sisoft and Schoolforge do not obtain it since they do not state that their data has open license. For this reason, although data is on the Web it cannot be legally used by third parties. The second and the third stars are obtained when data is

---

[6] http://www.sisoft.ucm.es/

published as machine-readable structured data using non-proprietary formats. These requirements are also satisfied by CoolToolsForSchools and Ontoolsearch; thus, third-parties can automatically obtain their data. However, it is convenient not only to allow the data extraction by third parties, but also to make this task as easy as possible. In this regard, the fourth star is obtained when publishing data using open standards from the W3C, which is done by Ontoolsearch but not by CoolToolsForSchools. Finally, the fifth star requires the dataset to be federated to others in the Web of Data, which is not done by Ontoolsearch.

Table 1 sums up the characteristics previously analyzed. It can be seen that none of these examples is able to obtain data from third parties, which severely limits its sustainability since each party has to do all the authoring effort. In addition, although some of these datasets allow their data to be retrieved, none of them are linked to the Web of Data. This is an important disadvantage that hinders the federation of these datasets, since it significantly increases the effort of accessing their data [8, chap. 2]. In order to further motivate the federation of educational datasets to the Web of Data, next subsection proposes a scenario that shows its advantages and underlines the requirements that such a registry should fulfill.

Table 1. Main characteristics of four educational ICT tool repositories

Characteristic	Sisoft	SchoolForge	CoolTools ForSchools	Ontoolsearch
Data authoring	Expert	Communities of educators and tool providers	Community of educators	Expert
Third-party data	No	No	No	No
Berners-Lee's stars	0	0	3	4

2.2   Motivating Scenario

Software Engineering (SE) is a third-year course in Telecommunication Engineering in the University of Valladolid (Spain) in which a software project is designed. One exercise is to individually generate a UML diagram. Then, groups of four students are formed to carry out a peer-review activity where each student obtains a text generated by its partners criticizing his UML model.

The educator responsible of SE does not know which ICT tools could support the aforementioned activity, so he uses an ICT tool search system stating that he needs a `tool for modeling UML models` and a `tool that allows a group to write a text document`. Several months before, someone published in Wikipedia a description of Umbrello[7], saying that it is a UML modeling tool; similarly, another one published in Freebase a description of Google Docs[8], stating that it is an on-line word processor that allows collaboration. These two tools

---

[7] http://en.wikipedia.org/wiki/Umbrello_UML_Modeller

[8] http://www.freebase.com/view/m/02ffncb

can well support the students of SE in their activity; so that, when the educator submits the abovementioned educational requests, the search system provides Umbrello and Google Docs as well as their descriptions.

This scenario can be achieved by a search facility that allows the educator to **make queries using educational abstractions** [11] while relying on an **infrastructure capable of obtaining data from different sources of the Web of Data** (F1), including DBpedia [10], which publishes information from Wikipedia, and Freebase[9]. With these data sources the infrastructure has to make the necessary **translations to provide educational data of ICT tools** (F2). So that, this infrastructure needs to relate to an educational vocabulary all the data retrieved from third-party data sources, which may not be educational. In order to allow the data consumption by third-parties' applications, the infrastructure **should offer a public data API**, using **W3C standards** to facilitate interoperation and to contribute to the growth of the Web of Data (F3). Finally, since educators are supposed to use an interactive user interface when searching for tools, **response time when retrieving the data should not be too high** even if they make complex queries. Figure 1 graphically sums up the data flow of this scenario.

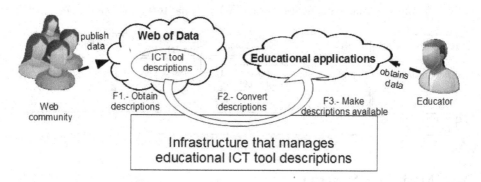

Fig. 1. Educational consumption of ICT tool descriptions from the Web of Data

3   SEEK-AT-WD Architecture and Prototype

Taking into account the requirements previously collected, this section briefly depicts the most important decisions when designing and implementing SEEK-AT-WD, both focusing on its data layer and on its software infrastructure.

3.1   SEEK-AT-WD Data Layer

The main decision when defining SEEK-AT-WD data layer is to follow the Linked Data principles [7]. This consideration has two main advantages: first, the

---

[9] http://www.freebase.com/

publication methodology and the data interface are standardized, thus facilitating the consumption of data; second, it allows the federation of SEEK-AT-WD to already existing datasets in the Web of Data, as well as to other educational registries that may appear in the future.

Best practices to publish data on the Web [8, chap. 3-4] have been followed. They highly recommend to reuse already existing vocabularies that are in used by third parties, since it reduces the effort of creating a vocabulary and facilitates the federation of this dataset to others in the Web. After analyzing the vocabularies used in the Web of Data some of them were found to be useful to describe the administrative parameters of the ICT tools (e.g. license or last version), their functionality (e.g. word processor) or some technological characteristics (e.g. the Operating System were they can be deployed). These characteristics can be defined using concepts from several well-known vocabularies, such as Dublin Core[10], FOAF[11], DBpedia Ontology [10] or RDFS[12].

However, none of these vocabularies define the educational capabilities of ICT tools. For this purpose, the vocabularies used in the learning domain were studied and Ontoolcole [11] was selected since it is the only one specifically developed to describe educational ICT tools. Ontoolcole defines three taxonomies: one of educational tools (i.e. "text editor"), another one of educational tasks that can be supported by these tools (i.e. "writing"), and finally another taxonomy of artifacts that can be managed by the tools (i.e. "text documents"). Thus, Ontoolcole can express complex educational descriptions, such as "a tool that allows students to draw collaboratively". Note that the educational specificity of Ontoolcole is important, since other generic vocabularies, such as WSMO [12] or OWL-S [13], do not provide the required expressiveness to describe the educational functionality of the tools; instead they mainly focus on their technical characteristics (see [11]).

However, Ontoolcole is a very complex vocabulary since it formally defines the relationships between the three taxonomies it states. Nevertheless, all the expressiveness it provides is not necessary in most situations [4]. So that, a simplified vocabulary called "SEEK ontology" was developed, defining these same taxonomies but without relating them. This way, describing a tool is much simpler while its type, the educational tasks it supports and the artifacts it handles can still be stated.

3.2   SEEK-AT-WD Software Architecture

Moving to the architecture design, its aim is to offer educational descriptions of ICT tools (F3) retrieving all the information from the Web of Data (F1). Several patterns can be followed in order to design such infrastructure [8, chap. 6]. In this case, the crawling pattern [14] was chosen, which separates the tasks of obtaining data (F1) and prepare it for its educational consumption (F2), and

---

[10] http://dublincore.org/2010/10/11/dcelements.rdf
[11] http://xmlns.com/foaf/spec/
[12] http://www.w3.org/1999/02/22-rdf-syntax-ns

offering this data to third-parties (F3). For this reason, when retrieving data from SEEK-AT-WD there is no need to wait for the crawler to obtain data from external sources. Thus, tool descriptions can be obtained from different sources while the response time is low enough to satisfy the requirements of educational applications. On the other hand, it is possible that the cache may contain stale data, so the crawler should periodically crawl the Web of Data.

Figure 2 shows the devised architecture. It mainly consists on a `web crawler` and a data repository (`SEEK-KB`) that provides a `SPARQL interface` [15] for the educational data consumption, as recommended by the Linked Data principles [7]. In this architecture the crawler plays an important role since it is responsible of gathering data from the Web and enriching it with educational concepts. Its key idea is to access several known data sources from the Web of Data that provide information about ICT tools, and then crawl the Web to obtain richer descriptions of these tools.

For each ICT tool description retrieved, the crawler relates it to SEEK vocabulary (F2) using well-known techniques of ontology mapping [6]. Thus, an educational value is added to the data from the Web. For example, the tool `Umbrello` is defined as a `UML tool` in DBpedia, and the crawler automatically infers that this tool is a `UML editor` that supports the tasks of `Modeling` and `Viewing UML models`, as defined by SEEK vocabulary. This way, SEEK-AT-WD offers educational descriptions of ICT tools that can be consumed both submitting SPARQL queries or browsing the dataset out of a `Linked Data interface`. More details about these mappings can be seen at [16].

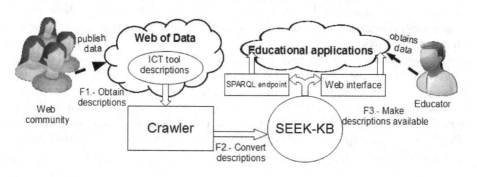

Fig. 2. SEEK-AT-WD architecture

Finally, the data provided by SEEK-AT-WD is expected to be used by different educational applications. Specifically, authors expect two different type of applications to be developed: search clients and annotation applications. Indeed, we have already developed a graphical search client that consumes data form SEEK-AT-WD and is briefly presented in the following subsection. Further, we are also devising a social annotation system that will allow educators to enrich the SEEK-AT-WD dataset. Nevertheless, we expect third-party applications that freely consume data from SEEK-AT-WD to appear in the future.

3.3   SEEK-AT-WD Prototype and Data Consumption

After designing the infrastructure architecture, a prototype has been developed. Firstly, the crawler was implemented using the Jena ontology API[13]. Current version of the crawler obtains data from DBpedia [10] and Factforge[14] and then crawls the Web using the *follow your nose* principle [8, chap. 6], meaning that if these datasets state that more information about a tool is published in another source, the crawler accesses that source to obtain more data. On March 2012 this crawler retrieved 3556 tool descriptions from the Web of Data. Then, SEEK-KB was deployed using RKBExplorer[15], an already working data publishing infrastructure that was developed under the Resist Network of Excelence[16]. SEEK-KB can be accessed at http://seek.rkbexplorer.com/.

SEEK-AT-WD data can currently be browsed (e.g see Umbrello at http://seek.rkbexplorer.com/id/tool/Umbrello_UML_Modeller) and queried through a SPARQL interface at http://seek.rkbexplorer.com/sparql/. However, appropriate applications should be developed for educators in order to take profit of SEEK-AT-WD. As a proof of concept, an educational ICT tool search client has been developed that hides the complexity of the SPARQL query language. The user interface is inspired on the one developed for Ontoolsearch, which has already been evaluated by educators [4]. Using a graph abstraction, educators can compose their queries by adding graph nodes corresponding to tool types, supported tasks or mediating artifacts. In addition, educators can use keywords and include additional restrictions referred to operating systems, licenses and so on. Figure 3 shows a screenshot of this application that can be tested at http://www.gsic.uva.es/seek/useek/.

4   Evaluation

Before developing a range of applications that use SEEK-AT-WD data it is convenient to evaluate whether this infrastructure supports the functionality required by its users. In addition, a functional comparison between this infrastructure and other educational tool registries described in Section 2.1 is interesting in order to show the advantages and the limitations of SEEK-AT-WD. With this aim, an evaluation based on the well-known feature analysis method [17] is proposed. Specifically, some use cases are considered and it is discussed how they are carried out by the infrastructure under evaluation in comparison to other approaches. The flexibility and the facility to carry out the evaluation are its main advantages. On the other hand, this methodology presents a certain degree of subjectivity, and different results may be obtained depending on the assessors.

---

[13] http://jena.sourceforge.net/ontology/
[14] http://factforge.net/
[15] http://www.rkbexplorer.com
[16] http://www.resist-noe.org/

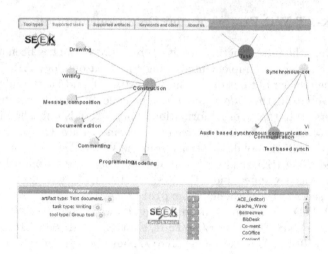

Fig. 3. Screenshot of SEEK search client when retrieving ICT tools that allow a group to write a text document

In order to decrease this subjectivity, use cases were proposed out of the experience of two different research groups. One of them is the GSIC[17] group, from the University of Valladolid, which is devoted to Computer Supported Collaborative Learning; the other is the WAIS[18] group, from the University of Southampton, which focuses on Web Science. They represent two different research groups who have experience in developing and providing educational ICT tools, as well as managing educational repositories of ICT tools.

4.1   Proposed Use Cases

Section 2.1 shows that the stakeholders that manipulate the data contained in a registry of educational ICT tools are tool providers, registry administrators and educators. Some experiences with educators were carried out in 2011. Educators where asked to propose some queries that they would ask to a search system of educational ICT tools and they filled in several questionnaires specifying their requirements. Altogether, 28 university educators from 3 different Spanish universities and 4 different areas of knowledge participated in these experiences. They expressed several requirements for ICT tool search systems that are represented in the following scenarios. Additionally, in February 2012 some other questionnaires were sent to four tool providers and three registries administrators from both research groups in order to understand their needs when publishing data and maintaining registries of educational ICT tools. All the information gathered can be looked up at http://www.gsic.uva.es/seek. These questionnaires were analyzed and, out of them, the main use cases were proposed to summarize the main requirements of each stakeholder, which can

---

[17] http://www.gsic.uva.es/

[18] http://www.wais.ecs.soton.ac.uk/

be seen in Table 2. This section shows how these use cases are supported by SEEK-AT-WD in comparison to the registries summarized in Table 1.

Table 2. Main findings in the experiences with the stakeholders related to the educational registries of ICT tools

Stakeholder	ID	Finding	Use case
Tool provider	ID 1	They want to increase the visibility of their tools	A
	ID 2	The size of the community of users should worth the effort of publishing their tools in a registry	A
	ID 3	It is a problem to discover educational ICT tool datasets	A
	ID 4	Updating the description should be as easy as possible	B
Registry administrator	ID 5	Vocabularies do not change frequently	-
	ID 6	Updating tool descriptions requires a significant cost	B
Educator	ID 7	They want to share their experience.	C
	ID 8	The impact in the educational community should worth the effort of publishing data in the registry	C
	ID 9	They want to obtain updated information of a big and comprehensive collection of ICT tools	D

A.- A Tool Provider Wants Its New Tool to Appear in Educational Tool Registries

Tool providers want to increase the visibility of their tools as much as possible (ID 1) in order to make their potential users to know them. In this regard, if they consider education as a potential market they will be interested in publishing their tools in registries of educational tools.

Some registries, such as Sisoft and Ontoolsearch, do not give any kind of support for this use case since they only allow an expert inside their organization to publish data. In these cases, the only possibility for the tool provider is to contact the administrators of the registry and convince them to include his tool. Other registries, such as CoolToolsForSchools and Schoolforge, trust their community of users to include new tool descriptions. Tool providers can add descriptions of their tools in these registries, but they need to be motivated to do it. Concerning this, they need to realize that the effort related to the tool publication is compensated by a big community of users that access their data (ID 2). For these reasons very few tool providers spend their time advertising their tools in educational registries that follow a Web 2.0 approach: as tools have to be manually introduced in each one and their communities are not very big, the effort is not compensated. In addition, several tool providers were not aware of the existence of educational-specific registries of ICT tools, so their discovery is another related problem (ID 3).

The approach followed by SEEK-AT-WD to support this scenario is completely different: it only requires providers to publish a description of their tools in the Web of Data (for example publishing them in Wikipedia). Once the descriptions are in the Web of Data, the crawler will automatically discover and publish them in SEEK-KB. As a result, SEEK-AT-WD is able to offer tool descriptions even if its tool provider does not have an explicit intention of publishing it in an educational tool registry. For this reason, there is no effort related to the registry discovery and the tool publication.

B.- A Stale Tool Description Has to Be Updated

Once a tool description is published in a registry it may need to be updated because of two reasons: either because a new version appears and the description becomes out-dated or because the vocabulary managed by the registry changes. According to the interviewees, registry vocabularies do not change frequently (ID 5), and when they do it is not problematic to update the descriptions. However, providing up-to-date data is seen as a problem for registry administrators (ID 6) and tool providers (ID 4).

Those registries which do not give support to social annotation (i.e. Sisoft or Ontoolsearch) require experts to be aware of new versions of the tools and to manually update the data. On the other hand, registries where the community of users can modify their data (i.e. CoolToolsForSchools or Schoolforge) it is either educators or tool providers who update the descriptions. For the educators to update the descriptions they should be aware of the appearance of new versions and they should be motivated enough to spend their time in this task. Also note that these registries can not share data; so that, a user of CoolToolsForSchools will not be aware if a description has been updated in Schoolforge. Finally, tool providers should manually update the description of their tools in each registry were they have published each time a new version appears. It can be seen that updating data in Web 2.0 registries is a hard task that should be done by each isolated community. This limitation makes these registries to be rarely updated.

These problems are alleviated by the approach followed by SEEK-AT-WD. As the Web of Data is periodically crawled it is only required that the tool provider updates the description of his tool in its origin of the Web of Data. Next time the crawler retrieves the description of the tool it will be updated in SEEK-KB. Note that this same description can also be crawled and imported to other registries of the Web of Data, but the data publisher does not need to worry about updating them manually.

C.- An Educator Enriches a Tool Description

Educators find very interesting the possibility of sharing knowledge about the use of ICT tools in real educational situations (ID 7). However, only the registries that allow educators to add data (i.e. CoolToolsForSchools or Schoolforge) give support to this scenario. In these cases, educators who have previous experience with ICT tools should be motivated to share their experience publishing it in a registry. In this regard, several educators admit that the larger the community of users is the more motivated they will be in sharing their experiences, since the time they spend publishing their experience is rewarded by a bigger impact in the educational community (ID 8).

As an infrastructure, SEEK-AT-WD supports this scenario. However, it requires a social annotation tool to be available, which is proposed as future work. In this regard, this social annotation tool not only will allow educators to share their experience to SEEK-AT-WD users, but it also will be published in the Web of Data, so third-parties can make use of them.

D.- An Educator Searches for ICT Tools

The most basic functionality of an educational registry of ICT tools is to allow educators to obtain the data it contains. The way the data is extracted highly depends on the user interface [4], which is out of the discussion of this paper, but it also depends on the way the registry structures the tool descriptions.

All the examples summarized in Table 1 count with a taxonomy of ICT tools to organize the descriptions they contain. These taxonomies allow educators not only to make queries based on keywords, but also to discover tools when retrieving descriptions that belong to a specific category. In this regard, Ontoolsearch requires a special mention since the data it contains is based on taxonomies of educational concepts. For this reason, this registry is capable of answering complex questions about ICT tools that involve educational-specific concepts, thus providing better results to educators' needs [11].

From this point of view, SEEK-AT-WD offers the same functionality as Ontoolsearch, since it is based on the same vocabulary. Nevertheless Ontoolsearch can only access data provided manually by an expert. So that, SEEK-AT-WD provides a higher amount of descriptions that are expected to be up to date, which is considered very positive by educators (ID 9).

4.2 Discussion

Previous subsection shows that SEEK-AT-WD is capable of supporting the most important use cases required for a registry of educational ICT tools. Regarding the management of data, SEEK-AT-WD follows an approach that differs from the rest of the registries analyzed: it is federated to the Web of Data; so that, although ICT tools are not described using educational abstractions, these descriptions can be automatically retrieved from third-party repositories and then converted to some extend to description with educational abstractions by making inferences based on an educational vocabulary.

As SEEK-AT-WD is able to retrieve data from third-party repositories it has a collection of 3556 ICT tool descriptions, as on March 2012, that was automatically created. This is an important feature since educators retrieve a wide range of ICT tools that can potentially be used to support their learning tasks. In addition, the automatic retrieval of information makes SEEK-AT-WD data very easy to maintain, since the Web of Data is periodically crawled and tool descriptions are automatically updated when they change in third-party repositories. Nevertheless, current version of SEEK-AT-WD does not provide information about the educational relevance of the tools published. As authors consider this a critical aspect, future work will deal on how to collect this information.

Finally, note that the federation of SEEK-AT-WD to the Web of Data provides an important added-value. All the data managed by SEEK-AT-WD is publicly available from a SPARQL endpoint and third parties can make use of it. External developers might use this data to build their own applications. Thus the data created by SEEK-AT-WD community might also be reached by other communities, increasing its educational impact.

5   Conclusions and Future Work

This paper presented SEEK-AT-WD as an infrastructure that federates a registry of educational ICT tools to the Web of Data. This feature allows SEEK-AT-WD to take advantage of the data already published in the Web, thus creating an initial collection of ICT tools that will be automatically updated when new relevant information appears on the Web. SEEK-AT-WD has already been built and its data can be publicly accessed either browsing the dataset or submitting queries using a SPARQL endpoint. In addition, an educational ICT tool search client has been developed in order to facilitate educators to submit queries to SEEK-AT-WD.

A feature evaluation was carried out to compare SEEK-AT-WD to other educational registries of ICT tools. It could be seen that the most important use cases related to these registries can be supported while taking advantage of the dataset federation to the Web of Data. Tool providers are only require to publish and maintain updated a description of their tools in a source of the Web of Data. Then, SEEK-AT-WD will crawl these descriptions and make them available to educational applications without any additional effort from the publisher. In this regard, a collection of 3556 descriptions of ICT tools was retrieved from the Web, which is periodically updated. Thus, educators can find a large quantity of tools that were published in SEEK-AT-WD without any human intervention.

The automatic retrieval of ICT tool descriptions from the Web and the inference of their educational capabilities are an important added value of SEEK-AT-WD. However, an evaluation with educators is needed to assess that the data collected and inferred satisfies their information needs. Our near future work will focus on this evaluation. Then, a social annotation tool will be developed, envisioning a scenario where experiences about the use of ICT tools in real educational situations can be shared and published on the Web of Data.

Acknowledgements. This work has been partially funded by the Spanish Ministry of Economy and Competitiveness (projects TIN2008-0323, TIN2011-28308-C03-02 and IPT-430000-2010-054) and by the Government of Castilla y Leon, Spain (VA293A11-2 and VA301B11-2). Authors also want to thank Ian Millard for his helpful support.

References

1. Conole, G., Alevizou, P.: A literature review of the use of Web 2.0 tools in higher education. Technical report, The Open University (2010)
2. Richardson, W.: Blogs, wikis, podcasts, and other powerful web tools for classrooms, 3rd edn. Corwin Press, Thousand Oaks (2010)
3. Madden, A., Ford, N., Miller, D.: Using the Internet in teaching: the views of practitioners (A survey of the views of secondary school teachers in Sheffield, UK). British Journal of Educational Technology 36(2), 255–280 (2005)

4. Vega-Gorgojo, G., Bote-Lorenzo, M.L., Asensio-Pérez, J.I., Gómez-Sánchez, E., Dimitriadis, Y.A., Jorrín-Abellán, I.M.: Semantic search of tools for collaborative learning with the Ontoolsearch system. Computers & Education 54(4), 835–848 (2010)
5. O'Reilly, T.: What is Web 2.0. Design patterns and business models for the next generation of software (2005), http://oreilly.com/web2/archive/what-is-web-20.html (last visited March 2012)
6. Choi, N., Song, I.Y., Han, H.: A survey on ontology mapping. ACM Sigmod Record 35(3), 34–41 (2006)
7. Berners-Lee, T.: Linked Data - Design Issues (2006), http://www.w3.org/DesignIssues/LinkedData.html (last visited March 2012)
8. Heath, T., Bizer, C.: Linked Data: Evolving the Web into a Global Data Space, 1st edn. Synthesis Lectures on the Semantic Web: Theory and Technology. Morgan & Claypool (2011)
9. Tiropanis, T., Davis, H., Millard, D., Weal, M.: Semantic Technologies for Learning and Teaching in the Web 2.0 Era. IEEE Intelligent Systems 24(6), 49–53 (2009)
10. Auer, S., Bizer, C., Kobilarov, G., Lehmann, J., Cyganiak, R., Ives, Z.G.: DBpedia: A Nucleus for a Web of Open Data. In: Aberer, K., Choi, K.-S., Noy, N., Allemang, D., Lee, K.-I., Nixon, L.J.B., Golbeck, J., Mika, P., Maynard, D., Mizoguchi, R., Schreiber, G., Cudré-Mauroux, P. (eds.) ASWC 2007 and ISWC 2007. LNCS, vol. 4825, pp. 722–735. Springer, Heidelberg (2007)
11. Vega-Gorgojo, G., Bote-Lorenzo, M.L., Gómez-Sánchez, E., Asensio-Pérez, J.I., Dimitriadis, Y.A., Jorrín-Abellán, I.M.: Ontoolcole: Supporting educators in the semantic search of CSCL tools. Journal of Universal Computer Science (JUCS) 14(1), 27–58 (2008)
12. Roman, D., Keller, U., Lausen, H., de Bruijn, J., Lara, R., Stollberg, M., Polleres, A., Feier, C., Bussler, C., Fensel, D.: Web Service Modeling Ontology. Applied Ontology 1(1), 77–106 (2005)
13. Martin, D., Burstein, M., Hobbs, J., Lassila, O., et al.: OWL-S: Semantic markup for web services. White paper OWL-S 1.1, DARPA Agent Markup Language Program (November 2004), http://www.daml.org/services/owl-s/1.1/overview/ (last visited March 2012)
14. Hartig, O., Langegger, A.: A database perspective on consuming Linked Data on the Web. Datenbank-Spektrum 10(2), 1–10 (2010)
15. World Wide Web Consortium. SPARQL Query Language for RDF. Recommendation, W3C (2008), http://www.w3.org/TR/rdf-sparql-query/ (last visited March 2012)
16. Ruiz-Calleja, A., Vega-Gorgojo, G., Gómez-Sánchez, E., Alario-Hoyos, C., Asensio-Pérez, J.I., Bote-Lorenzo, M.L.: Automatic retrieval of educational ICT tool descriptions from the Web of Data. In: Proceedings of the Twelfth International Conference on Advanced Learning Technologies (in press, 2012)
17. Kitchenham, B.A.: Evaluating software engineering methods and tools. Parts 1 to 9. SIGSOFT Software Engineering Notes (1996-1998)

Key Action Extraction for Learning Analytics

Maren Scheffel[1], Katja Niemann[1], Derick Leony[2], Abelardo Pardo[2],
Hans-Christian Schmitz[1], Martin Wolpers[1], and Carlos Delgado Kloos[2]

[1] Fraunhofer Institute for Applied Information Technology FIT,
Schloss Birlinghoven, 53754 Sankt Augustin, Germany
{maren.scheffel,katja.niemann}@fit.fraunhofer.de,
{hans-christian.schmitz,martin.wolpers}@fit.fraunhofer.de
[2] Universidad Carlos III de Madrid,
Avenida Universidad 30, E-28911, Leganés (Madrid), Spain
{abel,dleony,cdk}@it.uc3m.es

Abstract. Analogous to keywords describing the important and relevant content of a document we extract key actions from learners' usage
data assuming that they represent important and relevant parts of their
learning behaviour. These key actions enable the teachers to better understand the dynamics in their classes and the problems that occur while
learning. Based on these insights, teachers can intervene directly as well
as improve the quality of their learning material and learning design.
We test our approach on usage data collected in a large introductory C
programming course at a university and discuss the results based on the
feedback of the teachers.

Keywords: Usage data, learning analytics, self-regulated learning, activity patterns.

1 Introduction

In order to support students within a course – guiding them when they are on
the wrong track, giving advice even when they cannot ask precise questions and
general troubleshooting – teachers must be aware of what the students are doing
and have been doing so far. In other words: the teachers need information on
the students' activities. Such information is also needed for the evaluation of
a course, its didactic concept and the materials provided, including contents,
tools and tests. Information on the students' activities gives insights into which
materials actually have been used and which have not, when troubles occurred,
when the students started their work and in which parts of the course they got
stuck, etc. Such information can be referred to for optimising the course and
thus supporting the learning process.

One can monitor the students' activities and list them in one large file. This
file, however, will contain more information than teachers and students can effectively evaluate. It will therefore meet neither the needs of the teachers nor those
of the students. This is where learning analytics comes in as Siemens and Long

A. Ravenscroft et al. (Eds.): EC-TEL 2012, LNCS 7563, pp. 320–333, 2012.

explain [1]. A distillation of the recorded data is required, so that irrelevant information is filtered out and information overload is avoided. The question now is which means of data distillation are useful and help students and teachers in mastering their tasks.

This paper deals with one particular means of data distillation, namely the extraction of key actions and key action sequences. We claim the hypothesis that key action extraction is a very useful form of data distillation. We prove our hypothesis by implementing the approach in a larger test bed, namely an introductory programming course with theoretical lectures and practical lab sessions. The course was held in the fall semester of 2011 at the Universidad Carlos III de Madrid (UC3M), Spain.

The rest of this paper is structured as follows: in section 2, we will describe background, related work and previous experiments. We will then report on the evaluation of the approach within a larger test bed, namely the already mentioned course on C programming (section 3). Section 4 will deal with the implementation of key action extraction and the setting of parameters for the test bed. In section 5, we will report on the insights the teachers got from the key actions and thus qualitatively evaluate the approach. Finally, in section 6 we will summarise and give an outlook on future work.

2 Related Work and Linguistic Background

2.1 User Monitoring and Learning Analytics

Many university courses consist of self-regulated learning activities. The students take over responsibility for planning and reflecting these activities. Being aware of one's own activities is a prerequisite for reflection. In [2], we have shown that one way of helping a learner to become aware of his actions and learning processes is to record and store his interaction with his computer and to then analyse the collected data. Results of these analyses can on the one hand be used to foster his self-reflection processes or on the other hand to give recommendations, e.g. of further steps in his current learning scenario. The same applies to teachers.

If the number of students in a course is high and the tasks the students are engaged in are not trivial, then the teachers need assistance for keeping track of the students' activities. They cannot constantly observe their students themselves. It will be of great advantage for them if monitoring is supported automatically. A survey conducted by Zinn and Scheuer [3] about the requirements of student tracking tools showed that aspects such as competencies, mastery level of concepts, skills, success rate and frequent mistakes are seen as highly important to teachers. Many teachers said that employing student tracking would allow them (the teachers) to be able to adapt their teaching to the behaviour of the students and to identify problems in the students' learning processes.

Several approaches deal with the creation of feedback for the teacher to enable him to improve the quality of his courses. Kosba et al. [4] for example developed the TADV (Teacher ADVisor) framework which uses data collected by a course management system to build student models and a set of predefined rules to recognise situations that require teachers' intervention. Based on the student models and the rules, the framework creates advice for the teacher as well as a recommendation for what is best to be sent to the students. Similar to our approach, the goal of this framework is to enable the instructors to improve their feedback and guidance to students. However, our approach does not use predefined rules as we do not want to force the instructor to perform a specific action but to enable him to get new insights into the learning behaviour of his students and thus to rethink and improve his teaching.

The CourseVis system also visualises data taken from a courses management system [5]. It addresses social, cognitive and behavioural aspects of the students' interaction with the course management system. While the discussion graph visualises the number of threads started and their number of follow-ups per student, the discussion plot shows originator, date, topic and number of follow-ups in a scatterplot. The visulisations of the cognitive aspects maps students' performance to the course's concepts and another visualisation deals with the students' access times. While these details can be very useful for teachers, CourseVis only depicts the students' interaction with the course management system and does not take other tool interactions into account.

Another relevant tool is the LOCO-Analyst by Jovanovic et al. [6] which is an educational tool that can be embedded in various LCMSs to analyse the usage data of learners. It has the goal to provide teachers with meaningful feedback about the users' interaction with the learning content. The teacher is for example informed about the average time the students spent on each learning object. When quizzes about the learning content are provided, the teacher gets detailed feedback about the incorrect answers per question and which questions are the toughest ones, as they have an error rate above the average. As with our approach, the LOCO-Analyst does not provide real time help, but tries to help the teacher to improve the quality of his learning content and learning design. Even though the LOCO-Analyst considers a lot of event types for its analysis, it does not create and use sequences of events so far, which we assume give the teachers of a course a better insight into the learning behaviour.

2.2   User Observation and Key Action Extraction

For our approach we make use of the contextualised attention metadata (CAM) schema that allows for describing a user's interaction with digital content [7]. The schema is event centered and is thus well suited for the evaluation and analysis of a user's entire computer usage behaviour. The schema has evolved over years and can be subject to further change in the future.[1] Analysing collected CAM can for example result in an overview of actions taking place in the environment or the discovery of changes, trends, etc. in usage behaviour. In controlled environments, e.g. formal learning, where activities are often scheduled, it can be useful to know what takes place when. In less controlled environments, e.g. informal or blended learning, CAM analyses can help to understand when users are active.

---

[1] The latest version is available at https://sites.google.com/site/camschema/

The concept of a key action is best described in analogy to a key word: a list of key words is a superficial but albeit highly useful semantic representation of a text [8]. The keywords of a text are those content words (or sequences of words) that are significant for the content of the entire text and by which the text can be distinguished from other texts. They do not exhaustively describe what the text is about but still give a clear impression of its theme. Knowing a text's keywords, one can capture the essence of a text's topic and grasp the essential information the text is trying to pass along. In analogy, key actions are those actions that are significant for an underlying set of actions and that give an impression of what has happened. We deem key actions to represent the session they are taken from (with a session being anything from a few minutes or hours up to days, weeks, months, etc.). Key actions (or key action sequences as a parallel to key phrases) indicate what a user has been doing. They give an overview of the essential activities. By no means can they be exhaustive but they provide a superficial yet almost noise-less impression.

Let it be given that we recorded all actions of one student in one practical session. From these we extract the key actions which, so the idea, gives us an impression of what the student essentially did in this session. The approach we apply for the analysis of usage data in order to detect meaningful patterns is the so called n-gram approach [9], followed by a ranking approach, namely *tf*idf* weighting, with *tf* being the term frequency and *idf* the inverse document frequency [10]. The following formula shows how the weight of a word can be calculated in more detail, with $w_{i,j}$ being the weight of word j in document i, $t_{i,j}$ being the term frequency of word j in document i, f_i being the number of documents containing word j and N being the number of documents in the collection: $w_{i,j} = t_{i,j} \cdot log\frac{N}{f_j}$.

We use the *tf*idf* algorithm to weight extracted key actions based on two assumptions: First, if a collection of sessions contains a specific key action more often than another collection of sessions, then this key action is more relevant to the first set of sessions and gets a higher weight. Second, if a key action does not appear as often in the collection of sessions as other actions, it should also get a higher weight in the session it does appear in.

In the fall semester 2010 we ran a first round of tests to extract key actions from CAM collected during a C programming course at UC3M [11]. The collected interactions were gathered from a virtual machine with programming tools and the forum interactions of the learning management system at the time. All calculations were done on the basis of the whole course, i.e. taking all sessions from all students into account at the same time. These first results were deemed useful by the teaching staff and thus some events were analysed in a more detailed way to find frequent error patterns.

The promising results motivated us to continue with our approach. Instead of only analysing the whole course at the same time, we wanted to make the available results more diverse and more detailed. We therefore again deployed our approach during a second year C programming course at the Universidad Carlos III de Madrid. This time, however, we look at the whole course as well as the different lecture degrees, lab sections, teams and individual students in order to provide teachers with an even better idea of where their students are at and how their learning processes can be supported.

3   Application of Key Action Extraction in Detail

3.1   A C Programming Course

We use the introductory C programming course at the Universidad Carlos III de Madrid as a testbed to test our approach more diversely. The course has four main objectives: The students are supposed to learn to write applications in C with non trivial data structure manipulation and they are told to use industry-level tools such as compiler, debugger, memory profiler and version control to get familiar with it. Additionally, they work in teams and need to create their own plans, divide tasks, solve conflicts etc. Finally, the course is assumed to increase the self-learning capability of the students.

The course is split into two halves. In the first half of the course, the students attend theoretical lectures to get a basic understanding of the programming language and start to program in teams of two in supervised lab sessions. In the second half they are split into larger groups of 4 - 5 people and work together on larger projects. Altogether, 10 instructors work on supervising the course, all of them having at least a master's degree and 4 of them being professors. For the theoretical lecture units, the students are divided into 5 groups that are taught by the professors. For the lab sessions, they are divided into 11 groups that are supervised by the 10 instructors.

At the beginning of the course, the students are offered a Virtual Machine that is already pre-configured. It is a UNIX-based system having the compiler, text editor, debugger and memory profiler – that the students are supposed to use – already installed among additional standard tools, e.g. a browser. Within the Virtual Machine, the main events we consider to be important for further analysis are logged and whenever a student uses the subversion system to download a document or upload own material, the collected events are uploaded and stored in a central database (see [12] and [13] for further details). It is important to mention that the students are well informed about the data collection process and are reminded of it every time they open the Virtual Machine. If they do not want to be monitored, they are able to easily stop it without stopping the use of the Virtual Machine.

For registered students, the course material, i.e. documents, exercises and C files, is accessible via an Apache Server. Every time a page is served, a new log entry is created that can be analysed to extract the events. Additionally, the forum functionalities of the learning platform Moodle are used to offer the students a place to discuss problems and ideas. Given that Moodle offers the teacher the possibility to download all events that took place within a course in Excel format, the forum events can easily be extracted. For the Apache and

Moodle Logs, the students are not able to stop the logging themselves but they are informed about the analysis and can write an e-mail to their teacher stating that they do not want to have their data stored which are then not considered for the analysis.

Altogether, we were able to monitor 33 different event types from three different sources. For the complete course, we collected approximately 340,000 (156,000 first half / 184,000 second half)[2] conducted by 332 distinct students in the period from September 5, 2011, until December 18, 2011.

3.2   Events That Are Monitored

Accessing Web Pages. The course material is accessible to logged in and authorised students. Every time a page is served, a new log entry is written which comprises the time stamp, the user identifier, the IP address, the URL served and the HTTP code. For the following analysis only the successful events are considered, as most of the failed access attempts are due to the reason that a student was not logged in or tried to log in with a false user name or password. Certainly, the students do not only access the course material while learning or programming. They also search for related material, forums that answer their questions or code snippets they can use; additionally, they browse the web for private purposes during breaks. This data is collated using the Virtual Machine that is offered at the beginning of the course and captures the use of the browser as well as of other main tools. Altogether, we were able to collect 131,071 (68,130 / 62,941) web page accesses.

Program, Compile and Debug. The students were told to use the text editor KATE to write their programs as they are supposed to learn to code from scratch in the introductory course without any help from a development environment. Within the Virtual Machine it is stored whenever KATE is opened or closed, but no further information, e.g. about the file that is opened or actions that are performed within KATE, are logged. The students are also expected to learn how to use gdb as debugger to find problems in the code and Valgrind as memory profiler to find memory leaks. As for KATE, it is only stored when the debugger or memory profiler are opened, respectively closed. When the students compile their code using gcc, the compile command is extracted from the command line in the shell including the file that is compiled as well as the resulting warning and error messages. These messages are stored as well, as they keep a lot of insight for the students themselves, e.g. for self-reflection, as well as for the teachers, e.g. to understand with which errors the students get stuck or which warnings they just ignore [11]. With 87,148 (23,298 / 63,850) occurrences, the compile event is the most frequent one in this category, followed by the text editor KATE which was invoked 27,363 (6,875 / 20,488) times. Interestingly, with

---

[2] If not otherwise indicated, the format of giving the numbers for the whole course followed by the numbers for the first and the second half of the semester in brackets is continued for the rest of this paper.

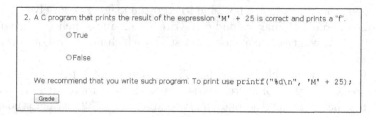

Fig. 1. Screen capture of a Quiz in the Learning Material

13,591 (1,081 / 12,510) accesses, the memory profiler is more often used than the debugger that was only accessed 2,838 (583 / 2,255) times.

Answering Quizzes in the Learning Material. Some of the learning resources contain small quizzes after important sections to support the students in testing their understanding of what they just read, see Figure 1. The students can select an answer and click on *"Grade"* to find out if the answer was correct or not. If the answer was wrong they can try again or display the correct answer. Overall, 21,865 (19,380 / 2,485) times a test was carried out and 5,306 (4,908 / 398) times the students choose to show the correct results after a test and did not try it again.

Communicating in the Course Forum. The Moodle forum used in this course contains 219 discussions which comprise 439 posts. The larger part of these were created during the first half of the semester, namely 148 discussions with 334 posts, while only 71 discussions with 105 posts were created in the second semester. All events conducted in Moodle can be downloaded by the teachers as an Excel file. Each event comprises the name of the course, the time stamp, the IP address, the full name of the student, the event type, e.g. *course view* or *forum add post* and a field for further information, e.g. the name of the viewed course or the title of the post added to the forum. Due to privacy reasons the user names are mapped to the respective user ids before the events are considered for further analyses. In total, there were 52,087 (31,605 / 20,482) LMS-related events. 20,828 (12,886 / 7,942) of them were forum-related.

3.3   Stock Taking: Descriptive Statistics on Tool Usage

The activity of the students is not distributed equally. It is important to take into account that the students work in tandem teams (first half) and small groups (second half) during the lab sessions. Therefore, it is possible that they always work on the computer of one student. However, most students (76,3%) conducted between 100 and 2,000 events. 8,7% conducted less than 100 events. This can have several reasons: (1) they do not use the virtual machine (i.e. configure entirely their personal computer); (2) they use the virtual machine but disable the event recording; (3) they work with somebody else and it is this other person that generates events; (4) they do not actively participate in the

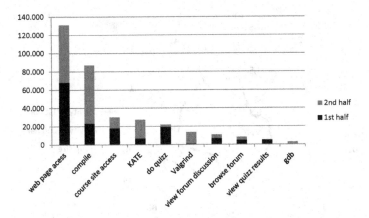

Fig. 2. Ten Most Frequent Events of the Whole Course, for the first and the second half of the semester

course.[3] The rest of the students (15%) can be said to be very active with 2,000 to 6,507 events per student.

Figure 2 shows the distribution of the ten most frequent events. Note that we subsumed some events into one event for this diagram, e.g. *start KATE* and *close KATE* are presented as *KATE*. The top event is *web page access*. This is not surprising in the first and more theoretical half of the course as the students need to access a lot of learning material and need to get familiar with the system and the tools. We could for example observe that the students often need to look up Linux commands. The graph also shows that the number of compile events rose drastically in the second and more practical half of the course.

Figure 3 shows the normalised activity per group for the two parts of the semester and the normalised average group size in comparison. The average amount of events conducted by a single student is 1,019 (470 / 586). Group B-2 is the group with the most active students with on average when looking at the whole year (2063 events per student). This results in an activity score of $2063/1019 = 2.02$. In the first half, group Group A-2 is the most active one with an activity score of 1.55 and in the second half, Group B-2 is the most active one with an activity score of 2.32. The least active group for the whole year as well as the two halves is Group E-2 with an activity score of 0.61 (0.55 / 0.64). This means, wherever the activity score of a course is higher than 1, its students are more active than the average and vice versa. It is noticeable that in many cases the lab groups that belong to the same theoretical group have a similar activity score although the lab groups are not supervised by the same teachers.

Each lab group comprises on average 28.7 students. Group B-2 is not only the most active but also the smallest group with only 10 students and a group size score of $10/28.7 = 0.35$. Group E-1 is the biggest group with 38 students and a

---

[3] Given the way the data are collected, there is no simple way to distinguish the causes. Clarifying which one would require some extreme, highly unfeasible measures such as taping them in and out of class.

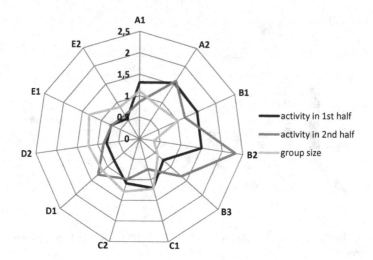

Fig. 3. Web Graph Comparing All Activities per Group to the Group Size

group size score of 1.32. The figure shows that in many cases in which the group size is below the average the activity of the students is above the average and vice versa.

4   Extracting Key Actions

The n-gram approach has been applied to the complete data set as well as to the several sub-sets of theoretical units and lab sessions. When extracting n-grams for a user or a group, we first gather all actions using the userId(s) and order a user's actions depending on the actions' time stamps. Additionaly we take the sessionIds into account to not combine actions conducted in different sessions or by different users.

The base of an activity is usually a complete CAM instance, i.e. time stamp, student ID, and event type (e.g. visitUrl, start, send, etc.) together with an item (e.g. a URL, a terminal command, a file, etc.) and a tool (e.g. editor, browser, etc.). All activities have a time stamp and a student ID and most of these activities have tool, event type and item as well. Some however, such as starting or ending the editor or the debugger, do not have an item.

By specifically taking some aspects out of the calculation, results can be made more general. The less information is taken into account, the more general the results are. In our experiments the calculations were done for several granularity levels: (a) tool, event type and item, (b) tool, event type and item where URLs were shortened to their domain and C files ignored, (c) tool and item, (d) event type only and (e) item only. As explained, shortening the URLs to their domain allows a broader, more general combination of actions. Several students might use Google to find information about the same task, their query term though might differ. The action they execute, however, is essentially the same. This principle also applies to actions where the names of the C files are taken out as students may use any name for their files, the fact that they compile a C file, however, stays the same. During all calculations key actions of length one and two were discarded as they had not been deemed meaningful enough by the teaching staff of the course.

After extracting the key actions as described above *tf*idf* weights were calculated for the theoretical units as well as the lab groups with the whole course serving as the corpus. Then the 10 most frequent key actions of each calculation were juxtaposed with the 10 highest weighted key actions of the *tf*idf* results. The original key action extraction results and the *tf*idf* results were then given to the course's teachers for a qualitative evaluation.

5   Qualitative Evaluation of the Results

As we claim the hypothesis that key action extraction is a very useful form of data distillation, we gave all analysis results (visualised as in Figure 4 and as text files) to the course's teaching staff and asked for feedback. We especially wanted to know what they can deduce from the results and whether they think they are useful or not and why. The following paragraphs summarise the feedback of the different teachers about their lecture and lab groups as well as the whole course.

In the calculations for the whole course, many extracted key actions dealt with the quizzes. The quizzes are directly embedded in the course notes and are not mandatory to fill and can thus be deployed to measure the level of spontaneous engagement with the course material. Very often students attempted to take a quiz, presumably got a wrong answer and, instead of attempting the quiz again, subsequently clicked on *show result*. For the teachers, this indicates that many students not only did not know the answer of the first two questions of a quiz but also that they needed – or wanted – to "be told" the answer instead of just trying with a second option. Figure 4 shows a visualisation of such extracted key actions.

The key actions reveal those questions that had to be repeated more often, and therefore posed a more complex problem to students. This was deemed very valuable information by the teachers. They concluded the following: The key actions point to those questions that were interacted with the most and thus should be reviewed in class, covered in more detailed or even discussed in the forum. Another aspect of the quizzes is that they show the level of engagement of students with the material. If key actions do not reflect quiz usage, then students are not taking enough quizzes. The high frequency of pointer related questions in the key actions, for example, denotes to the teachers that this is a block that requires special treatment such as reviewing the material in an extra class, complementary course notes, additional exercises, etc. In the key actions, not only the reflection of answering quizzes, but also the rate of failures and redoing of quizzes helped the teachers to understand what they can change in upcoming courses. Also, the differences between doTest-key action-frequencies among the different groups gave the teachers a reference point to see if they should improve student engagement within their groups.

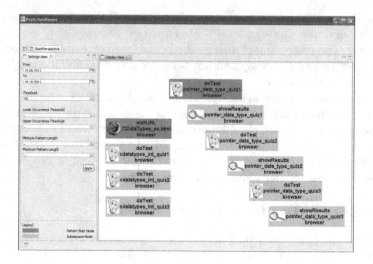

Fig. 4. Visualisation of Some Key Actions from the Whole Course

From the calculations where the C files had not been taken to account, the teachers noticed that there are fairly long and frequent compilation chains. As one of the main objectives of the course is for the students to be able to write applications in C, this was deemed a good result. Another interesting factor was the comparison of the students' compile behaviour from the different theoretical units and lab groups. For three of the five theoretical units the longest key action sequences were compile activities. For the other two the longest sequences came from the browser. The teachers interpreted this as a possible lower engagement in the course for these two units.

Looking at the different lab groups gave them even more detailed insights. For five groups, the longest sequences were compilation activities. These groups were deemed "on track". Another five groups had such activitites as second largest sequences after browser related sequences. That was seen as accaptable. The one group that had the compile activities only in the fourth largest sequences was seen as lacking behind and needed to improve. The teachers agreed that in general they would like to see compile actions as the longest sequences for every lab group. Figure 5 shows the normalised compile activity per group. As can be seen in comparison to Figure 3, the number of compile activities and all activities correlate as do the observations with the longest key action sequences.

The calculated key actions supported what was also reflected in the statistic analysis: the debugger was not used as frequently as it should have been, according to the teachers. For them, without the need of a much deeper analysis, these results reveal that the debugging tool is used significantly below expectations. A straightforward consequence of these results is that changes need to be introduced on the activities to lower the adoption threshold for this tool.

Many lab groups generated most of their key actions related to programming in the second half of the course. This behaviour coincides with the assignment of

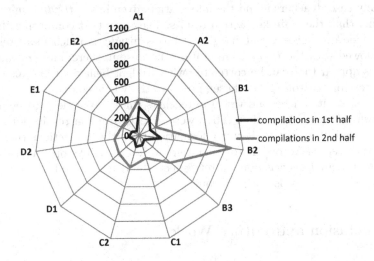

Fig. 5. Web Graph Showing Compile Activities per Group

the course's main project that is due at the end of the second half. The key actions revealed that a relevant amount of students do not really start working on actual programming tasks until they have the obligation to work on the project. The teachers, however, would have liked to see more programming activity already during the weeks before the assignment.

One teacher noticed that the use of the memory profiler in his lab group had a high $tf*idf$ weight as did viewing forum discussions which was a good sign for him. Other sequences that had rather some $tf*idf$ weight included accesses to the SVN material of the course. From this the teacher guessed that the students did not put enough effort into learning the command syntax but just copied and pasted it when they needed it. He appreciated this insight planning to work on this topic more in his class.

Looking at $tf*idf$ results calculated for the ten students with the highest final mark revealed that they very frequently use both the compiler and the memory profiler. When looking at the results for the ten students with the lowest final mark, such key actions are not weighted high, instead they check the LMS, use the text editor, surf the web and so forth. Teachers noticed a striking absence of the compiler. The $tf*idf$ results also showed that the students with higher scores check the content of the forum more often, whereas students with lower scores tend to consult course notes, notices, etc., but not so much the course forum. What seems meaningless but was found remarkable by the teachers was the fact that students with good final marks regularly checked the file containing the course scores (as was reflected in their key actions), whereas students with low final marks barely did so. They deduced that checking for the scores more frequently denotes a higher level of commitment to the course.

In many cases teachers found the information given in the *tf*idf* results more interesting than the plain key action results. They said that computing the key actions alone does give a first insight, but this insight is much more relevant when followed with the *tf*idf* computations because in there, more meaningful sequences appear that can be correlated with student behaviour and also allow for a better comparison of the chosen group to the whole course.

Summing up it can be said that the teachers think key actions to be a very useful form of data distillation. They were able to use the results for course evaluation and liked getting better information from the key actions than from the logs themselves. According to them there were "lots of eye-opening things in there". As the analysis was done in retrospect, the teachers are looking forward to employing the analysis during a course.

6   Conclusion and Future Work

In this paper we presented our approach of key action extraction as a means of data distillation for a collection of contextualised attention metadata. The approach was implemented in a large test bed as part of a C programming course at a university. Student activity data have been recorded, from these key actions have been extracted which in turn have been used by the teachers for the evaluation of the course, that is, for understanding how the students actually dealt with their exercises and how the course can be improved according to their behaviour. As the reactions of the teachers show, the key actions and especially their subsequent *tf*idf* weighting give interesting insights into the course that have not been provided otherwise.

Although we managed to answer the question whether key action extraction results are useful for the teachers at all, new questions arise: Will it be possible to apply the analysis during the course instead of in retrospect? Will this help teachers to react directly and adapt their teaching? Another important question is how to provide the students with their own data. Can it help them to self-reflect on their learning activities? Do they perceive the system's and the teachers' feedback as useful?

In order to answer these new question, the key action extraction will be implemented into the running system of the next fall semester's C programming course. From this we expect to gain even more insight of the usefulness of key action extraction. Apart from providing students and teachers with the results throughout the course and collecting general feedback impressions during several stages, we will also employ questionnaires with which we especially want to address privacy and security issues that are wont to arise.

Acknowledgments. Work partially funded by the European Community's Seventh Framework Programme (FP7/2007-2013) under grant agreement no 231396 (ROLE project), the Learn3 project (TIN2008-05163/TSI), the eMadrid project (S2009/TIC-1650), and the Acción Integrada DE2009-0051.

References

1. Siemens, G., Long, P.: Penetrating the Fog: Analytics in learning and Education. Educause Review 46(5) (September/October2011)
2. Schmitz, H.C., Scheffel, M., Friedrich, M., Jahn, M., Niemann, K., Wolpers, M.: CAMera for PLE. In: Cress, U., Dimitrova, V., Specht, M. (eds.) EC-TEL 2009. LNCS, vol. 5794, pp. 507–520. Springer, Heidelberg (2009)
3. Zinn, C., Scheuer, O.: Getting to Know Your Student in Distance Learning Contexts. In: Nejdl, W., Tochtermann, K. (eds.) EC-TEL 2006. LNCS, vol. 4227, pp. 437–451. Springer, Heidelberg (2006)
4. Kosba, E., Dimitrova, V., Boyle, R.: Using Student and Group Models to Support Teachers in Web-Based Distance Education. In: Ardissono, L., Brna, P., Mitrović, A. (eds.) UM 2005. LNCS (LNAI), vol. 3538, pp. 124–133. Springer, Heidelberg (2005)
5. Mazza, R., Dimitrova, V.: CourseVis: A Graphical Student Monitoring Tool for Supporting Instructors in Web-based Distance Courses. Int. J. Hum.-Comput. Stud. 65(2), 125–139 (2007)
6. Jovanovic, J., Gasevic, D., Brooks, C.A., Devedzic, V., Hatala, M.: LOCO-Analyst: A Tool for Raising Teachers' Awareness in Online Learning Environments. In: Duval, E., Klamma, R., Wolpers, M. (eds.) EC-TEL 2007. LNCS, vol. 4753, pp. 112–126. Springer, Heidelberg (2007)
7. Schmitz, H.C., Kirschenmann, U., Niemann, K., Wolpers, M.: Contextualized Attention Metadata. In: Roda, C. (ed.) Human Attention in Digital Environments, pp. 186–209. Cambridge University Press (2011)
8. Rose, S., Engel, D., Cramer, N., Cowley, W.: Automatic Keyword Extraction from Individual Documents. In: Berry, M.W., Kogan, J. (eds.) Text Mining, pp. 1–20. John Wiley & Sons, Ltd. (2010)
9. Hulth, A.: Improved Automatic Keyword Extraction Given More Linguistic Knowledge. In: Proceedings of the 2003 Conference on Empirical Methods in Natural Language Processing, EMNLP 2003, Stroudsburg, PA, USA. Association for Computational Linguistics, pp. 216–223 (2003)
10. Dörre, J., Gerstl, P., Seiffert, R.: Volltextsuche und Text Mining. In: Carstensen, K.U., Ebert, C., Endriss, E., Jekat, S., Klabunde, R., Langer, H. (eds.) Computerlinguistik und Sprachtechnologie, pp. 425–441. Spektrum, Heidelberg (2001)
11. Scheffel, M., Niemann, K., Pardo, A., Leony, D., Friedrich, M., Schmidt, K., Wolpers, M., Delgado Kloos, C.: Usage Pattern Recognition in Student Activities. In: Delgado Kloos, C., Gillet, D., Crespo García, R.M., Wild, F., Wolpers, M. (eds.) EC-TEL 2011. LNCS, vol. 6964, pp. 341–355. Springer, Heidelberg (2011)
12. Niemann, K., Schmitz, H.C., Scheffel, M., Wolpers, M.: Usage Contexts for Object Similarity: Exploratory Investigations. In: Proceedings of the 1st International Conference on Learning Analytics and Knowledge, LAK 2011, pp. 81–85. ACM, New York (2011)
13. Romero Zaldívar, V.A., Pardo, A., Burgos, D., Delgado Kloos, C.: Monitoring Student Progress Using Virtual Appliances: A Case Study. Computers & Education 58(4), 1058–1067 (2012)

Using Local and Global Self-evaluations to Predict Students' Problem Solving Behaviour

Lenka Schnaubert[1], Eric Andrès[2], Susanne Narciss[1], Sergey Sosnovsky[2], Anja Eichelmann[1], and George Goguadze[2]

[1] Technische Universität Dresden, Dresden
[2] Deutsches Forschungszentrum für Künstliche Intelligenz, Saarbrücken

Abstract. This paper investigates how local and global self-evaluations of capabilities can be used to predict pupils' problem-solving behaviour in the domain of fraction learning. To answer this question we analyzed log-files of pupils who worked on multi-trial fraction tasks. Logistic regression analyses revealed that local confidence judgements assessed online improve the prediction of post-error solving, as well as skipping behaviour significantly, while pre-assessed global perception of competence failed to do so. Yet, for all computed models, the impact of our prediction is rather small. Further research is necessary to enrich these models with other relevant user- as well as task-characteristics to make them usable for adaptation.

1 Introduction

The Network of Excellence Stellar identified affective and motivational aspects of technology-enhanced learning (TEL), as well as personalized learning to be among the eleven most important core research areas for TEL in the next ten years [1]. An important aspect of personalization for learning is the design and evaluation of adaptation strategies selecting instructional elements appropriate for the learners individual characteristics. To design such adaptation strategies, it is important to understand learners' behaviour within the system and identify the important factors the system needs to adapt to [2].

In literature, static and dynamic adaptation strategies are differentiated. The former uses as the source of adaptation global learner characteristics, while the latter is usually based on the learners' features derived locally, from their interaction with the environment [2]. Static adaptation strategies can use data from questionnaires or pre-tests administered before the learning starts, in order to set the global parameters of the learning environment to the needs of an individual learner. Dynamic adaptation strategies, on the other hand, take into account learners' behaviour within the system to optimize the next learning step. Therefore the system has to make decisions "on the go" and dynamically model needs/skills/knowledge of learners.

Our study will focus on the processes to help finding indicators enabling the prediction of learning behaviour on the basis of global, as well as local learner

A. Ravenscroft et al. (Eds.): EC-TEL 2012, LNCS 7563, pp. 334–347, 2012.

characteristics. In particular, we are interested in the impact of global perception of competence and local response confidence on problem solving behaviour. In line with previous research [3], we distinguish between two dimensions of behaviour within multi-trial learning tasks: quantity of input (did a learner provide a solution to a task?) and quality of input (was his/ her solution correct?). Quantity of input helps us to measure learners' engagement in a task, while quality of input is a performance measure. The long-term goal of this research is to establish the set of learners' parameters important to predict their problem solving behaviour and to design effective adaptation strategies facilitating the learning process.

The rest of the paper is structured as follows: We will first provide the theoretical and empirical background of our research, which includes an overview of research on our independent variables perception of competence and response confidence and their impact on learning behaviour (section 2). Section 3 states the research questions addressed in this paper. In section 4, we describe in detail the methodology we used to approach these questions. This includes a brief description of the learning environment and the empirical study, which data we re-analyzed in order to answer our research questions. It further describes the operationalization and assessment of the independent and dependent variables involved as well as the statistical analyses used. We report our results in section 5 before discussing possible interpretations and consequences in section 6.

2   Self-evaluation and Its Relation to Behaviour

Several lines of research indicate that students' self-evaluations of capabilities and/ or performance influence different aspects of their motivation and/ or behaviour (e.g., goal setting and effective self-regulated learning [35], feedback processing [25,33], choice and application of adequate learning strategies [34], satisfaction with performance [14]).

Self-evaluations can range from global self-concepts (e.g., academic self-concept) to domain-specific competence perceptions, up to task-specific self-evaluations (e.g., prospective self-efficacy or retrospective confidence). Although these concepts are interrelated [4], the former ones are more stable over time and focus on the learners' abilities or competencies, while the latter are more dynamic and therefore focus on the current interaction of the learner with the learning environment (e.g., specific learning task). While it seems plausible that local self-evaluations are more appropriate to predict (local) behaviour, there is research suggesting that global measures might also be indicative in this respect [5,3]. Global self-evaluations do not require online collection of dynamic data, thus do not force learners to switch between content-related and content-unrelated tasks [6].

The aim of this work is to compare the predictive power of global perception of competence (PoC) and local confidence judgements (response confidence, RC) on critical behaviour (e.g., giving up on a problem or solving a problem). Since PoC and RC are different concepts, but yield overlapping information about the learner (Zhao & Linderholms anchoring and adjustement framework of metacomprehension [7] even suggests that initially assessed global self-evaluations may serve as an anchor for local self-evaluations), we will briefly describe these variables and show why they might both carry important information for adaptation.

Perception of Competence (PoC). Perception of competence combines domain-specific beliefs about own capabilities [8] and, therefore, includes knowledge that individuals have about their skills and abilities [9,7]. Thus, PoC is more specific than representations of global abilities (like academic self-concept), yet more general than task-specific evaluations of performance (e.g., self-efficacy). Domain-specific perception of competence stems from individual experiences of success and failure within a certain domain, but is also influenced by more global beliefs about the self [9] and social comparison processes [10]. PoC is usually assessed via questionnaire data. Since it is thought to be relatively stable it can be assessed pre- or post-performance, although it, undoubtedly, is influenced by performance itself. Existing theoretical and as empirical work suggests that domain-specific perception of competence is related to domain-specific performance and motivation, all of which mutually reinforce each other [11,12,13]. Furthermore, PoC has been shown to influence learners' reactions to negative feedback [5,3].

Response Confidence (RC). Response confidence is the product of a metacognitive monitoring process [15]. It is the confidence a learner assigns to a particular (local) task performance (response) and is therefore usually assessed item-by-item post-performance, but prior-feedback. While some theories of confidence stress the importance of task-inherent cues, task-processing cues and memory cues in the formation of RC [16,17,18,19], others try to explain the inter-individual variance by stressing global self-confidence traits [20]. RC has been a core concept of cognitive feedback research since the seminal work by Kulhavy and colleagues [21]. Its role for improving performance assessment and for tailoring feedback to learners' needs has been investigated in several studies, e.g., [22,23,24]. Still, little is known about the predictive value of RC for concrete critical behavioral events when learners work with an ITS. Although, one robust finding was that errors committed with high confidence in the correctness of the solution are followed by a higher feedback study time, than errors committed with low confidence [21]. Attention and time allocated to feedback can account for some benefits learners with high confidence have in correcting errors [26]. However, another prominent hypothesis is that confidence reflects domain familiarity [27,28,29]. Thus, the hypercorrection of errors committed with high confidence is due to different types of errors learners make (e.g., slips vs. errors caused by lacks of knowledge) and/ or prior knowledge they have in the domain of interest. Both the motivational, and the cognitive approach are supported by empirical evidence and, most probably, both contribute to this hypercorrection phenomenon.

Common Ground: RC and PoC. As we can see, RC and PoC influence each other and produce overlapping information. The main differences are their scope, granularity and stability. Table 1 compares and contrasts the two concepts.

Table 1. Comparison of PoC and RC characteristics

	PoC	RC
focus	domain-specific (global)	tasks-specific (local)
referres to	competence (global)	performance (local)
assessment	pre-assessed	online-assessed
measure	more static	more dynamic

Even though they are interrelated, correlation between domain-specific PoC and mean RC is often weak [30].

3 Research Questions and Goals

In our previous work [3], we presented a method of analyzing behavioral data of learners solving multi-trial problems. We differentiated between system-provided trials and learners' reactions on two dimensions:

1. quantity of input: try vs. skip
2. quality of input: fail vs. succeed

Further, we compared behavioural patterns within multi-trial error correction tasks of learners with low vs. high prior-assessed perception of competence (PoC) and found global differences in success, as well as skipping behaviour.

This work extends our previous research by analyzing the predictive capability of global and local self-evaluations for critical post-error behaviour. Our research questions are:

1. Can critical post-error behaviour be predicted using PoC or RC?
2. Is RC or PoC more suitable for predicting critical post-error behaviour?

Following the research on hypercorrection [27], we assume that high RC is a good predictor of success while lower RC might lead to skipping. We assume that both of these predictions are also true for PoC but to a smaller extent, since this measure does not project the dynamics of actual performance and, thus, can only account for inter-individual (not intra-individual) differences in behaviour.

4 Method

To answer our research questions we re-analyzed a partial sample of data collected in 2009. The selected data sample has been balanced according to learner and task-requirements (the data selection is explained in section 4.1). For a more detailed description of the original study and the learning tasks used see [3].

338   L. Schnaubert et al.

4.1   Study Description

Learning Environment. The experiment has been conducted using Active-Math [31], an adaptive intelligent learning environment that offers a broad spectrum of features, including the ability to run complex exercises with various feedback strategies. Figure 1 presents a screenshot of a typical exercise used in this experiment.

159 pupils from German schools attending the 5th to 7th grade participated in the study that consisted of five phases: pre-questionnaire, pre-test, treatment, post-test and post-questionnaire (see Figure 2). The learners were randomly assigned to one of four feedback-strategies, which combined procedural and conceptual hints and explanations. During the pre-questionnaire phase, learners were required to answer several questions measuring their PoC for the domain of fraction arithmetics.

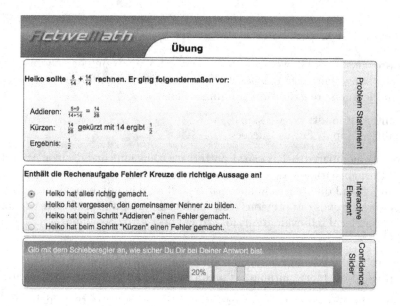

Fig. 1. Screenshot of an exercise within ActiveMath

During the treatment, learners had 45 minutes to work self-paced on learning tasks. Within each task, they had a maximum of three trials to provide a correct solution. They were allowed to skip a trial by leaving the input field blank. If they failed to provide a correct solution in their initial attempt (by either providing an incorrect solution or no solution), they received a tutoring hint and were allowed to re-try. If they failed (or skipped) again, they received a tutoring explanation and were given a final chance to solve the task. If they failed a third time, they were given a worked out solution and were directed to the next task. Each learner had to provide a mandatory confidence judgement (RC) with the first of the three trials.

Task-Sample. The treatment consisted of 4 blocks of 8 tasks each (2 of each of the following task requirements: represent a fraction, expand a fraction, add fractions with and without a common denominator). Most of the learners fully completed only the first block of tasks and partially completed the other three. In order to produce a data sample balanced with regard to learners and task requirements, we reduced the number of tasks to the first block. Learners who did not complete all the tasks form the first block due to system failures or experiment time constraints have not been included in the data sample. The remaining data consisted of 101 learners each represented by 8 tasks (N_{tasks} = 808). Within the tasks we differentiated between the initial trial to solve the task and all behaviour following an erroneous attempt to solve the task (post-error behaviour). 478 tasks were solved within the initial trial, the remaining 330 tasks contained post-error behaviour.

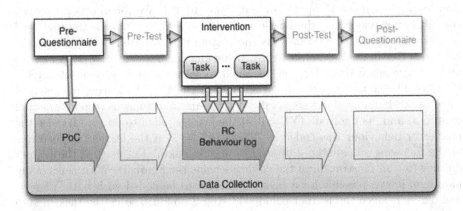

Fig. 2. Overview of the study workflow and data collection relevant to this article

4.2   Data Coding

Predictor 1 (X1): RC - Response Confidence. RC was assessed within each task. The learners had to indicate via a slider from "not sure at all" (0%) to "absolutely sure" (100%) how confident they were that they had solved the task correctly on the first trial. This is a common procedure in confidence research to ask the learner to make a probability judgement of a (binary) event [15]. To the left of the slider, a preview of the chosen degree of confidence was presented to the learners via a text field indicating the percentage value selected (cf. Fig. 1). Using this procedure, each of the eight tasks a learner completed was paired with a potentially different RC-score which ranged from 0 to 100.

Predictor 2 (X2): PoC - Perception of Competence. PoC was assessed with the following three items in the motivational questionnaire regarding fraction learning at the beginning of the study:

- I am very satisfied with my performance regarding fraction-tasks.
- I think my performance regarding fraction-tasks is very good.
- I think I am very skilled in solving fraction-tasks.

These items were adapted from [32]. The learners indicated their approval of the three statements on an anchored 6-point rating scale from "not true at all" (= 0%) to "totally true" (= 100%). The reliability in our sample was fairly high (Cronbachs $\alpha = .87$). We computed the arithmetic average to get individual PoC scores. These scores served as predictor variables for further computations and were static for each learner. That means that each of the eight task completions of a learner was paired with the same PoC score which ranged from 0 to 100.

Critical Behaviour: Solving (Y1) and Skipping (Y2). The system analyzed the input of the learners automatically according to pre-defined task-requirements and coded it "correct" or "incorrect". Additionally we re-coded if an entry was made ("input") or not ("no input"). We differentiate between the initial behaviour and the post-error behaviour. The former is the first trial to solve the task, this step informs the system if the learner potentially needs help to solve the task. The post-error behaviour includes all other steps, which are only presented if the learner has failed to solve the task on the initial trial. Post-error-behaviour was coded as "solved" (Y1 = 1) if the learner managed to solve the task in one of the steps following the initial erroneous response (step 2 or step 3) and "not solved" (Y1 = 0) if the learner failed to do so. Likewise the post-error behaviour was coded as "skip" (Y2 = 1) if the learner skipped one of the following steps (step 2 or step 3) and "no skip" (Y2 = 0) if the learner did not (for an illustration of the coding scheme see figure 3). With this coding paradigm it was possible for a task-completion to be coded with both Y1 = 1, and Y2 = 1, since the learner could skip step 2 and still solve the task in step

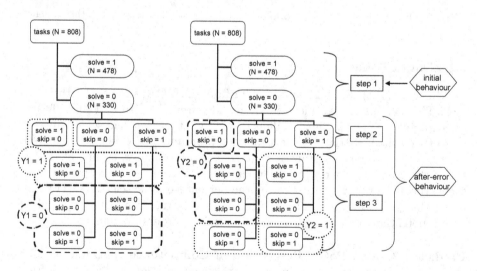

Fig. 3. Coding Scheme for Solving (Y1) and skipping (Y2)

3, although we know from our previous research [3] that this is highly unlikely, and in our sample it was only the case in 1 task completion.

4.3 Statistical Analyses

We conducted logistic regressions separately for the two factors: RC (X1) and PoC (X2), with the two predicted variables: solve (Y1) and skip (Y2). Y1=1 corresponds to a learner solving a task after the initial error, Y1=0 corresponds to failing all subsequent steps. Similarly, Y2=1 corresponds to the learner giving up on a task after the initial error, Y2=0 corresponds to the learner trying to solve the task. Level of significance was set at 5%.

5 Results

Out of the 808 included task-completions, an initial error occurred in 330 cases. For these (N = 330) cases, the regression models were computed. Equation (1) shows the null- (or intercept-only-) model for Y1 (solving), the equations (2) and (3) describe the logistic regression models if we include our predictors RC (2) and PoC (3).

$$logit(Y1) = -1.276 \tag{1}$$
$$logit(Y1) = -2.336 + .019 * RC \tag{2}$$
$$logit(Y1) = -1.517 + .005 * PoC \tag{3}$$

Respectively, the equations (4)-(6) describe the logistic regression models for Y2 (skipping).

$$logit(Y2) = 0.146 \tag{4}$$
$$logit(Y2) = 0.882 - .015 * RC \tag{5}$$
$$logit(Y2) = 0.407 - .005 * PoC \tag{6}$$

The null-models (or intercept-only models) only include a constant while the other models use either RC or PoC as additional predictors. Table 2 presents these logistic regression models (with the predictors RC and PoC included).

We can see that the -2LL value (log-likelihood) improves significantly for both criteria Y1 and Y2 if we include RC in our predictions, but not if we include PoC (cf. table 3). But even though this improvement is significant for RC, the value itself (which describes the deviance between predicted and observed values of Y) is quite high in both cases. Thus, even by including RC in the predictions, R^2 (which can be used as a measure of relative (but not absolut) effect-size) is quite low.

The Wald-statistic (cf. table 2) supports this observation; it indicates a significant difference from zero for the B-values of RC, but not for the B-values of PoC. The Hosmer-Lemshow Goodness of Fit Test (which indicates if the model prediction for Y differs significantly from the observed values of Y) does only

Table 2. Regression Statistics

	Regression Coefficient B	SD	Wald	df	p	Exp(B)	95% confidence interval for Exp(B)	
							lower bound	upper bound
Y1: solve								
constant	−2.336	.279	70.208	1	.000	.097		
RC	.019	.004	24.902	1	.000	1.019	1.011	1.026
constant	−1.517	.317	22.966	1	.000	.219		
PoC	.005	.005	.724	1	.395	1.005	.994	1.015
Y2: skip								
constant	.882	.187	22.171	1	.000	2.416		
RC	−.015	.003	25.408	1	.000	.985	.979	.991
constant	.407	.256	2.513	1	.113	1.502		
PoC	−.005	.004	1.277	1	.258	.995	.986	1.004

Table 3. Goodness of fit characteristics

	Hosmer-Lemshow			-2 log likelihood				R^2	
	HL	df	p	−2LL	Chi^2	df	p	Cox-Snell	Nagelkerkes
Y1: solve									
null-model				346.234					
RC-model	4.014	5	.547	318.382	27.852	1	.000	.081	.125
PoC-model	13.158	8	.107	345.506	.728	1	.393	.002	.003
Y2: skip									
null-model				455.730					
RC-model	5.519	6	.479	428.807	26.923	1	.000	.078	.105
PoC-model	23.185	8	.003	454.446	1.284	1	.257	.004	.005

become significant in the PoC-model for skipping, supporting the notion of a poor fit here.

Unfortunately, in the case of X1 (RC) → Y1 (solving), even though the RC-model seems to be more appropriate than the null-model, the value of the constant is too high for the low impact of the RC to take effect ($\forall x \in [0, 100]$: logit($Y1$) < 0 ⇒ predicted $Y1 = 0$). Therefore the classification tables (cf. table 4) of the null-model and the RC-model are identical (as well as the PoC-model, even though that is not surprising since the regression coefficient B_{PoC} does not differ significantly from zero) and can predict the criterion with 78.2% accuracy (sensitivity = 0, specificity = 1). In the case of Y2 (skipping), the classification tables differ (cf. table 4), leading to an increase in accuracy for the RC-model (61.5%, sensitivity = .68, specificity = .54), as compared to the null-model (53.6%, sensitivity = 1, specificity = 0), which is not significantly better than guessing, as the constant in the null-model is not significantly different from 0. In its turn, the PoC-model (52.1%, sensitivity .90, specificity .08) is performing even worse that the null-model. The Receiver-Operating-Characteristics

Table 4. Crosstables: predicted and observed values of Y1 (left) and Y2 (right)

solve		predicted values		$\sum_o$
		$Y1_p = 0$	$Y1_p = 1$	
	$Y1_o = 0$	null 258	null 0	258
observed		RC 258	RC 0	
values		PoC 258	PoC 0	
	$Y1_o = 1$	null 72	null 0	72
		RC 72	RC 0	
		PoC 72	PoC 0	

skip		predicted values		$\sum_o$
		$Y2_p = 0$	$Y2_p = 1$	
	$Y2_o = 0$	null 0	null 153	153
observed		RC 82	RC 71	
values		PoC 13	PoC 140	
	$Y2_o = 1$	null 0	null 177	177
		RC 56	RC 121	
		PoC 18	PoC 159	

(cf. figure 4) support the overall observation that even though the RC-models are in both cases more valuable than the null-models, with an Area Under Curve (AUC) of $AUC_{X1,Y1} = .69$ and an $AUC_{X1,Y2} = .67$ they still have to be considered poor. The AUC for PoC is even slightly below .50 in one case ($AUC_{X2,Y1} = .46$ & $AUC_{X2,Y2} = .54$).

6 Discussion

Our analysis of prediction models for skipping and solving behaviour has shown that dynamic (item-by-item) self-evaluation variables have a clear advantage in predicting item-by-item behaviour compared to static variables. While PoC did not add to the prediction of critical (post-error) behaviour, RC did. These results are not surprising since the static variable PoC could not account for intra-individual differences in behaviour (e.g. due to differing task-requirements), while RC judgements can be adjusted to upcoming obstacles or varying task-difficulties. While PoC might still account for overall behavioural strategies, it is not suitable for predicting local behaviour.

For all computed models the impact of our predictors was rather small. Since post-error success proved to be a rather rare event, the intercept-only model for solving was almost 80% accurate. Despite it's significant impact, RC was not able to add to this (rather high) accuracy. The situation is different with the model for skipping, where RC significantly contributes to the model and enhances the prediction accuracy.

Still, to use these models for decision making in adaptation algorithms, the predictive value of the present models has to be enhanced. The RC data therefore should be combined with other (dynamic) data about the learner (e.g., preceding behaviour, the learners approach to solve a task, types of errors or time data) and / or task-characteristics. Accounting for skills required to solve a task seems particularly interesting. The preliminary analysis shows that the ratio of learners skipping after failure ranges from 25% to almost 75% depending on the task requirements.

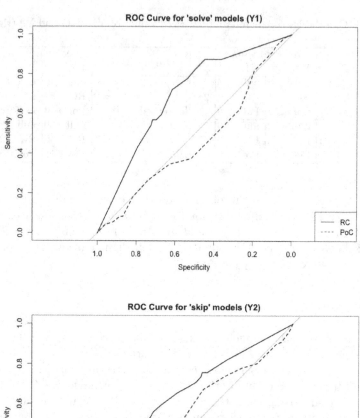

Fig. 4. ROC curve for the logistic regression models predicting Y1 (top) and Y2 (bottom)

Further research needs to be conducted to identify the relevant cognitive, motivational and metacognitive learner characteristics improving the prediction of potentially critical learning behaviour, as well as to design adaptive strategies utilizing these variables on order to improve learning. In particular, reliable prediction of skipping behaviour would allow a system to adjust not only the post-task feedback, but to take remedial action motivating the learner to engage in the task before he/ she skips. This is closely related to one of the Grand

Challenges in TEL identified by Stellar, GCP4: Increasing student motivation to learn and engaging the disengaged [1].

Despite some methodological shortcomings of this study (balancing original data-set but not post-error-data, partially-dependent data), the results indicate the importance of including dynamic self-evaluation measures in behavioural models. An interest potential direction of research is to design a methodology for soliciting (or deriving) such online confidence judgements in a less obtrusive way without the loss of predictive power.

Acknowledgements. This work was supported by the DFG - Deutsche Forschungsgemeinschaft under the ATuF project ME 1136/8-1 and NA 738/10-1.

References

1. Spada, H., Plesch, C., Wiedmann, M., Kaendler, C., Deiglmayr, A., Mullins, D., Rummel, N.: D1.6: Final report on the STELLAR Delphi study: Future directions for TEL and TEL research: Areas of Tension, Core Research Areas, and Grand Challenge Problems. The STELLAR Network of Excellence (2012)
2. Vandewaetere, M., Desmet, P., Clarebout, G.: The contribution of learner characteristics in the development of computer-based adaptive learning environments. Computers in Human Behavior 27, 118–130 (2011); Third International Cognitive Load Theory Conference on Current Research Topics in Cognitive Load Theory
3. Schnaubert, L., Andrès, E., Narciss, S., Eichelmann, A., Goguadze, G., Melis, E.: Student Behavior in Error-Correction-Tasks and Its Relation to Perception of Competence. In: Kloos, C.D., Gillet, D., Crespo García, R.M., Wild, F., Wolpers, M. (eds.) EC-TEL 2011. LNCS, vol. 6964, pp. 370–383. Springer, Heidelberg (2011)
4. Bouffard, T., Narciss, S.: Benefits and risks of positive biases in self-evaluation of academic competence: Introduction. International Journal of Educational Research 50, 205–208 (2011)
5. Eckert, C., Schilling, D., Stiensmeier-Pelster, J.: Einfluss des fähigkeitsselbstkonzepts auf die intelligenz- und konzentrationsleistung. Zeitschrift für Pädagogische Psychologie 20, 41–48 (2006)
6. Dempsey, J.V., Driscoll, M.P.: Error and feedback: Relationship between content analysis and confidence of response. Psychological Reports 78, 1079–1089 (1996)
7. Zhao, Q., Linderholm, T.: Adult metacomprehension: Judgment processes and accuracy constraints. Developmental Psychology 20, 191–206 (2008)
8. Harter, S.: A model of intrinsic mastery motivation in children: Individual differences and developmental changes. In: Collins, A. (ed.) Minnesota Symposia on Child Psychology, pp. 215–255. Erlbaum (1981)
9. Marsh, H.W., Shavelson, R.: Self-concept: Its multifaceted, hierarchical structure. Educational Psychologist 20, 107–123 (1985)
10. Zhao, Q., Linderholm, T.: Anchoring effects on prospective and retrospective metacomprehension judgments as a function of peer performance information. Metacognition and Learning 6, 25–43 (2011)
11. Bandura, A.: Self-efficacy. Encyclopedia of Human Behavior 4, 71–81 (1994)
12. Schunk, D.H.: Self-efficacy and academic motivation. Educational Psychologist 26, 207–231 (1991)

13. Zimmerman, B.J., Bandura, A., Martinez-Pons, M.: Self-motivation for academic attainment: The role of self-efficacy beliefs and personal goal setting. American Educational Research Journal 29, 663–676 (1992)
14. Narciss, S., Körndle, H., Dresel, M.: Self-evaluation accuracy and satisfaction with performance: Are there affective costs or benefits of positive self-evaluation bias? International Journal of Educational Research 50, 230–240 (2011); Knowledge: The Legacy of Competence, pp. 143–151. Springer, Netherlands (2008)
15. Dunlosky, J., Metcalfe, J.: Metacognition. Sage Publications (2009)
16. Dougherty, M.R.P.: Integration of the ecological and error model of overconfidence. Journal of Experimental Psychology: General 130 (2001)
17. Gigerenzer, G., Hoffrage, U., Kleinbölting, H.: Probabilistic mental models: A brunswikian theory of confidence. Psychological Review 98, 506–528 (1991)
18. Koriat, A., Lichtenstein, S., Fischhoff, B.: Reasons for confidence. Journal of Experimental Psychology: Human Learning and Memory 6, 107–118 (1980)
19. Koriat, A., Nussinson, R., Bless, H., Shaked, N.: Information-based and experience-based metacognitive judgments: Evidence from subjective confidence. In: Dunlosky, J., Bjork, R.A. (eds.) A Handbook of Memory and Metamemory, pp. 117–136. Lawrence Erlbaum Associates (2008)
20. Kröner, S., Biermann, A.: The relationship between confidence and self-concept towards a model of response confidence. Intelligence 35, 580–590 (2007)
21. Kulhavy, R., Stock, W.: Feedback in written instruction: The place of response certitude. Educational Psychology Review 1, 279–308 (1989)
22. Hancock, T.E., Stock, W.A., Kulhavy, R.W.: Predicting feedback effects from response-certitude estimates. Bulletin of the Psychonomic Society 30, 173–176 (1992)
23. Hancock, T.E., Thurman, R.A., Hubbard, D.C.: An expanded control model for the use of instructional feedback. Contemporary Educational Psychology 20, 410–425 (1995)
24. Vasilyeva, E., Pechenizkiy, M., De Bra, P.: Tailoring of Feedback in Web-Based Learning: The Role of Response Certitude in the Assessment. In: Woolf, B.P., Aïmeur, E., Nkambou, R., Lajoie, S. (eds.) ITS 2008. LNCS, vol. 5091, pp. 771–773. Springer, Heidelberg (2008)
25. Kulhavy, R.W., Wager, W.: Feedback in programmed instruction: Historical context and implications for practice. In: Dempsey, J.V., Sales, G.C. (eds.) Interactive Instruction and Feedback, pp. 3–20. Educational Technology Publications, Englewood Cliffs (1993)
26. Fazio, L., Marsh, E.: Surprising feedback improves later memory. Psychonomic Bulletin & Review 16, 88–92 (2009)
27. Butterfield, B., Metcalfe, J.: Errors committed with high confidence are hypercorrected. Journal of Experimental Psychology Learning Memory and Cognition 27, 1491–1494 (2001)
28. Butterfield, B., Metcalfe, J.: The correction of errors committed with high confidence. Metacognition and Learning 1, 69–84 (2006)
29. Metcalfe, J., Finn, B.: Peoples hypercorrection of high-confidence errors: did they know it all along? Journal of Experimental Psychology Learning Memory and Cognition 37, 437–448 (2011)
30. Stankov, L., Crawford, J.D.: Self-confidence and performance on tests of cognitive abilities. Intelligence 25, 93–109 (1997)
31. Melis, E., Goguadze, G., Homik, M., Libbrecht, P., Ullrich, C., Winterstein, S.: Semantic-aware components and services in activemath. British Journal of Educational Technology. Special Issue: Semantic Web for E-learning 37, 405–423 (2006)

32. Narciss, S.: Informatives tutorielles Feedback. Waxmann (2006)
33. Mory, E.H.: Feedback research revisited. In: Jonassen, D.H. (ed.) Handbook of Research on Educational Communications and Technology, 2nd edn., pp. 745–783. Lawrence Erlbaum Associates, Mahwah (2004)
34. Thiede, K.W., Anderson, M.C.M., Therriault, D.: Accuracy of Metacognitive Monitoring Affects Learning of Texts. Journal of Educational Psychology 95, 66–73 (1980)
35. Stone, N.J.: Exploring the relationship between calibration and self-regulated learning. Educational Psychology Review 12, 437–475 (2000)

Taming Digital Traces for Informal Learning:
A Semantic-Driven Approach

Dhavalkumar Thakker, Dimoklis Despotakis, Vania Dimitrova,
Lydia Lau, and Paul Brna

University of Leeds, Leeds LS2 9JT, UK
{D.Thakker,scdd,L.M.S.Lau,V.G.Dimitrova}@leeds.ac.uk,
paulbrna@mac.com

Abstract. Modern learning models require linking experiences in training environments with experiences in the real-world. However, data about real-world experiences is notoriously hard to collect. Social spaces bring new opportunities to tackle this challenge, supplying digital traces where people talk about their real-world experiences. These traces can become valuable resource, especially in ill-defined domains that embed multiple interpretations. The paper presents a unique approach to aggregate content from social spaces into a semantic-enriched data browser to facilitate informal learning in ill-defined domains. This work pioneers a new way to exploit digital traces about real-world experiences as authentic examples in informal learning contexts. An exploratory study is used to determine both strengths and areas needing attention. The results suggest that semantics can be successfully used in social spaces for informal learning – especially when combined with carefully designed nudges.

Keywords: Semantic Data Browser, Social Semantic Web, Semantic Augmentation, Adult Informal Learning.

1 Introduction

The dramatic rise of social media leads to the development of a vast amount of user-generated content which is radically transforming today's practice in many areas (e.g. policy making, disaster response, open government). Trends and predictions [1, 2] point out that social media will have a strong impact on learning in workplace, providing a huge resource of user-generated content for learning that may be unplanned and driven by circumstances – i.e. informal learning. Social spaces can offer a plethora of digital traces (DTs)[1] of real world experiences: people write reviews about their experiences with staff or services (e.g. in hotels); share their personal stories (e.g. in blogs); leave comments pointing at situations they have experienced (e.g. when watching videos). If selected carefully, these DTs can give broad, authentic and up-to-date digital examples of job activities, and can be a useful source for informal learning by adults who are self-regulated and shaped by experience [3], and are accustomed to social media and pervasiveness of technologies.

---

[1] The term "digital traces" is confined to traces of real world experiences from social media.

A. Ravenscroft et al. (Eds.): EC-TEL 2012, LNCS 7563, pp. 348–362, 2012.
© Springer-Verlag Berlin Heidelberg 2012

DTs can be particularly appealing as a source for informal learning of soft skills (e.g. communicating, planning, managing, advising, negotiating). Soft skills are highly demanded in the 21st century [4], and fall in the category of ill-defined learning domains [5], i.e. they are hard to specify and often require multiple interpretations and viewpoints. Modern informal learning environments for soft skills can exploit user-generated content to provide learning situations linked to real world experience. For instance, learners can browse through examples with different job situations, share personal stories, read stories or examples and comment on them, become aware of different perspectives, triggering self reflection and goal-setting for personal development.

To realise this vision, novel architectures of social spaces are needed which use robust and cost-effective ways to retrieve, create, aggregate, organise, and exploit DTs in social learning situations; in other words, to tame DTs for informal learning. Recent advances in social semantic web can offer the technical underpinning for taming DTs; allowing semantically augmented user generated content available via semantic data browsers. However, for learning environments, new intelligent techniques are needed to extend semantic data browsers with features that facilitate informal learning, yet preserving the exploratory nature of social environments.

In this work, we propose a novel approach for taming DTs in social spaces for informal learning by combining major advancements in semantic web (semantic augmentation, semantic query, relatedness, similarity, summarisation) with semantically driven 'nudges'[2]. We believe that these nudges can preserve the freedom for users to explore the social spaces and yet providing guidance to benefit from informal learning [6]. The paper addresses the following research questions:

Q1: How can semantics be used in social spaces to aggregate DTs on a specific activity and to generate nudges to facilitate exploration of DTs for informal learning?

Q2: What are potential benefits of using semantically augmented DTs and nudges in social spaces for informal learning, and what are the further issues to address?

The contributions of this paper are: (i) an intelligent content assembly framework which aggregates and organises DTs with user generated content in a social space for learning; (ii) implementation of the framework within a representative soft skills domain - interpersonal communication with the focus on non-verbal cues and emotion; (iii) an exploratory study which provides an initial, qualitative analysis of the benefits of the main features (augmented content and nudges); (iv) lessons learnt for the feasibility of the approach and further research directions in using semantic data browsers for learning. This work is part of the "Immersive Reflective Experience-based Adaptive Learning" (ImREAL) project which develops an approach for linking experiences in virtual environments with real-world context by following andragogy and self-regulated learning principles. Interpersonal communication skills is the learning domain.

The rest of the paper is organised as follows. Section 2 positions the work in the relevant literature and highlights the key contributions. The semantic data browser and its main components are described in Section 3. An exploratory study with the

---

[2] Henceforth, *"nudges"* is used to stand for "semantically-driven nudges".

objective of addressing the aforementioned research questions was conducted and is described in Section 4. The findings are presented in Section 5. Finally, Section 6 reflects on the lessons learnt from the study.

2   Related Work

The work presented in this paper is positioned in the emergent strand of exploiting user-generated content in learning environments; and pioneers an approach that utilises techniques from semantic augmentation, data browsers and nudges.

Social spaces for sharing personal experience in informal learning contexts have been developed in several projects. MATURE [7] uses a semantic media wiki and information extraction techniques to develop social spaces that capture organisational knowledge in career guidance from shared job experience. AWESOME [8] exploits a semantic media wiki and the emergent ontology to provide semantic social scaffolding for sharing dissertation writing experience in the form of examples, stories or tips. Videos with job-related activities have been used in KP-LAB [9] as "shared authentic objects" of work practices to trigger collaborative interactions for collective knowledge creation, discovery and exploitation of tacit knowledge, e.g. for teachers' professional development. A domain ontology and semantic tagging are used to support knowledge discovery in a shared space for experiential learning. Most recently, one of the applications being developed in MIRROR [10] organises and retrieves relevant descriptions of job situations from a pool with shared experiences by carers of elderly people, in order to facilitate contextualised learning. Similarly to these projects, we consider collective content as a source for authentic experience, and adopt semantic techniques for augmentation and search. The distinct aspects of our work are: (i) exploits several ontologies and state of the art techniques for semantic augmentation, to group content related to a job activity and present different viewpoints; (ii) adopts entity and ontology summarisation techniques to generate nudges that facilitates awareness of activity aspects; (iii) explores a new domain - interpersonal communication - and developing original semantic techniques to include social signals (non-verbal cues and emotion) in learning contexts.

Access to informal knowledge from social media (such as Delicious, Slideshare and YouTube) in the context of learning tasks is supported with iFSSS [11]. It exploits resources, tags and user social networks, to enable knowledge discovery for informal learning, using a domain ontology. Similarly, we consider social media content as a source for informal learning. In contrast to iFSSS, which searches for available learning resources, we exploit DTs that include user comments or stories related to aspects of a job activity, which provides authentic examples of different viewpoints and is crucial for learning in ill-defined domains. Moreover, we provide semantic enrichment of DTs and aggregate DTs from both an open social space (YouTube videos and comments) and a closed social environment (descriptive stories).

Semantic augmentation is being increasingly adopted as the main mechanism to aggregate and organise social content by linking it with ontology concepts. A review of semantic augmentation tools is available in [12]. Semantic units are extracted mainly by identifying named entities (e.g. people, organisations, locations). In this work, we have developed a semantic augmentation service by using a state-of-the-art information extraction engine and a semantic repository for automatically detecting important activity aspects from textual sources. This focuses on entities related to an interpersonal communication activity and enables converting DTs to conceptual spaces for exploration.

Semantic data browsers that combine semantically augmented data and ontological knowledge bases, are being utilised in various domains, such as sensemaking or statistical data analysis (see review in [13]). Semantic browsers can offer opportunities to build learning environments in which exploration of content is governed by ontologies that capture contextual aspects. Data browsers assume that the users are in charge of what they do when using the browser. This puts the cognitive onus on the user, and is particularly acute in the case of a user being a learner, i.e. not familiar with the conceptual space in the domain and may be unable to decide what is the best course of action for him/her. Hence, directly adopting semantic web browsers in learning contexts would not be sufficient for effective learning environments – new intelligent techniques are needed to extend these browsers with features that facilitate informal learning. The paper presents a novel approach to extend semantic browsers with nudges in order to influence the choices users make and benefit learning. Our technical implementation follows the pedagogical framework proposed in [6].

3   Intelligent Content Assembly Workbench (I-CAW)

An Intelligent Content Assembly Workbench (I-CAW) is a semantic data browser with intelligent features that facilitate informal learning. I-CAW utilises: (i) ontological underpinning and semantic augmentation and query services to aggregate and organise DTs, and (ii) *nudges* such as signposting and prompts to guide users through the browsing process. This section outlines the main components of I-CAW (Fig. 1).

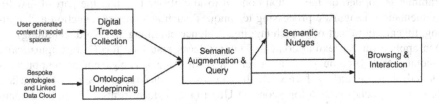

Fig. 1. The main components of the Intelligent Content Assembly Workbench

3.1   DTs Collection and Ontological Underpinning

The availability of social web APIs has made it possible to consume DTs from these sources and to build custom applications. I-CAW supports trainers by offering the options to browse the videos and comments from YouTube (linked within I-CAW) and personal stories collected from a blog-like story telling environment. Trainers can then select appropriate content for training material.

The Ontological Underpinning is crucial for aggregating DTs on an activity, as a semantic model describes the key aspects of that activity in the form of an ontology. An Activity Model Ontology (AMOn) is developed in ImREAL by a multi-disciplinary team of computer scientists and social scientists [14]. The ontological underpinning for aggregation of DTs in I-CAW utilises AMOn, DBPedia from Linked Data Cloud and public ontologies (WordNet-Affect and SentiWordNet). The ontologies are used by intelligent services for semantic augmentation, query and for designing semantic nudges as described below.

3.2   Semantic Augmentation and Query Services

Semantic augmentation uses ontologies to enrich unstructured or semi-structured data in relation to the context of an activity. This service takes any text and outputs semantically augmented text. It has two main components: the Information Extractor (IE) and the Semantic Indexer (SI). The IE component is implemented using the General Architecture for Text Engineering (GATE[3]) and outputs Annotation Sets of extracted entities with offsets, ontology URI and type information. The SI is designed using semantic repositories and converts the annotation sets to RDF triples. The augmentation process consists of two stages: set up (manual and offline) and processing (automatic and on-the-fly).

Set up: This includes selection of ontologies for the domain. These ontologies are utilised to build a list of known entities (called Gazetteers in GATE). The mapping from ontologies to gazetteers exploits the label properties (rdfs:label, skos:prefLabel and skos:altLabel) for each ontology class. A JAPE grammar component is also designed to define the rules for pre-processing of the content and post-processing of the extraction results.

Processing: The input to the processing algorithm is textual content. The JAPE grammar is applied on the textual content to identify and extract the part of text for augmentation. Linguistic processing techniques (such as sentence detection and splitting, tokenisation, part-of-speech tagging, sentence chunking and word stemming) are then applied on the textual content. This produces a surface form (i.e. linguistically annotated text with nouns, verbs, adverbs and adjectives). The gazetteer component is applied on these surface forms which matches the ontological labels to the surface forms and attaches an ontology concept URI to the surface forms. Then, the semantic indexer converts the annotation sets into set of triples, checks the existing index for the content and stores new or updated indexes into the semantic repository.

Semantic query provides a mechanism for querying and browsing the semantically augmented content in the repository. In the processing stage: this service takes a focus concept (C_f) as keywords and outputs information relating to the concept (e.g. triples $T_f{}^i$) and content ($CO_f{}^i$) in the SPARQL/XML format. This is achieved by concept lookup in the form of the triple (`<`C_f`><P><O>`), where P is any property and O is any object; and content lookup in the form of (`<O1>` `<media:hasEntity>` `<`C_f`>`) and (`<`CO_f`><media:hasAnnotation><O1>`).

---

[3] http://gate.ac.uk/

The query service also allows access to content related to a focus concept using a semantic content relatedness algorithm.

> *Semantic content relatedness for focus concept C_f:* First, the content CO_f tagged with C_f is retrieved using content lookup. Then the concepts C_r that have triple pattern ($<C_f><P_1><C_r>$) or ($<C_r><P_2><C_f>$), where P_1 and P_2 are any valid properties, are retrieved. Next, the content CO_r tagged with C_r is retrieved. $CO_f \cup CO_r$ gives the list of content for C_f.

3.3   Nudges

I-CAW proactively suggests areas of exploration to learners. We have developed a novel approach - *nudges* - based on Sunstein and Thaler's choice architecture [15] and a proposal for its adoption in learning [6]. In a choice architecture, a choice architect is responsible for organizing the context in which people make decisions. A nudge is any aspect of the choice architecture that alters people's behaviour in a predictable way without forbidding any options, and tries to influence choices in a way that will make choosers better off. Two types of *nudge* have been chosen: signposting and prompts.

Signposting. The first type of *nudge* is related to "default options". The choice architecture encourages careful design of default choices as usually most users will end up with these. The semantic data browsers generally have information on a focus concept with a list of known facts from ontological knowledge bases. The presentation (signposting) of these facts influences the navigational path that learners take and facts they can read. Three types of signposting made possible by semantic technologies are explained below.

"All Facts" signposting. All facts (T_a^i), where $i=1$ *to* n, about a focus concept (C_f) are made of two sets of triples: the first set where the C_f is the subject of a triple ($<C_f><P><C_o>$) and second set where the C_f is the object of a triple ($<C_s><P> C_f>$). As per the choice architecture, the "all choices" shall be made available to the learners if they wish, and "all facts" signposting allows achieving this in the browser.

"Key Facts" signposting. This is a summary of a focus concept with fewer facts, and yet containing sufficient information for learners to quickly identify the concept [16]. The key facts can be considered as providing immediate exploration space and can be implemented using entity summarisation techniques.

> *Entity summarisation algorithm for focus concept C_f:* The key facts (T_k^j), where $j=1$ to m and m is the number of facts required as part of the key facts and can be system dependent.   The key facts include:
> 1. fact($<C_s>$ $<$rdf:type$>$ $<C_f>$) when concept C_s is   the   direct   instance   of C_f
> 2. fact($<C_f>$ $<$rdf:type$>$ $<C_o>$) when C_f is the direct instance of the concept C_o
> 3. fact($<C_f>$ $<$rdfs:subClassOf$>$ $<C_o>$) when concept C_o is the direct super class of C_f.
> 4. If m is the number of facts required as part of key facts and q is the number of the facts with triple pattern ($<C_s>$ $<$rdf:type$>$ $<C_f>$) and the number of other qualifying facts (using above 2-3) are r and $(q+r) > m$. Then: Limit the number of facts with pattern ($<C_s>$ $<$rdf:type$>$ $<C_f>$) to $q-m-r$.

"Overview" signposting. This is the overview of the ontological knowledge base with an interaction focus. The overview can be considered as providing overall exploration space and is implemented using ontology summarisation techniques [17].

Ontology summarisation algorithm for focus concept C_f: Overview facts for a focus concept (C_f) are (T_o^i), where $i=1$ to m, m is the number of facts required as part of overview and can be system dependent; are derived using the following:

1. If (<C_f> <rdf:type> <C_{obj1}>) exists do the following:
    1a. Add (<C_f> <rdf:type> <C_{obj1}>) in overview, <u>where</u> C_f must be the direct instance of the concept C_{obj1}.
    1b. Add a fact (<C_f> <P> <C_{obj2}>) <u>when</u> (<C_f> <rdf:type> <C_{t1}>), (<C_{t2}> <rdfs:domain> <C_{t1}>), (<C_{t2}> <rdfs:range> <C_{obj2}>) and (<C_{obj2}> <rdf:type> <?>) exists.
2. If (<C_f> <rdfs:subClassOf> <C_{obj1}>) exists do the following:
    2a. Add (<C_f> <rdfs:subClassOf> <C_{obj}>) in overview, <u>where</u> C_f must be the direct subclass of the C_{obj1}.
    2b. Add a fact (<C_f> <P> <C_{obj2}>) <u>when</u> (<C_f> <rdfs:subClassOf> <C_{t1}>), (<C_{t2}> <rdfs:domain> <C_{t1}>) , (<C_{t2}> <rdfs:range> <C_{obj2}>) and (<C_{obj2}> <rdf:type> <?>) exists.

Prompts. This is second type of nudge which provides non-invasive suggestions based on similar and/or contradictory learning objects (factual knowledge). Following are two formats of a prompt, each with corresponding algorithm and example.

<u>*Similarity prompt Format:*</u> C_f is a C_t. There are further examples of <u>Similar</u> C_t to C_f which will be useful for you to explore. Click on one of the following to explore further. {C_{sim}^i}

Similarity measurement: If C_f is the focus concept then (<C_f><rdf:type> <C_t>) where C_f must be the direct instance of C_t. Then, (< C_{sim}^i ><rdf:type><C_t>) where C_{sim}^i is the direct instance of C_t and $i=1$ to n where n=number of similarity concepts and can be system specific.

Example: Nervousness is a negative emotion. There are further examples of <u>Similar</u> negative emotions to "Nervousness" which will be useful for you to explore. Click on one of the following to explore further.
NEGATIVE EMOTION » ANXIOUS » PANIC

<u>*Contradictory prompt Format:*</u> C_f is a C_t. There are <u>Contradictory</u> examples to C_f which will be useful for you to explore. Click on one of the following to explore further. {C_{con}^i}

Contradictory measurement: If C_f is the focus concept then (<C_f> <rdf:type> <C_t>) where C_f must be the direct instance of C_t. If there exists (<C_t> <owl:disjointWith> <C_d>), Then, (< C_{con}^i> <rdf:type> <C_d>) where C_{con}^i is the direct instance of C_d and $i=1$ to n where n=number of contradictory concepts and can be system specific.

Example: Aggression is a negative emotion. There are <u>Contradictory</u> examples to "Aggression" which will be useful for you to explore. Click on one of the following to explore further.   POSITIVE EMOTION » CALM » EMPATHY

3.4   An Illustrative User Interaction

A user interface was developed in I-CAW to incorporate the services and nudges; and to allow learners to search and browse relevant digital traces for interpersonal communications. Here is a usage scenario: Jane is commencing her first job in the human resources department. Her role involves conducting job interviews and she has been advised by some colleagues that non-verbal behaviour can play a crucial role in interpersonal communication and it is important that she is aware of it when dealing with the interviewees. Jane wants to learn more about this. She knows that there is a large volume of DTs (videos and comments, blogs, stories etc) on the Web but it is time consuming to search this content, especially when approaching them randomly.

Using I-CAW she searches the term "body language" and is offered information on various aspects of body language including an overview of body language (using <u>overview signposting</u>) along with several DTs such as videos on YouTube (with comments) on job interviews and stories about real experiences. While reading some comments, she learns that eye-contact can be important and that there are several possible interpretations of the same aspect of body-language. She clicks on eye-contact and is particularly interested in the link pointing to "Body Language Meanings" (derived from the ontology by I-CAW) which leads to a list of manifestations of eye-contact (gaze, stare, and its interpretation as meaning nervous etc). From there, she clicks on "Nervousness" and arrives at another page with a collection of related YouTube videos, personal stories and comments (Fig. 2). The diversity of such content is a result of the semantic augmentation service which makes such aggregation possible. From the information she finds out other types of body language signals (in additional to eye-contact) that may indicate nervousness. From the comments associated to each DTs, Jane also learns about different interpretations by people from different cultures or from different perspectives (i.e. interviewer vs. job applicant, novice vs. experienced individuals).

While reading a personal story on a nervous applicant, a <u>prompt</u> appears and suggests experiences for <u>similar</u> behaviour such as 'anxious'. Although not personally experienced with this behaviour, Jane feels that the different comments on these resources have opened her eyes and she is much better prepared for future interviews.

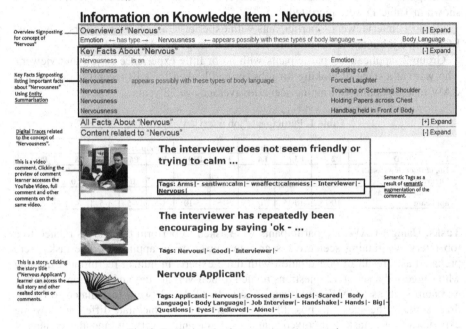

Fig. 2. Exploring the focus concept 'nervous'

4   Exploratory Study

An exploratory study was conducted following a qualitative approach in order to get an insight of user interactions with I-CAW, and the potential benefits of using semantic services and nudges for informal learning.

Domain. Job interview was selected as the setting for interpersonal communication (IC) activity, as (i) it is relevant to many people at different stage of their lives and is often learned by sharing experience; (ii) it consists of a mixture of well-defined elements (such as standard practices in job interviews) and ill-defined elements (such as interpersonal skills); and (iii) DTs of this activity are widely available in social media. Non-verbal cues and emotion were selected as the prime focus, as they play a key role in IC and are fairly well covered in the ontologies currently used by I-CAW.

DTs Used. DTs were sampled from the ImREAL collection, which included: (i) user comments from YouTube – relevant videos on job interviews or non-verbal cues were selected; cleaned comments were aggregated using the ImREAL YouTube noise filtration mechanism (described in [18]); (ii) user comments from an experimental environment which encouraged different viewpoints to be collected on fragments of job interview videos (comment collection and viewpoint analysis are presented in [19]); and (iii) personal stories from volunteers in the ImREAL project. Following the framework presented in Section 3, the sample of DTs was semantically augmented and used to seed I-CAW with relevant content for the study.

Participants. Ten participants with different level of experience in interviews (as shown in Table 1) were recruited:

Group 1: interviewers –.participants with experience as job interviewers, who used I-CAW in a job interviewer training scenario; average age 44 years.

Group 2: applicants – participants with no or little experience as job interviewers, who were at a stage of looking for jobs (or had looked for jobs recently), and used I-CAW in a job applicant training scenario; average age 28 years.

Table 1. Participants' job interview experience

Participant ID	Group 1: Interviewers					Group 2: Applicants				
	P2	P3	P4	P5	P10	P1	P6	P7	P8	P9
No. of interviews as an interviewer	10-15	10-15	10-15	>15	>15	0	0	0	0	1-5
No. of interviews as an applicant	10-15	10-15	10-15	>15	5-10	1-5	1-5	1-5	5-10	5-10

Tasks. Using I-CAW, the participants were asked to perform three tasks related to a job interview training scenario (as an interviewer or an applicant). The tasks were prepared as a result of consultation with three experts in human recruitment training who suggested a set of 12 questions related to non-verbal cues and emotion in IC that a system such as I-CAW could help. The tasks (Table 2) were formulated to (i) address some of the questions posed by the experts; and (ii) facilitate different browsing experience – i.e. task 1 involved mainly the semantic overview and the content it signposted to; task 2 required broader exploration of content; and task 3 led to a specific aspect with insufficient content (a likely situation with user generated content).

Table 2. Tasks in the study

Task1: *What nonverbal cues can be observed in job interview situations?* (same for both groups)
Task2: *What nonverbal cues show nervousness?* (same for both groups)
Task3: *How would an interviewer deal with an aggressive applicant?* (for group 1) and *How would an applicant deal with an aggressive interviewer?* (for group 2)

Procedure and Data Collection. In each session, a participant was firstly introduced to I-CAW [5 min] by following a script to perform a simple independent task. A standard script with the three tasks (Table 2) was then given to the participant which required the use of search box or signposting (All Facts, Key Facts and Overview) to find/browse relevant examples in I-CAW. When the participant finished a task, a semantic prompt was presented by the system when appropriate (e.g. task 2 included a similarity-based prompt, and task 3 included a contradiction prompt). After a participant completed all the tasks, the experimenter collected the participant's feedback on his/her experience with I-CAW (using a semi-structured interview and a questionnaire). The materials for the study are available online[4].

5   Results

The researchers' notes of individual sessions were analysed jointly by the authors to get an insight of the users' interaction with I-CAW and their feedback. The findings are summarised below, exploring benefits and issues that require further attention.

5.1   Semantically Augmented DTs

Feedback from the participants on the usefulness of these content was analysed.

Benefits. Firstly, the retrieved DTs provided valuable authentic examples, in particular the personal stories, and the comments to some extent. Two participants in group 1 stressed that examples would benefit both learners and tutors:

> *"Examples are the beauty of the system - I will learn more from examples"* [P10]
> *"Anything that facilitates the preparation of training materials and provides real world examples to backup the training is very helpful."* [P5]

Secondly, DTs offered different points of view (interviewer/applicant) on non-verbal cues. This refers mainly to the comments on YouTube videos. The participants valued the aggregated / filtered comments as a group as it enabled quick browsing to get a sense of the overall picture.

Thirdly, DTs provided a stimulus for participants to contribute content on cultural differences. For instance, comments about an aggressive applicant provoked P2 to point out that aggressiveness was not an expected behaviour at job interviews in China, while the same content led P10 to comment that at some job interviews the interviewer may deliberately be aggressive to test the applicant's ability in dealing with difficult IC situations. These additional experiences could be added to the pool of DTs, semantically augmented, and made available for browsing in I-CAW.

Issues Requiring Attention. Some participants found the resulting content confusing or misleading. This was caused mainly by the semantic relatedness algorithm (see section 3.2). For example, body language and eye body language were considered as related concepts to direct eye contact. However, as there was no content on direct eye contact, I-CAW offered the nearest content tagged with eye body language or body language. The relevance was not immediately obvious to the participants. This could be addressed by explaining to the learners why DTs have been suggested.

The two most experienced interviewers (P5 and P10) found some content could be mistaken as the norm. For instance, a comment associated with a video stated *"The*

---

[4] http://imash.leeds.ac.uk/imreal/icaw.html#evaluation

interviewer has his hands in front of him, which indicates that he is concentrating and not fidgeting...". P5 and P10 stressed that inexperienced users may see a comment in isolation and believe it would be valid in all situations. It was suggested that short comments could be augmented with contextual information to assist the assessment of the credibility of the different viewpoints (e.g. the interviewer or the applicant). There was a strong demand for a better way to compare and contrast different viewpoints on the same video or on similar videos. This requires that, in the next stage of development, a suitable pedagogical approach is coupled to both the content and the *nudges*.

5.2   Nudges –Signposting

Table 3 summarised the feedback from the participants on their usefulness. Further analysis was conducted on the reasons behind the perceived usefulness of these nudges and issues to be addressed.

Table 3. Summary of the participants' opinions on semantic nudges in I-CAW

	Informative	Relevant	Confusing	Overwhelming		Helpful	Not Helpful	Not Sure
Overview	6	9	3	0		6	2	2
Key Facts	9	9	2	1		9	0	1
All Facts	7	6	4	5		7	1	2
Similarity	10	8	1	2		8	0	2
Contradiction	7	9	1	0		7	1	2

Benefits of Overview. The overview signposting was found relevant by all participants except P4. It was seen by some as a helpful starting point for exploration when examining a specific concept, for examples:

"Gives you an indication of main items to look for." [P4]

"Provides concise definition of terms and links to other related terms" [P8]

"...shows the links between different things, for instance nervousness to body language"[P9]

Issues Requiring Attention. A participant felt that a summary of a concept was not sufficiently helpful, and wanted to see also a summary of the relevant content as part of the overview. Two participants felt that the overview was not intuitive:

"The overview is robotic at the moment and needs to be more human." [P10]

On some occasions participants commented that the overview did not present the right information for the task. Task-based adaptation in browsers is generally hard to address without using a pre-defined set of tasks and limiting the exploratory nature.

P4 was unsure of the overview helpfulness but made an interesting statement:

"...[the overview] does not say much,   but if it's not there, it would be missed." [P4]

Benefits of Key Facts. Key facts were found informative, relevant and helpful by all participants except P5. The participants often used the Key Facts as an anchor for exploring related concepts or as a quick summary for understanding a concept:

"I will just go to the key facts, they are most useful. The quality and relevance is good." [P10]

"[key facts] show the most relevant links/information relevant to my query." [P9]

"Helps see areas that may have been missed and should look at." [P4]

Issues Requiring Attention. Confusion could be caused when key facts did not lead to relevant content due to the lack of user generated content (as discussed above), A possible improvement could be by adding an indicator of the number of DTs available for a particular concept. This is a crucial lesson for entity summarisation techniques used in learning environments, as a learner would intuitively assume that when a fact (concept) is signposted, relevant examples would be present. Another issue highlighted by P1 was the lack of explanation of the reasoning behind the selection of key facts and their concepts and how this relates to the user's task.

Benefits of All Facts. Opinion on the usefulness of the all facts signposting was mixed. Participants who found it helpful, informative and relevant considered all facts as a provider of diversity of concepts, additional information and details.

Issues Requiring Attention. Half of the participant found all facts overwhelming and suggested that a better layout and presentation was needed. The lack of content for a concept was also stated as a reason for confusion and misleading.

Overall Comment on Signposting. A main purpose of signposting is to lead a learner to find something new to learn. When the participants were asked at the end of the three tasks: "Did you find anything in summary that surprised you or you were not aware of?", most of them responded positively (e.g. For task 1 - 9 participants; task 2 – 8; and task 3 – 5). Example answers include:

"Did not know kinesics, clicked on it and the definition helped."[P4]; *"Quite a coverage of body language."*[P8]; *"Had not thought of some of the aspects before."*[P3].

5.3   Nudges –Prompts

Table 3 summarised the feedback from the participants on prompts' usefulness.

Benefits of Similarity Prompts were unanimously found informative, and was seen as relevant and helpful in most cases (notably, all interviewers found the similarity prompt helpful). Similarity prompts were also seen as task setting since they pointed at aspects participants might have not thought about (e.g. shyness or anxiousness experienced at interviews could be related emotions to nervousness). **Contradiction Prompts** were unanimously found as relevant, and were considered as complementary knowledge, prompting participants to *"look at the task in a holistic way"*, *"pointing at alternatives"*, and *"helping see the big picture"*.

Issues Requiring Attention. The participants who found the similarity prompts misleading or confusing thought that prompts per se were not sufficiently helpful without proper content. A way to improve prompts is to combine ontology summarisation with user browsing history and available content. The participants who found semantic prompts overwhelming felt that the text was not intuitive; to improve this a consultation with trainers could be sought. It was also pointed out that emotion similarity could be situation dependent and could have subjective/cultural interpretation (e.g. P5 commented that shyness and nervousness in some cultures could be seen as respect, and should not be marked as negative emotions). This refers to a broader aspect about subjective views and agreed ontologies (e.g. we used WordNet-Affect and Senti-WordNet which are broadly used as generic resources but do not take encounter of subjective or cultural aspects). The participants who were not positive about contradiction prompts felt that they did not lead to interesting examples, as the content linked to the prompt was quite limited.

Overall Comment on Semantic Prompts. All participants followed the prompts and clicked on at least one concept from the suggestions, which led to further browsing of content and seeing other signposts. Hence, the prompts were found effective for further exploration of the conceptual space under examination.

6   Discussion and Conclusions

We have demonstrated with semantic technologies how DTs can be aggregated and utilised for generating nudges in social spaces for informal learning. The exploratory study reported in this paper shows potential benefits of our approach. There are also promising avenues for further work. Key aspects of our findings are discussed below.

DTs as Authentic Examples and Stimuli. The participants in the study particularly liked the authenticity of the content, which probed them to: (i) further reflect on their experiences, and in some cases help participants articulate what they had been doing intuitively; (ii) contribute their different viewpoints (e.g. due to culture, environment, or their tacit knowledge); (iii) engage deeper with the learning resource; and (iv) sense the diversity or consensus of opinions on the selected topic. We plan to carry out deeper analysis to extract patterns of DTs usage in a variety of contexts.

Capturing Viewpoints from DTs. Aggregation of content from heterogeneous sources in I-CAW strongly supported the requirement for further improving capturing and grouping multiple viewpoints. Examining DTs in a group (related content shown in I-CAW), different people saw different aspects of an activity; in some cases these aspects were complementary, in others contradictory. In many occasions, learners wanted to compare different viewpoints and considered them crucial to raising their awareness. Appropriate techniques for capturing, aggregating and comparing viewpoints are needed. Some of our ongoing work addresses these issues, e.g. [19].

Nudges **to Support Informal Learning.** We demonstrated the use of *nudges* as a step in the right direction for turning a semantic browsing experience into an informal learning experience. Overall, the signposting *nudges* were considered a fruitful way to provide a quick summary for understanding a concept and for exploration which leads to something new to learn. The prompts were seen as task setting for the learners to browse further to understand the bigger picture. Different strategies for signposting and different types of prompts could be explored further. For examples, adding 'complementary' prompts to suggest complementary learning objects or 'reflection' prompts to ask learners to link their previous experiences to content they have seen.

Contextualising *Nudges*. The three key decisions a choice architect makes are what, when and how to nudge. The approach presented in this paper only deals with "what" (semantic data and DTs) and "how" (signposts and prompts). Deciding an appropriate time to nudge is an interesting challenge and requires taking the context into account. This context can come from the learner's interaction with the system (i.e. what learner has seen so far), the learner's profile, the competency of the learner, pedagogical goal, the interaction focus and availability of content. Experimental studies are planned to explore these issues.

New Opportunities for Learning. Using social content brings in new sources for learning, e.g. the diverse range of real-world experiences. Further work is needed to capitalise on the new opportunities brought by the broad availability of social content. Our approach is just a step in this direction, and we expect that there will be further research to exploit the reuse of social content for learning.

Acknowledgements. The research leading to these results has received funding from the European Union Seventh Framework Programme (FP7/2007-2013) under grant agreement no ICT 257831 (ImREAL project).

References

1. Cara Group Plc., How Informal Learning is Transforming the Workplace, A Pulse Survey – Social Media's Impact on Informal Workplace Learning (2010)
2. Redecker, C., Ala-Mutka, K., Punie, Y.: Learning 2.0 - The Impact of Social Media on Learning in Europe, Policy Brief, European Commission, Joint Research Centre, Institute for Prospective Technological Studies (2010)
3. Knowles, M.S., Holton, E.F., Swanson, R.A.: The Adult Learner: The Definitive Classic in Adult Education and Human Resource Development. Elsevier (2005)
4. Ananiadou, K., Claro, M.: 21st Century Skills and Competences for New Millennium Learners in OECD Countries. OECD Education Working Papers, No. 41 (2009)
5. Lynch, C., Ashley, K., Pinkwart, N., Aleven, V.: Concepts, structures, and goals: Redefining ill-definedness. Int. Journal of AI in Education 19, 253–266 (2009)
6. Kravcik, M., Klamma, R.: On Psychological Aspects of Learning Environments Design. In: Kloos, C.D., Gillet, D., Crespo García, R.M., Wild, F., Wolpers, M. (eds.) EC-TEL 2011. LNCS, vol. 6964, pp. 436–441. Springer, Heidelberg (2011)
7. Weber, N., Schoefegger, K., Bimrose, J., Ley, T., Lindstaedt, S., Brown, A., Barnes, S.-A.: Knowledge Maturing in the Semantic MediaWiki: A Design Study in Career Guidance. In: Cress, U., Dimitrova, V., Specht, M. (eds.) EC-TEL 2009. LNCS, vol. 5794, pp. 700–705. Springer, Heidelberg (2009)
8. Dimitrova, V., Lau, L., O'Rourke, R.: Semantic Social Scaffolding for Capturing and Sharing Dissertation Experience. IEEE-TLT 4, 74–87 (2011)
9. Markkanen, B.: Schrey-Niemenmaa: Knowledge Practices Laboratory Overview (2008)
10. Karlsen, K.: Supporting reflection and creative thinking by carers of older people with dementia. In: Proceedings of PervasiveHealth 2011, pp. 526–529 (2011)
11. Westerhout, E., et al.: Enhancing the Learning Process: Qualitative Validation of an Informal Learning Support System Consisting of a Knowledge Discovery and a Social Learning Component. In: Wolpers, M., Kirschner, P.A., Scheffel, M., Lindstaedt, S., Dimitrova, V. (eds.) EC-TEL 2010. LNCS, vol. 6383, pp. 374–389. Springer, Heidelberg (2010)
12. Rizzo, G., Troncy, R.: NERD: Evaluating Named Entity Recognition Tools in the Web of Data. In: WEKEX 2011 @ ISWC 2011, Bonn, Germany (2011)
13. Popov, I.O., Schraefel, M., Hall, W., Shadbolt, N.: Connecting the Dots: A Multi-pivot Approach to Data Exploration. In: Aroyo, L., Welty, C., Alani, H., Taylor, J., Bernstein, A., Kagal, L., Noy, N., Blomqvist, E. (eds.) ISWC 2011, Part I. LNCS, vol. 7031, pp. 553–568. Springer, Heidelberg (2011)
14. Thakker, D., Dimitrova, V., Lau, L., Karanasios, S., Yang-Turner, F.: A Priori Ontology Modularisation in Ill-defined Domains. In: I-Semantics 2011, pp. 167–170 (2011)

15. Sunstein, C., Thaler, R.: Nudge: Improving Decisions about Health, Wealth, and Happiness. Penguin Books, New York (2009)
16. Cheng, G., Tran, T., Qu, Y.: RELIN: Relatedness and Informativeness-Based Centrality for Entity Summarization. In: Aroyo, L., Welty, C., Alani, H., Taylor, J., Bernstein, A., Kagal, L., Noy, N., Blomqvist, E. (eds.) ISWC 2011, Part I. LNCS, vol. 7031, pp. 114–129. Springer, Heidelberg (2011)
17. Zhang, X., Cheng, G., Ge, W., Qsssu, Y.: Summarizing Vocabularies in the Global Semantic Web. Journal of Computer Science and Technology 24(1), 165–174 (2009)
18. Ammari, A., Lau, L., Dimitrova, V.: Deriving Group Profiles from Social Media to Facilitate the Design of Simulated Environments for Learning. In: LAK 2012, Vancouver (2012)
19. Despotakis, D., Lau, L., Dimitrova, V.: A Semantic Approach to Extract Individual Viewpoints from User Comments on an Activity. In: AUM@UMAP 2011 (2011)

Part III
Short Paper

Analysing the Relationship between ICT Experience and Attitude toward E-Learning

Comparing the Teacher and Student Perspectives in Turkey

Dursun Akaslan[1] and Effie Lai-Chong Law[2]

Department of Computer Science, University of Leicester, Leicester, UK
info@dursunakaslan.com, elaw@mcs.le.ac.uk

Abstract. This paper analyses the relationship between the experience of using different ICT and attitudes towards e-learning. We have conducted two surveys with teachers and students from the academic institutions associated with the subject of electricity in Turkey. Both surveys have been built on our conceptual models of readiness for e-learning. 280 and 483 valid responses from teachers and students have been collected, respectively. Overall, the findings indicate that the more experiences the teachers and students have of using different ICT the more positive their attitudes towards e-learning and that e-learning should be integrated into campus-based education and training.

Keywords: E-learning, Experience, Higher education, Electricity, Survey.

1 Introduction and Background

Several barriers hinder the integration of e-learning into higher education institutions (HEIs). To address this concern in the context of HEIs associated with the subject of electricity in Turkey, we have conducted two surveys with representative samples of teachers and students from these HEIs in 2010 and 2011, respectively. Specifically, we employed questionnaires and semi-structured interviews to measure the two target groups' readiness for e-learning and to investigate how to implement e-learning in these HEIs. Results of the surveys have been published ([1] [2] [3] [4]). However, no systematic comparison between the responses of the two groups has been made. It is intriguing to know if there is any gap between the target groups, which may hamper the implementation of e-learning. With the questionnaires two major aspects were investigated: First, both students and teachers were asked with several close-ended items to evaluate their usage of different ICT and their attitudes towards e-learning. Second, were asked with an open-ended item to elaborate their past or current experiences with e-learning, if any, and attitudes towards e-learning. In summary, the main goal of this paper is to analyse teachers' and students' self-reported evaluations to find out whether they are different significantly from each other in different aspects pertaining to e-learning.

A. Ravenscroft et al. (Eds.): EC-TEL 2012, LNCS 7563, pp. 365–370, 2012.
© Springer-Verlag Berlin Heidelberg 2012

2   Theoretical Background

Grounded in the meticulous reviews of the relevant literature, we have developed a conceptual model on teachers' (without the component "traditional skills") and students' readiness for e-learning in HEIs (Fig. 1). The notion of 'readiness for e-learning' can be defined as the ability of an individual or organization to benefit from e-learning [5].

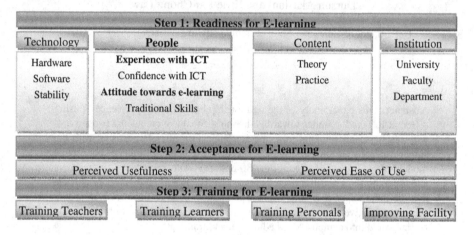

Fig. 1. Model on readiness for e-learning

As shown in Fig. 1, there are many factors affecting the ability of teachers or students to take the advantage of e-learning in their own working or studying context. As we aim to find out whether individuals tend to embrace or ostracize e-learning when they have more or less experiences of using ICT, in this paper we focus on two attributes of the factor *People*, namely *Experience with ICT* and *Attitude towards e-learning*.

Attitudes towards E-learning. Attitudes of individuals towards e-learning are emphasised as an important aspect of predicting and improving e-learning usage [6]. Hence, it is deemed relevant to find out potential stakeholders' attitudes towards e-learning *before* implementing it. Different researchers (e.g. [7]) measure people's attitudes with different approaches with Technology Acceptance Model (TAM) [8] being a common one. TAM is used to measure two constructs: *perceived usefulness* and *ease of use*, which denote the degree to which people believe using a system would be useful and free of effort, respectively. We also adopted TAM to measure teachers' and students' beliefs whether e-learning would be free of effort and useful for their respective tasks. However, rather than using the TAM, based on the literature (for details see [1] [3]), we have identified five sub-factors to measure attitudes for e-learning: *Knowledge, ICT competencies, Time, Feeling of readiness*, and *Thinking about others*.

Experience with ICT. Earlier research studies indicate that the usage of a system is significantly affected by previous experiences of other systems (e.g. [9]). Based on the related literature (for details see [1] [3]), we identified six sub-factors to study individuals' experiences of deploying various ICT for e-learning: *the Internet, E-mail, Office software, Engineering software, Instant messaging* and *Social network sites.*

3   Methodology

All the close-ended items in the two questionnaires were evaluated with a five-point Likert-scale with the leftmost and rightmost anchors being "Strongly Disagree" and "Strongly Agree" respectively. An option "Not applicable / Do not know" was also included into the questionnaire, given the relatively short history of e-learning in Turkey. The five-point Likert-scale were also coded in a way where 1 indicates the lowest readiness while 5 the highest one. Based on [10], we categorized the mean scores into three groups: (i) 3.4 and above as the expected level of sufficient experience with ICT and positive attitudes; (ii) 2.6 and below as the expected level of insufficient experience with ICT and negative attitudes towards e-learning; (iii) between 2.6 and 3.4 (exclusive) as the expected level of medium experience and neutral attitudes towards e-learning. There were altogether 11 items in the questionnaires addressing the two factors (Table 1). 417 and 456 departments of universities in Turkey were involved in the surveys. After filtering invalid response, 280 and 483 completed responses from the teachers and students respectively were included in the analyses.

Table 1. Items on experiences with ICT and attitudes towards e-learning

Factor 1: Experiences with ICT	
I1	I use the Internet as information source.
I2	I use e-mail as the main communication tool (/with my peers)
I3	I use the office software for (content delivery and demonstration/my coursework)
I4	I use social network sites.
I5	I use instant messaging software.
I6	I use electrical/engineering software, e.g. AutoCAD.
Factor 2: Attitudes towards E-learning	
I7	I have enough information about what e-learning is.
I8	I have enough ICT competencies to prepare (e-learning materials./ my coursework in electronic format.)
I9	I feel that I am ready (to integrate e-learning in my teaching. /for e-learning.)
I10	I have enough time to prepare (e-learning materials./my coursework in electronic format.)
I11	I believe my students will like e-learning.
Part 2: Open-ended Item	
I12	Can you elaborate your personal experiences of and attitudes towards e-learning?

Note: The words in brackets are alternative formulations of the item, separated by '/', for teachers and students, respectively.

4   Results

Results of the analysis are reported as two major parts: Eleven close-ended items and the open-ended item on the two attributes.

Analysis of Close-Ended Items. Mean scores of individual items of the two attributes for both teachers and students are shown in Table 2. It also indicates the results of independent-sample t-test to verify statistical significance of differences between the teachers and students. A Pearson product-moment correlation coefficient was also used to examine the relationship between the participants' scores on their experiences with ICT and attitudes towards e-learning. Significant positive correlations were found in the case of student sample ($r_{(398)} = 0.340$, $p < 0.000$) and of the pooled sample of participants ($r_{(641)} = 0.237$, $p < 0.000$), who had sufficient experiences in using various ICT (i.e. their mean scores across the items I1-I6 were above 3.40).

Table 2. Statistics for the items on the two attributes

Number and Mean of the Items											
Attribute 1: Experiences with ICT						Attribute 2: Attitudes towards E-learning					
I	Teac.	Stud.	Pool.	t	P	I	Teac.	Stud.	Pool.	t	p
I1	4.65	4.25	4.40	-7.332	0.000	I7	3.70	3.29	3.44	-5.366	0.000
I2	4.53	3.80	4.07	-9.429	0.000	I8	3.72	3.65	3.68	-0.825	0.409
I3	4.46	4.04	4.19	-6.080	0.000	I9	3.70	3.74	3.73	0.538	0.591
I4	**2.95**	3.90	3.55	9.845	0.000	I10	**2.81**	3.54	3.27	9.445	0.000
I5	3.40	3.70	3.59	3.124	0.002	I11	3.64	3.60	3.61	-0.629	0.530
I6	4.40	3.90	4.08	-6.443	0.000	-	-	-	-	-	-
M_0^*	**4.07**	**3.93**	**3.98**	**-2.659**	**0.008**	M_0^*	**3.51**	**3.56**	**3.55**	0.870	0.370

Note: **I**: Items; **Teac**: Teacher (N=280); **Stud: Student** (N=483); **Pool: Pooled** (N=763).
*overall mean scores across all the items related to the respective factor.

Analysis of the Open-Ended Item. 147 of the participating teachers and students responded to the item I12 (Table 1). We analysed the responses based on two major parts of I12. Due to the space limit, here we highlight some salient points.

Experience of e-learning. The responses were roughly categorized into five groups: The first group consisted of twenty teacher participants, who delivered some modules partially or entirely through e-learning. The second group comprised eleven teacher participants, who had experiences of developing e-learning materials. The third group was composed of twenty student participants, who had experience of downloading e-learning materials (e.g. video, lecture note, presentation) over the Internet to enhance their knowledge about the respective modules in their department or to develop their personal skills. The fourth group consisted of fifteen student participants who studied some modules through e-learning. The fifth group comprised two teachers who gained e-learning experiences through working in European projects.

Views on e-learning. The responses were categorized into three groups (Note: 'S' and 'T' denote student and teacher; numbers are identifiers): (a) *The negative comments:* e-learning would lead to the increase of unemployment rate (S626); e-learning was boring (S269) and distracting as it contains lots of information (S269, S60); e-learning is neither better nor secure than using an e-mail (S219); electricity was a subject based on practice and hence e-learning was not suitable because students could not conduct real experiments (T312, T352, S435, and S437); e-learning was weaker than the classroom environment (SP646, S280). (b) *The neutral comments*: Half of the participants stated that they did not have enough information about the notion of e-learning and hence avoided taking a stance on the value of e-learning. Some participants were ambivalent towards e-learning because it could be both positive and negative. For instance, e-learning could help students learn quickly, but it could be distracting (T357). Other concerns were difficulty in motivating students to learn on their own pace (T121) and inadequate infrastructural support for delivering e-learning material such as simulations of experiments (T114). (c) *The positive comments*: The potential benefits of e-learning include providing flexibility (S455, S32), widening access (S260, S601), developing information skills (S526), saving time (S598, S574, S644), more detailed information (S360), and bringing innovation to the institution.

5   Discussion

The above analyses identify several factors that potentially influence the effectiveness of e-learning. First, the risk that the access to the web-based resources such as social network sites may distract students from online learning can be aggravated by their low level of traditional learning skills such as note-taking and time management skills [3]. Second, our findings also manifest the importance of accessibility in delivering e-learning. It seems that the biggest challenge is to design e-learning materials available to as many students as possible with various connectivity speeds at anytime and anywhere. This is highly important because we examined teachers' and students' access to the Internet at their home and university as well as the stability of such an access [1,3]. The data indicated that while the majority of them had access to the Internet at home and university, they were not satisfied with the speed. This highlights that we need to think about the connectivity speed when we are preparing e-learning materials. Another key technical factor may also affect access to e-learning is different operating system, especially when we consider the mobile context of using smartphone and tablets. Third, remote labs are deployed to help students increase their practical skills, which can further be enhanced with practices in real-life lab settings. Hence, e-learning should be integrated with the campus-based training to substantiate learning efficacy as well as to meet social needs. Besides, students appreciate the flexibility of learning packages such as AutoCAD, MATLAB and programming languages by viewing videos with their own computers outside the lab.

6   Concluding Remark

Our analysis results indicate that both teachers and students tend to embrace e-learning for their teaching and learning when they have more experience in using different ICT and that maximising potential benefits of e-learning entails integration of campus-based education and training into web-based learning environments. The next major step of our research plan is to identify strategies, based on empirical data, how we should develop e-learning materials and train individuals to implement e-learning successfully. Accordingly, we propose a three-stage e-learning delivery model (Fig. 2). This approach can enhance students' motivation and pose challenges for them. We plan to elaborate this model and validate it with future research studies.

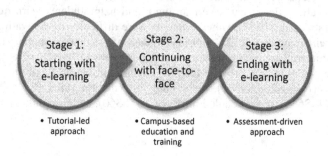

Fig. 2. Three-stage blended e-learning delivery

References

1. Akaslan, D., Law, E.L.-C.: Measuring Teachers' Readiness for E-learning in Higher Eduation Institutions associated with the Subject of Electricity in Turkey. In: Proceedings of the 2011 IEEE Global Engineering Education Conference, Amman (2011)
2. Akaslan, D., Law, E.L.-C., Taskin, S.: Analysis Issues for Applying E-learning to the Subject of Electricity in Higher Education in Turkey. In: Proceedings of the 17th International Conference on Engineering Education, Belfast (2011)
3. Akaslan, D., Law, E.L.-C.: Measuring Student E-Learning Readiness: A Case about the Subject of Electricity in Higher Education Institutions in Turkey. In: Leung, H., Popescu, E., Cao, Y., Lau, R.W.H., Nejdl, W. (eds.) ICWL 2011. LNCS, vol. 7048, pp. 209–218. Springer, Heidelberg (2011)
4. Akaslan, D., Law, E.L.-C., Taskin, S.: Analysis of Issues for Implementing E-learning: The Student Perspective. In: Proceedings of the 2012 IEEE Global Engineering Education Conference, Marrakesh (2012)
5. Lopes, C.T.: Evaluating E-learning Readiness in a Health Sciences Higher Education Institution. In: Proceedings of IADIS International Conference e-Learning, Porto (2007)
6. Liaw, S.-S., Huang, H.-M., Chen, G.-D.: Surveying instructor and learner attitudes toward e-learning. Computers & Education 49, 1066–1080 (2007)
7. Rosenberg, M.J.: E-learning, strategies for delivering knowledge in the digital age. McGraw-Hill, New York (2001)
8. Davis, F.D., Bagozzi, R.P., Warshaw, P.R.: User acceptance of computer technology: A comparison of two theoretical models. Management Science 35, 982–1003 (1989)
9. Park, N., Roman, R., Lee, S., Carver, J.E.: User acceptance of a digital library system in developing countries. International Journal of Information Mangement 29, 196–209 (2009)
10. Adyin, C.H., Tasci, D.: Measuring Readiness for E-learning. Educational Technology & Society 8, 244–257 (2005)

Integration of External Tools in VLEs with the GLUE! Architecture: A Case Study[*,**]

Carlos Alario-Hoyos, Miguel Luis Bote-Lorenzo, Eduardo Gómez-Sánchez,
Juan Ignacio Asensio-Pérez, Guillermo Vega-Gorgojo, and Adolfo Ruiz-Calleja

School of Telecommunication Engineering, University of Valladolid,
Paseo de Belén 15, 47011 Valladolid, Spain
{calahoy@gsic,migbot@tel,edugom@tel,juaase@tel,guiveg@tel,
adolfo@gsic}.uva.es

Abstract. This paper presents a case study of the usage of GLUE!, a loosely-coupled architecture that enables the integration of external tools in VLEs. The case study is a collaborative learning situation carried out through a VLE, but involving several external tools. GLUE! is used to instantiate and enact this situation in two authentic experiments. Evaluation results show that GLUE! alleviated educators in the instantiation process, and facilitated the effective collaboration among students. These results are relevant as this is a real situation with a complex structure and groups that change over time.

Keywords: integration, collaboration, GLUE!, external tools, VLEs.

1 Introduction

Virtual Learning Environments (VLEs) (also referred to as Learning Management Systems) [2], such as Moodle, LAMS or Blackboard, are representative examples of broadly adopted learning systems that support the design, instantiation and enactment of collaborative learning situations. VLEs typically provide a shared workspace in which educators can design and instantiate individual and collaborative activities. These activities can later be enacted by students that aim at reaching the learning objectives. As interacting with peers fosters collaborative learning [3], most VLEs include features that allow the creation of group structures, the arrangement of students in groups, and the assignment of different resources and tools to each group in each activity.

Main VLEs generally include a limited number of built-in tools (e.g. forums, chats), that can support both individual and collaborative activities. For example, Moodle includes 14 built-in tools (2.2 version). Nevertheless, research studies like [4] found that many educators agree on the fact that VLE built-in tools are

---

[*] This work has been partially funded by the Spanish Ministry of Economy and Competitiveness (TIN2008-03023, TIN2011-28308-C03-02 and IPT-430000-2010-054) and the Autonomous Government of Castilla y Leon (VA293A11-2 and VA301B11-2). Authors also thank David A. Velasco-Villanueva and Javier Aragón for their development work.

[**] This paper is complemented by [1], also published in the EC-TEL 2012.

A. Ravenscroft et al. (Eds.): EC-TEL 2012, LNCS 7563, pp. 371–376, 2012.

not enough for the support of their learning activities, specially those activities involving specific purpose tasks (e.g. simulations, drawings).

The integration of external tools in VLEs [5,6] is a research trend that aims at offering educators a larger set of alternatives when designing and instantiating learning activities within VLEs. Researchers working in this line are particularly encouraged by the recent spread of web technologies and the growth of Web 2.0, which brought an explosion of software tools employed by practitioners (see http://c4lpt.co.uk/top-100-tools-for-learning-2011), in theory, outside of VLEs. However, the integration problem cannot be easily tackled, mainly due to the wide variety of integration contracts [7] imposed by VLEs and tools. A deep analysis of the main generic approaches tackling the integration problem can be found in [7].

In this context, authors proposed the GLUE! (Group Learning Uniform Environment) architecture [8], which intends to facilitate the instantiation and enactment of collaborative activities that require the integration of external tools in VLEs. This paper assesses whether GLUE! actually facilitates the instantiation and enactment of collaborative activities by means of a case study. Significantly, software systems in general, and CSCL systems in particular, are commonly evaluated using case studies that involve end-users [9]. Here, the case study is a collaborative learning situation, put into practice thanks to GLUE! in two different experiments with real educators and students at university level.

2   Overview of GLUE!

GLUE! is a middleware architecture that enables the lightweight integration of multiple existing external tools in multiple existing VLEs (many-to-many integration) [8]. Figure 1 presents the structure of GLUE!, which is composed by three kinds of loosely-coupled distributed services where heterogenous VLE and tool contracts are adapted through an intermediate software layer, namely *GLUE! core*, and a set of adapters (*VLE adapters* and *tool adapters*). This structure of distributed elements fosters that every new integrated tool could be used within every available VLE and vice versa; this is a remarkable feature in many-to-many integration approaches [8].

The GLUE! core defines two integration contracts (one for the VLE-side and one for the tool-side) that homogenize existing VLE and tool contracts, enabling a common way of communication between VLEs and tools through the GLUE! core. The GLUE! contracts are characterized using popular and loosely-coupled

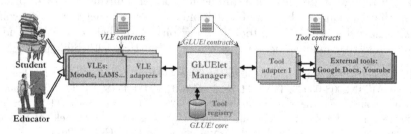

Fig. 1. Overview of the GLUE! architecture

web technologies; particularly, REST interfaces [10] are defined for the communication between the GLUE! core and the adapters. The selection of these popular technologies facilitates the integration of many external tools, due to the growing trend of web applications, in main VLEs, typically offered as web-based platforms. Further details on these contracts can be found in [8].

GLUE! supports practitioners in managing the *life cycle of external tool instances*[8], which includes the creation, configuration, retrieval, update, and deletion of external tool instances (*educators* can request all these actions, while *students* can only retrieve instances). The GLUE! core partially assumes this life cycle (in the GLUElet Manager), receiving requests from VLE adapters, and routing them to tool adapters, which translate them to the specific tool contract. The GLUE! core also persists information about the external tools (in the internal tool registry). VLE adapters embed instances in the VLE graphical interface, mapping them to VLE activities, users and groups. This is relevant in the processes of instantiating and enacting collaborative activities, since users belonging to the same group in the same activity need to share the same instances.

Reference implementations for the GLUE! core and some adapters have been developed (see http://gsic.uva.es/glue). These adapters currently enable the integration of at least 17 external tools in Moodle, LAMS and MediaWiki.

3   A Collaborative Learning Situation

Advanced Networking (AN) is a third-year course in the Telecommunication Engineering curriculum at the University of Valladolid (UVa). One of its core practical exercises targets the development of a message server. Prior to that, AN educators want their students to reflect and discuss about how the message server should be designed. To facilitate this discussion, AN educators designed a blended collaborative learning situation with different group settings that involves the realization of five collaborative activities.

In the first three activities students work in pairs and must: a) draw the sequence diagram and the flowchart of the message server; b) justify the decisions made on these drawings; c) review the drawings of some other pairs. In the last two activities, students are arranged in bigger groups (supergroups), and must: d) agree on the final sequence diagram and flowchart; e) make a presentation with the challenges and final results. A shared whiteboard, a collaborative text editor and a collaborative presentation tool may serve to carry out these duties. Each group setting (pairs, supergroups) must separately work using its specific tool instances in each activity. This is a collaborative situation that helps students achieve a gradual agreement on the proposed learning objectives.

Moodle is the institutional VLE at the UVa, being thus the preferred VLE to carry out this situation (although others could be chosen instead). However, Moodle built-in tools do not include shared whiteboards, nor presentation tools, while collaborative text editors (e.g. wiki tool) are quite limited; there are neither official Moodle plugins to make presentations, nor drawings. Trying to integrate these tools with tight coupling approaches like IMS LTI [6] could undertake a significant development effort. Among loosely-coupled approaches, [5] does not currently integrate presentation tools, and those tools intended for collaborative writing or drawing are very simple. Finally, Basic LTI, the subset of IMS LTI for lightweight integrations, does not allow the creation of separate tool instances for the different groups, hindering the instantiation of this situation. Thus, AN educators used GLUE! to integrate the kinds of tools expected, with similar conditions that Moodle built-in tools as far as collaborative settings concerns.

4   Evaluation

Software systems that promote collaboration are frequently evaluated through authentic experiments involving real users [9]. In this context, the aforementioned situation was successfully instantiated and enacted with GLUE! by real educators and students in the first semester of the 2010-2011 and 2011-2012 school years (from now on AN-2010 and AN-2011). Both experiments lasted one week, starting and ending with a two hours face-to-face session. Apart from Moodle, three external tools were employed: Dabbleboard (shared whiteboard), Google Documents (collaborative text editor) and Google Presentations (collaborative presentation tool). Evaluation data from these experiments were gathered from optional questionnaires to educators and students that included open text questions and Likert scales[1]. Besides, interviews with educators and focus groups with students were intended to confirm or discard those trends detected in the questionnaires, as suggested in the mixed method proposed in [11].

4.1   Instantiation of the Situation

Two AN educators participated in the instantiation of the situation presented in section 3 in both AN-2010 and AN-2011. Data collected from the questionnaires filled out by these educators indicated that they all agreed or completely agreed that "*GLUE! facilitated the instantiation of the collaborative learning situation*". Interestingly, they also agreed or completely agreed that "*this collaborative learning situation was significant in the context of the AN course*" and that "*the centralization of the activities in a single interface facilitated that students could achieve the learning goals*" established for this situation.

These opinions were contrasted with quantitative measurements of the instantiation time. These measurements were obtained from two experiments that replicated the instantiation process in AN-2010 and AN-2011 with and without GLUE!. The latter was feasible here because the three tools were web tools, although it required the manual creation and configuration of instances within the tool graphical interface, and also the manual copying and pasting of the representation of these instances (URLs) in individual Moodle resources. Interestingly, 62 instances were created in AN-2010 (31 for Dabbleboard, 24 for Google Documents, and 7 for Google Presentations) and 72 in AN-2011 (36, 28, and 8). The time demanded for the creation, configuration and assignment of instances with GLUE! was about 6.5 and 7.5 minutes in AN-2010 and AN-2011.

---

[1] The raw answers to these questionnaires can be consulted at http://gsic.uva. es/glue

These values contrasted with the instantiation time without GLUE! which was about 37.5 and 42.5 minutes respectively.

Thus, GLUE! facilitated the instantiation of this collaborative learning situation, which otherwise could not have been done or would have been much more demanding. Apart from the reduction of time, GLUE! also reduces the instantiation complexity, since external tool instances are managed within VLEs.

4.2 Enactment of the Situation

This situation was enacted by 47 and 51 students in AN-2010 and AN-2011, although only 38 answered the optional questionnaires. Aggregated results from those questions showing evidences of GLUE! facilitating the collaboration are shown in Table 1. Significantly, 84.2% of students collaborated much or very much with their partners during the experiments. Besides, 68.4% of students thought that the technological support (Moodle, the three external tools, and GLUE!) facilitated much or very much the performance of the activities in collaboration. Some comments supporting this finding are: *"Having the required tools integrated in one platform made our work easier"*; *"I think it [the performance of the activities in collaboration] was facilitated because of the ease with which we could share and visualize our partners' work"*. The latter argument can also be applied to triangulate the answers regarding the 68.4% of students that considered easy or very easy to see their partners' contributions. Students also highlighted this idea in the open text questions: *"It was very easy to see my partners' contributions"*; *"Everything was integrated in the same platform, and you could access both your work and your partners' work in a similar way."*

Negative answers came from a pair of AN-2010 students who did not see the built-in Moodle option to see the instances of their supergroup partners in the review activity: *"We did not see our partners' work because we did not know there was an option to change the group displayed"*; educators took this comment into account warning students in AN-2011, who did not find this problem.

It can be thus concluded that the kind of integration promoted by GLUE! facilitated AN students the realization of the proposed activities in collaboration, and allowed them to see their partners' work when required.

Table 1. Aggregated answers from the students that enacted the AN experiments

Options	Question 1: "How much did you collaborate with your partners?"	Question 2: "How much did the technological support facilitate the performance of the activities in collaboration with your partners?"	Options	Question 3: "How easy / difficult was to see your partners' contributions along the activities?"
Very much	2/38 (5.3%)	7/38 (18.4%)	Very easy	11/38 (28.9%)
Much	**30/38 (78.9%)**	19/38 (50%)	**Easy**	**15/38 (39.5%)**
Some	5/38 (13.2%)	8/38 (21.1%)	A bit easy	9/38 (23.7%)
A Little	0/38 (0%)	3/38 (7.9%)	A bit difficult	2/38 (5.3%)
Little	0/38 (0%)	0/38 (0%)	Difficult	0/38 (0%)
Very little	0/38 (0%)	1/38 (2.6%)	Very difficult	1/38 (2.6%)
No answer	1/38 (2.6%)	0/38 (0%)	No answer	0/38 (0%)

5 Conclusions

This paper has presented a preliminary evaluation of the GLUE! architecture employing a Moodle-based authentic collaborative learning situation, which was instantiated and enacted by real practitioners in two consecutive years (2010 and 2011). Evaluation results from both experiments showed that GLUE! facilitated educators the instantiation of the collaborative activities, greatly reducing the instantiation time and complexity. Besides, these results also showed that GLUE! enabled the enactment of these activities, also facilitating the collaboration among students, and allowing them to achieve the learning goal of this collaborative learning situation. Significantly, this and other similar situations could have been carried out within other VLEs and employing other tools [8], since GLUE! promotes a many-to-many integration. For instance, a demonstration of the instantiation of this situation in LAMS can be seen in [1].

References

1. Alario-Hoyos, C., et al.: Demonstration of the Integration of External Tools in VLEs with the GLUE! Architecture. In: Ravenscroft, A., Lindstaedt, S., Delgado Kloos, C., Hernández-Leo, D. (eds.) EC-TEL 2012. LNCS, vol. 7563, pp. 465–470. Springer, Heidelberg (2012)
2. Dillenbourg, P., et al.: Virtual Learning Environments. In: Proceedings of the 3rd Hellenic Conference "Information & Communication Technologies in Education", Rhodes, Greece, pp. 3–18 (2002)
3. Dillenbourg, P. (ed.): Collaborative Learning: cognitive and computational approaches. Elsevier Science, Oxford (1999)
4. Bower, M., Wittmann, M.: A Comparison of LAMS and Moodle as Learning Design Technologies - Teacher Education Students' Perspective. Teaching English with Technology, Special Issue on LAMS and Learning Design 11(1), 62–80 (2011)
5. Wilson, S., et al.: Distributing education services to personal and institutional systems using Widgets. In: Proceedings of the 1st International Workshop on Mashup Personal Learning Environments (MUPPLE 2008), Maastricht, The Netherlands, pp. 25–32 (2008)
6. IMS GLC. IMS GLC Learning Tools Interoperability Implementation Guide (2012), http://imsglobal.org/lti (last visited: June 2012)
7. Alario-Hoyos, C., Wilson, S.: Comparison of the main alternatives to the integration of external tools in different platforms. In: Proceedings of the International Conference of Education, Research and Innovation (ICERI 2010), Madrid, Spain, pp. 3466–3476 (2010)
8. Alario-Hoyos, C., et al.: GLUE!: An Architecture for the Integration of External Tools in Virtual Learning Environments. Computers & Education (submitted, 2012)
9. Dewan, P.: An integrated approach to designing and evaluating collaborative applications and infrastructures. Computer Supported Cooperative Work 10(1), 75–111 (2001)
10. Richardson, L., Ruby, S.: RESTful Web Services. O'Reilly Media, Inc., Sebastopol (2007)
11. Martínez, A., et al.: Combining qualitative evaluation and social network analysis for the study of classroom social interactions. Computers & Education 41(4), 353–368 (2003)

Mood Tracking in Virtual Meetings

Angela Fessl[1], Verónica Rivera-Pelayo[2], Viktoria Pammer[1], and Simone Braun[2]

[1] Know-Center, Graz, Austria
[2] FZI Research Center of Information Technologies, Karlsruhe, Germany

Abstract. Awareness of own and of other people's mood are prerequisite to a reflective learning process which allows people to consciously change their perception of, attitude towards and behaviour in future work situations. In this paper we investigate the usage and usefulness of the MoodMap App – an application for tracking own mood and creating awareness about the mood of team members in virtual meetings. Our study shows that especially the possibility to compare own mood to the mood of others' is perceived as useful and therefore enhances interpersonal communication in virtual team settings. Whilst users express an interest in tracking own mood, they need to relate their mood to the current context, and wish to receive feedback or other helpful input from the app in order to achieve propitious reflective learning.

Keywords: mood tracking, reflective learning, workplace, empirical study.

1 Introduction

Learning from own experiences by reflection (re-visiting and re-evaluating past experiences) is a powerful way of learning, and especially relevant in work-integrated learning where more often than not there is no teacher, mentor or coach available. Emotions can help or hinder people when thinking back and critically reviewing past experiences. Additionally, it is necessary to consider which emotions were present during an experience and why, in order to be able to re-evaluate the past experience (see e.g.[2]). Within this work we investigate mood tracking in virtual meetings. We base our work on the assumptions that mood tracking i) raises awareness about emotions, ii) provides the possibility to consciously consider emotions during past experiences at time of reflection, and iii) improves interpersonal communication in virtual team meetings since it provides for additional non-verbal communication clues. In this paper we report on a user study involving a team of 12 people throughout 4 virtual team meetings. The goal of the user study was to find out whether people are interested in mood tracking in their normal work surroundings, to evaluate the usability of the Moodmap App (the App used in the study for mood tracking), and to investigate which benefits study participants perceive from mood tracking in virtual meetings.

A. Ravenscroft et al. (Eds.): EC-TEL 2012, LNCS 7563, pp. 377–382, 2012.

2   Related Work

A lot of research has been conducted at defining, modeling and measuring emotions in the area of psychology, leading to numerous different theoretical models of assessing and representing emotional states [1,11,9]. The word 'emotion' is often applied to a wide variety of phenomena, such as moods, sentiments, temperament, and passions, which in fact refer to different affective states. In our work we will focus our understanding of emotion and mood on the work of Scherer [11] and Fridja [4], who define emotions as affective reactions to an event, typically short-lived and directed at a specific object. More recent work has been done about the relationship between emotional components and mood as well as the relation between behaviour and emotions in [1]. The authors present an emotional theory consisting of feedback and retrospective appraisal of past situations, which leads to a model on how emotions shape behaviours including reflection and learning through reflection. Latest work includes technology for tracking and representing emotions through user-initiated approaches. Studies in health or psychological settings use emotion capturing to enhance self-awareness by illuminating data trends and self-regulation through some tool interventions offered by the applications [3,8]. Furthermore, there are several tools for mood tracking in the context of The Quantified Self[1], a community of users using tools to collect personally relevant information for self-knowledge about one's behaviours, habits and thoughts.

All these approaches address mood awareness and self-regulation for mood improvement mainly in Human Computer Interaction (HCI) context and in the health sector. Nonetheless, some work has been conducted in work-related settings [6,7,10], but none of them considers mood awareness to promote reflective learning and improve future experiences.

3   The MoodMap App

In extension to existing research, we set our investigation with the MoodMap App into the context of work, and specifically into virtual team meetings. We combine individual mood-tracking, which is expected to raise emotional self-awareness, with mood sharing, which is expected to improve interpersonal communication through raising awareness of the others' mood, and provide an additional feedback loop to the individual by enabling a comparison of mood. Through this combination, users may create new perspectives about shared experiences.

The MoodMap App[2] enables users i) to note and review their own moods over time to support self-awareness and self-reflection and ii) to get an insight into moods of their team members to support non-verbal communication and offer other perspectives that may trigger reflective processes.

---

[1] www.quantifiedself.com

[2] The MoodMap App as described in this paper is available online at http://moodmap. apps.mirror-demo.eu

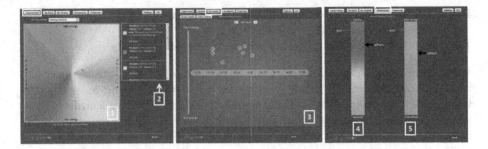

Fig. 1. Users track their own mood in the two-dimensional colour coded mood map (1) and review it in the moodlist (2) or along a timeline (3). Every user can compare their own latest valence (4) and arousal (5) with the average values of the team.

Mood Representation. The mood representation within the MoodMap App is based on Russell's Circumplex Model of Affect [9], which distinguishes between valence (negative to positive feelings) and arousal (low to high energy). Both values are coded in the range 0 to 1. The mood map has as background a gradient of colours, with colours associated to the main moods, based on Itten's colour system [5]. A theoretical derivation of the mood representation in the MoodMap App is given in [7].

Capturing Mood. In the MoodMap App, this two-dimensional colour coded representation is used directly for mood capturing and is carried on throughout all representations of recorded mood as well. During the meeting, the user can express her mood by clicking at the appropriate area of the mood map (Fig. 1, point 1). Additionally, it appears as entry within the moodlist (Fig. 1, point 2), where the user can contextualize the current mood with the help of notes, which will facilitate the subsequent reflective learning process.

Individual Views. Apart from the moodlist, users can review their own mood by replaying the sequence of mood entries (My Mood, not depicted) or along a timeline (My Timeline, see Fig. 1 point 3).

Collaborative Views. Every user can compare her own mood with the mood of other team members (Compare Me, see Fig. 1, point 4) according to the valence and arousal values. In the Collaborate view (not depicted), a large red cross shows again the average of the latest mood entries of all users placed on the mood map. By clicking this cross, the distribution of the anonymous individual entries (red dots) is shown.

4 Study Design

We designed a study about usage of the MoodMap App in virtual team meetings to answer the following three research questions:

RQ1: *Are participants interested in tracking their own mood and that of others?*

RQ2: *Do participants appreciate the usability and features of the MoodMap App as it is implemented?*

RQ3: *What benefits do participants perceive from the MoodMap App?*

With these research questions we want to be able to differentiate the user acceptance and attitude of mood tracking in a real work setting as well as to investigate if mood tracking and sharing may i) lead to reflective learning about work experiences by raising emotional self-awareness and ii) improve their team communication by raising awareness of others' mood.

The user study was carried out within a spatially distributed team in a European telecommunication company during their weekly team meetings. The MoodMap App was used in four meetings and users were asked to enter their mood at the beginning and at the end of the meeting. They could also do it during the meeting and add notes to the moods. After each meeting the participants were invited to reflect about their mood development during the meeting. The participating team consisted of 12 people.

Short Questionnaire. It consisted of 6 questions about the atmosphere of each meeting, the participant's role in the meeting, the current mood and what insights they gained during the meeting. It was answered after each meeting.

Final Questionnaire. The final questionnaire consisted of 23 questions (with 5-Likert scale or free text) concerning general interest in mood tracking and aspects like usefulness, usage or perceived benefit regarding the MoodMap App.

Interviews. Six interviews were conducted online via Microsoft Live Meeting and lasted about 30 minutes. We based the interviews in an adaptation of the main points addressed in the final questionnaire.

5   Results

As a first step, we analyzed all data captured by the participants using the MoodMap App to compare the four meetings according to valence and arousal values. In *Meeting 1* the average values changed rapidly, which may reveal excitement and nervousness among participants. In *Meeting 2*, the participants seemed to get used to the MoodMap App, as the values followed a more rational tendency. In *Meeting 3*, valence and arousal values showed several changes, which may be referred to the low number of meeting participants. In *Meeting 4*, the average values stayed slightly stable. This could be either explained that the participants are getting used to the application or they have not already seen any benefit or insights for themselves. Unfortunately we have no information about the topics discussed in the meetings.

RQ1: Are Participants Interested in Tracking Their Own Mood and That of Others? Regarding this first question, we obtained two opposite results. On the one hand, the majority of the participants agreed or were at least neutrally interested in capturing their mood, found the app interesting and useful e.g. a participant mentioned *'it made me aware of my mood and if there was* *a shift in my mood...*'. On the other hand, we received statements that they were not very interested in capturing their own mood, because they already know how they feel. Most of the participants were more interested in the mood captured by their colleagues and to compare themselves with others.

RQ2: Do Participants Appreciate the Usability and Features of the MoodMap App as it is Implemented? Most of our participants liked the look and feel of the MoodMap App, especially the mood representation in a bi-dimensional map of valence and arousal. Regarding the aspects of liking the MoodMap App or having fun to use it, even though the results were ambiguous, there was consensus that the MoodMap App was very easy to use, no user guide was necessary and the interface was kept simple. Regarding potential improvements, the participants stated that the application should give directly feedback. Unfortunately for one participant the bi-dimensional representation of the mood was very difficult to understand, because she could not match her mood to a corresponding color.

RQ3: What Benefits Do Participants Perceive from the MoodMap App? Regarding the perceived benefits, our evaluation shows mixed results and different opinions. Several participants already identified some benefits, became aware of their mood and saw some influences on it during the meeting. The possibility of comparing the mood of oneself with the team's mood was emphasized as the most obvious benefit and it led to initial reflections of the participants. Additionally, the awareness concerning the team's mood was raised, e.g. *'I think it helps to develop the emotional intelligence of the team'*.

On the other hand, there were also participants, who did not perceive any benefits. They mentioned some issues e.g. technical problems, missing features or the short evaluation time. Nevertheless they see great potential for the MoodMap App and made suggestions for improvements, like mood trend analysis over a longer period of time or direct feedback for the participants after the meeting on the individual as well as collaborative level. Such features would help them to start reflective learning processes.

6   Discussion and Outlook

The results of our study show quite diverging opinions. People's reactions to mood tracking and sharing in virtual meetings range from being very interested to not being able to deal with such an application. Currently the benefits of tracing and sharing mood are not directly recognizable for the users, which therefore needs further discussion and investigation.

Analysing the study results from the perspective of reflective learning, we conclude that most of the participants agree that the collaborative views have major potential to trigger reflection. Unfortunately, the team did not reflect together on the past meetings neither alone nor in a subsequent meeting, which could have led to more benefits or insights. This could be taken into account in further research. Reviewing the individual moods e.g. on the timeline are not seen as very useful to re-experience the meeting. Comparing one's own mood with the average mood of others in e.g. the timeline as well as comparing all collaborative moods of the four meetings on the other hand was seen as useful. The participants see further potential to reflect especially on critical topics, which might occur during a meeting when the mood changes significantly. Similarly, the participants see a high potential for additional context information. For instance, more information from outside and not only what is captured within the MoodMap application ('I think the best way to reflect is when you receive new input from outside and this is quite simple input...') or a better contextualisation like replaying the recorded meeting in comparison to the moods.

Acknowledgement. The project "MIRROR - Reflective learning at work" is funded under the FP7 of the European Commission (project number 257617). The Know-Center is funded within the Austrian COMET Program - Competence Centers for Excellent Technologies - under the auspices of the Austrian Federal Ministry of Transport, Innovation and Technology, the Austrian Federal Ministry of Economy, Family and Youth and by the State of Styria. COMET is managed by the Austrian Research Promotion Agency FFG.

References

1. Baumeister, R.F., Vohs, K.D., DeWall, C.N., Zhang, L.: How emotion shapes behavior: Feedback, anticipation, and reflection, rather than direct causation. Personality and Social Psychology Review 11, 167–203 (2007)
2. Boud, D., Keogh, R., Walker, D.: Reflection: turning experience into learning, 1st edn. Routledge Falmer (1985)
3. Church, K., Hoggan, E., Nuria, O.: A study of mobile mood awareness and communication through mobimood. In: Proceedings of the 6th Nordic Conference on Human-Computer Interaction: Extending Boundaries, pp. 128–137 (2010)
4. Frijda, N.: Emotions and episodes. Moods and sentiments, pp. 59–67. Oxford University Press (1994)
5. Itten, J.: Kunst der Farbe, 1st edn. Otto Maier Verlag, Ravensburg (1971)
6. Matic, A., Papliatseyeu, A., Gabrielli, S., Osmani, V., Mayora-Ibarra, O., Via Alla Cascata, D.: Happy or Moody? Why so? Ubihealth (2010)
7. Mora, S., Rivera-Pelayo, V., Müller, L.: Supporting mood awareness in collaborative settings. In: Georgakopoulos, D., Joshi, J.B.D. (eds.) CollaborateCom., pp. 268–277. IEEE (2011)
8. Morris, M.E., Kathawala, Q., Leen, T.K., Gorenstein, E.E., Guilak, F., Labhard, M., Deleeuw, W.: Mobile therapy: case study evaluations of a cell phone application for emotional self-awareness. Journal of Medical Internet Research 12, 10 (2010)
9. Russell, J.A.: A circumplex model of affect. Journal of Personality and Social Psychology 39, 1161–1178 (1980)
10. Saari, T., Kallinen, K., Salminen, M., Ravaja, N., Yanev, K.: A mobile system and application for facilitating emotional awareness in knowledge work teams. In: Proceedings of the 41st Annual Hawaii International Conference on System Sciences, HICSS 2008, p. 44 (2008)
11. Scherer, K.R.: What are emotions? and how can they be measured? Social Science Information 44(4), 695–729 (2005)

Teachers and Students in Charge

Using Annotated Model Solutions in a Functional Programming Tutor

Alex Gerdes[1], Bastiaan Heeren[1], and Johan Jeuring[1,2]

[1] School of Computer Science, Open Universiteit Nederland
P.O. Box 2960, 6401 DL Heerlen, The Netherlands
agerdes@me.com, {bhr,jje}@ou.nl
[2] Department of Information and Computing Sciences, Universiteit Utrecht

Abstract. We are developing ASK-ELLE, a programming tutor that supports students practising functional programming exercises in Haskell. ASK-ELLE supports the stepwise construction of a program, can give hints and worked-out solutions at any time, and can check whether or not a student is developing a program similar to one of the model solutions for a problem. An important goal of ASK-ELLE is to allow as much flexibility as possible for both teachers and students. A teacher can specify her own exercises by giving a set of model solutions for a problem. Based on these model solutions our tutor generates feedback. A teacher can adapt feedback by annotating model solutions. A student may use her own names for functions and variables, and may use different, but equivalent, language constructs. This paper shows how we track intermediate student steps in ASK-ELLE, and how we avoid the state space explosion we get when analysing intermediate, incomplete, student answers.

Keywords: Functional programming, tutoring, Haskell.

1 Introduction

Learning to program is challenging. The results of a first course in programming are often disappointing [11]. Learning by doing through developing programs, and learning through feedback [7] on these programs are essential aspects of learning programming. To support learning programming, many intelligent programming tutors have been developed. Intelligent programming tutors support the development of programs, and can give immediate feedback to the student. There exist programming tutors for Lisp, Prolog, Java, Haskell, and many more programming languages. Some of these tutors are well-developed tutors extensively tested in classrooms, others haven't outgrown the research prototype phase yet, and quite a few have been abandoned. Evaluation studies have indicated that working with an intelligent tutor supporting the construction of programs may have positive effects. For example, using the LISP tutor is more effective when learning how to program than doing the same exercise "on your own" using only a compiler [3].

A. Ravenscroft et al. (Eds.): EC-TEL 2012, LNCS 7563, pp. 383–388, 2012.

The following aspects are relevant for programming tutors:

- *Development process*: does the tutor support the incremental development of programs, where a student can obtain feedback or hints on incomplete programs, can a student follow her preferred way to solve a programming problem, can a student submit a complete solution to a problem in the tutor?
- *Correctness*: does the tutor guarantee that a student solution is correct, can it check that a student has followed good programming practices, can it verify that a solution has the desired efficiency, does it give an explanation why a program is incorrect, does it give counterexamples for incorrect programs, and does it detect at which point of a program a property is violated?
- *Adaptivity*: can a teacher add her own exercises to a tutor, and can she adapt the behaviour so that particular solutions are enforced or disallowed?

No existing programming tutor addresses all of the above aspects. In particular, it is usually quite hard for teachers to add programming exercises to, and adapt the feedback given by, tutors that support the incremental development of programs. Anderson et al. [2] mention the lack of adaptability as one of the main reasons for the slow uptake of their tutors outside their own teaching environment. Teacher adaptability is of fundamental importance for the uptake of learning environments.

Another important aspect of a programming tutor is that it offers sufficient freedom to students: a student should be able to use her own names, to use her own favourite programming style, her own refinement step-size, etc. Similar to the Lisp tutor [3], the refinement rules in our tutor model Haskell at the finest grain size that has functional meaning in Haskell, but we want to offer students the possibility to make larger steps than these small steps.

This paper shows how we address the above aspects, and investigates how we can develop a programming tutor:

- in which a student incrementally develops a program that is equivalent (modulo syntactic variability) to one of the teacher-specified model solutions for a programming problem,
- that gives feedback and hints on intermediate, incomplete, and possibly buggy programs, based on teacher-specified annotations in model solutions,
- to which teachers can easily add their own programming exercises, and in which teachers can adapt feedback,
- and in which a student can use her preferred step-size in developing a program: from making a minor modification to submitting a complete program in a single step.

In particular, we address some of the technical challenges that need to be solved to develop such a tutor.

This paper is accompanied by a paper demonstrating our programming tutor ASK-ELLE [10]. In the demonstration paper, which is best read before this paper, we show a hypothetical session of a student with the tutor, how a teacher adds an exercise to the tutor, and how a teacher adapts the feedback of the tutor. A more extensive description of the ideas discussed in this paper can be found in an accompanying technical report [5].

This paper is organised as follows. Section 2 discusses how we can combine teacher-annotated model solutions to both give hints to students as well as diagnose partial student programs. Section 3 shows how we recognise student steps where step size doesn't matter. Section 4 discusses related work and concludes.

2   The Teacher in Charge

Our tutor takes a set of teacher annotated model solutions for a programming exercise as input. Using these solutions, it constructs a programming strategy [8,4], which it uses to follow a student when incrementally solving the programming exercise. The strategy is interpreted as a recogniser that recognises program refinement steps of students. This section discusses how we construct a recogniser from several possible model solutions, such that teacher annotations in model solutions are used when giving feedback, hints, or worked-out solutions.

2.1   Strategy Recogniser

We interpret a strategy as a context-free grammar. The language generated by a strategy can be used to determine whether or not a sequence of rules applied by a student follows a strategy. The sequence of rules should be a sentence in the language, or a prefix of a sentence, since we solve exercises incrementally. A recogniser for a context-free grammar recognises refinement steps that are applied to some initial term, usually the empty program. The recogniser maintains the current location within the strategy at which the student has applied a refinement rule, to give precise feedback. Using the information about the progress of a student, we can calculate which steps are allowed next, and check whether or not a student deviates from a path towards a model solution.

The recogniser maintains the active labels, which contain the texts that are used when a student asks for a hint. The interpretation of a strategy with a label introduces the special rules ENTER and EXIT, parameterised by the label. These rules are only used for tracing positions in strategies, and delivering feedback texts when necessary. A label is active when we have recognised the ENTER rule of that particular label, but not yet its corresponding EXIT rule.

Most of the feedback is derived from the grammar functions *empty* and *firsts*. The *empty* function determines whether or not the language described by a strategy contains the empty sentence. The *firsts* function determines the set of rules with which a sentence in the language of a strategy can start. These grammar functions are used to give feedback to students about whether or not an exercise is solved, possible next steps, or complete solutions.

2.2   Parallel Top-Down Recogniser

The recogniser recognises prefixes, and hence also accepts intermediate (incomplete) solutions. It cannot use backtracking, since this would imply that it accepts steps that do not lead to a solution, and hence guides a student into the wrong direction. It follows that the recogniser needs to choose between the various model solutions on the basis of a single refinement step. This is problematic when multiple model solutions share a first step, i.e., when we encounter a left-factor in the strategies generated for the model solutions. Note that combining model solutions almost always leads to left-factors. The introduction of a declaration, and a function name is very often shared between the different model solutions. The standard method to deal with this problem is to apply left-factoring, a grammar transformation that removes left-factors. However, the presence of labels makes it hard to use left-factoring, since moving or merging labels leads to scrambling annotations of model solutions, making it very hard if not impossible to give the intended hints. We need to defer committing to a particular path in the strategy.

To deal with left-factors, we fork the recogniser whenever we run into a left-factor. If any of these recognisers fails to recognise the student solution, we discard it. Thus we obtain a top-down variant of a parallel recogniser. Using a top-down parallel recogniser we allow a teacher to specify model solutions that have common components.

3   The Student in Charge

We have performed several experiments with ASK-ELLE and asked students to evaluate the programming tutor [6]. Students were generally positive about using the tutor; their main comment was that the tutor is of no help when performing many refinement steps in a single step. At the time of the experiments, the tutor could only recognise a limited number of steps towards a solution when a student submitted a (possibly partial) program. The enhancements described in this section lift this restriction.

3.1   Pruning

We allow students to interleave most of the refinement steps from an incomplete program to a complete solution. Different students develop programs in different ways, and we support this. However, if a student takes multiple steps before checking with the tutor, the number of different sequences of refinements is enormous. This makes recognising multiple steps hard.

We constrain the search space of intermediate answers to determine whether or not a student submission follows a strategy. First, we observe that the first steps of the different strategies for model solutions may be the same, but they diverge after a number of steps. Since we use refinement rules, a student can no longer refine her program towards model solutions that do not include the performed refinement. This reduces the number of interleavings significantly. We filter out these intermediate answers by determining whether or not the normalised abstract syntax trees of the model solution and the student submission overlap, where a hole in the program overlaps with any tree. We use depth-first search to find matching solutions, since it is more likely that a student first finishes a particular part of the program, such as a case alternative, than doing refinements at arbitrary places.

3.2 A Search Mode for the Interleave Combinator

Although pruning is a step forward, it is not good enough. Even with pruning, the search space remains too large, due to the amount of possible interleavings. To reduce the number of interleavings, we observe that when recognising multiple steps, the *order* of refinements of holes that may be interleaved is irrelevant. We use the irrelevance of refinement order when recognising multiple steps by introducing a *search mode* for the interleave combinator used in our strategy language. The semantics of the original interleave combinator chooses between the left-interleave of both sub-strategies: either start with the first step of the left sub-strategy, or with the first step of the right sub-strategy. The search mode for interleave chooses between left-interleaving the left sub-strategy with the right sub-strategy, or taking the (non-interleaved) right sub-strategy. This implies that *if* the left sub-strategy has been applied in the refinement, it is performed before the rest of the strategy. This is safe because the order of refinement steps does not matter. Using the search mode for interleave, all sequences of refinement steps leading to the same intermediate program are replaced by a single sequence, drastically reducing the search space.

Our approach can be applied in the functional programming domain because we use refinement rules. If we would also use rewrite rules, we would need to prove that the rewriting system is Church-Rosser before we can use the alternative semantics of interleave.

4 Conclusions and Related Work

We have discussed two important issues for Ask-Elle, a programming tutor for Haskell. In the accompanying demonstration paper [10], we have shown how teachers can add programming exercises to our programming tutor by means of annotated model solutions. In this paper we show that we cannot use backtracking or problem compilation [3] to track students in our framework since do not want to give hints that do not lead to a solution. Instead we introduce parallel top-down recognition. Second, we have shown how we recognise almost arbitrary many student steps on the way to a solution. A student may take refinement steps in any order, but when recognising student steps we fix the order to reduce the search space.

The concepts we have introduced to deal with these issues are not specific for Haskell. We can use the approach described in this paper to develop similar programming tutors for other functional programming languages, such as Lisp or OCaml. We believe that we have not made assumptions that exclude imperative programming languages, but we would have to further investigate this. We have not yet performed experiments with teachers using our system, except for ourselves using the system. We want to perform experiments to test the new functionality of our tutor.

Our tutor resembles the Lisp tutor [1] in that it supports the stepwise development of programs, and gives hints at intermediate steps. By generating strategies from model solutions we think it is easier to add programming exercises to our tutor. Moreover, teachers can easily fine-tune the generated feedback. J-Latte [9] verifies complete student Java programs against constraints. In the future we want to add the possibility to check constraints on an incomplete student program to our tutor.

References

1. Anderson, J.R., Conrad, F.G., Corbett, A.T.: Skill acquisition and the LISP tutor. Cognitive Science 13, 467–505 (1986)
2. Anderson, J.R., Corbett, A.T., Koedinger, K.R., Pelletier, R.: Cognitive tutors: Lessons learned. Journal of the Learning Sciences 4(2), 167–207 (1995)
3. Corbett, A.T., Anderson, J.R., Patterson, E.J.: Problem compilation and tutoring flexibility in the LISP tutor. In: Proceedings of ITS 1988: 4th International Conference on Intelligent Tutoring Systems, pp. 423–429 (1988)
4. Gerdes, A., Heeren, B., Jeuring, J.: Constructing Strategies for Programming. In: Cordeiro, J., Shishkov, B., Verbraeck, A., Helfert, M. (eds.) Proceedings of the First International Conference on Computer Supported Education, pp. 65–72. INSTICC Press (March 2009)
5. Gerdes, A., Heeren, B., Jeuring, J.: Teachers and students in charge — using annotated model solutions in a functional programming tutor. Technical Report UU-CS-2012-007, Utrecht University, Department of Computer Science (2012)
6. Gerdes, A., Jeuring, J., Heeren, B.: An interactive functional programming tutor. In: Proceedings of ITICSE 2012: the 17th Annual Conference on Innovation and Technology in Computer Science Education (to appear, 2012); Also available as technical report Utrecht University, UU-CS-2012-002
7. Hattie, J., Timperley, H.: The power of feedback. Review of Educational Research 77(1), 81–112 (2007)
8. Heeren, B., Jeuring, J., Gerdes, A.: Specifying rewrite strategies for interactive exercises. Mathematics in Computer Science 3(3), 349–370 (2010)
9. Holland, J., Mitrovic, T., Martin, B.: J-Latte: a constraint-based tutor for Java. In: Proceedings of ICCE 2009: the 17th International on Conference Computers in Education, pp. 142–146 (2009)
10. Jeuring, J., Gerdes, A., Heeren, B.: ASK- ELLE: A Haskell Tutor Demonstration. In: Ravenscroft, A., Lindstaedt, S., Delgado Kloos, C., Hernández-Leo, D. (eds.) EC-TEL 2012. LNCS, vol. 7563, pp. 453–458. Springer, Heidelberg (2012)
11. McCracken, M., Almstrum, V., Diaz, D., Guzdial, M., Hagan, D., Kolikant, Y.B., Laxer, C., Thomas, L., Utting, I., Wilusz, T.: A multi-national, multi-institutional study of assessment of programming skills of first-year CS students. In: Working Group Reports from ITiCSE on Innovation and Technology in Computer Science Education, ITiCSE-WGR 2001, pp. 125–180. ACM (2001)

The Effect of Predicting Expertise in Open Learner Modeling

Martin Hochmeister[1], Johannes Daxböck[1], and Judy Kay[2]

[1] Electronic Commerce Group, Vienna University of Technology
Favoritenstraße 9-11/188-4, 1040 Vienna, Austria
martin.hochmeister@ec.tuwien.ac.at,
johannes.daxboeck@student.tuwien.ac.at
[2] School of Information Technologies, University of Sydney
1 Cleveland Street, NSW, Australia
judy.kay@sydney.edu.au

Abstract. Learner's self-awareness of the breadth and depth of their expertise is crucial for self-regulated learning. Further, of learners report self-knowledge assessments to teaching systems, this can be used to adapt teaching to them. These reasons make it valuable to enable learners to quickly and easily create such models and to improve them. Following the trend to open these models to learners, we present an interface for interactive open learner modeling using expertise predictions so that these assist learners in reflecting on their self-knowledge while building their models. We report study results showing that predictions (1) increase the size of learner models significantly, (2) lead to a larger spread in self-assessments and (3) influence learners' motivation positively.

Keywords: Prediction, Expertise, Open Learner Model, Self-assessment, Metacognition, Adaptive Educational Systems.

1 Introduction

Self-regulated learning is the ability to understand and control one's learning environments. Metacognition is an important part of self-regulated learning because it enables learners to scrutinize their current expertise levels and to plan and allocate scarce learning resources [15]. Being aware of one's own expertise is referred to as self-knowledge [8], which includes knowledge of one's strengths and weaknesses. In recent years, learner models have been increasingly opened to learners allowing them to scrutinize and update information stored in adaptive educational systems [6,14,5]. One of the potential benefits of this approach is to gain more accurate and extensive learner models allowing these systems to provide more effective personalization. Furthermore, the active involvement of learners in building and maintaining their models may contribute to learning [11,7].

To use open learner models to elicit learner's expertise, we need to find ways to support learners in estimating their expertise. In this paper, we hypothesize that expertise predictions have the potential to serve an important role in

A. Ravenscroft et al. (Eds.): EC-TEL 2012, LNCS 7563, pp. 389–394, 2012.
© Springer-Verlag Berlin Heidelberg 2012

guiding learners in self-assessing their knowledge to quickly create rich learner models. While learner self-assessment may not necessarily be accurate, there is considerable evidence that bias may be systematic [13] and so it can be valuable.

We understand expertise predictions as representations of topics paired with score values ranging from 0 to 100 points such as *programming:75*. Predictions are calculated based on learners' self-assessments as they are reported to the system. While learners perform self-assessment, expertise predictions are promptly recalculated and displayed to the learners. Given these expertise predictions, we address the following questions: (1) *Will expertise predictions affect the size of learner models?* and (2) *Will expertise predictions motivate learners to focus on their strengths and weaknesses equally?* To answer these questions we propose a user interface featuring predictions and conduct an experimental study for evaluation.

Even though self-knowledge constitutes an important aspect of metacognitive behavior, it is important to emphasize that the validity of self-knowledge seems to be most crucial for learning per se. However, to determine the accuracy of learners' self-assessments goes beyond the scope of this paper.

2   Related Work

This work aims to *elicit a rich user model* as a basis for subsequent personalization in a learning environment. It builds on the growing body of work on *Open Learner Models* (OLMs). Open learner modeling research has shown that OLMs can play several roles, including improving the accuracy of the model, navigation within an information space and supporting metacognitive processes such as setting goals, planning, self-monitoring, self-reflection and self-assessment [5]. Our work builds on the last of these, so that we can quickly create a learner model. At the same time, the process of self-assessment support self-reflect, a valuable way to improve learning [2]. There are many forms of interfaces to open learner models [5] and the ways they can be part of an application or play a use independent role [10,4]. especially for supporting reflection [4]. Our work continues this trend, as we explore the creation of an interface to support self-assessment.

For large learner models, there are interface challenges for OLMs. The VlUM interface tackled this with a novel exploration interface tool [1] that could be incorporated into a system [11]. It showed an overview of the learner model. Each concept was color coded, green indicating a concept was known and red that it was not known. The colour intensity indicated the knowledge score, with learner control on setting the threshold for these colours. A later version, called SIV had ontological inference [12] so that data about fine-grained model concepts could be used to infer values for more general ones, and vice-versa. In this paper, our work explores a different approach to creation of the interface to a learner model because we aim to support learner self-assessment rather than reflection.

3   User Interface

Figure 1 depicts the proposed interface leveraging expertise predictions. In the upper part, learners select topics from a hierarchically structured domain ontology (giving 454 concepts), estimate their expertise scores and add the expertise to their model shown in the table below. To support learners' self-assessment, we provide expertise predictions by employing a score propagation algorithm [9]. Learner models are stored as ontology overlays [3]. The algorithm exploits models' structures to propagate expertise scores amongst ontology topics. The algorithm's scores are integrated with the learner model as shown in the bottom right part of Figure 1. The top left shows the selection of the topic (1). Learners can either enter a topic in the top text box (1a) or select one of the hierarchy of topics, such as Programming (1b). A selected topic then appears on the top right, where the learners assign their self-assessments (2) and Add/Update their scores to the model illustrated at the bottom. The prediction engine dynamically calculates scores based on the scores shown in column *self* and updates the model table. The learner can customize the model's display (3) by filtering the model according to a specific string and by setting a score threshold (ranging from 10 to 100 points in steps of 5) to restrict the display of predicted scores below the threshold value. Learners can now scrutinize (4) their model by inspecting its structure and scores. They can alter their self-assessments by clicking on a topic

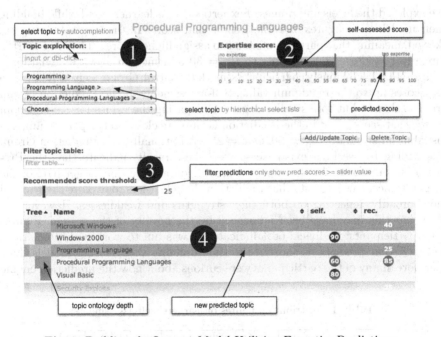

Fig. 1. Building the Learner Model Utilizing Expertise Predictions

in the model, which loads the topic in the top view as it is the case in Figure 1 for the topic *Procedural Programming Languages*.

4   Evaluation and Results

We conducted a user study with master students in a computer science program. Participants were randomly separated into two groups: The Control Group (using the proposed interface but without the prediction feature) and the Prediction Group (working with the same interface but with predictions). We put both interface variants online and notified the participants to start building their learner models from the scratch within two-weeks time. The prediction group was required to self-assess five topics in advance, then the prediction feature started to operate and adapted its results to the growing set of self-assessments. After both groups constructed their models, we asked them for feedback. Importantly, participants completed the given task as a one-off with no consequences (either benefits nor negative effects) for poor self-assessments. With 21 students in the Control Group and 29 students in the Prediction Group completing the task, we observed significant larger learner models in the Prediction Group (avg 38 topics vs. 20 topics).

4.1   Levels and Range of Self-assessments

We explored the levels and ranges of expertise scores learners used while building their models. The figures in Table 1 show that scores in the Control Group are skewed, meaning that participants tend to assign higher scores. The interquartile range ($iqr = Q_3 - Q_1$) amounts to $iqr = 30$ and median average deviation is $mad = 14.83$. Obviously, participants in the Control Group were reluctant to use scores up to the maximum value. Self-assessments in the Prediction Group are also skewed but to a smaller degree. Comparing the values for iqr and mad we see that scores used in the Prediction Group are closer to the perfect uniform distribution standard ($iqr = 50, mad = 25$). Additionally, the Prediction Group was willing to use high expertise scores. These results indicate that the Prediction Group focused their expertise scoring on a somewhat larger part of the model. Hence, this suggests that predictions help learners to explore their model more broadly, reflecting on both their strengths and weaknesses. However, we note that this may have been influenced by the novelty of the system. Figures from participants' feedback: 66% indicated it was fun to work with predictions and 62% that predictions shorten the time building their learner models. Furthermore, many of the participants were curious about how the prediction engine

Table 1. Distribution figures of learners' self-assessments

	n	min	Q_1	median	mean	sdev	Q_3	max	mad
Control group	411	5	40	60	52.94	21.31	70	90	14.83
Prediction group	1115	5	40	60	58.09	24.29	80	100	29.65

works. Together, this suggests that predictions may have led to a higher level of motivation to use the system. This could be very important for maintaining the model over a longer period.

4.2 Feedback

We asked the participants of the Prediction Group to complete an online questionnaire after building their models. For closed questions, 62% of participants liked the predicted scores although 38% rated them mostly useless. 83% found the slider element to be useful to limit the display of predicted scores. 62% believe that a prediction feature shortens the time to build a learner model. And finally, 66% said that it was fun to work with predictions.

From the open questions about likes, dislikes and improvements, it seems that participants found it challenging to decide what it means to be an expert. Selected quotes: *"When is someone an expert and when not?"*, *"I got a very good in Artificial Intelligence. But am I an expert in this topic?"*, *"Someone else might say that he has used Java for 10 years but he still feels that there are better people than him, so he gives himself 80%."*, *"Further I don't the reference point of the scores. (e.g. 'all people', students of informatics, ...?)"*. Even though we declared the expert level as having problem-solving capability in the respective topic, participants experienced difficulties. This is part of a broader challenge in defining what an expert level means.

Another finding concerns self-reflection. Selected quotes: *"It was interesting to think about questions i did not have in mind before (what is my expertise)."*, *"It helps to find mistakes and makes me rethink my self-assessment."*, *"Was interesting to see how the software thinks my expertise is."*. These statements suggest that predictions can trigger mechanisms to think about one's expertise in more detail as well as scrutinize one-selves believes.

Lastly, participants expressed the wish after a more transparent prediction process: *"I dislike the present interface because I don't understand how the predicted score is calculated."*, *"The system should reason (comment) its predictions."*, *"It would be nice to be able to get a short explanation from the system on how the score was derived."*, *"Scores were irritating, because I don't know how they are determined"*.

5 Conclusions and Future Work

We examined the effects of expertise predictions in supporting learners during self-assessment. Our study results indicate that predictions can have a positive influence on learners' motivation. This appears to be one reason that models for the Prediction Group were almost double the size of those for the Control Group. Furthermore, predictions appear to help learners to broaden their focus to include both their strengths and weaknesses. This may indicate that expertise predictions facilitate learners' reflection on their self-knowledge. The majority of participants appreciated the system's expertise predictions and also think that they shorten the time effort in building their models. Although we have not tested the validity of participants' self-assessments, our study represents a critical precursor before incorporating this class of interfaces into broader contexts, e.g., long term learner modeling. Moreover, tendencies to bias in self-assessments are likely to be consistent [13] and over the long term, changes in these self-assessment could be valuable for learners' reflection on their progress.

In future work, we will explore enhancing our predictions with a collaborative filtering approach, based on learners' similar expertise. This will introduce new topic areas to learners since they are not based on topics learners explicitly stated. We will explore if this helps learners explore more new areas over familiar ones.

References

1. Apted, T., Kay, J., Lum, A., Uther, J.: Visualisation of ontological inferences for user control of personal web agents. In: IV 2003, pp. 306–311. IEEE (2003)
2. Boud, D.: Reflection: Turning experience into learning. Routledge (1985)
3. Brusilovsky, P., Millán, E.: User Models for Adaptive Hypermedia and Adaptive Educational Systems. In: Brusilovsky, P., Kobsa, A., Nejdl, W. (eds.) Adaptive Web 2007. LNCS, vol. 4321, pp. 3–53. Springer, Heidelberg (2007)
4. Bull, S., Gardner, P.: Highlighting learning across a degree with an independent open learner model. In: Artificial Intelligence in Education, pp. 275–282 (2009)
5. Bull, S., Kay, J.: Student Models that Invite the Learner In: The SMILI:() Open Learner Modelling Framework. IJAIED 17(2), 89–120 (2007)
6. Bull, S.: Supporting learning with open learner models. In: Proceedings of the 4th Hellenic Conference in Information and Communication Technologies in Education, Athens, Greece, pp. 47–61 (2004)
7. Bull, S., Kay, J.: Open Learner Models. Springer (to appear, 2012)
8. Flavell, J.: Metacognition and cognitive monitoring. American Psychologist 34(10), 906–911 (1979)
9. Hochmeister, M.: Spreading expertise scores in overlay learner models. In: Proceedings of CSEDU (to appear, 2012)
10. Kay, J.: Lifelong learner modeling for lifelong personalized pervasive learning. IEEE Transactions on Learning Technologies 1(4), 215–228 (2008)
11. Kay, J., Li, L., Fekete, A.: Learner reflection in student self-assessment. In: Proceedings of the Ninth Australasian Conference on Computing Education, vol. 66, pp. 89–95. Australian Computer Society, Inc. (2007)
12. Kay, J., Lum, A.: Exploiting readily available web data for reflective student models. In: Proceedings of AIED 2005. Artificial Intelligence in Education, pp. 338–345. IOS Press, Amsterdam (2005)
13. Kleitman, S.: Metacognition in the Rationality Debate: self-confidence and its Calibration. VDM Verlag (2008)
14. Mabbott, A., Bull, S.: Student Preferences for Editing, Persuading, and Negotiating the Open Learner Model. In: Ikeda, M., Ashley, K.D., Chan, T.-W. (eds.) ITS 2006. LNCS, vol. 4053, pp. 481–490. Springer, Heidelberg (2006)
15. Schraw, G., Crippen, K., Hartley, K.: Promoting self-regulation in science education. Research in Science Education 36(1), 111–139 (2006)

Technology-Embraced Informal-*in*-Formal-Learning

Isa Jahnke

Umeå University
Department of Applied Educational Science
Interactive Media and Learning (IML), Sweden
isa.jahnke@edusci.umu.se

Abstract. A characteristic of informal learning is that a person has an unsolved issue and starts searching for answers. To what extent can we transfer such a 'motivation to learn' into formal education? In 2002, an online, open, free forum at a university has been launched for around 2,000 students at a study program (CS). Users got the opportunity to co-construct new knowledge about issues what they want (e.g., course content, how to study successfully). Designed in that way, the online forum provides an informal learning space. Studying it from a sociological theory of social roles, one conclusion is that the iForum activates the conative level of learning. The term conation by K. Kolbe in 1990 refers to a concrete action; the learner does not only know, s/he really acts, s/he is willing to do sth. This rubric of learning is neglected in *designs for* formal schooling where cognitive learning 'textbook knowledge' is more focused.

Keywords: Conative learning, Online Forum, Role theory.

1 Informal Learning, Unplanned Learning

Informal learning usually takes place when a learner has unsolved issues outside of a formal instruction given by a teacher. Sometimes these informal unsolved issues are clear problems and conscious to an individual; sometimes they are less clear and less obvious. Imagine, a person who wants to know something and starts searching for an answer; planned informal learning. Such 'unsolved problems' are, for instance, improving a swim style by watching YouTube videos, checking information, observing keynote speakers, discussing newspaper articles. When facts are discussed offline at least one person takes her smartphone and 'googles' the information – unplanned informal learning takes place. Such forms of informal learning can lead to a deeper understanding and a different quality of a learning outcome; it enables the learner to expand her thinking beyond a receptive behavior at formal schooling and beyond a traditional reproduction of existing knowledge. A combination of both informal learning added to formal education might be a win-win situation for learners.

The research question is: To what extent, how, can formal schooling create structures and spaces to foster informal learning; for what purposes is informal-*in*-formal learning meaningful? To answer this question, an online forum for a computer science study program has been analyzed. The research aim is to deepen the knowledge about student's behavior in an online forum that represents Technology-Enhanced informal-*in*-formal-learning, to understand the students' motivations and expectations.

A. Ravenscroft et al. (Eds.): EC-TEL 2012, LNCS 7563, pp. 395–400, 2012.

2   Designing for Linking Informal and Formal Learning

Informal learning can be described by the concepts of "incidental learning" [14] and "experiential learning" by D. Kolb [11]. A person is doing Kolb's four learning steps by contrasting her experiences with the experiences of others [3]. Sometimes, these forms are unplanned learning situations that also occur in formal or non-formal learning situations. Formal, non-formal, informal learning differs in a) the degree of organization b) formal certificates, c) the criterion of 'who triggers learning' [1]:

- Formal learning is triggered by an instructor or teacher, organized by such a person or educational institution, the learner get credits or a formal degree;
- Non-formal learning is also a form of planned learning, and structured with regard to learning objectives, time, support; it is organized by an external person, but it usually takes place outside of educational institutions (e.g. community programs);
- Informal learning is a self-directed learning situation, or not-organized at all, triggered by the learner instead by an external teacher, no degree included.

A difference between the learning forms is the external organizer. Formal and non-formal learning is related to a teacher and tutor, who give instructions and rules; informal learning is related to an inspiring environment, reflections by the learner and supporting structures [15]. Supporting structures can be created through an online forum; learning through peer-reflection. Online forums, blogs using comments by readers, Facebook and LinkedIn groups are just few examples where informal learning can take place; detailed information is online in Jahnke [6].

Context of the Study and Description of Mixed Methods
In 2002, an online forum at a Computer Science faculty has been launched [9]. The free and open online forum has been offered to support students in doing their studies. The forum works as an informal learning approach in which learning is defined as the co-construction of knowledge among new and senior students, study advisors as well as faculty members [4]. The forum covers two fields, sub-forums for a) teaching like lectures, seminars, and b) planning and organizing the study from the students' perspectives. The forum is called iForum (InPUD). The data collection, analysis and redesign were conducted in iterative cycles of research and development from 2002 to 2009. The data gathering included mixed methods like open-ended interviews, standardized questionnaires, user statistics, content analysis and log-files.

The Advanced Role Theory as a Framework for Studying the iForum
Our theoretical underpinning is the expansion of the role theory in sociotechnical systems; read [5] where we describe the term "role" and its long tradition. To make it short here, the two paradigms "symbolic interaction" (Mead 1934) and the "functionalistic perspective" (Linton 1936), elaborated by many others to the mid of the 1990s, attempt to explain the relationship between the individual and society, between a person and the system. The functionalistic approach suggests the existence of objective structures made by the society that determine the individuals' behavior. In contrast, the symbolic interaction approach emphasizes that roles are formed more on the subjective will by actors. Both influence each other ([5], [10], [8]):

Social actors <-> Situated in co-constructed Roles <-> On-/Offline system/network.

Using the role approach is helpful to explore structures of group interactions within sociotechnical systems – the actors, the group and the system/network are parts of that theory. Our advanced role theory explains the social co-construction of online and offline *reality* by social actors situated in roles. Following our prior work [5], a role is socially constructed by

- a formal position within a system (online community, network, group etc.) created by someone, e.g., designers or managers
- a formal function as well as tasks related to that position
- explicit and implicit behavior expectations of different people towards the position which change over time
- dynamics of role-playing (e.g., same role but different role-playings by different actors)

Roles can be visible but can also follow a hidden agenda. A role is not a static phenomenon; it is rather a socially co-constructed formal-informal complex and (un-) conscious negotiation of actors embedded into broader social systems influencing each other, based on a historical body ("role-mechanisms" [5]).

Mixed Methods
For this paper, we focus on the analysis of log files and content analysis in addition to online questionnaires in 2002 and 2008/9. The quantitative survey included 24 standardized and open-ended questions, was four weeks online and 345 questionnaires returned (response rate of ca. 20 percent of all enrolled students of 2,000). The quantitative data have been analyzed with SPSS 7. Table 1 shows the elements of the role theory, operalizationed and connected to the forms of data collections [2].

Table 1. Advanced role theory & data collection

Role Theory elements	iForum (operationalized)	Data collection
position in the iForum	What kinds of members (self-perception)	Questionnaires, Log files
function/tasks	Degree of contribution (self-perception)	Questionnaires
behavior expectations	What members expect towards the others	Open ended questionnaires
role-playing (co-constr.)	What do the members really do?	Log files; content analysis

Description of the Open, Free, Online iForum
The iForum is a PHP technical system. Users need only an Internet access to read the iForum. To post, a registration with a free username is required. iForum supports a public communication based on the anonymity of its users. This is different to traditional LMS, which require registration given by the university administration desk and the real name of the users. In 2008, iForum had more than 30 sub-boards. The sub-boards exist for a) courses like lectures and seminars (e.g., to discuss exercises or content of lectures) and b) study organization, for example, users share knowledge about the CS degree, information about 'how to manage a study for a degree'. The decision about the topics mainly depends on what the students want to discuss. The iForum is characterized by a large size (1,500 users) and an extended lifespan, started in 2002, lives until today, providing a space for interactions, usually asynchronously.

3   Results

Positions and Functions in the iForum: Self-perception & the Other Reality
In December 2008, around 1,500 individuals had an iForum account. Usually, the core group has fewer members than readers and lurkers (e.g., [13]).

Table 2. What the users say *and* what they do

iForum	What the users *say* (questionnaire) n=345	What the users *do* (logfiles) n=1,478
Core group	8	18
Registered lurking (Newbies)	30	22
Regular/peripheral members	62	60
	100 percent	100 percent

According to the log files, a total of almost 1,500 members from 2,000 students were registered but 22 percent did not contribute actively. They are only registered. What the members say and what they really do is shown in table 2. The core group and registered lurkers differ in around 8 and 10 percent. To the question "Do you label yourself as a community-member?" more than 70% of the students agreed (n=188). This is a surprise since it means that not only the core members but also active and peripheral members rate themselves as part of iForum.

Learning Activities: Behavior Expectations Towards Contribution/Participation
More than 71% of the respondents use the iForum *"to ask subject-specific questions about courses"*. They do this often, once a week and more than once week. Around 66% use the forum for a) sharing information about lectures and tutorials b) solving exercises online collaboratively, and c) learning to handle different opinions [6].

Non-active contributors (registered lurkers): More than 300 registered members did not contribute but where registered (contribution=0). These registered iForum-lurkers are about 15% of all 2,000 students. According to Preece et al. ([13]), there are various reasons for why users do not post (e.g., no motivation, curiosity). To understand the reasons for non-contribution, we had an open question that 113 students answered (coded afterwards). The survey collected different motivations, why iForum-users have an account but do not actively contribute, read table 3.

Table 3. Motivations for reading and lurking

Motivations of non-active contributors in iForum (open question, coded)	% (n=113)
"Questions, I have, already there in the iForum"; "answers already available" (F1)	31.8
Communication problems/weakness: "difficulties with language", "shy", "I'm afraid of asking sth.", "I do not want to ask stupid/dumb questions" (F2)	16.8
Forum as information source: "an account has the advantages to get information what happens in which sub boards"; "automatic notification via email" (F3)	15.9
No motivation: "no interests", "I'm too lazy", "I have no time" (F4)	15.4
Questions can be clarified on other ways: "Face-to-face is better"; different contact points available; no need for information online, "I see no necessity" (F5)	12.4
"No special topics available where I can say something" (F6)	8.0

Contradictory Forms of Role-Playing?

We studied the role-playing of iForum-users. Most of the answers are "*I ask uncleared and unsolved questions*", and "*I need answers or solutions*". Some interviewees also mentioned they like to help other students: "*I help other students since I hope they will help me later, when I need help*", "*That's the sense of a community, we help each other*", and "*Only active members affect active, vivid forums*". Other interviewees did like the opportunity to get in contact with others at unusual time slots, "*direct contact possibilities at unusual time in the night*", and stress the anonymity: "*because of the anonymity, I can ask 'stupid' questions*". During the data analysis two new aspects for active contributions came up. These are (1) criticizing deficiencies and (2) gaining attention out of a huge group:

Ad 1) Students use the iForum to criticize shortcomings within the study program. The respondents said: "*I want to show my opinion*", "*I can show my anger by using anonymity*", „*I can scarify deficiencies*", and "*When I'm annoyed about something or somebody, I can say it in the forum*".

Ad 2) The users use the forum to get out from the huge group of learners. Students perceive those large groups as 'anonymous mass'. So, when writing something in the iForum, they want to show their individual faces and voices, and try to gain attention: "*I post because I have to say something*", and "*Sometimes, I even want to say something*". Some users stressed especially the factor of awareness: "*I think the professor will be better aware of me when I'm active in the iForum. So, I'm not just a pure number for him but become an individual*".

The interesting result is that anonymity has a contrary function. Because of the anonymity, some students use the forum to show their anger or to reveal aspects they do not agree with. On the other hand, some other members use the iForum to gain more attention and getting out of the anonymity of large groups by saying something and by creating a voice. By participating online, some members expect that other people would perceive their individual voices better than without the iForum. Additional data supports this. Almost 55 percent (n=133) agreed "*the iForum (digital life) has a positive impact on my offline life*".

4   Implications – Preparing for Informal-*in*-Formal Learning

Designs for the conative level of learning: The results about the iForum indicate a special feeling of a membership. This feeling is expressed in terms like "*That's the sense of a community, we help each other*" (interviewees). It seems that the group feeling within the forum activates a) the user's perception of having a specific form of social proximity triggered by technology and b) activates the conative level of learning. The term "conation" refers to a concrete action conducted by a learner; s/he does not only know but s/he really acts, s/he is willing to do sth. and really *does* [12], The concept of conation stresses what a learning outcome really is, a changed behavior of the learner. This level of learning is often neglected in formal schooling where the cognitive learning 'learning what' and 'textbook knowledge' is focused without supporting the learners to practice this in action or to reflect practices. Traditional teaching neglects the designs for learning as an active process including reflective action (students as pro-sumers) but also neglects to create *designs for* social relations among students as well as teacher to students.

Conclusion. This short paper illustrated a differentiated picture of informal learning added to a formal education; *read the long version online* [6]. Online boards can be a differentiator that supports the individual needs of users. Such a flexibility of a "just-in-time-communication" ([8]) is useful to engage students in learning and can also link weakly coupled learners. The addition of informal learning expands formal education and leads to an all-embracing learning experience that activate learners on all levels such as the cognitive, affective and on the *conative level* as well; this is what we call *designing for* technology-embraced informal-*in*-formal learning.

References

1. Ainsworth, H.L., Eaton, S.: Formal, Non-formal and Informal Learning in the Sciences. Onate Press, Calgary (2010)
2. Bryman, A.: Social research methods, 3rd edn. Oxford University Press, New York (2008)
3. Daudelin, M.: Learning from experience through reflection. Organizational Dynamics 24(3), 36–48 (1996)
4. Duffy, T.M., Cunningham, D.J.: Constructivism: Implications for the design and delivery of instruction. In: Handbook of Research for Educational Communications and Technology, p. 171 (1996)
5. Herrmann, T., Jahnke, I., Loser, K.-U.: The Role Concept as a Basis for Designing Community Systems. In: Darses, F., Dieng, R., Simone, C., Zackland, M. (eds.) Cooperative Systems Design, pp. 163–178. IOS Press, Amsterdam (2004)
6. Jahnke, I.: Technology-Embraced Informal-*In*-Formal Learning extended. Long version including additional research results (2012), http://isajahnke.webnode.com/publications
7. Jahnke, I., Bergström, P., Lindwall, K., Mårell-Olssen, E., Olsson, A., Paulsen, F., Vinnervik, P.: Understanding, Reflecting and Designing Learning Spaces of Tomorrow. In: Sánchez, A., Isaías, P. (eds.) Proceedings of IADIS Mobile Learning, Berlin, pp. 147–156 (2012)
8. Jahnke, I.: Dynamics of social roles in a knowledge management community. Computers in Human Behavior 26, 533–546 (2010), doi:10.1016/j.chb.2009.08.010
9. Jahnke, I.: A Way out of the Information Jungle – a Longitudinal Study About a Sociotechnical Community and Informal Learning in Higher Education. Journal of Sociotechnology and Knowledge Development (4), 18–38 (2010), doi:10.4018/jskd.2010100102
10. Jahnke, I., Ritterskamp, C., Herrmann, T.: Sociotechnical Roles for Sociotechnical Systems: a perspective from social and computer science. In: AAAI Fall Symposium: Roles, an Interdisciplinary Perspective, Arlington, Virgina, November 3-6 (2005)
11. Kolb, D.: Experiential Learning: Experience as the Source of Learning and Development. Prentice-Hall, Inc., Englewood Cliffs (1984)
12. Kolbe, K.: The conative connection. Addison-Wesley Publishing, Reading (1990)
13. Preece, J., Abras, C., Maloney-Krichmar, D.: Designing and evaluating online communities. International Journal of Web Based Communities 1, 2–18 (2004)
14. Reischmann, J.: Learning "en passant": The Forgotten Dimension. In: Proceedings of the Conference of Adult and Continuing Education (October 1986)
15. Watkins, K., Marsick, V.: Towards a Theory of Informal and Incidental Learning in Organisation. International Journal of Lifelong Education 11(4), S287–S300 (1992)

Towards Automatic Competence Assignment
of Learning Objects

Ricardo Kawase[1], Patrick Siehndel[1], Bernardo Pereira Nunes[2,1],
Marco Fisichella[1], and Wolfgang Nejdl[1]

[1] L3S Research Center, Leibniz University Hannover, Germany
{kawase,siehndel,nunes,fisichella,nejdl}@L3S.de
[2] Department of Informatics - PUC-Rio - Rio de Janeiro, RJ - Brazil
bnunes@inf.puc-rio.br

Abstract. Competence-annotations assist learners to retrieve and better understand the level of skills required to comprehend learning objects. However, the process of annotating learning objects with competence levels is a very time consuming task; ideally, this task should be performed by experts on the subjects of the educational resources. Due to this, most educational resources available online do not enclose competence information. In this paper, we present a method to tackle the problem of automatically assigning an educational resource with competence topics. To solve this problem, we exploit information extracted from external repositories available on the Web, which lead us to a domain independent approach. Results show that automatically assigned competences are coherent and may be applied to automatically enhance learning objects metadata.

Keywords: Metadata Generation, Competences, e-Learning, Automatic Competence Classification.

1 Introduction

Understandability of resources by learners is one essential feature in the learning process. To measure it, a common practice is the use of competence metadata. A competence is the effective performance in a domain at different levels of proficiency. Educational institutions apply competences to understand whether a person has a particular level of ability or skill. Thus, an educational resource, enriched with competence information, allows learners to identify, on a fine-grained level, which resources to study with the aim to reach a specific competence target.

With the catch up of the Open Archives Initiative, plenty of learning materials are freely available. Through the utilization of the OAI-PMH protocol[1], a learning environment can list the contents of several external repositories. Although this open content strategy provides numerous benefits for the community, new challenges arise to deal with the overload of information. For example, every time a new repository is added to a library, thousands of new documents may come at once. This makes the experts' task of evaluating and assigning competences to the learning objects impossible.

---

[1] http://www.openarchives.org/pmh

A. Ravenscroft et al. (Eds.): EC-TEL 2012, LNCS 7563, pp. 401–406, 2012.

Table 1. The compentence classification of the OpenScout repository and the respective examples of most relevant keywords

Competences	Relevant Keywords
Business and Law	law,legal,antitrust,regulation,contract,formation,litigation...
Decision Sciences	decision,risk,forecasting,operation,modeling,optimization...
General Management	planining,plan,milestone,task,priority,management,evaluation...
Finance	finance,financial,banking,funds,capital,cash,flow,value,equity,debt...
Project Management	management,monitoring,report,planning,organizing,securing...
Accounting and Controlling	accounting,controlling,balance,budgets,bookkeeping,budgeting...
Economics	economics,economy,microeconomics,exchange,interest,rate,inflation...
Marketing and Sales	marketing,advertising,advertisement,branding,b2b,communication...
Organizational Behavior and Leadership	organizational,behavior,leadership,negotiation,team,culture...
Management Information Systems	management,information,system,IT,data,computer,computation...
Human Resource Management	resources,management,career,competence,employee,training,relation...
Entrepreneurship	entrepreneurship,entrepreneurs,start-up,opportunity,business...
Technology and Operations Management	technology,operation,ebusiness,egovernment,ecommerce,outsourcing...
Strategy and Corporate Social Responsibility	strategy,responsibility,society,sustainability,innovation,ethics,regulation...
Others	-

In this paper, we present our work towards an automatic competence assignment tool, taking into account the speed of educational resources development, exchange, and the problem of ensuring that these materials are easily found and understandable. Our goal is to provide a mechanism that facilitates learners in finding relevant learning materials and to enable them to better judge the required skills to understand the given material through the interpretation of competence levels.

2 Competences

Our work is contextualized within the OpenScout learning environment[2]. The Open-Scout portal is the outcome of an EU co-funded project[3], which aims at providing skill-and-competence based search and retrieval Web services that enable users to easily access, use, and exchange open content for management education and training. Therefore, the project not only connects leading European Open Educational Resources (OER) repositories, but also integrates its search services into existing learning suites [2]. As the platform integrates different content repositories, many learning materials are daily added to the environment without the experts' annotations regarding competence levels. To tackle this problem we proposed a novel approach to automatically annotate the educational resources in OpenScout with competences.

Within the project, a management-related competence classification was developed (see Table 1), in order to support learners and teachers while searching for appropriated educational resources that meet a specific competence level. In a first major step, a

---

[2] http://learn.openscout.net
[3] http://openscout.net

focus group was organized consisting of a sample of ten domain experts from Higher Education, Business Schools, and SMEs, including two professors, six researchers and two professionals with the aim to generate an initial competence classification from experience and academic literature.

In addition to the competence classification, within the OpenScout project we created a list of keywords that are mostly relevant to each competence (see Table 1). These descriptions are essential for our automatic competence assignment tool, further explained in Section 3.

In order to build the competence descriptions, eight researchers from the ESCP Europe Business School[4] with different research focuses and knowledge about certain domains were asked to provide a list of terms that best fit their domains (competences). Participants had completed different diploma studies in Germany, the US, UK, Australia, or China and had an average of two years of work experience at the university; three of them had also been employed full-time in several industries before. All experts have emphasized that they provided a subjective assessment creating the keyword list related to each competence. Thus, due to their long years of experience and ongoing education in their respective field of knowledge, these experts fulfilled the necessary criteria for providing the most relevant keywords.

3   Automatically Assigning Competences

In order to solve the problem of automatically assigning competence annotations to learning objects, we developed an unsupervised method that can be applied to any repository of documents where the competences involved are known in advance. The method is a tag-based competence assigner. To better understand the proposed method, in the next subsection we briefly introduced the methodology involved to extract tags from learning objects, followed by the actuall competence assigning method.

3.1 α-TaggingLDA

Our proposed competence annotation method is an extension of the α-TaggingLDA. This method is a state-of-the-art LDA based approach for automatic tagging introduced by Diaz-Aviles et al. [1]. α-TaggingLDA is designed to overcome new item cold-start problems by exploiting content of resources, without relying on collaborative interactions. The details involving the technical aspects of the automatic tagger are out of the scope of this paper and we refer to [1] for more details. The important abstraction to be considered is that, for a given LO, α-TaggingLDA outputs a ranked list of most representative tags.

3.2   Tag-Based Competences

On top of the automatic tagging method presented in Section 3.1, we added a new layer to identify which is the most probable competence a document includes. The

---

[4] http://www.escpeurope.eu

classification layer uses two different inputs; (i) a ranked list of keywords that describes the resource to be classified (tags) and (ii) a list of competences that a document can belong to with a list of keywords describing each competence (see Table 1).

With these two inputs, the classification method assigns scores for each match found between the document's list of keywords and the competences' keywords. Since the document's tags are already properly ranked, we apply a linear decay on the matching-score. It means that the competences' keywords that matches the first document's keywords have a greater score. In the other hand, the higher a document's keywords is positioned in the ranking, the lower is the final score. After the matching process, we compute the sum of the scores for each competence and the document is assigned with the top scoring competence. The pseudo-code (Algorithm 1) depicts the matching method. It is important to remark that all keywords involved are first submitted to a stemming process.

Algorithm 1: Pseudocode for keyword-term matching method.

```
1  begin
2      for each document do
3          Get top N α-TaggingLDA keywords;
4          for each keywords do
5              KeywordIndex++; for each competence do
6                  Get competence's terms;
7                  for each competence's terms do
8                      if keyword == terms then
9                          competence-score += 1/KeywordIndex;

10     return scoring competences;
```

3.3 Evaluation

To evaluate our method we used the OpenScout dataset containing 21,768 learning objects. We pruned these data to consider only objects that are in English, with the description with a minimum length of 500 characters, which resulted in a set of 1,388 documents. Thus, on these documents we applied the competence assignment method. Since the dataset is relatively new and very few items have been assigned with competences, we propose an automatic method to evaluate the outcomes of the automatic competence assignments.

Our evaluation method considers the similarity among the learning objects and a set of assumptions/cases that we believe can validate whether the automatic competence assigner produces optimum results or not. To measure the similarity among the documents, in our study, we used MoreLikeThis, a standard function provided by the Lucene search engine library[5]. MoreLikeThis calculates similarity of two documents by computing the number of overlapping words and giving them different weights based on

---

[5] http://lucene.apache.org/core/old_versioned_docs/versions/3_4_0/api/all/org/apache/lucene/search/similar/MoreLikeThis.html

TF-IDF [3]. MoreLikeThis runs over the fields we specified as relevant for the comparison - in our case the description of the learning object - and generates a term vector for each analyzed item (excluding stop-words).

To measure the similarity between documents, the method only considered words that are longer than 2 characters and that appear at least 2 times in the source document. Also, words that occur in less than 2 different documents are not taken into account for the calculation. For calculating the relevant documents, the method used the 15 most representative words, based on their TF-IDF values, and generated a query with these words. The ranking of the resulting documents is based on Lucene's scoring function which is based on the Boolean model of Information Retrieval and the Vector Space Model of Information Retrieval [4].

Let $c(LO_i)$ be a function returning the competence for a specific learning object LO_i and let $s(LO_i, LO_j)$ be a function measuring the similarity between two resources LO_i and LO_j. Then, given the set of learning objects, the similarity scores $s(LO_i, LO_j)$ and the competence assignments $c(LO_i)$, we evaluate the results through four given cases:

- **Case 1:** If two LOs have the same competence and are similar to some extent, it is resonable to assume that the compentece assigner is coherent. If $c(LO_1) = c(LO_2)$ and $s(LO_1, LO_2) >= 0.7$
- **Case 2:** If two LOs have been assigned with the same competence but are not similar, it is not completely implausible and means that the competence is broad. If $c(LO_1) = c(LO_2)$ and $s(LO_1, LO_2) < 0.7$
- **Case 3:** If two LOs have been assigned with different competences and are very similar, it suggests that the automatic competence assigner committed a fault. Thus, the lower the assignments that fall in this case the better the results. If $c(LO_1) \neq c(LO_2)$ and $s(LO_1, LO_2) >= 0.7$
- **Case 4:** Finally, for the cases where two LOs have been assigned with different competences and the LOs are not similar, correctness can not be derived but a high value also demonstrates the coherence of the method. If $c(LO_1) \neq c(LO_2)$ and $s(LO_1, LO_2) < 0.7$

3.4 Evaluation Results

In this section, we present the results of the proposed automatic competence assigner method. In Table 2, we plot the results discerning the number of occurrences (in percentage) that fall in each case. Additionally, we alternate the number of competences considered in the evaluation. First, we used only the top scoring competence for a given LO and, in a second round of the evaluation, we considered the top two scoring competences.

The results show that very few items fell in the cases 1 and 3, meaning that most of the items did not meet the minimum threshold value of similarity, thus, showing that most of the classified documents are dissimilar. The low similarities are also caused by the short textual descriptions available. Regarding the documents that are similar (>= 0.7), only around 1% of the items fell into case 3; given our assumptions in Section 3.3, we consider this 1% as a false assignment.

Table 2. Results of the automatic comepetence assignments according to the cases defined in Section 3.3, considering the top one and top two competences with similarity threshold at 0.7

Rule	Tags(1)	Tags(2)
1) Same Competence Sim. >= 0.7	0.24	0.24
2) Same Competence Sim. < 0.7	9.60	10.63
3) Dif. Competence Sim. >= 0.7	1.02	0.95
4) Dif. Competence Sim. < 0.7	89.12	88.16

4 Conclusion

In this work, we proposed a methodology to automatically assign competences to learning objects. Our proposed method is based on an automatic tagging tool that does not require a training set or any previous users' interaction over the resources. We also proposed an automatic methodology to evaluate the given competences through a set of cases that considers objects' textual similarities. Although the cases cannot guarantee the correctness of an assignment, the third case indeed exposes misassigned items. The results obtained showed very few occurrences where different competences were assigned to very similar items. We interpret that as evidence of the coherence and effectiveness of the proposed method that may be applied to effectively enhance competence metadata for learning objects.

As future work, we plan to improve the competence classifier by including the whole content of documents and representing it through a weighted term vector. Additionally we plan to automatically quantify necessary competence levels according to the European Qualification Framework (EQF).

Acknowledgement. This research has been co-funded by the European Commission within the eContentplus targeted project OpenScout, grant ECP 2008 EDU 428016 (cf. http://www.openscout.net) and by CAPES (Process n° 9404-11-2).

References

1. Diaz-Aviles, E., Georgescu, M., Stewart, A., Nejdl, W.: Lda for on-the-fly auto tagging. In: Proceedings of the Fourth ACM Conference on Recommender Systems, RecSys 2010, pp. 309–312. ACM, New York (2010)
2. Niemann, K., Schwertel, U., Kalz, M., Mikroyannidis, A., Fisichella, M., Friedrich, M., Dicerto, M., Ha, K.-H., Holtkamp, P., Kawase, R., Parodi, E., Pawlowski, J., Pirkkalainen, H., Pitsilis, V., Vidalis, A., Wolpers, M., Zimmermann, V.: Skill-Based Scouting of Open Management Content. In: Wolpers, M., Kirschner, P.A., Scheffel, M., Lindstaedt, S., Dimitrova, V. (eds.) EC-TEL 2010. LNCS, vol. 6383, pp. 632–637. Springer, Heidelberg (2010)
3. Salton, G., McGill, M.J.: Introduction to Modern Information Retrieval. McGraw-Hill, New York (1983)
4. Salton, G., Wong, A., Yang, C.S.: A vector space model for automatic indexing. Commun. ACM 18(11), 613–620 (1975)

Slicepedia: Automating the Production of Educational Resources from Open Corpus Content

Killian Levacher[1], Seamus Lawless[2], and Vincent Wade[3]

Trinity College, Dublin, Ireland

Abstract. The World Wide Web (WWW) provides access to a vast array of digital content, a great deal of which could be ideal for incorporation into eLearning environments. However, reusing such content directly in its native form has proven to be inadequate, and manually customizing it for eLearning purposes is labor-intensive. This paper introduces Slicepedia, a service which enables the discovery, reuse and customization of open corpus resources, as educational content, in order to facilitate its incorporation into eLearning systems. An architecture and implementation of the system is presented along with a preliminary user-trial evaluation suggesting the process of slicing open corpus content correctly decontextualises it from its original context of usage and can provide a valid automated alternative to manually produced educational resources.

1 Introduction

Educational Adaptive Hypermedia systems (EAH) have traditionally attempted to respond to the demand for personalized interactive learning experiences through the support of adaptivity, which sequences re-composable pieces of information into personalized presentations for individual users. While their effectiveness and educational benefits have been proven in numerous studies [1], the ability of EAH systems to reach the mainstream audience has been limited [2]. This is in part due to their reliance upon large volumes of educational resources available at high production costs, incurred by labor-intensive work [3].

This dependency has been extensively studied by the research community, which has addressed this issue mainly by improving either the discovery [4] or the reuse [5] of existing educational resources. Solutions proposed so far however, do not address the fundamental problem which is the labor intensive manual production of such resources.

In parallel with these developments, the field of Open Adaptive Hypermedia (OAH) has attempted to leverage the wealth of information, which has now become accessible on the WWW as open corpus information. However, open corpus reuse and incorporation has been achieved so far, using either manual [6] or at best traditional information retrieval (IR) approaches [7]. Even when retrieving relevant open web information, these IR techniques suffer because they

A. Ravenscroft et al. (Eds.): EC-TEL 2012, LNCS 7563, pp. 407–412, 2012.

only provide one-size-fits-all, untailored, document level, delivery of results, with limited control over topics, granularity, content format or associated meta-data.

This results in limited and restricted reuse of such resources in OAH. Open corpus material, in its native form, is very heterogeneous. It comes in various formats, languages, is generally very coarse-grained and contains unnecessary noise such as navigation bars, advertisements etc. Hence, there remains a significant barrier to automatically convert native open corpus content into reusable educational resources meeting specific content requirements (topic covered, style, granularity, delivery format, annotations) of individual EAH.

We believe that the cost intensive, manual production and/or adaptation of educational resources must be augmented or replaced by the automated repurposing of open corpus content into such resources. This transformation will make it possible to provide on-demand automated production and right-fitting of educational resources in large volumes to support re-composition and personalization within EAH systems.

Contribution: This paper presents Slicepedia, a service that leverages content[1] from open corpus sources to produce large volumes of right-fitted educational resources at low cost. This novel approach leverages complementary techniques from IR, Content Fragmentation, Information Extraction (IE) and Semantic Web to improve the reuse of open corpus resources by converting them into information objects called slices.

- An implementation of the system architecture is presented, which has been applied in an authentic educational user-trial scenario.
- Initial results, of an evaluation currently underway, investigating the quality of automated open corpus reuse and its suitability within an educational context, are presented in this paper.

2   The Web Converted as Slices

The Slicepedia service enables the automated reuse of open corpus content through slicing. Slicing [8] is the process of automatically harvesting, fragmenting, semantically annotating and customizing original web resources into re-composable information objects called slices, tailored for consumption by individual EAH systems. The system is available as a fully autonomous service, composed of successive and easily pluggable components, and provides slices according to the formats (LOM, SCORM etc...) and meta-data requested.

Harvesting: The first component of a slicer pipeline acquires open corpus resources, from the web, in their native form. Standard IR systems[2] or focused

---

[1] In order to deal with the significant technical challenges of right sizing and reuse, some specific aspects are deemed beyond the scope of this paper; namely copyright and digital rights management issues.

[2] http://developer.yahoo.com/search/web/V1/webSearch.html

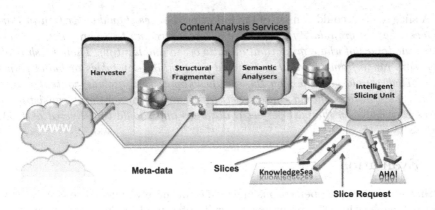

Fig. 1. Slicepedia Architecture

crawling techniques [9] are used to gather relevant documents, which are then cached locally for further analysis.

Structural Decontextualisation: Resources harvested are then fragmented into individual structurally coherent sections (such as menus, advertisements, main article). Structural meta-data, such as the location of each fragment within the original resource, is extracted and stored in the meta-data repository. This phase is critical since, maximising the reuse potential of a resource involves the ability to identify specific reusable parts of pages from any clutter content present within the original document. Densitometric fragmentation [10] was selected for this slicer implementation.

Semantic Analyser: Once decontextualised, each resulting fragment, available as linked-data in a Sesame store[3], is annotated with rdf semantic labels, using a list of pre-selected algorithms. A boilerplate detection algorithm[4] annotates to what degree a fragment of a page can be reused or not. Concepts mentioned within each fragment are identified using the AlchemyApi concept tagging service[5] and tagged as Dbpedia concepts[6]. Reading level difficulty[7] of fragments, expressed as Flesh Reading scores and finally, the part of speech, noun phrase and verb phrases are also identified and annotated[8] with their relevant linguistic attributes.

Slice Creation: The fourth step finally analyses slice requests received from EAH slice consumers, which are then matched with all possible fragments/meta-data combinations available in the system. Fragment combinations obtaining the closest match are delivered to EAH systems in their format of choice.

---

[3] http://www.openrdf.org/
[4] http://code.google.com/p/boilerpipe/
[5] http://www.alchemyapi.com/api/
[6] http://dbpedia.org/About
[7] http://flesh.sourceforge.net/
[8] http://gate.ac.uk/

A slice request could consist of the following: *"Slices should be written in Portuguese, have a granularity ranging from 3 sentences up to 3 paragraphs, should cover the topics of whale migration, atlantic ocean and hunting. Their Flesh reading score range from 45 to 80 and should not contain any tables or bullet points lists. They should be scoped on the specified topics (i.e. exclude content not on these topics). Slices should contain between 7 and 15 annotations consisting of verbs conjugated in the past perfect continuous and should be delivered as LOM objects".*

3 Evaluation

Although the reuse of open corpus material is ultimately aimed at re-composition and personalization, the preliminary results presented in this paper focus, as a first step, in evaluating the automated reuse of individual open corpus slices. The purpose of the evaluation presented below hence was to investigate whether the approach to automate the production of educational resources via open corpus content reuse using slicing techniques could:

- H1: Correctly decontextualise preliminary open corpus resources from original associated clutter
- H2: Offer a suitable alternative to manually generated educational resources from a user perspective.

The experiment compared automated and manual reuse of arbitrary selected open corpus content using a real life user-trial. A simple online e-assessment application (available in English, Spanish and French), built for this experiment, presented users with traditional gap filler grammar exercises, built using different sets of open corpus resources converted (manually or automatically) into grammatical e-assessment exercises. Verb chunks conjugated at specified tenses were removed and replaced by gaps, which users had to fill according to particular infinitives and tenses specified for each gap. The answers provided were compared to the original verb chunks and users were assigned a score for each grammar point.

In order to guarantee open corpus resources harvested, represented a truly random set of resources, a group of five independent English teachers were asked to arbitrarily select a total of 45 pages (content batch CBN) of their choice from the web, from any source/topic of their choice. These teachers were then asked to arbitrarily select fragments (of any size) of pages harvested, which they felt were adequate for grammar exercises, and manually annotate tenses encountered within these extracts, to produce content batch CBM. The entire collection of pages was then harvested from the web in their original form by the slicer and sliced in order to produce a set of automatically generated resources CBO, with similar characteristics as their manual counterparts, This resulted in 3 content batches consisting of CBN (open corpus pages in their native form), CBM (annotated fragments manually produced) and CBO (annotated fragments automatically produced). All of the extracts produced were subsequently converted into grammar e-assessment pages.

The slice consumer application represents an excellent evaluation platform for the purpose of this experiment since it was necessary to select a reuse vehicle where the user needs were very sensitive to: (i) the accuracy of annotations (i.e.: verbal annotations) and (ii) the visual layout (i.e.: content formatted correctly). After performing exercises on each content batch (presented using a Latin Squared Distribution), users were asked to rate each set of pages presented to them using a 10 point Likert scale.

4 Results

A total of 41 users, divided into two groups consisting of Experts (63%) and Trainees (37%), performed the experiment.

H1: As pointed out in Section 2, a part of maximizing the reuse potential of a previously published resource requires the ability to decontextualize this content from its original setting. Hence, users were asked directly, for each content, whether *in addition to the main content, a lot of material displayed on the page was irrelevant to the task (such as advertisement, menu bar, user comments..).* Results obtained for both the manual and automatically produced content were very similar (Mean CBM=2.13, Mean CBO=2.53) with paired t-tests considering mean differences as insignificant ($p=0.423$). These results indicate that although users did notice a difference in average between the decontextualisation of preliminary open corpus resources carried out by the slicer, the difference was statistically insignificant.

H2: Following a correct decontextualisation of open corpus resources, the overall re-purposing performance of both content batches, with respect to their ability to provide adequate e-assessments, was measured. The number of grammar mistakes (E=29.74%, T=42.61%) measured upon content created automatically, appears to be higher than for the content produced manually (E=23.92%, T=35.13%) for both groups of users. This appears to suggest that automatically generated content occasioned users to perform more errors during the e-assessments tasks. Although the difference in errors between content batches was slightly higher for the trainees in comparison to the experts group, an independent t-test considers this difference as insignificant (Mean Percentage Difference: E=5.80%, T=7.48%, $p=0.891$, Equal Variances Assumed), which would indicate that although the automatically generated content did induce users to answer erroneously some assessment units, users from the expert group didn't appear to use their language skills to compensate differences between content batches used. When trainees were asked whether, for content batch CBO, *"the number of erroneous assessment units presented was tolerable"*, a mean score of 7 on the liker scale was measured. When asked whether *"Overall, I felt this content was adequate to perform a grammar exercise"* both content achieved very similar scores (CBM=8.33, CBO=8.57, $p=0.536$) with t-tests suggesting any difference observed as insignificant. Hence, these result appear to indicate that although users appear to achieve lower performances on assessments automatically generated in comparison to those manually produced, this tendency didn't appear to affect Trainees more than the Experts group of users, nor did it appear to decrease the perceived usefulness of the content for the assessment task performed.

5 Conclusion

The preliminary results presented in this paper appear to indicate that the process of automatically slicing content can correctly decontextualise individual portions of open corpus documents into structurally coherent and independent information objects.

Initial results obtained with respect to the suitability of content automatically generated by this approach, as an alternative to manually produced educational resources, suggest a slight difference in appropriateness. However, user perception appears to consider both content types as interchangeable. Taking into account the low production cost and high volume of educational objects such a slicing service could provide, a slight decrease in content quality could more than likely be tolerable in many educational use cases. An experiment investigating the reading quality decrease, re-composition and personalization of slices in an independent third party AHS is currently in progress.

References

1. Lin, Y.L., Brusilovsky, P.: Towards Open Corpus Adaptive Hypermedia: A Study of Novelty Detection Approaches. In: Konstan, J.A., Conejo, R., Marzo, J.L., Oliver, N. (eds.) UMAP 2011. LNCS, vol. 6787, pp. 353–358. Springer, Heidelberg (2011)
2. Armani, J.: VIDET: a Visual Authoring Tool for Adaptive Tailored Websites. Educational Technology & Society 8, 36–52 (2005)
3. Meyer, M., Hildebrandt, T., Rensing, C., Steinmetz, R., Ag, S.A.P., Darmstadt, C.E.C.: Requirements and an Architecture for a Multimedia Content Re-purposing Framework. In: Nejdl, W., Tochtermann, K. (eds.) EC-TEL 2006. LNCS, vol. 4227, pp. 500–505. Springer, Heidelberg (2006)
4. Drachsler, K.R.: ReMashed - An Usability Study of a Recommender System for Mash-Ups for Learning. Int. Journal of Emerging Technologies in Learning 5 (2010)
5. Meyer, M., Rensing, C., Steinmetz, R.: Multigranularity reuse of learning resources. In: Multimedia Computing, Communications, and Applications (2011)
6. Henze, N., Nejdl, W.: Adaptation in Open Corpus Hypermedia. International Journal (2001)
7. Zhou, D., Truran, M., Goulding, J.: LLAMA: Automatic Hypertext Generation Utilizing Language Models. In: Int. Conf. on Hypertext and Hypermedia (2007)
8. Levacher, K., Lawless, S., Wade, V.: Slicepedia: Providing Customized Reuse of Open-Web Resources for Adaptive Hypermedia. In: HT 2012: Proc. of the 23rd Conf. on Hypertext and Social Media (2012)
9. Lawless, S., Hederman, L., Wade, V.: OCCS: Enabling the Dynamic Discovery, Harvesting and Delivery of Educational Content from Open Corpus Sources. In: Int. Conf. on Advanced Learning Technologies (2008)
10. Kohlschütter, C., Nejdl, W.: A Densitometric Approach to Web Page Segmentation. In: Proceeding of the 17th International Conference on Information and Knowledge Management, CIKM, pp. 1173–1182 (2008)

Fostering Multidisciplinary Learning through Computer-Supported Collaboration Script: The Role of a Transactive Memory Script

Omid Noroozi[1,*], Armin Weinberger[2], Harm J.A. Biemans[1], Stephanie D. Teasley[3], and Martin Mulder[1]

[1] Wageningen University (Netherlands)
[2] Saarland University (Germany)
[3] University of Michigan (United States)

Abstract. For solving many of today's complex problems, professionals need to collaborate in multidisciplinary teams. Facilitation of knowledge awareness and coordination among group members, that is through a Transactive Memory System (TMS), is vital in multidisciplinary collaborative settings. Online platforms such as ICT tools or Computer-supported Collaborative Learning (CSCL) have the potential to facilitate multidisciplinary learning. This study investigates the extent to which establishment of various dimensions of TMS (specialization, coordination, credibility) is facilitated using a computer-supported collaboration script, i.e. a transactive memory script. In addition, we examine the effects of this script on individual learning satisfaction, experience, and performance. A pre-test, post-test design was used with 56 learners who were divided into pairs based on disciplinary background and randomly divided into treatment condition or control group. They were asked to analyse, discuss, and solve an authentic problem case related to their domains. Based on the findings, we conclude that a transactive memory script in the form of prompts facilitates construction of a TMS and also improves learners' satisfaction with learning experience, and performance. We provide explanations and implications for these results.

Keywords: CSCL, multidisciplinary, transactivity, transactive memory system, transactive memory script.

Introduction

The main advantage of multidisciplinary learning is that learners from different backgrounds take advantage of one another's complimentary expertise and bear on a problem from various perspectives and viewpoints. Although considering various

---

* Correspondence concerning this article should be addressed to Omid Noroozi, Chair Group of Education and Competence Studies, Wageningen University, P.O. Box 8130, NL 6700 EW Wageningen, the Netherlands. E-mail: omid.noroozi@wur.nl.

A. Ravenscroft et al. (Eds.): EC-TEL 2012, LNCS 7563, pp. 413–418, 2012.
© Springer-Verlag Berlin Heidelberg 2012

viewpoints can be productive, multidisciplinary group members need to establish a common ground, which is vital to group performance but difficult to achieve. Learners may thus engage in non-productive discussing of such pieces of information that may already be known to all members from the start (Stasser & Titus, 1985). Hence, they make work together for extended periods, before starting to efficiently pool their unshared knowledge. Speeding up the process of pooling unshared information is best achieved when group members have meta-knowledge about the expertise and knowledge of their learning partners, that is, a Transactive Memory System (TMS).

According to Wegner (1987), learning partners work best when they encode, store, and retrieve information distributed in the group. In the encoding process, the initiation of a TMS begins with the process of getting to know "who knows what" in the group. Encoding is followed by information allocation in the storage process group members allocate new information on a topic to the relevant expert(s) in the group on that topic. Next, group members need to retrieve required information from the expert who has the stored information on a particular topic (Wegner, 1987 & 1995). The TMS comprises group members' views in terms of awareness of one another's knowledge, the accessibility of that knowledge, and the extent to which they take responsibility for providing knowledge in their own areas of expertise (Lewis, 2003). In scientific literature three dimensions of TMS, i.e. specialization, coordination, credibility, have been studied. Specialization represents awareness and recognition of expertise distributed among members of a group. Coordination represents group members' ability to work efficiently on a task with less confusion, fewer misunderstandings but a greater sense of collaboration. Credibility or trust

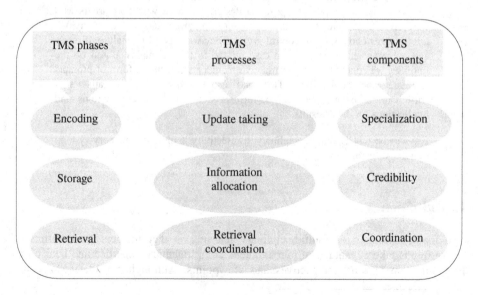

Fig. 1. Relation between TMS processes and components

represents the degree to which group members trust and rely on each other's expertise while collaborating (Michinov & Michinov, 2009). Figure 1 shows a graphical representation of various processes and dimensions of TMS.

Much research has reported positive impacts of TMS on group performance (see Michinov & Michinov, 2009). Despite the vast research on TMS, no study has explicitly investigated the role of TMS in online learning environments. This is striking since online support systems for collaborative learning, such as CSCL platforms, play a key role in terms of potential for establishment of TMS (see Noroozi & Busstra et al., 2012; Noroozi & Biemans et al., 2011 & 2012). Various forms of CSCL scripts are manifested as stand-alone instructional tools or scaffolds to guide learners to engage in specific activities. Scripts provide explicit guidelines for learners to clarify what, when and by whom certain activities need to be executed. Therefore, we implement a particular transactive memory script in an online learning platform to facilitate the establishment of TMS. We also examine the effects of this script on learners' satisfaction, experiences, and performance.

Method

Participants were 56 students from Wageningen University. Each pair (two learners with complementary disciplinary backgrounds) was randomly assigned to one of the treatment conditions or the control group. The subject was Community-Based Social Marketing (CBSM) and its application in Sustainable Agricultural Water Management (SAWM). The task of the pairs was to apply the concept of CBSM in fostering sustainable behaviour among farmers in terms of SAWM.

The experiment took 3.5 hr. and consisted of 4 phases including (1) introduction and pre-test phase (35 min.), (2) individual phase (40 min.), (3) collaborative phase (90 min.), (4) post-test and debriefing phase (45 min.). An asynchronous text-based discussion board called "SharePoint" was customized for this study. In the control group, learning partners received no further support beyond being asked to analyse and discuss the problem case on the basis of the conceptual space and to type their arguments into a blank text box. Building on Wegner (1987), we expanded a transactive memory script over three phases: encoding, storage, retrieval. For each phase, specific types of prompts were embedded in the platform. All the prompts were delivered at the collaborative phase of the experiment. In the encoding phase, learners were given 10 minutes to introduce themselves, compose a portfolio of their expertise, and indicate what aspects of their expertise applied in the given case. In the storage phase, they were given 15 minutes to read the portfolios and discuss the case with the goal of distributing parts of the task. In the retrieval phase, the group members were supposed to analyse and solve part of the task from their expertise perspective in 15 minutes. They were subsequently given 40 minutes to reach an agreement by discussing and sharing their individual solutions.

Measurement of various aspects of TMS: We adapted a questionnaire from Lewis (2003) to assess TMS. This questionnaire included three dimensions of TMS with 15 items on a five-point Likert scale ranging from "strongly disagree" to "strongly agree". The reliability and validity of this scale have been reported to be adequate in various contexts. In this study, the reliability was sufficiently high for all dimensions of TMS.

Measurement of learners' satisfaction and experience: We adapted a questionnaire from Mahdizadeh (2008) to assess these items. This questionnaire comprised five sections and 32 items on a five-point Likert scale ranging from "almost never true" to "almost always true". The first section (10 items) assessed Perceived Effects of Learning. The second section (4 items) captured Attitude toward Computer-Assisted Learning. The third section (3 items) collected information on the Ease of Use of the Platform. The fourth section (4 items) assessed Satisfaction with Learning Effects. The last section (11 items) collected information on Appreciation of the Learning Materials. Cronbach Alpha was sufficiently high (around .90) for all five categories.

Measuring learning performance. We analysed individual written analyses in the pre-test and post-test based on expert solutions on a five-point Likert scale ranging from "very poor solution" to "very good solution". The inter-rater agreement and intra-coder test-retest reliability were reported to be satisfactory. The difference in the scores from pre-test to post-test served as an indicator for learning performance. Inter-rater agreement between two coders (Cohen's $\kappa > 85$) were sufficiently high.

Results and Discussions

The effects of transactive memory script on TMS. The average scores for all dimensions of TMS were higher for scripted than unscripted learners. The difference between specialization means was significant at the 0.05 level. This was also true for credibility and coordination. Instructional prompts helped learners to label information as being in one another's expertise domains, to store information with the appropriate individuals who had the expertise and to discover and retrieve needed information by each individual during collaboration. Scripted learners could coordinate the process of problem-solving by assigning responsibility to the individual who had the most relevant expertise in group. If learners could trust one another to take responsibility for parts of the task for which they had the most expertise, they could be sure that no information would be missed by the group as a whole. In the final stage, prompts helped learners to retrieve required information by discovering and associating the label of the information with sources of expertise from the expert who had the stored information (Wegner, 1987).

The effects of transactive memory script on learners' satisfaction and experience. The average scores for all dimensions of these items were higher for learners in the treatment condition than for learners in the control group, except for the Ease of Use of Platform. For all variables, the differences between scripted and unscripted learners were significant at the 0.05 level. No difference was reported for the Ease of Use of Platform. In the scripted condition, learners received detailed instructions during the collaborative phase on what, when, how, and by whom certain activities needed to be executed for accomplishing the collaborative learning task. The positive effect of this information was reflected in the learners' satisfaction and experience at the end of the experiment.

The effects of transactive memory script on learning performance. Learners in both conditions did not differ significantly regarding their pretest scores. The scores of all learners improved significantly at the 0.05 level from pre-test to post-test. Gain of knowledge for learners in the treatment condition was significantly higher at the 0.05 level than that of learners in the control group (see Figure 2). In collaborative settings, groups whose members are aware of one another's knowledge and expertise develop a shared understanding of who knows what in the group (Wegner, 1987) and thus perform better than groups whose members do not possess such knowledge. The significance of shared knowledge for collaborative learning activities especially among heterogonous groups of learners has been widely acknowledged in the scientific literature.

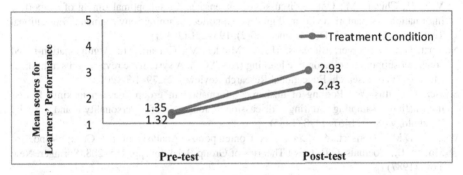

Fig. 2. Mean score of learners' performance in the treatment condition and control group

Conclusion

No research on multidisciplinary context has yet reported the use of transactive memory script for facilitation of TMS in online learning platforms. This study was conducted in a laboratory setting to investigate the effects of a transactive memory script on TMS, learning satisfaction, experience, and performance in a multidisciplinary CSCL setting. Based on our study, we conclude that using a transactive memory script can positively foster the establishment of dimensions of TMS (specialization, credibility, and coordination) in a multidisciplinary CSCL setting. This facilitation of TMS in the treatment condition not only resulted in higher scores for learners' satisfaction than in the control group condition, but also for learning experience and performance.

References

Lewis, K.: Measuring transactive memory systems in the field: scale development and validation. Journal of Applied Psychology 88(4), 587–604 (2003)

Mahdizadeh, H.: Student collaboration and learning. Knowledge construction and participation in an asynchronous computer-supported collaborative learning environment in higher education. PhD dissertation, Wageningen University, The Netherlands (2007)

Michinov, N., Michinov, E.: Investigating the relationship between transactive memory and performance in collaborative learning. Learning & Instruction 19(1), 43–54 (2009)

Noroozi, O., Biemans, H.J.A., Busstra, M.C., Mulder, M., Chizari, M.: Differences in learning processes between successful and less successful students in computer-supported collaborative learning in the field of human nutrition and health. Computers in Human Behavior 27(1), 309–318 (2011)

Noroozi, O., Biemans, H.J.A., Busstra, M.C., Mulder, M., Popov, V., Chizari, M.: Effects of the Drewlite CSCL platform on students' learning outcomes. In: Juan, A., Daradoumis, T., Roca, M., Grasman, S.E., Faulin, J. (eds.) Collaborative and Distributed E-Research: Innovations in Technologies, Strategies and Applications, pp. 276–289 (2012)

Noroozi, O., Busstra, M.C., Mulder, M., Biemans, H.J.A., Tobi, H., Geelen, M.M.E.E., van't Veer, P., Chizari, M.: Online discussion compensates for suboptimal timing of supportive information presentation in a digitally supported learning environment. Educational Technology Research & Development 60(2), 193–221 (2012)

Noroozi, O., Weinberger, Biemans, H.J.A., Mulder, M., Chizari, M.: Argumentation-based computer supported collaborative learning (ABCSCL). A systematic review and synthesis of fifteen years of research. Educational Research Review 7(2), 79–106 (2012)

Stasser, G., Titus, W.: Pooling of unshared information in group decision making: Biased information sampling during discussion. Journal of Personality and Social Psychology 48(6), 1467–1478 (1985)

Wegner, D.M.: Transactive Memory: A Contemporary Analysis of the Group Mind. In: Mullen, B., Goethals, G.R. (eds.) Theories of Group Behavior, pp. 185–208. Springer, New York (1987)

Wegner, D.M.: A computer network model of human transactive memory. Social Cognition 13(3), 1–21 (1995)

Mobile Gaming Patterns and Their Impact on Learning Outcomes: A Literature Review

Birgit Schmitz[1,2], Roland Klemke[2,1], and Marcus Specht[2]

[1] Humance AG, Goebenstr. 10-12, 50672 Köln, Germany
{bschmitz,rklemke}@humance.de
[2] Center for Learning Science and Technology, Valkenburgerweg 177,
Heerlen, The Netherlands
{birgit.schmitz,roland.klemke,marcus.specht}@ou.nl

Abstract. Mobile learning games have increasingly been topic of educational research with the intention to utilize their manifold and ubiquitous capabilities for learning and teaching. This paper presents a review of current research activities in the field. It particularly focuses is on the educational values serious mobile games provide. The study results substantiate their generally assumed motivational potential. Also, they indicate that mobile learning games may have the potential to bring about cognitive learning outcomes.

Keywords: Mobile games, mobile learning, serious games, game design patterns, learning outcomes.

1 Introduction

The interest in learning games has considerably grown within the last decade. This is not only due to the growing number of people playing games. Games seem to enable students to gain skills needed in an information-based culture and to learn innovatively [16]. Investigations into educational games centre on their motivational potential and their low-threshold learning opportunities [5][10]. Games on mobile devices open up new target groups and new access to learning [26][19]. The Mobile Learning NETwork's (MoLeNET) review on learning game technologies suggests that mobile learning games provide potential for learning and teaching in terms of 'assessment', 'learner performance and skills development' or 'social and emotional well-being' [11]. In order to determine the mechanisms and design elements that make the use of novel learning scenarios successful and transferrable, it is necessary to explore how these technologies can be used for teaching and learning [12][15][17].

Therefore, the objective of this paper is to scrutinize the learning effects of mobile games and to understand the game mechanisms that have led to it. The results could provide valuable insight into the working mechnisms of mobile learning games that may positively influence future design decisions.

A. Ravenscroft et al. (Eds.): EC-TEL 2012, LNCS 7563, pp. 419–424, 2012.

2   Theoretical Framework

The conceptual framework for our analysis comprises two main components: the game design patterns for mobile games by Davidsson et al. [8] and the taxonomy of learning outcomes by Bloom [4].

We decided to base our study on patterns because especially in the context of educational games, the traditional categorization of games according to genres has proved to be of little use [8]. As an expansion to the already existing set of Game Design Patterns by Björk and Holopainen [3], [8] introduced 74 new patterns that describe the unique characteristics of mobile games. Each pattern is identified by a core definition, a general definition, example(s), descriptions of how to use it (by listing related patterns or patterns that can be linked to it), the description of its consequences, relations with regard to instantiation (patterns causing each other's presence) and modulation (patterns influencing each other), as well as references. The pattern *Physical Navigation* for example "forces players of a mobile game to move or turn around in the physical world in order to successfully play the game" [8, p.18]. The MLG 'Frequentie 1550' makes use of this pattern. Players have to move around to find sources of information and to complete tasks [1].

On the other hand, learning games, as any educational measure, can be classified according to learning outcomes. Well advanced in years but notwithstanding adequate is Bloom's taxonomy [4] which sorts learning outcomes into the *affective, cognitive* and *psychomotor domain.* The affective domain encompasses attitudes and motivation. The cognitive domain deals with the recall or recognition of knowledge and the development of intellectual abilities and skills. For this domain, Bloom distinguishes six successive levels: Knowledge, Comprehension, Application, Analysis, Synthesis and Evaluation. The psychomotor domain encompasses manual or physical skills or the performance of actions. Learning outcomes in relation to this domain, e.g. exer-games [29] we did not consider, as they have a different didactic approach.

3   Basis for the Review

For the review, we focused on 42 empirical research articles and practical papers. The following keywords were used: mobile educational game, mobile serious game, mobile learning game, mobile game-based learning, (location-based, ubiquitous, mixed reality, augmented reality, pervasive) learning game. We included practical papers (publicly available journal paper and conference proceedings) that (a) report evaluation results from pilot studies with mobile learning games, (b) have a clear focus on affective and/or cognitive learning outcomes, (c) allow identification of mobile game design patterns and (d) report on concrete learning outcomes where the learning outcomes can be correlated with a pattern used in the game.

Due to the educational focus of our analysis, we excluded 4 papers because they reported on games other than serious games, e.g. [14]. Also, we excluded 12 technical reports that focused on innovation, functionality, playability and/or usability testing, e.g. [2][9][23][22] or [30]. For our purpose, an explanation of effects in relation to individual game play mechanisms was crucial. We excluded another 9 papers that provided evaluation data on a very general level, thus no pattern – effect correlation was possible. We did not take into consideration a specific age group. The research we reviewed was conducted mainly on pupils and young adults (age range: 10 – 25 years). Possible variations in effect due to that range of age were not considered.

4   Results

In the following, we present the most significant results of the survey. First, we scrutinized what games impact motivation (affective learning outcomes) and/or knowledge (cognitive learning outcomes). We thenwent into detail, focusing on theindividual patterns used in the game and analysed how they impact affective and/or cognitive learning outcomes.

Our analysis reveals that game mechnisms such as *Collaborative Actions, Augmented Reality* and *Roleplaying* are vital motivational factors providing an incentive to get engaged with a learning environment and/or a certain topic.

The 'Virus Game' [24] for example integrated the pattern *Collaborative Actions* by providing different roles with distinct abilities. 'Each of the roles is dependent on the others both for information and for action. This fosters collaboration through jig sawing' [24, p. 40]. The study indicates that *Collaborative Actions* can bring about a change in students' attitude by providing insight into the working mechanisms of interpersonal communication. In the course of the 'Virus Game', students depend on each other for information and for action to reach their goal. The 'jig sawing of complementary information' (p. 35) brought about 'an understanding of the interdependence of the roles' (p. 43). Students 'grasped the resulting importance of communication and collaboration for success in the game' (p. 40).

Through the integration of *Roleplaying* in the game, students become more involved. Students felt personally embodied in the game and became tightly associated with the tasks they were responsible for 'like a real occupation' [24, p.40]. In the 'Virus Game', players take on the roles of doctors, medical technicians, and public health experts to contain a disease outbreak. The personal embodiment enabled by these roles motivated students' actions in the game [24].

Though empirical evidence on cognitive learning outcomes is inconsistent, some evaluations report on positive interrelations between mobile learning games and cognitive learning outcomes. Liu and Chu [20] for example investigate the potential of the context-aware, ubiquitous learning game HELLO (Handheld English Language Learning Organization). To measure possible cognitive effects, they evaluate students' English listening and speaking skills. Playing HELLO improves students' learning outcomes as they collaborated in their tasks in real conditions (pattern: *Collaborative Actions*). The collaborative learning activity was a story relay race. In the beginning, students could listen to several sample stories after which they were asked to edit a story collaboratively [20].

Table 1. Learning outcomes of mobile game patterns

Pattern Definition	Learning Outcome	Domain
Augmented Reality (AR) Players' perception of the game world is created by augmenting their perception of the real world.	Students feel "personally embodied" in the game. Their actions in the game are intrinsically motivated [24]. Learners are attentive [27].	Affective
	Students can discuss geometrical aspects [27]. They can describe and illustrate a disease model [24] and reflect on the process of learning [7].	Cognitive - Comprehension
Collaborative Actions Several players meeting at a location or attacking a target simultaneously.	Students are engaged in the game [7][12][20][24]. They exchange and discuss game progress [18]. Participants are driven by a good team spirit [7].	Affective
Cooperation Players have to work together to progress.	Students memorize their knowledge [28]. Students can explain and rewrite the knowledge learned [20].	Cognitive - Knowledge
Pervasive Games Play sessions coexists with other activities, either temporally or spatially.	Participants are exceptionally activated. Their attitude towards learning material improves [21].	Affective
	Students are able to transfer the learned material [21]. They reflect on their learning [6].	Cognitive - Comprehension
	Students can solve problems related to the object of learning. They can create new problems related to the object of learning [6].	Cognitive - Synthesis
	Students can judge and evaluate the material for a given purpose - critical thinking skills [6].	Cognitive - Evaluation
	Students are able to analyse and classify the learned material [6].	Cognitive - Analysis
Physical Navigation Players have to move around in the physical world to play the game.	Students are highly motivated [12]. Participants are interested and moved [25]. Students's are exited [13].	Affective
Roleplaying Players have characters with at least somewhat fleshed out personalities. Play is about deciding on how characters would take actions in staged imaginary situations.	Learners are involved in the game [13]. They feel highly engaged and identify with their roles in the game [7]. They are tightly associated with their tasks in the game [24][27]. They take on an identity and are eager to work together [12].	Affective
	Students can give examples for the importance of communication and collaboration [24].	Cognitive - Comprehension

5   Discussion and Future Work

This paper reports the results of a practical research paper review focussingon the affective and cognitive learning outcomes mobile learning games may have. The review identified patterns within mobile learning games that positively influence motivation and knowledge gain. With regard to 'hard learning' [25], empirical evidence is fragmented though, e.g. the diverse studies had different statistical bases (dependent/independent variables) and applied different research methods. Also, the studies did not explicitly focus on the effects of isolated patterns but on a set of diverse patterns embedded in the games. Therefore, the impact of one particular pattern on learning is difficult to determine. Further research on the correlations between patterns and learning outcomes has thus to focus on a limited number of the patterns in existence [3][8].

To comprehensively support future design decisions, a comprehensive investigation of the effects of individual patterns has yet to follow. It will seek to understand which pattern impacts motivation and which knowledge. Future study settings have to comprise (a) an experimental variation of patterns, i.e. game settings that enable/disable individual patterns and (b) an in-depth variation of patterns, i.e. game settings that allow different instances for the same pattern. This way, measurable and feasible results can be obtained that may serve as a base for design guidelines which define (a) patterns which support the achievement of a desired learning outcome and (b) ways of how to apply the different patterns.

References

1. Akkerman, S., Admiraal, W., Huizenga, J.: Storification in History education: A mobile game in and about medieval Amsterdam. Computers & Education 52(2), 449–459 (2009)
2. Ballagas, R., Walz, S.: REXplorer: Using player-centered iterative design techniques for pervasive game development. In: Magerkurth, C., Röcker, C. (eds.) Pervasive Gaming Applications - A Reader for Pervasive Gaming Research. Shaker Verlag, Aachen (2007)
3. Björk, S., Holopainen, J.: Patterns in Game Design. River Media, Boston (2004)
4. Bloom, B.S.: Taxonomy of educational objectives, Handbook 1: Cognitive domain. David McKay, New York (1956)
5. Carstens, A., Beck, J.: Get Ready for the Gamer Generation. TechTrends 49(3), 22–25 (2010)
6. Conolly, T.M., Stansfield, M., Hainey, T.: An alternate reality game for language learning: ARGuing for multilingual motivation. Computers & Education 57(1), 1389–1415 (2011)
7. Costabile, M.F., De Angeli, A., Lanzilotti, R., Ardito, C., Buono, P., Pederson, T.: Explore! Possibilities and challenges of mobile learning. In: Proceeding of the Twenty-Sixth Annual SIGCHI Conference on Human Factors in Computing Systems, pp. 145–154. ACM (2008)
8. Davidsson, O., Peitz, J., Björk, S.: Game design patterns for mobile games. Project report, Nokia Research Center, Finland (2004), http://procyon.lunarpages.com/~gamed3/docs/Game_Design_Patterns_for_Mobile_Games.pdf (accessed August 15, 2010)
9. Diah, N.M., Ehsan, K.M., Ismail, M.: Discover Mathematics on Mobile Devices using Gaming Approach. Procedia - Social and Behavioral Sciences 8, 670–677 (2010), doi:10.1016/j.sbspro.2010.12.093
10. Douch, R., Savill-Smith, C.: The Mobile Learning Network: The Impact of Mobile Game-Based Learning. In: Proc. of the IADIS Int'l Conf. Mobile Learning 2010, Porto, Portugal, pp. 189–197 (2010)
11. Douch, R., Attewell, J., Dawson, D.: Games technologies for learning. More than just toys (2010), https://crm.lsnlearning.org.uk/user/order.aspx?code=090258 (April 13, 2011)

12. Dunleavy, M., Dede, C., Mitchell, R.: Affordances and limitations of immersive participatory augmented reality simulations for teaching and learning. Journal of Science Education and Technology 18(1), 7–22 (2009)

13. Facer, K., Joiner, R., Stanton, D., Reid, J., Hull, R., Kirk, D.: Savannah: mobile gaming and learning? Journal of Computer Assisted Learning 20(6), 399–409 (2004)

14. Falk, J., Ljungstrand, P., Bjork, S., Hannson, R.: Pirates: Proximity-triggered interaction in a multi-player game. In: Extended Abstracts of Computer-Human Interaction (CHI), pp. 119–120. ACM Press (2001)

15. Huizenga, J., Admiraal, W., Akkerman, S., Dam, G., Ten: Mobile game-based learning in secondary education: engagement, motivation and learning in a mobile city game. Journal of Computer Assisted Learning (2009), doi:10.1111/j.1365-2729.2009.00316.x

16. Johnson, L., Levine, A., Smith, R., Stone, S.: The 2011 Horizon Report. The New Media Consortium, Austin (2011)

17. Klopfer, E.: Augmented Learning. Research and Design of Mobile Educational Games. The MIT Press, Cambridge (2008)

18. Klopfer, E., Squire, K.: Environmental Detectives—the development of an augmented reality platform for environmental simulations. Educational Technology Research & Development 56(2), 203–228 (2008), doi:10.1007/s11423-007-9037-6

19. Liao, C.C.Y., Chen, Z.-H., Cheng, H.N.H., Chen, F.-C., Chan, T.-W.: My-Mini-Pet: a handheld pet-nurturing game to engage students in arithmetic practices. Journal of Computer Assisted Learning 27(1), 76–89 (2011), doi:10.1016/j.compedu.2011.01.009

20. Liu, T.-Y., Chu, Y.-L.: Using ubiquitous games in an English listening and speaking course: Impact on learning outcomes and motivation. Computers & Education 55(2), 630–643 (2010)

21. Markovic, F., Petrovic, O., Kittl, C., Edegger, B.: Pervasive learning games: A comparative study. New Review of Hypermedia and Multimedia 13(2), 93–116 (2007)

22. Martin-Dorta, N., Sanchez-Berriel, I., Bravo, M., Hernandez, J., Saorin, J.L., Contero, M.: A 3D Educational Mobile Game to Enhance Student's Spatial Skills. In: Advanced Learning Technologies (ICALT), pp. 6–10 (2010), doi:10.1109/ICALT.2010.9

23. Moore, A., Goulding, J., Brown, E., Swan, J.: AnswerTree – a Hyperplace-based Game for Collaborative Mobile Learning. In: Proceedings of 8th World Conference on Mobile and Contextual Learning - mLearn, pp. 199–202 (2009)

24. Rosenbaum, E., Klopfer, E., Perry, J.: On Location Learning: Authentic Applied Science with Networked Augmented Realities. Journal of Science Education and Technology 16(1), 31–45 (2006)

25. Schwabe, G., Göth, C.: Mobile learning with a mobile game: design and motivational effects. Journal of Computer Assisted Learning 21(3), 204–216 (2005)

26. Unterfauner, E., Marschalek, I., Fabian, C.M.: Mobile Learning With Marginalized Young People. In: Proc. of IADIS Int'l Conf. Mobile Learning 2010, Porto, Portugal, pp. 28–36 (2010)

27. Wijers, M., Jonker, V., Drijvers, P.: MobileMath: exploring mathematics outside the classroom. ZDM 42(7), 789–799 (2010)

28. Winkler, T., Ide-Schoening, M., Herczeg, M.: Mobile Co-operative Game-based Learning with Moles: Time Travelers in Medieval Ages. In: Mc Ferrin, K., Weber, R., Carlsen, R., Willis, D.A. (eds.) Proc. of SITE, pp. 3441–3449. AACE, Chesapeak (2008)

29. Yang, S., Foley, J.: Exergames get Kids moving. In: Gray, T., Silver-Pacuilla, H. (eds.) Breakthrough Teaching and Learning, pp. 87–109. Springer, New York (2011)

30. Yiannoutsou, N., Papadimitriou, I., Komis, V., Avouris, N.: "Playing with" museum exhibits: designing educational games mediated by mobile technology. In: Proc. of 8th Int'l Conf. on Interaction Design and Children, pp. 230–233. ACM, New York (2009)

Adaptation "in the Wild": Ontology-Based Personalization of Open-Corpus Learning Material

Sergey Sosnovsky[1], I.-Han Hsiao[2], and Peter Brusilovsky[2]

[1] Center for e-Learning Technology, DFKI,
Campus D3.2, D-66123 Saarbrücken, Germany
sosnovsky@gmail.com
[2] University of Pittsburgh, School of Information Sciences,
135, N. Bellefield ave., Pittsburgh, PA, 15260, USA
{peterb,hyl12}@pitt.edu

Abstract. Teacher and students can use WWW as a limitless source of learning material for nearly any subject. Yet, such abundance of content comes with the problem of finding the right piece at the right time. Conventional adaptive educational systems cannot support personalized access to open-corpus learning material as they rely on manually constructed content models. This paper presents an approach to this problem that does not require intervention from a human expert. The approach has been implemented in an adaptive system that recommends students supplementary reading material and adaptively annotates it. The results of the evaluation experiment have demonstrated several significant effects of using the system on students' learning.

Keywords: Open-Corpus Personalization, Adaptive Educational System.

1 Introduction

From the educational perspective, the WWW can be viewed as a very large collection of learning material. For many subjects, one can find online tutorials, textbooks, examples, problems, lectures slides, etc. Nowadays, teachers often do not have to create most of course materials themselves, instead they can reuse the best content available online. For example, a teacher developing a course on Java programming might decide to use a web-based Java tutorial, an electronic version of the course book, an existing Web-based assessment system, and online code examples. Although, all these resources are useful, students might get lost in this large volume of content without additional guidance. Organizing adaptive access to the course materials would help solving the problem. Appropriate tutorial pages can be recommended to students based on their progress; an adaptive navigation technique can be implemented to facilitate the choice of the most relevant example; an intelligent tutoring system can adaptively sequence learning problems. A teacher might be able to find a system implementing one of these technologies and providing adaptive access to one of the collections of learning material. However this system will not be aware of the rest of the available content, unless it supports Open-Corpus Personalization (OCP).

A. Ravenscroft et al. (Eds.): EC-TEL 2012, LNCS 7563, pp. 425–431, 2012.

OCP is one of the classic problems of adaptive information systems, in general, and adaptive educational systems (AES), in particular. Many research projects tried to propose a solution for it with different degrees of completion (e.g. [1], [2], [3], [4]). Brusilovsky and Henze in [5] presented a comprehensive overview of the problem and draw the evolution of research addressing it. This paper focuses on the OCP based on semantic content models, as the dominant personalization approaches in the field of e-Learning rely on representation of student knowledge and learning activities in terms of domain semantics. Therefore automatic extraction of domain knowledge from Web-content becomes an important component of the problem we address here. We propose a novel approach towards a *fully automated* OCP in the context of e-Learning. It is based on harvesting coarse-grained models from semi-structured digital collections of open-corpus education material (such as tutorials and textbooks) and mapping them into the pre-defined domain ontology serving as the main domain model and the reference point for multiple open-corpus collections. Once the mapping is established, the content from the processed open-corpus collection can be presented to students in adaptive way, according to their student models computed in terms of the central ontology. The rest of this paper describes the details of the approach, the adaptive e-learning service implementing it, and the results of the evaluation experiment demonstrating several learning effects of the developed service.

2   Ontology-Based OCP Approach

Information on the Web is not without structure. Authors of many online resources create them as a reflection of their own internal organization of related knowledge. They encode this organization by formatting the text with lists and headings, breaking documents into sections and pages, linking pages together, creating tables of contents, etc. The approach proposed in this paper attempts to utilize this hidden semantic layer of well-formatted content collections to achieve fully automated OCP. The entire procedure consists of the three steps presented below.

Step 1: Modeling of Open-Corpus Content in Terms of its Structure. An author creating an instructional resource tries to make it more readable and understandable by structuring it into chapters and sections. Every section is intended to represent a coherent topic. It is given a title conveying the meaning of the topic and contains the text explaining it. Their main purpose is to structure content, but they inescapably structure the knowledge, as well. A topic-based structure of such a resource can be parsed automatically and represented formally, e.g. as an RDF model. This model will have some drawbacks: (1) subjectivity; (2) poor granularity; (3) undefined semantics of topics and relations between them; (4) incompleteness. Yet, such model provides means to access the material of the collection in terms of topics, reason about the material in terms of topics and adapt the material in terms of topics.

Step 2: Mapping Extracted Model into the Central Domain Ontology. Extraction of the hidden semantic layer is not enough for two reasons. First, coarse-grained domain and content models can be effective when delivering the adaptation to students, but cannot maintain student modeling of good quality [6]. Second, a model extracted from a single collection can be used to adaptively present only the content of this collection. The learning material from different collections will be isolated. The solution is to use the central domain ontology as a reference model. It will help to model students' knowledge and to translate between the topic-based structures of individual open-corpus collections. The connection between the harvested models and the central ontology is established based on the automatic mapping of these models.

Step 3: Mediated Personalization of Open-Corpus Learning Material. Once the two models are mapped the systems can reason across them. The mapping bridge between the central ontology and the tutorial model enables two principle procedures: (1) tracing student's actions with the tutorial's topics, representing these actions in terms of the ontology concepts and updating the ontology-based student model; (2) requesting the current state of student model expressed in terms of ontology concepts, translating it into the open-corpus topics, and adapting students' access to the open-corpus material. Fig. 1 summarizes the principle relation between the components of the central ontology and the open-corpus material, as well as the information flow across these relations.

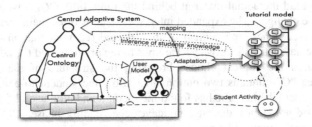

Fig. 1. Meditated personalization of open-corpus learning material

3   The Ontology-Based Open-Corpus Personalization Service

The proposed approach has been implemented in the Ontology-based Open-corpus Personalization Service (*OOPS*). It has been developed as a value-added service used in parallel with a central exercise system and augmenting it with adaptive access to supplementary reading material. As a central system we used *QuizJET* – the system serving parameterized exercises for Java programming language [7]. Both *OOPS* and *QuizJET* have been integrated with the *CUMULATE* user modeling server [8]. As an open-corpus collection of instructional material we used the electronic version of an introductory Java textbook. *QuizJET* is responsible for objective assessment of students' knowledge. Its exercises are indexed in terms of the central ontology and it report students' activity to the central user modeling component – *CUMULATE*, which models students' knowledge in terms of the central domain ontology and reports it to *OOPS*. As a result, student practicing with *QuizJET* exercises and struggling with a difficult topic receives recommendations of relevant open-corpus reading material from *OOPS*.

The student interface of *OOPS* has two interaction phases: recommendation (when a student is presented with a list of recommended pages) and reading (when a student is studying recommended material). Left part of Fig. 2 presents a screenshot of the recommendation phase. Area "B" is the interface of the central system – *QuizJET*. Area "A" presents a list of recommendations produced by *OOPS* for the current exercise of *QuizJET*. Every item in the list is a topic label from the harvested open-corpus content collection. The order of an item in the list is determined by its relevance to the current *QuizJET* exercise computed based on the aggregated similarity of the topic and the concepts indexing the exercise. The similarity values are calculated by the ontology mapping algorithm. The recommended items are provided with adaptive annotation in form of human-shaped colored icons. The coloring of an icon annotating a topic represents the amount of knowledge a student has demonstrated for the learning material behind this topic measured in terms of central ontology concepts mapped to the topic and provided by the central student model. The annotation level is computed as a weighted aggregate of knowledge levels for all concepts mapped into the topic. Once a student decides to accept a recommendation by clicking on a topic link, s/he goes into the reading phase of the *OOPS* interface (right part of Fig. 2). In this phase, *OOPS* provides a student with an opportunity to read the actual material behind the topic link. *OOPS* widget expands, and its interface changes. The expanded interface contains three main areas. Area "A" is the content area, where the content of the selected recommendation is presented. Area "B" is the navigation area, where the links to the previous and the next topics are presented, should the student choose browsing the structure of the open-corpus collection. Area "C" contains two buttons that allow the student to exit the reading phase and to report whether s/he has found the recommendation useful for the current learning task or not. Once the student leaves the reading phase, *OOPS* interface switches to the recommendation phase again.

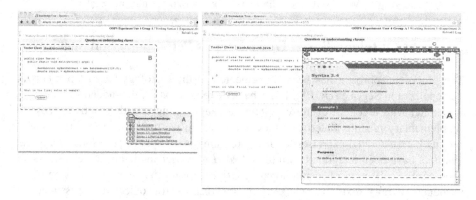

Fig. 2. Interface: *Left*: recommendation phase; *Right*: reading phase

4   The Evaluation

This section presents the results and the procedure of the *OOPS* service evaluation. It was organized as a controlled balanced experiment comparing the developed system against two control conditions. The experimental system (*open-corpus*) provided students solving *QuizJET* exercises with open-corpus recommendation of reading material. Another version of the system (*closed-corpus*) had the identical interface and generated recommendations from the same pool of reading material, but used traditional closed-corpus adaptation approach based on the manual indexing of recommended pages. The last configuration of the system (*textbook*) did not recommend any reading material. Instead, students using this version had a hard copy of the textbook, which was the source of reading material for the first two versions. The experiment consisted of two sessions corresponding to two sets of introductory Java topics. First set included simpler topics: from basics of variable and object handling to conditional statement and Boolean expressions. The second set covered more advanced topics: from loops to arrays and ArrayLists. Each session started with a pretest, continued with the 30 minutes work with the system and ended with the posttest. Forty subjects with limited Java programming experience participated in the experiments. Subjects were randomly assigned to one of the four groups:

A. Session $1_{easy\ topics}$ – *open-corpus*; Session $2_{complex\ topics}$ – *closed-corpus*;
B. Session $1_{easy\ topics}$ – *open-corpus*; Session $2_{complex\ topics}$ – *textbook*;
C. Session $1_{easy\ topics}$ – *closed-corpus*; Session $2_{complex\ topics}$ – *open-corpus*;
D. Session $1_{easy\ topics}$ – *textbook*; Session $2_{complex\ topics}$ – *open-corpus*;

General Learning Effect. In order to verify that work with the system actually leads to learning, pair-wise comparisons of scores on the pre-test and the post-test have been made (Table 1). Significant learning has been registered for all groups and conditions during Session 1. For Session 2, the *open-corpus* condition and the *closed-corpus* condition resulted in significant (or bordering on significance) learning. At the same time, the *textbook*-condition led to no learning.

Table 1. General learning effect statistics ($Score_{post-test}$ VS. $Score_{pre-test}$)

Group	Session 1		Session 2	
	t(9)	p-value	t(9)	p-value
A	3.787	0.004	1.941	0.084
B	4.409	0.002	0.0	1.0
C	8.213	<0.001	2.250	0.051
D	4.077	0.03	3.361	0.008

Effect on Learning Complex Material. The main difference between Session 1 and Session 2 is material complexity. The analysis of the general learning effect suggests that the recommendation of supplementary reading material can have a positive influence on learning the complex learning material. During Session 1 (easy topics), none of the comparisons resulted in significant difference in Knowledge Gain. However, once the learning material became more complex (Session 2), the *open-corpus* system significantly outperformed the *textbook*: Knowledge Gain for the *open-corpus* condition (M=1.55; SD=1.23) is significantly higher than Knowledge Gain for the *textbook* condition (M=0.60; SD=0.97): t(28) = 2.124; p = 0.043 during Session 2 (complex topics). At the same time, no difference was observed between the closed-corpus and the open-corpus system when students were learning complex material. This is an important effect with a reasonable explanation. When learning easy material, students need less support from the system. They learn just by practicing with *QuizJET* exercises. And if they need extra reading, it is easier for them to find a relevant chapter in the textbook. On the other hand, when the material becomes complex, students can benefit from the recommendations and the adaptive annotations guiding them to the most important piece of reading material. Thus, personalized learning support results in better learning when support is needed. The comparison of open-corpus and closed-corpus conditions show that they are equally effective, which indicates that OCP produced by *OOPS* has similar quality to the traditional closed-corpus personalization.

Effect on Learning Conceptual Material. The personalization implemented by *OOPS* is aimed at achieving two instructional goals: (1) Support students solving self-assessment exercises by bringing them the most relevant reading material; (2) Balance students' learning by giving them the opportunity to read instructional texts in addition to practicing. The second means that *OOPS* should contribute better to the knowledge of important concepts and fact in the domain. The pre- and post-tests of both sessions contained two kinds of questions: those evaluating students' practical skills in code understanding and manipulation and those checking their factual and theoretical knowledge. In order to measure the conceptual knowledge gain, only the second kind of questions was taken into account. The comparison of conceptual knowledge gain between the *open-corpus* and the *textbook* conditions shows that the hypothesis is partially confirmed. During Session 1, the conceptual knowledge gain for the *open-corpus* condition (M=2.61; SD=1.75) was higher, than for the *textbook* (M=1.75; SD=1.15), but not significantly: $t(28)=1.762$; $p=0.089$. However, during Session 2 (complex topics), the conceptual knowledge gain for the open-corpus condition (M=0.73; SD=0.47) was significantly higher than the one for the textbook condition (M=0.30; SD=0.42): $t(28)=2.403$; $p=0.023$. No significant effect was observed when comparing the open-corpus and the closed-corpus conditions.

5   Conclusion

In this paper, we have addressed the problem of OCP in the context of e-Learning and proposed a solution for it. As a proof-of-concept the adaptive e-Learning service *OOPS* has been implemented. It adaptively recommends and annotates pages for supplementary reading to students solving self-assessment exercises. The evaluation of *OOPS* has shown that students were able to achieve significant learning while using the open-corpus version of the system. *OOPS* significantly improved students' knowledge gain when they work with more challenging learning material. In comparison, students using the textbook demonstrated no significant learning while working with complex topics. *OOPS* helped to maintain a more balanced learning by significantly improving gain in conceptual knowledge, no such effect was observed for the textbook. At the same time, on no tests, we could statistically distinguish between the results of the proposed fully-automated open-corpus approach and a conventional closed-corpus technique based on a carefully handcrafted content model.

References

1. Henze, N., Nejdl, W.: Adaptation in open corpus hypermedia. Int. J. of Artificial Intelligence in Education, 12(4), 325–350 (2001)
2. Carmona, C., Bueno, D., Guzmán, E., Conejo, R.: SIGUE: Making Web Courses Adaptive. In: De Bra, P., Brusilovsky, P., Conejo, R. (eds.) AH 2002. LNCS, vol. 2347, pp. 376–379. Springer, Heidelberg (2002)
3. Jovanović, J., Gasevic, D., Devedzic, V.: Ontology-based automatic annotation of learning content. Int. J. on Semantic Web and Information Systems 2(2), 91–119 (2006)
4. Apted, T., Kay, J.: MECUREO Ontology and Modelling Tools. Int. J. of Continuing Engineering Education and Lifelong Learning 14(3), 191–211 (2004)
5. Brusilovsky, P., Henze, N.: Open Corpus Adaptive Educational Hypermedia. In: Brusilovsky, P., Kobsa, A., Nejdl, W. (eds.) Adaptive Web 2007. LNCS, vol. 4321, pp. 671–696. Springer, Heidelberg (2007)
6. Sosnovsky, S., Brusilovsky, P.: Layered Evaluation of Topic-Based Adaptation to Student Knowledge. In: Workshop on Evaluation of Adaptive Systems at UM 2005, Eduburgh, UK, pp. 47–56 (2005)
7. Hsiao, S., Brusilovsky, P., Sosnovsky, S.: Web-based Parameterized Questions for Object-Oriented Programming. In: E-Learn 2008, Las Vegas, NV, USA, pp. 3728–3735 (2008)
8. Brusilovsky, P., Sosnovsky, S., Shcherbinina, O.: User Modeling in a Distributed E-Learning Architecture. In: Ardissono, L., Brna, P., Mitrović, A. (eds.) UM 2005. LNCS (LNAI), vol. 3538, pp. 387–391. Springer, Heidelberg (2005)

Encouragement of Collaborative Learning Based on Dynamic Groups

Ivan Srba and Mária Bieliková

Slovak University of Technology in Bratislava
Faculty of Informatics and Information Technologies
Ilkovičova 3, 842 16 Bratislava, Slovakia
{srba,bielik}@fiit.stuba.sk

Abstract. We propose a method for creating different types of study groups with aim to support effective collaboration during learning. We concentrate on the small groups which solve short-term well-defined problems. The method is able to apply many types of students' characteristics as inputs, e.g. interests, knowledge, but also their collaborative characteristics. It is based on the Group Technology approach. Students in the created groups are able to communicate and collaborate with the help of several collaborative tools in a collaborative platform called PopCorm which allows us to automatically observe dynamic aspects of the created groups. The results of these observations provide a feedback to the method for creating groups. In the long term experiment groups created by our method achieved significantly better results in the comparison with the reference method (k-means clustering).

Keywords: CSCL, Collaboration, Group Technology, Groups.

1 Introduction

Research in Computer-Supported Collaborative Learning (CSCL) domain can be grouped into systematic and dialogical approaches [4]. The systematic approach concerns the creating of models describing how the specific features of technological systems support or constrain collaboration, reasoning, knowledge representation, and structure of discourse [3]. On the other hand, the dialogical approach considers learning as a social-based activity. Therefore, we should pay appropriate attention to the group formation process which can significantly influence collaboration and thus, it is possible source of many improvements how to support effective collaboration.

In this paper, we deal with the dialogical approach, especially with the encouragement of students in collaborative learning by creating dynamic short-term study groups and design a collaboration platform which allows these groups to collaborate efficiently. The reason to follow this goal is the fact that we do not know what makes collaboration really effective and therefore how to join the students into effective groups. Thus, if we want students to collaborate effectively we should help them find appropriate collaborators.

A. Ravenscroft et al. (Eds.): EC-TEL 2012, LNCS 7563, pp. 432–437, 2012.

2   Method for Creating Dynamic Groups

Recently, several methods and techniques were applied to group formation, e.g. ontologies, genetic algorithms, agent-based methods or methods for socially intelligent tutoring [8]. These methods usually use only one source of information about students and do not consider actual context, i.e. characteristics of the collaboration. Also they suppose that a teacher knows which attributes make collaboration more effective.

One prospective approach to group formation is based on Group Technology. According to Selim, et al. [6] Group Technology (GT) is an approach to manufacturing and engineering management that helps manage diversity by capitalizing on underlying similarities in products and activities. One application of the GT approach in manufacturing is a so-called Cellular Manufacturing. Groups of machines should be located in close proximity in order to produce a particular family of similar parts and thus minimize production and transfer time [2]. Several types of methods are described in [6] to solve the problem of cell formation. The most appropriate for us are procedures based on cluster analysis, especially array-based clustering techniques.

The basic idea of our method is derived from the GT approach because it solve similar problem as we have to solve to reach our goal. Analogy between domain entities can be easily found. It is possible to replace a machine with a student, a part with a characteristic, assignment of parts to the machine with assignment of characteristics to the student, and a family of similar parts with a set of related characteristics. Moreover, we can find this analogy also in goals; instead of optimizing machine production we need to optimize collaboration process.

The proposed method consists of two main processes:

1. *Group Formation* takes different personal or collaborative characteristics as inputs and creates study groups. Personal characteristics can be student's knowledge, interests, or any other personal characteristics (e.g. age, gender). We can obtain these characteristics from many sources, such as existing user models, social networks or questionnaires. Furthermore, characteristics can include collaborative aspects, such as students' collaborative behavior;
2. *Collaboration* allows students of created groups to participate on task solving via a collaboration platform which provides appropriate collaboration tools together with functionality for observation groups' dynamic aspects which are used as one of inputs in the method for creating groups.

Input data to our method are composed of two matrices: a matrix of related characteristics and a matrix of assignments of characteristics to students. We consider characteristics related if their combination leads to positive influence on collaboration.

The matrix of related characteristics is defined as follows. Let C be the set of all characteristics $C = \{c_j\}$, $j = 1,2,...,n$. Every characteristic can be represented as a n-dimensional vector $c_j = (c_j^1, c_j^2, ..., c_j^n)$, where:

$$c_j^i = \begin{cases} 1 & \text{if characteristic } c_j \text{ should be combined with characteristic } c_i \\ 0 & \text{if characteristic } c_j \text{ should not be combined with characteristic } c_i \end{cases} \quad (1)$$

The matrix of assignment of characteristics to students is defined as follows. Let L be the set of all learners $L = \{l_k\}$, $k = 1,2,...,m$. Every learner can be represented as a n-dimensional vector $l_k = (l_k^1, l_k^2, ..., l_k^n)$, where:

$$l_k^i = \begin{cases} 1 & \text{if characteristic } c_j \text{ is typical for learner } l_k \\ 0 & \text{if characteristic } c_j \text{ is not typical for learner } l_k \end{cases} \tag{2}$$

Calculation of clusters of learners and characteristics is performed in several steps. First of all, three values are defined for each learner vector $l_k \in L$ and characteristic vector $c_j \in C$:

1. Value a is a number of characteristics contained in both vectors.
2. Value b is a number of characteristics which are typical for the current student but should not be connected with the current characteristic.
3. Value c is a number of characteristics which are not typical for the current student but should be connected with the current characteristic.

Then similarity (SC) and relevance coefficient (RC) can be defined as follows:

$$SC(l_k, c_j) = \frac{a}{a+b+c} \tag{3}$$

$$RC(l_k, c_j) = \frac{a}{a+b} \tag{4}$$

Afterwards *Group Compatibility Matrix*, $GCM = (a_{ij})$, $i \in [1,n]$, $j \in [1,m]$, is calculated as:

$$a_{ij} = \begin{cases} 1 & \text{if } SC \geq \theta^{SC} \text{ and } RC \geq \theta^{RC} \\ 0 & \text{else} \end{cases} \tag{5}$$

Values $\theta^{SC}, \theta^{RC} \in \langle 0,1 \rangle$ represent minimal thresholds for similarity and relevance coefficient. Algorithm set thresholds to ones and continuously decreases them until a valid Group Compatibility Matrix (GCM) matrix is found. A GCM matrix is valid as soon as each student has at least one assigned characteristic. Finally, it is necessary to perform clustering on a GCM matrix with any array-based clustering algorithm. We used Modified Rank Order Clustering (MODROC) for our purpose.

Output data from our method is a GCM matrix in which the clusters of the students and the characteristics are concentrated along the main diagonal (see Table 1, as characteristics are used activities which are typical for particular students). Assignment of a student to a cluster of characteristics means that this student has these characteristics or these characteristics should combine with characteristics which are typical for this student. Particular study groups can be created with any combination of students from the same cluster.

We apply our method iteratively which allows us to use several matrices of related characteristics. Each matrix can represent different requirements how to combine characteristics together, i.e. a matrix of complementary characteristics or a dynamic matrix based on achieved results. The dynamic matrix can solve the problem of absence of information about attributes (in our proposal characteristics' combinations) which make collaboration effective and successful. After each group finishes task solving, its collaboration and achieved result is evaluated. Afterwards each combination between characteristics which are typical for members of this group is strengthened according to the achieved evaluation. Equally the dynamic matrix of assignment of characteristics to students can be updated according to the number of performed activities which contribute to these characteristics.

Table 1. An example of clustered GCM matrix acquired in the first phase of evaluation

Characteristic activity	Student 1	Student 2	Student 3	Student 4	Student 5
Warn of mistake	1	1	0	0	0
Accept warn of mistake	1	1	0	0	0
Write general message	0	0	1	0	0
Ask for explanation	0	0	0	1	1
Give explanation	0	0	0	1	1
Propose action	0	0	0	1	1
Accept action	0	0	0	1	1
Write praise	0	0	0	1	0

3   Evaluation

Evaluation of our method for group formation cannot be accomplished without a collaborative environment where it is applied. Therefore, we have designed and realized the collaboration platform called *Popular Collaborative Platform – PopCorm* which is integrated within Adaptive Learning Framework ALEF [7]. It consists of four collaborative tools which are suitable for task solving in CSCL: a text editor, a graphical editor, a categorizer, and a semi-structured discussion. The categorizer is a special tool developed for solving different types of tasks in which the solution consists of one or more lists (categories). The semi-structured discussion represents a generic communication tool independent of a particular type of a task being solved. It provides 18 different types of messages (e.g. propose better solution). These different message types allow us to automatically identify student's activities. Recorded activities are used to measure the collaboration by set of seven dimensions designed rooted in studies in psychology: sustaining mutual understanding, information exchanges for problem solving, argumentation and reaching consensus, task and time management, sustaining commitment, shared task alignment and fluidity of collaboration.

We performed evaluation of our method and the collaboration platform in two phases. Firstly, we realized in February 2012 a short-term controlled experiment. The purpose of this experiment was to evaluate preconditions of the proposed method; namely, the precondition whether activities form natural clusters which influence collaboration in the positive or, on the contrary, in the negative way. Moreover, the experiment was also an opportunity to get valuable comments on the implementation of the collaboration platform. Five participants in total took part in the experiment and solved 12 tasks. The precondition was confirmed and our method was able to identify three clusters of students and activities at the end of the experiment with grouping efficacy more than 88% (see Table 1).

The second phase consisted of a long-term experiment which was realized during summer term as a part of education on the course Principles of Software Engineering at the Slovak University of Technology in Bratislava. 106 students in total participated in 208 created groups. 3 613 activities are recorded during task solving. Each activity corresponds to one sent message in the semi-structured discussion.

Table 2. Comparison of achieved results during the second phase of the experiment

Groups created	Average evaluation	Feedback
By the proposed method	0.459	4.01
By the reference method (k-means clustering)	0.392	3.55
Randomly	0.422	3.29

The 8-dimensional evaluation of the groups created using our method was compared with a reference method (k-means clustering) and randomly created groups (see Table 2). Groups created by our method achieved the most effective and successful collaboration in comparison with the other two types of groups. We employ ANOVA statistical model to evaluate significance of achieved results and we got p-value 0.0048. Thus, the achieved results can be considered as highly significant. Additionally, students have provided a higher explicit feedback in these groups.

4   Related Work and Conclusion

Several works employing Group Technology (GT) approach in CSCL domain exist. Pollalis, et al. [5] proposed a method for learning objects recommendation to student groups according to students' knowledge of relevant domain terms. Two input matrices were used. The first one represented student's knowledge; the second one represented similarity or mutual dependency of relevant domain terms which was derived from their common occurrence in the same learning object. The output was clusters of students and learning objects which were suitable for these students to learn.

Similar approach is described in [2]. The main goal of this research was to identify sets of students which use similar strategies to solve mathematical exercises. Similarly to the previous work, two matrices were calculated: the dynamic matrix representing assignment of strategies to students and the static matrix representing mutual similarity of strategies. The output was clusters of students and assigned groups of strategies. The identified clusters can be used to assign new task to particular group of students according to strategies which are familiar to the members of the group and which are suitable to solve this task as well.

As opposed to previous two works, authors in [1] considered only one matrix as input. This matrix represents teachers and subjects they teach. A hybrid grouping genetic algorithm was used to identify groups of similar subjects.

Our method considers its iterative application in contrast to the existing methods for group formation based on GT approach. This allows us to take into consideration already achieved students' results in collaboration and adjust input parameters to encourage better collaboration between students. It means that we can start the group formation process with no or minimal information about students and related characteristics. Our method then automatically learns which collaborative characteristics are typical for students and which characteristics should be combined together to achieve more effective collaboration. Moreover, automatic evaluation by seven dimensions defined according psychological studies provides immediate feedback to students and advices how to collaborate more effectively.

Our method is not limited only to the CSCL domain. It can be easily applied in other domains where dynamic groups should be created according to different user characteristics. We have successfully applied the proposed method during the experiment in collaborative learning by creating dynamic short-term study groups, which showed high potential of proposed method. It would not be possible to evaluate our method for group creation without the collaborative platform PopCorm which provides students the appropriate environment for effective task solving and automatic identification of their activities.

Acknowledgement. This work was supported by grants No. VG1/0675/11, VG1/0971/11 and it is a partial result of the Research and Development Operational Program for the projects SMART, ITMS 26240120005 and SMART II, ITMS 26240120029, co-funded by ERDF.

References

1. Blas, L., et al.: Team formation based on group technology: A hybrid grouping genetic algorithm approach. Comput. Oper. Res. 38(2), 484–495 (2011)
2. Cocea, M., Magoulas, G.D.: Group Formation for Collaboration in Exploratory Learning Using Group Technology Techniques. In: Setchi, R., Jordanov, I., Howlett, R.J., Jain, L.C. (eds.) KES 2010, Part II. LNCS, vol. 6277, pp. 103–113. Springer, Heidelberg (2010)
3. Dillenbourg, P.: What do you mean by collaborative learning? In: Dillenbourg, P. (ed.) Collaborative Learning: Cognitive and Comp. Approaches, pp. 1–19. Elsevier, Oxford (1999)
4. Ludvigsen, S., Mørch, A.: Computer-supported collaborative learning: Basic concepts, multiple perspectives, and emerging trends. In: McGaw, B., Peterson, P., Baker, E. (eds.) The Int. Encyclopedia of Education, 3rd edn., Elsevier (2009)
5. Pollalis, Y.A., Mavrommatis, G.: Using similarity measures for collaborating groups formation: A model for distance learning environments. European Journal of Operational Research 193(2), 626–636 (2009)
6. Selim, H.M., et al.: Cell formation in group tech.: review, evaluation and directions for future research. Comput. Ind. Eng. 34(1), 3–20 (1998)
7. Šimko, M., Barla, M., Bieliková, M.: ALEF: A Framework for Adaptive Web-Based Learning 2.0. In: Reynolds, N., Turcsányi-Szabó, M. (eds.) KCKS 2010. IFIP AICT, vol. 324, pp. 367–378. Springer, Heidelberg (2010)
8. Tvarožek, J.: Bootstrapping a Socially Intelligent Tutoring Strategy. Information Sciences and Technologies Bulletin of the ACM Slovakia 3(1), 33–41 (2011)

Part IV
Demonstration Paper

An Authoring Tool to Assist the Design
of Mixed Reality Learning Games

Charlotte Orliac, Christine Michel, and Sébastien George

Université de Lyon, CNRS,
INSA-Lyon, LIRIS, UMR5205
F-69621 Villeurbanne Cedex, France
{charlotte.orliac,christine.michel,
sebastien.george}@insa-lyon.fr

Abstract. Mixed Reality Learning Games (MRLG) provide new perspectives in learning. But obviously, MRLG are harder to design than traditional learning games environments. The main complexity is to cope with all the difficulties of learning design, game design and mixed reality design at the same time, and with the integration of all aspects in a coherent way. In this paper, we present existing tools and methods to design learning games or mixed reality environments. Then, we propose a model of MRLG design process. In the last part, we present an authoring tool, MIRLEGADEE, and how it supports this process.

Keywords: Learning Game, Mixed Reality, Design, Authoring tool.

1   Introduction

Milgram and Kishino [1] defined the Mixed Reality (MR) environments as real world and virtual world objects, presented together within a single display. MR has already been exploited in learning or game fields and has proven positive outcomes. Cook *et al.* [2] found that body gesturing improves long term learning, and in gaming area, MR preserve communication between players while adding functionalities such as saving the game [3]. Mixed Reality Learning Games (MRLG) combine the assets of game-based learning and the new perspectives provided by mixed reality. They offer real benefits for teaching: they enable active pedagogy trough the physical immersion of learners, "in situ" information while practicing and authentic context. Besides, the learner is strongly implicated and has a better motivation. Despite the huge opportunities brought by the MRLG, not so many exist. The little number of MRLG can be explained by the innovative, unusual and non-mature aspect of mixed reality technologies. Moreover, creating a MRLG is long and difficult.

In this paper, we focus on the design process of MRLG (only the pre-production, not its implementation). If some methods or tools exist for learning games or mixed reality design, there is no specific one for MRLG design. Our aim is to make MRLG creation easier, by providing useful tools or methods to support MRLG design.

A. Ravenscroft et al. (Eds.): EC-TEL 2012, LNCS 7563, pp. 441–446, 2012.

2   Existing Methods and Tools for MRLG Design

MRLG refer to three domains: mixed reality, games, and pedagogy. Each of these fields has its own tools and methods.

In MR area, task models [4] aim at structuring the task with a tree and interaction models [6] describe the MR systems for a given task. The latter enable the representation of the complexity of a MR system, especially with the difference between the real and digital elements. As both models are at a very high level of detail, we can only use them in addition to another model to enable a workflow description.

In game area, the existing tools support essentially the game development and not the conception. The specific step of design is not really assisted by tools or methods as it depends on the game designer itself. Even so, some game designers identified the main elements to consider in order to design a game, for instance the goal and the topic [7].

In educational field, some teachers already use learning scenario writing as a way to describe a unit of learning. Based on the theory of instructional design, some tools and models intend to support the description of learning scenario. We previously analyzed and compared 3 formalisms and found out that these models are not complete enough to describe a whole MRLG [8]. In particular, they cannot specify the devices used in MRLG: they can neither express a distinction between tangible and digital parts nor describe the interface of digital devices nor the physical positions of devices.

At the intersection of pedagogy and games, some tools aim at assisting learning game design, such as EDoS (now called LEGADEE) [9] or ScenLRPG [10]. But they do not allow the description of other types of interactions than classical human-computer interactions. As for educational models, the description of MR activities is not possible with these tools.

The methods and tools are quite different from one field to another. The main designers' concerns are really dissimilar and each field use a specific vocabulary. For these reasons, none of the existing tools or methods can be used directly to design a MRLG. A combination of these tools is also difficult because they are too different from each other, and they could not be linked easily into a whole global tool. Furthermore, except for the game design, the previously presented methods and tools only support the formalization and not the whole design process. Therefore, there is a real need for specific method and tools.

3   MRLG Design Process

We divide MRLG design process in four phases: definition of the project, creativity, formalization and the final specifications. For the creativity phase, where ideas are explored, decided, and changed, we rely on the process from Wallas [11], which includes 4 steps: preparation, incubation, illumination and verification (see Table 1).

The MRLG design process is included in a larger MRLG production process. After the design step, a prototype will be implemented and tested. With an iterative method, results of testing may then lead to the refinement of the former project.

Table 1. The seven steps of the MRLG design process

Project definition		The design process starts with a first idea, a main purpose or a motivation. This step leads to the definition of the needs, constraints, and main ideas of the project
Creativity	Preparation	Preparation consists in exploring data from project first ideas. A designer needs to gather information about what already exists (what are the existing devices, or game types?) and how s/he can combine these elements (which game type for which competencies?). S/he may also look for heuristics or best practices.
	Incubation	According to Wallas [11], incubation is an "interval free from conscious thought on the particular problem". During this step, MRLG designer's mind works unconsciously.
	Illumination	Several ideas emerge from the incubation step. They constitute a set of possible solutions.
	Verification	Each idea or possible solution (from the set of solution) is verified, according to the first needs, the constraints, and the other chosen solutions. When a first choice is made, whatever the choice is, it restricts the other possibilities.
Formalization		Once chosen, the solutions must be described. Creativity and formalization steps are actually bound: the formalization of ideas is also a way to verify, evaluate and refine the possible solutions because they constitute a support for discussion in a team.
Final specifications		The ideas are refined until a complete MRLG is designed and described in the final specifications for developers.

4   An Authoring Tool to Assist MRLG Design

We designed an authoring tool, named MIRLEGADEE (MIxed Reality LEarning GAme DEsign Environment), to support the MRLG design. This tool is an extension of LEGADEE [9], which already supports the design of learning games using a computer. With MIRLEGADEE, the goal is mainly to support the *formalization*, in order to provide complete and understandable specifications for the developers, but it also intends to assist the *project definition* and the *creativity*, as it is a lack of the existing tools.

4.1   Supporting the Project Definition

MIRLEGADEE includes an Ideas Wall to support the first step of MRLG design process, the *project definition*. Designers are encouraged to write their ideas, motivations or constraints using post-it notes. This step enables the expression of ideas freely and without influence of the model described below (way of describing the idea, order...). Then the identified constraints or important ideas can be specified in a more formal way with the model.

4.2   Supporting the Creativity

MIRLEGADEE supports the *creativity* by providing information about some elements the designer may use and describe for the MLRG (game types, games principles, devices for MR, etc.). It may also provide good practices or advices. The idea is not to include in the authoring tool all the information that the designer could need, but to help him to have a statement about all the fields linked with the *global specifications* (see part (1) in Fig. 1). Obviously, designers are not limited to our proposals and can also express their own ideas. Currently, we propose some lists of possibilities or examples for some elements. In addition, we provide a tool to support specifically the choice of devices, which constitute a crucial design issue for MRLG. This tool, still under development, aims at providing a better knowledge of the devices which support MR, along with their possibilities for learning or game aspects, and their constraints. It is based on identified uses of MR in existing environments.

4.3   Supporting the Formalization

MIRLEGADEE is structured by the underlying model f-MRLG. This model is based on a usual process in instructional design: the description of learning scenario. In our case, the MRLG designer has to integrate several aspects (learning, game and mixed reality) into a same scenario in a coherent way. In order to make the design activity easier and also to help the designer to focus attention on every aspect to create a well-balanced MRLG, we add a first stage. In this phase, the designer defines separately each element into a special part called *global specifications* (part (1) on Fig.1).

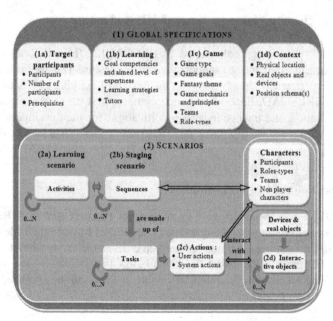

Fig. 1. Overview of the f-MRLG model

In MIRLEGADEE, *global specification's* elements are filled trough forms. The ideas previously written on the "ideas wall" are also displayed in this part.

Then, the designer describes the *scenarios* (part (2) in Fig 1.) as a sequence of events. According to Marfisi-Schottman et al. [9] the workflow for a learning game can be described from two points of view: the usual *learning scenario* (2a) and a fun scenario (that we call *staging scenario* (2b)). At the lowest level of the *staging scenario*, abstract *tasks* describe more precisely each *sequence*. *Actions* (2c) refine abstract *tasks* into concrete interactions. Through *user actions*, the participants interact with other participants or *interactive objects*, which are parts of the *devices* or *real objects* defined in context (2d). LEGADEE already allows the creation of both scenarios and the links to characters through palettes. In MIRLEGADEE, we use the palettes to display important elements from the *global specifications*, so the designer can include them in the scenarios.

In order to remain flexible, we consider that no element of the model is mandatory. Depending on the designer's profile, s/he can also complete the elements in any order (and may even start with the description of the scenarios in some cases).

4.4   Supporting the Export of the Final Specifications

MIRLEGADEE, like LEGADEE, includes a validation step. It checks for instance if all characters created really intervene in the scenario. At last, final specifications can be exported in an XML (machine readable) or HTML format (better looking).

5   First Evaluation and Discussion

An evaluation of a paper prototype of MIRLEGADEE was organized during the Game Based Learning Summer School 2011[1]. The 49 participants were gathered by groups, from 4 to 6 persons. They were asked to design a MRLG by using the formalism f-MRLG and lists of possibilities (to support the exploration step) on a structured poster on a table.

A questionnaire was filled at the end of the workshop (results in Table 2). Taken as a whole, the model and the authoring tool provide useful support for MRLG design. The participants highlighted that the model helped them to structure their ideas and to consider every aspect of a MRLG. Regarding the final posters, 9 groups on 10 performed completely the MRLG description and correctly understood the elements of f-MRLG. Only the interactive objects were sometime mixed up with devices. The computer version should prevent this error: an interactive object must belong to a device or real object already defined in the *global specifications*.

A second evaluation of the authoring tool is planned with the computer prototype. We will evaluate the usability of MIRLEGADEE but also its utility. In particular, we wish to know if the tool provided to choose device could help the MRLG designers to make a decision and to justify it, and if developers can use the final specifications.

---

[1] http://gbl2011.univ-savoie.fr/

Table 2. Synthesis of questionnaire answers

Criteria	Positive answers	Negative answers	NA
Expertise in learning or task modelling	26	3	17
Interest of using MR in learning games	37	3	9
Utility of the model for designing MRLG	30	7	6
Utility of the model for mutual understanding	23	1	5
Effectiveness of MRLG design using the model	22	5	17

Acknowledgements. This research is undertaken within the framework of the SEGAREM (SErious GAmes and Mixed Reality) project. The authors wish to thank both the DGCIS (Direction Générale de la Compétitivité, de l'Industrie et des Services) for the fund and the partners of this project, Symetrix and Total Immersion, for their collaboration.

References

1. Milgram, P., Kishino, F.: A taxonomy of mixed reality visual displays. IEICE Transactions on Information and Systems E Series D 77, 1321–1329 (1994)
2. Cook, S.W., Mitchell, Z., Goldin-Meadow, S.: Gesturing makes learning last. Cognition 106, 1047–1058 (2008)
3. Nilsen, T., Linton, S., Looser, J.: Motivations for augmented reality gaming. Proceedings of FUSE 4, 86–93 (2004)
4. Paternò, F., Mancini, C., Meniconi, S.: ConcurTaskTrees: A diagrammatic notation for specifying task models. In: Proceedings of the IFIP TC13 International Conference on Human-Computer Interaction, pp. 362–369 (1997)
5. Chalon, R., David, B.T.: Irvo: an interaction model for designing collaborative mixed reality systems. In: Présenté à HCI International Conference 2005, Las Vegas, United States (2007)
6. Crawford, C.: The art of computer game design. Osborne/McGraw-Hill (1982)
7. Orliac, C., George, S., Michel, C., Prévot, P.: Can we use Existing Pedagogical Specifications to Design Mixed Reality Learning Games? In: Proceedings of the 5th European Conference on Games Based Learning, Athens, Greece, pp. 440–448 (2011)
8. Marfisi-Schottman, I., George, S., Tarpin-Bernard, F.: Tools and Methods for Efficiently Designing Serious Games. In: Proceedings of 4th Europeen Conference on Games Based Learning, Copenhagen, Denmark, pp. 226–234 (2010)
9. Mariais, C., Michau, F., Pernin, J.-P., Mandran, N.: Supporting Learning Role-Play Games Design: A Methodology and Visual Formalism for Scenarios Description. In: Proceedings of the 5th European Conference on Games Based Learning, Athens, Greece, pp. 378–387 (2011)
10. Wallas, G.: The art of thought. Harcourt, Brace and Company (1926)

An Automatic Evaluation of Construction Geometry Assignments

Šárka Gergelitsová and Tomáš Holan

Charles University in Prague, Faculty of Mathematics and Physics
Malostranské nám. 2, Prague, Czech Republic
sarka@gbn.cz, tomas.holan@mff.cuni.cz

Abstract. A new method for evaluating assignments on constructions in geometry is proposed. Instead of comparing drawn objects and their parameters, a process of construction is evaluated as a procedure of getting from the given input to the required output. The method was implemented and used by approximately 10 teachers and 300 students of secondary schools with positive results.

Keywords: construction, geometry, task, evaluation, GeoGebra.

1 Introduction

An individual and unassisted activity of the student is an important part of the educational process. There are a lot of forms of these activities – tests, homework etc. – and each of them requires some feedback evaluating the correctness of students' work.

This paper deals with a new method of an effective automatic evaluation of students' solution of plane geometry constructions tasks.

1.1 Types of Answers

The evaluation of the student's answer depends on the type of the answer.

In the simplest case the answer is simply one number, a name or a date and evaluation of such answer consists in comparing the given answer with the correct one.

On the contrary, in the most complex and complicated cases precise and strict evaluation is not possible at all, e.g. if the answer is in the form of an essay or work of art. In such cases it is possible to evaluate spelling or other formal prerequisites, but it is not enough to give the correctness of the answer. But such cases obviously are not frequent at elementary or secondary schools.

Mathematics is somewhere between these extremes because students learn procedures and algorithms besides definitions and formulas. For testing and homework students get a problem with specific data and their task is to know and choose the right procedure and use it to get the required results. To evaluate such results you only need to compare the given results and the correct ones.

A. Ravenscroft et al. (Eds.): EC-TEL 2012, LNCS 7563, pp. 447–452, 2012.

The more complex cases are assignments where the student is not expected to say the result given by applying the procedure but the procedure itself. An example of such a case is the assignment in plane geometry, geometric constructions.

The following is an example of such an assignment: "Given three vertices of a triangle. Find the circle circumscribed of this triangle."

With the fixed points given the result is one particular circle and the assignment could be solved by selecting the right procedure and accomplishment of all its steps.

More interesting than an ability to "memorize the list of procedures, choose the right one and execute it" would be an ability "find or create the correct procedure", but this ability is hard to test or measure.

1.2   Construction Geometry Assignments

Construction geometry assignments in their generic form are represented by particular data, such as coordinates, lengths and angles and the result that the student gives to the teacher is a drawing.

The teacher can then assess the student's thoroughness, the existence of correct elements, including the auxiliary ones, the correct placement of these elements, but this does not give the evidence, that the student solved the assignment correctly. And vice versa, some solution can arrive at the correct result using different procedures.

It is also possible to ask students to prepare the construction protocol and to compare it with the correct one, but some students fail to produce the correct protocol and moreover the same task can be solved by different procedures and therefore simple comparing the protocols is not enough.

2   An Automatic Evaluation of Construction Geometry Assignments

We propose a new way of automatic evaluation of constructions used in geometry assignments.

Our way uses the dynamic geometry system GeoGebra ([1], [2]), in which students construct their solution in the computer and a part of the solution is also the whole procedure of construction.

The novelty of our evaluation method is based on the fact that we do not compare elements of construction nor the procedure (as e.g. [3]) but that we exploit the procedure of the solution to be evaluated as a way to get from elements given to elements to be constructed.

The correctness of the construction is evaluated by executing the procedure of the construction to a number of different collections of input objects and we test whether the procedure gives the correct result elements in all cases. If the evaluated procedure gives correct result for all cases of input data, we consider it right.

That means that the student can solve the task by using whatever procedure, if the submitted answer always matches the data, it is considered right.

3   Implementation

We have implemented the described principle in the form of web application named GeoTest ([4], [5]).

In GeoTest a logged-in teacher can create accounts for his/her students, create groups corresponding to one test or one homework and assign them tasks from the list of prepared tasks.

A logged-in student can see a list of his groups, e.g. last week's assignments, tests in the school, today's assignments and can select one of them.

In the group students can see a list of tasks, select one of them and solve it inside the GeoGebra applet.

After pressing a button, the applet sends the answer to the server where it is evaluated and application instantly displays whether the solution is correct or not. If the solution is not correct, the student can try to find the mistake or look for other ways to solve it and submit his/her new solution.

The teacher can see the table of all students and tasks in the given group and also the list of the last evaluated answers.

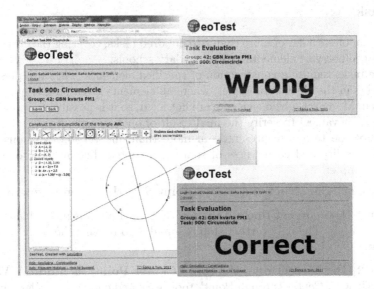

Fig. 1. Students' view on the GeoGebra applet containing the construction with two possible results of the evaluation

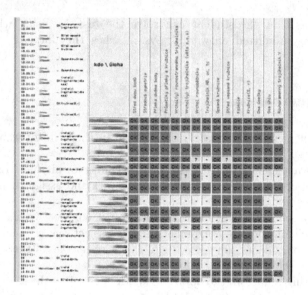

Fig. 2. The teacher's overview of one group of tasks. The rows of table belong to students, the columns to tasks, the teacher can see the solved tasks, the tasks with wrong answers and the unsolved tasks. On the left side there is a special column that presents a list of the last evaluated constructions with the name of the student, the task name, the time and the result of the particular submission.

4   Tasks

Our system offers a list of more than two hundred tasks, covering the curriculum of plane geometry at elementary or secondary schools.

These tasks require a construction of specified elements. They are presented to students in the GeoGebra applet, where all elements given have been already visualized. Moreover, given elements are not fixed, so it is possible to move them (change their values) to verify the final result of one's construction.

Some of the tasks are presented in more than one version. Versions differ in the generality and correctness of solution accepted. Let us show one quite simple example:

Assignment: Construct the center point S of the given segment AB.

Answer 1 (the correct one): If the tool "Center Point" is not hidden, it can be used directly. The only further step required is to rename the point acquired properly.

Answer 2 (the correct one, the tool "Center Point" is hidden): We use the "Circle with Center through Point" tool twice and construct circles with the center A through the point B and vice versa. Then the "Intersect Two Objects" tool is used to construct intersection points (e.g. P, Q) of the circles, then the "Line through Two Points" twice again and, finally, the result point S is the intersection point of the lines AB, PQ.

Answer 3 (not the correct one, in the easier category accepted, in the basic one rejected): We use the "Circle with Center and Radius" tool twice and construct circles with the centers A, B and with the equal radii, "long enough" so the circles intersect each other. Then we follow the sequence above.

Reasoning: If the points A, B are dragged it is possible to reach the configuration, where the circles don't intersect and the final point vanishes though the correct answer-point of the task exists. We haven't used the correct algorithm.

Why accept this answer? In case the construction was pencil-and-paper drawn, the teacher would accept it without hesitation.

5 Experience with Using the System

Our application has until now been used by (approximate numbers) 10 teachers and 300 students in 10 classes. The students were assigned 800 tasks (all students in the class were assigned the same tasks, in average approx. 80 plane geometry tasks for each student!), 25000 of constructions were submitted and evaluated and 93 % of the tasks that the students begun to solve was finished with the correct answer (in average approx. two evaluated answers for one correct).

We observe that:

- Students do solve the tasks. They solve them in the evenings, at weekends, on holidays.
- When students cannot solve some task, they come back later.
- ... but students come back also to the tasks they have already solved!
- While using the system during lessons every student can progress in his/her own pace, there is no need why the slow-working students should be put under pressure and the fast-working should be bored, while the teacher can exactly monitor each student's progress and recent answers.
- The teacher spends almost no time and effort on evaluation of her/his students' answers.
- The teacher has a good overview over the students and the tasks, he/she can see better which tasks are simple for students and which of them are difficult and perhaps need a further explanation.
- There is no difference between "small" and "big" mistake. Students accept that for example naming the result point "B" instead of "C" is an error and they do not try to argue (there is no one to argue with).
- Students can correctly solve more tasks, because of instant feedback.
- Probably the greatest benefit we observe is that students can continue with the task until it is correctly solved. When students see their answer is not correct, they simply try some other way. In case the assignment is in the form of drawing, the first day students produce the drawing, the second day they hand in it to the teacher and only on the third day they find out whether their solution was correct. It is most probable that they do not remember the task by then.

– The teacher would never be able to evaluate the assignments so quickly nor in such an amount as the computer is. Not to mention the work and time spent.

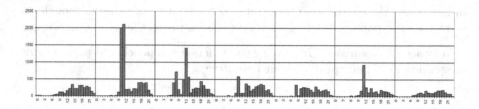

Fig. 3. Number of evaluated solutions by hours, from Monday 0:00 to Sunday 24:00

6   Conclusion

We have proposed a new way of automatic evaluation of construction geometry assignments. This way was implemented in the form of web application and used by more than 300 students of secondary schools. From the evaluation of submitted data of constructions and by using the described system we can see that the students are interested, the teachers are interested, and also that with instant feedback students carry on with their tasks until they find the correct solution. From the mentioned above we conclude that described solution has a positive impact on learning process.

References

1. GeoGebra, http://www.geogebra.org
2. Hohenwarter, J., Hohenwarter, M., Lavicza, Z.: Introducing Dynamic Mathematics Software to Secondary School Teachers: the Case of GeoGebra. Journal of Computers in Mathematics and Science Teaching 28(2), 135–146 (2009)
3. Kreis, Y., Dording, C., Keller, U., Porro, V., Jadoul, R.: Dynamic mathematics and computer-assisted testing: GeoGebra inside TAO. ZDM The International Journal on Mathematics Education 43(3), 1–10 (2011)
4. GeoTest, http://geotest.geometry.cz
5. Gergelitsová, Š., Holan, T.: Úlohy z planimetrie... teacher-friendly! In: Proceedings of the 31st Conference on Geometry and Graphics, pp. 95–102. VŠB – TU Ostrava, Ostrava (2011)

Ask-Elle: A Haskell Tutor
Demonstration

Johan Jeuring[1,2], Alex Gerdes[1], and Bastiaan Heeren[1]

[1] School of Computer Science, Open Universiteit Nederland
P.O.Box 2960, 6401 DL Heerlen, The Netherlands
{jje,age,bhr}@ou.nl
[2] Department of Information and Computing Sciences, Universiteit Utrecht

Abstract. In this demonstration we will introduce Ask-Elle, a Haskell tutor. Ask-Elle supports the incremental development of Haskell programs. It can give hints on how to proceed with solving a programming exercise, and feedback on incomplete student programs. We will show Ask-Elle in action, and discuss how a teacher can configure its behaviour.

1 Introduction

Ask-Elle[1] is a programming tutor for the Haskell [5] programming language, targeting bachelor students at the university level starting to learn Haskell. Using Ask-Elle, a student can:

- develop a program incrementally,
- receive feedback about whether or not she is on the right track,
- ask for a hint when she is stuck,
- can see how a complete program is stepwise constructed.

Ask-Elle is an example of an intelligent tutoring system [6] for the domain of functional programming.

In this demonstration we will show Ask-Elle in action, by means of some interactions of a hypothetical student with the tutor. Furthermore, we will show how a teacher can configure the behaviour of Ask-Elle. This demonstration accompanies our paper introducing the technologies we use to offer flexibility to teachers and students in our tutor [1,2]. Jeuring et al. [4] give more background information about Ask-Elle.

2 Ask-Elle in Action

We will start our demonstration of Ask-Elle with showing some interactions of a hypothetical student with the functional programming tutor. A screenshot of Ask-Elle is shown in Figure 1. It sets small functional programming tasks,

---

[1] http://ideas.cs.uu.nl/ProgTutor/

A. Ravenscroft et al. (Eds.): EC-TEL 2012, LNCS 7563, pp. 453–458, 2012.

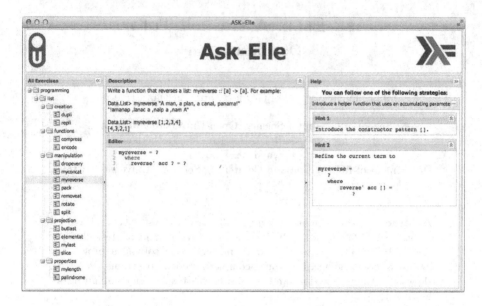

Fig. 1. The web-based functional programming tutor

and gives feedback in interactions with the student. We assume that the student has attended lectures on how to write simple functional programs on lists.

At the start of a tutoring session the tutor gives a problem description. Here the student has to write a program to construct a list containing all integers within a given range.

> Write a function that creates a list with all integers between
> a given range:

$$range :: Int \rightarrow Int \rightarrow [Int]$$

> For example:

$$> range\ 4\ 9$$
$$[4, 5, 6, 7, 8, 9]$$

and displays the name of the function to be defined, along with its parameters:

$$range\ x\ y = \bullet$$

The task of a student is to refine the holes, denoted by •, of the program. After each refinement, a student can ask the tutor whether or not the refinement is bringing him or her closer to a correct solution. If a student doesn't know how to proceed, she can ask the tutor for a hint. A student can also introduce new declarations, function bindings, and alternatives.

Suppose the student has no idea where to start and asks the tutor for help. The tutor offers several ways to help. For example, it can list all possible ways to proceed solving an exercise. In this case, the tutor would respond with:

```
You can proceed in several ways:
- Implement range using the unfoldr function.
- Use the enumeration function from the prelude².
- Use the prelude functions take and iterate.
```

We assume a student has some means to obtain information about functions and concepts that are mentioned in the feedback given by the tutor. This information might be obtained via lectures, an assistant, lecture notes, or even via the tutor at some later stage. The tutor can make a choice between the different possibilities, so if the student doesn't want to choose, and just wants a single hint, she gets:

```
Implement range using the unfoldr function.
```

Here we assume that the teacher has set up the tutor to prefer the solution that uses *unfoldr*, defined by:

$$unfoldr :: (b \rightarrow Maybe\ (a, b)) \rightarrow b \rightarrow [\,a\,]$$
$$unfoldr\ f\ b = \textbf{case}\ f\ b\ \textbf{of}\ Just\ (a, b') \rightarrow a : unfoldr\ f\ b'$$
$$Nothing \quad \rightarrow [\,]$$

The higher-order function *unfoldr* builds a list from a seed value b. The argument f is a producer function that takes the seed element and returns *Nothing* if it is done producing the list, or *Just* (a, b'), in which case a is prepended to the output list and b' is used as the seed value in the recursive call.

The student can ask for more detailed information at this point, and the tutor responds with increasing detail:

```
Define function range in terms of unfoldr, which takes two
arguments: a seed value, and a function that produces a new
value.
```

with the final bottom-out hint:

```
Define: range x y = unfoldr • •
```

At this point, the student can refine the function at two positions. We do not impose an order on the sequence of refinements. Suppose that the student chooses to first implement the producer function:

$$range\ x\ y = unfoldr\ f\ \bullet \quad \textbf{where}\ f\ i\ |\ \bullet = \bullet$$

Note that the student has started to define the producer function in a **where** clause. She continues with the introduction of the stop criterion:

---

² The prelude is the standard library for Haskell containing many useful functions.

$$range\ x\ y = unfoldr\ f\ \bullet \quad \textbf{where}\ f\ i\ |\ i == y + 1 = \bullet$$

There are several ways in Haskell to implement a condition. Here the student has chosen to define the function f with a so-called guarded expression; the predicate after the vertical bar acts as a guard. The student continues with:

$$range\ x\ y = unfoldr\ f\ \bullet \quad \textbf{where}\ f\ i\ |\ i == y + 1 = Just\ \bullet$$

The tutor responds with:

`Wrong solution:` $range\ 4\ 6$ `provides a counterexample.`

The partial definition of f does not match any of the correct solutions, and by means of random testing the tutor can find an example where the result of the student program differs from a model solution. Correcting the error, the student enters:

$$range\ x\ y = unfoldr\ f\ \bullet \quad \textbf{where}\ f\ i\ |\ i == y + 1 = Nothing$$

which is accepted by the tutor. If the student now asks for a hint, the tutor responds with:

`Introduce a guarded expression that gives the output value and the value for the next iteration.`

She continues with

$$range\ x\ y = unfoldr\ f\ \bullet \quad \textbf{where}\ f\ i\ |\ i == y + 1 = Nothing$$
$$|\ otherwise = Just\ \bullet$$

which is accepted, and then

$$range\ x\ y = unfoldr\ f\ \bullet \quad \textbf{where}\ f\ i\ |\ i == y + 1 = Nothing$$
$$|\ otherwise = Just\ (n, i + 1)$$

which gives:

`Error: undefined variable` n

This is a syntax-error message generated by the Helium [3] compiler, which we use in our tutor. The student continues with:

$$range\ x\ y = unfoldr\ f\ x \quad \textbf{where}\ f\ i\ |\ i == y + 1 = Nothing$$
$$|\ otherwise = Just\ (i, i + 1)$$

which completes the exercise.

A student can develop a program in any order, as long as all variables are bound. For example, a student can write

$$range\ x\ y = \bullet \quad \textbf{where}\ f\ i\ |\ \bullet = \bullet$$

and then proceed with defining f. This way, bottom-up developing a program is supported to some extent.

These interactions show that our tutor can give hints about which step to take next, in various levels of detail, list all possible ways in which to proceed, point out errors, and pinpoint where the error appears to be, and show a complete worked-out example.

3 Configuring the Behaviour of ASK-ELLE

In this part of the demonstration we show how a teacher adds a programming exercise to the tutor by specifying model solutions for the exercise, and how a teacher adapts the feedback given by the tutor.

3.1 Adding an Exercise

The interactions of the tutor are based on *model solutions* to programming problems. A model solution is a program that an expert writes, using good programming practices. We have specified three model solutions for *range*. The first model solution uses the enumeration notation from Haskell's prelude:

$$range\ x\ y = [x \mathrel{..} y]$$

The second model solution uses the prelude functions *take*, which given a number n and a list xs returns the first n elements of xs, and *iterate*, which takes a function and a start value, and returns an infinite list in which the next element is calculated by applying the function to the previous element:

$$range\ x\ y = take\ (y - x + 1)\ (iterate\ (+1)\ x)$$

The last model solution uses the higher-order function *unfoldr*:

$$range\ x\ y = unfoldr\ f\ x\ \ \textbf{where}\ f\ i\ |\ i \mathrel{==} y + 1 = Nothing$$
$$|\ otherwise = Just\ (i, i + 1)$$

The tutor uses these model solutions to generate feedback. It recognises many variants of a model solution. For example, the following solution:

$$range\ x\ y = \textbf{let}\ f = \lambda a \to \textbf{if}\ a \mathrel{==} y + 1\ \textbf{then}\ Nothing\ \textbf{else}\ Just\ (a, a + 1)$$
$$g = \lambda f\ x \to \textbf{case}\ f\ x\ \textbf{of}\ Just\ (r, b) \to r : g\ f\ b$$
$$Nothing\ \ \to [\,]$$
$$\textbf{in}\ g\ f\ x$$

is recognised from the third model solution. To achieve this, we not only recognise the usage of a prelude function, such as *unfoldr*, but also its definition. Furthermore, we apply a number of program transformations to transform a program to a normal form.

Using a *class* a teacher groups together exercises, for example for practicing list problems, collecting exercises of the same difficulty, or exercises from a particular textbook.

3.2 Adapting Feedback

A teacher adapts the feedback given to a student by *annotating* model solutions. The description of the entire exercise is given together with the model solutions in a configuration file for the exercise. Using the following construction we add a *description* to a model solution:

$$\{-\#\ DESC\ \ Implement\ range\ using\ the\ unfoldr...\ \#-\}$$

The first hint in Section 2 gives the descriptions for the three model solutions for the range exercise.

A teacher allows an *alternative* implementation for a prelude function by:

$$\{-\#\ ALT\ \ iterate\ f = unfoldr\ (\lambda x \to Just\ (x, f\ x))\ \#-\}$$

Using this annotation we not only recognise the prelude definition (*iterate f x = x : iterate f (f x)*), but also the alternative implementation given here. Alternatives give the teacher partial control over which program variants are allowed.

A teacher may want to enforce a particular implementation method, for example, use higher-order functions and forbid their explicit recursive definitions, for which we use the *MUSTUSE* construction:

$$range\ x\ y = \{-\#\ MUSTUSE\ \#-\}\ unfoldr\ f\ x$$

Specific feedback messages can be attached to particular locations in the source code. For example:

$$range\ x\ y = \{-\#\ FEEDBACK\ \ Note...\ \#-\}\ take\ (y - x + 1)\ \$\ iterate\ (+1)\ x$$

Thus we give a detailed description of the *take* function. These feedback messages are organised in a hierarchy based on the abstract syntax tree of the model solution. This enables the teacher to configure the tutor to give feedback messages with an increasing level of detail.

References

1. Gerdes, A., Heeren, B., Jeuring, J.: Teachers and Students in Charge — Using Annotated Model Solutions in a Functional Programming Tutor. In: Ravenscroft, A., Lindstaedt, S., Delgado Kloos, C., Hernández-Leo, D. (eds.) EC-TEL 2012. LNCS, vol. 7563, pp. 383–388. Springer, Heidelberg (2012)
2. Gerdes, A., Heeren, B., Jeuring, J.: Teachers and students in charge — using annotated model solutions in a functional programming tutor. Technical Report UU-CS-2012-007, Utrecht University, Department of Computer Science (2012)
3. Heeren, B., Leijen, D., van IJzendoorn, A.: Helium, for learning Haskell. In: Haskell 2003: Proceedings of the 2003 ACM SIGPLAN Workshop on Haskell, pp. 62–71. ACM (2003)
4. Jeuring, J., Gerdes, A., Heeren, B.: A Programming Tutor for Haskell. In: Zsók, V., Horváth, Z., Plasmeijer, R. (eds.) CEFP. LNCS, vol. 7241, pp. 1–45. Springer, Heidelberg (2012)
5. Peyton Jones, S., et al.: Haskell 98, Language and Libraries. The Revised Report. Cambridge University Press (2003); A special issue of the Journal of Functional Programming, http://www.haskell.org/
6. VanLehn, K.: The behavior of tutoring systems. International Journal on Artificial Intelligence in Education 16(3), 227–265 (2006)

Backstage – Designing a Backchannel for Large Lectures

Vera Gehlen-Baum[1], Alexander Pohl[2], Armin Weinberger[1], and François Bry[2]

[1] Departement of Educational Technology
Saarland University, Saarbrücken
{v.gehlen-baum,a.weinberger}@mx.uni-saarland.de
[2] Institute for Informatics
University of Munich
alexander.pohl@pms.ifi.lmu.de, bry@lmu.de
http://pms.ifi.lmu.de

Abstract. Students and lecturers use computers in lectures. But, the standard tools give a rather insufficient structure and support for better learning results. Backstage is an adjustable backchannel environment where students can communicate by microblogs, which they can link to the presenter slides. The lecturer can get feedback by Backstage and place quizzes with an Audience Response System. Backstage is designed to facilitate specific sequences of learning activities and to enhance student motivation with different functions, like asking questions anonymously via microblogs or to rate other students' questions.

Keywords: Lectures, Backchannel, Taking Notes, ARS.

1 Technology Supported Learning Fostering Activities in Large Lectures

Both lecturers and students frequently use computers in large lectures today. Whereas there is some understanding how lecturers use computers for presentations [1], little is known about how students can actively and productively use mobile devices, such as notebooks, tablets or smart phones in large lectures to better engage in learning activities and cognitively process what is being taught. Here, we present a learning environment called Backstage that allows learners to represent slides on their personal mobile devices, to take notes on those slides, and to post and answer questions to and from their peers and lecturers. We included also a feedback function as well as the possibility that students answer lecturer questions.

2 Computer-Supported Learning Activities in Large Classrooms

There are a couple of ways students in large lectures use computer technology: displaying slides, taking notes, browsing the internet, using Facebook or sending IMs

A. Ravenscroft et al. (Eds.): EC-TEL 2012, LNCS 7563, pp. 459–464, 2012.

and emails [2, 3]. The most frequent lecture-related way students use computers is to display the slides of the lecture on one's personal screen [4]. Lecturers often provide their slides in pdf for download before the lecture. To a lesser degree, students use computers for taking notes, especially since formats like pdf do not generally allow adding notes [4]. Taking notes has shown to be a good strategy for encoding and storing knowledge, especially when there are some additional strategies trained in how to take notes [5, 6] . Also additional learning material could be either provided for download by the lecturer or searched for online by students themselves, which may foster learners' understanding [6].

Specific educational technology is needed for supporting and enhancing specific social interactions in lectures that are conducive to learning. Traditional lectures have been criticized for being limited to a specific interaction pattern initiated by the lecturer only, namely asking a question, which only one student can answer directly, and commenting or evaluating the student's answer [7, 8]. There is some evidence, however, that students are more actively elaborating what is being taught and retain more knowledge when they initiate discourse and ask questions themselves (e.g. [9]). But in a large lecture, students often resist to do so due to fear of losing face or lack of metacognitive skills to identify knowledge gaps or to generate meaningful and critical questions (e.g. [10-13]).

To address these challenges, we have developed Backstage to enable all learners in large lectures to ask and reply to questions anonymously.

3   Backstage – An Environment to Foster Social Learning Activities in Large Lectures

Backstage runs on any online mobile device to provide carefully conceived student-student as well lecturer-audience interactions to support active participation and learning. Simultaneously, Backstage is customizable by the lecturer to set number of questions to be received or posed at what times, which results to mirror back to the students, or what interactions to allow between students. Backstage includes different functions, like pre-structured and peer-rated microblogging, displaying and connecting microblog messages to the slides, and an Audience Response System (ARS).

3.1   Pre-structured and Peer-rated Microblogs

Backstage supports microblogging that allows students to post questions to the whole class [14]. Microblogs are short messages with a fixed small number of characters, e.g. used in Twitter. These messages and the possibility for students to talk to each other during the lecture is characteristic of so called backchannels [15], which has yet met little acceptance by lecturers and students [14]. It is likely that this lack in acceptance is due to the fact that it is more difficult and time consuming for students to ask questions in written form.

3.2   Pre-structured Microblogging

Backstage pre-structures its microblog by featuring different communication modes and self-selectable predefined message categories. Backstage allows for two different communication modes, which can be enabled or disabled by the lecturer: anonymous communication, i.e. the author is not shown along with messages, and private communication, i.e. the messages are only visible to some users specified in the text bodies. Moreover, Backstage requires messages to be assigned to predefined categories (e.g., Question, Answer, Remark, Too Fast) to foster specific interaction patterns and encourage students to reflect on what they want to express.

3.3   Peer-rated Microblogging

Furthermore, students may rate their peers' microblog messages to assess relevance with respect to the lecture discourse. Only those messages may be selected that are rated and recommended by the students, similar to using Facebook's "like-Button" or Amazon's 5-star recommender system. Only those which reach a certain value will be passed on to the lecturer who can see how the messages rank by relevance in real-time in the lecturer display (see Figure 1). Backstage can also display an up-to-date overview of what kind of the predefined message categories is currently exchanged among the students at what rate. For instance, an increase in messages of the question or the Too Fast category may indicate to the lecturer that learners are getting lost.

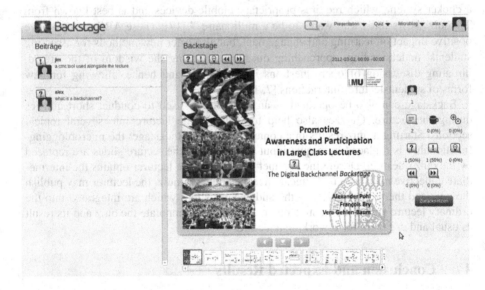

Fig. 1. Dashboard of Backstage as displayed to the lecturer

The lecturer can customize the number of messages to be displayed in the ranking. These features serve to reduce the volume of messages which have to be checked by the lecturer during a lecture and is different from some systems that suggest lecturers to additionally rate students' post (e.g. [16]). Also, the answering of any particular question by peer students and this answer could be rated by the audience as well, should reduce the workload of the lecturer. Both features aim to reduce workload and to foster sharing of interesting questions and answers.

3.4   Connecting Messages with the Slides

To align the backchannel communication with the lecture, the lecture slides are displayed in the Backstage interface and each microblog message is assigned to a distinct slide and is filtered accordingly. Furthermore, a message is placed on a certain location on a slide, i.e., messages are used to annotate slides.

The alignment with the slides allows navigating the message stream in a top-down fashion: the relevance of messages may be recognized foremost by the location on a slide and hence, peer students can select to view those messages linked to the slides they want to learn more about.

3.5   Audience Response System for Quizzes during the Lecture

In the last couple of years there has been a development of different tools aiming to engage students during large lectures. One of these tools is an Audience Response System (ARS), which offers the possibility that lecturers post questions to all attending students and all students may answer questions anonymously. This is often called a clicker system, which requires proprietary mobile devices and is best known from TV quiz shows like "who wants to be a millionaire?" [17]. These ARS seem to have positive impact on learning and engagement, but also offer new methods for assessing students' understanding, approaching questions by using the wisdom of the crowd, initiating discussion of open questions in the lecture and hence, allowing for new forms of student-lecturer interactions [7, 18].

Backstage can also be operated as an ARS, i.e. be used to conduct short quizzes during the lecture. Quizzes also help to structure the lecture into several topical sections. Starting a quiz results in a context switch on Backstage: the microblogging-functionality is replaced by a quiz input interface and the lecture slides are replaced by the quiz question. During the conduct of the quiz, the lecturer obtains the intermediate collective answer in real-time. When finishing the quiz, the lecturer may publish the quiz and the answer given by the audience as slides which are integrated into the ordinary lecture slides. Thus, students may review and annotate the quiz and its result as usual and use them for reworking a lecture.

4   Conclusion and Expected Results

Backstage is a highly customizable learning environment to support student –student as well as student –lecturer interaction in addition to facilitate learning in a lecture through different tools and included instructions. In contrast to many proprietary hardware systems, e.g. clicker systems, Backstage runs on any mobile devices students bring themselves to the lecture.

There are certain aspects, which should foster learning activities in Backstage in comparison to other learning environments. The combination of different features and tools in one learning environment aims to facilitate question asking and elaboration of the learning material. Usability studies show that students actively used the Backstage features in contrast to earlier use of microblogs [14]. Every one of them gave at least one comment or asked a question [19]. Especially the implementation of new features like giving feedback on motivation or learning results to the students should increase acceptance and use of Backstage.

We expect that the possibility for students to give feedback and to answer questions may enhance the use of Backstage. During and after the lecture, students can see which questions or answers are rated highly by themselves and other students, which may enhance a feeling of efficacy and social relatedness [20]. Future research will focus on how different types of feedback can make students and lecturers aware of vital learning processes in large lectures and how effective interaction patterns can be scripted onto large groups of students. Ultimately, Backstage aims to facilitate students' learning and future studies may inquire how and what kind of knowledge acquisition in large lectures can be facilitated with this tool.

We plan to conduct and report on an in-vivo study to identify which kind of support should be given within Backstage to formulate critical questions and to what extent learners participate and benefit more homogeneously when using Backstage.

References

1. Igel, C., Somson, T., Meiers, R.: Integration neuer Medien in universitäre Tätigkeitsfelder: Vergleich der Ergebnisse zweier Situationsanalysen. In: Informationszentrum Sozialwissenschaften & Gesellschaft Sozialwissenschaftlicher Infrastruktureinrichtungen (2005)
2. Fried, C.B.: In-class laptop use and its effects on student learning. Computers & Education 50(3), 906–914 (2008)
3. Kraushaar, J.M., Novak, D.C.: Examining the Effects of Student Multitasking with Laptops during the Lecture. Journal of Information Systems Education 21(2), 11 (2010)
4. Gehlen-Baum, V., Weinberger, A.: Notebook or Facebook? How Students Actually Use Mobile Devices in Large Lectures. In: Ravenscroft, A., Lindstaedt, S., Delgado Kloos, C., Hernández-Leo, D. (eds.) EC-TEL 2012. LNCS, vol. 7563, pp. 103–112. Springer, Heidelberg (2012)
5. Di Vesta, F.J., Gray, G.S.: Listening and note taking. Journal of Educational Psychology 63(1), 8–14 (1972)
6. Grabe, M.: Voluntary use of online lecture notes: Correlates of note use and note use as an alternative to class attendance. Computers & Education (2005)
7. Nassaji, H., Wells, G.: What's the Use of ' Triadic Dialogue '? An Investigation of Teacher - Student Interaction. Applied Linguistics 21(3), 376–406 (2000)
8. Scardamelia, M., Bereiter, C.: Computer support for knowledge-building communities. Journal of the Learning Sciences 3(3), 265–283 (1994)
9. King, A.: Guiding Knowledge Construction in the Classroom: Effects of Teaching Children How to Question and How to Explain. American Educational Research Journal 31(2), 338–368 (1994)

10. Aleven, V., Stahl, E., Schworm, S., Fischer, F., Wallace, R.: Help seeking in interactive learning environments. Review of Educational Research 73(3), 277–320 (2003)
11. Nelson-LeGall, S.: Help-seeking: An understudied problem-solving skill in children. Developmental Review 1, 224–226 (1981)
12. Rosenshine, B., Meister, C., Chapman, S.: Teaching Students to Generate Questions: A Review of the Intervention Studies. Review of Educational Research 66(2), 181–221 (1996)
13. Zhang, D., Zhao, J., Zhou, L.: Can e-learning replace classroom learning? Communications of the 47(5), 74–79 (2004)
14. Ebner, M., Schiefner, M.: Microblogging - more than fun? In: IADIS Mobile Learning Conference 2008, pp. 155–159 (2008)
15. McCarthy, J., Boyd, D.: Digital backchannels in shared physical spaces: experiences at an academic conference. In: Extended Abstracts, Human Factors in Computing Systems (CHI), pp. 550–553. ACM, Portland (2005)
16. Doroja, G.S., Ramos, S.M.L., Sabal, J.A.C., Fernandez, H.B.: Augmenting Teacher-Student Classroom Interaction Using Mobile Messaging. In: Proceedings of the 19th International Conference on Computers in Education (2011)
17. Mazur, E.: Farewell, Lecture? Science 323, 50–51 (2009)
18. Kay, R.H., LeSage, A.: Examining the benefits and challenges of using audience response systems: A review of the literature. Computers & Education 53(3), 819–827 (2009)
19. Gehlen-Baum, V., Pohl, A., Bry, F.: Assessing Backstage—A Backchannel for Collaborative Learning in Large Classes. In: Proceedings of the International Conference ICL, pp. 154–160 (2011)
20. Deci, L., Edward, Ryan, R.M.: Intrinsic motivation and self-determination in human behavior. Plenum, New York (1985)

Demonstration of the Integration of External Tools in VLEs with the GLUE! Architecture*,**

Carlos Alario-Hoyos, Miguel Luis Bote-Lorenzo, Eduardo Gómez-Sánchez,
Juan Ignacio Asensio-Pérez, Guillermo Vega-Gorgojo, and Adolfo Ruiz-Calleja

School of Telecommunication Engineering, University of Valladolid,
Paseo de Belén 15, 47011 Valladolid, Spain
{calahoy@gsic,migbot@tel,edugom@tel,juaase@tel,guiveg@tel,
adolfo@gsic}.uva.es

Abstract. The main objective of this paper is to illustrate the instantiation and enactment of a collaborative learning situation that requires the integration of three external tools, in two different VLEs (Moodle and LAMS), with GLUE!. GLUE! facilitates the instantiation of collaborative activities, reducing the time and effort educators need for the creation, configuration and assignment of external tool instances for each group. In particular, this paper details how GLUE! allows for each VLE to use its own group management routines with external tools, in the same seamless way as done with built-in tools. The GLUE! web site complements this paper, showing videos of these and other integrations.

1 Description of GLUE!

The GLUE! (Group Learning Uniform Environment) architecture [2] enables the integration of multiple existing external tools in multiple existing Virtual Learning Environments (VLEs), facilitating the instantiation and enactment of individual and collaborative activities that require the integration of external tools. As part of the instantiation process, educators can request GLUE! the creation, configuration, update and deletion of external tool instances within their commonly-used VLE. These instances can automatically be assigned to VLE users, so that those belonging to the same group share the same instance in each activity. As part of the enactment process, participants can request the retrieval of created instances in order to achieve the learning objectives proposed in each activity. These instances are delivered following the group structure defined by educators within VLEs.

Authors have incrementally developed reference implementations for the elements composing the GLUE! architecture [2], enabling, at the moment, the integration of about seventeen external tools in Moodle, LAMS and MediaWiki.

---

* This work has been partially funded by the Spanish Ministry of Economy and Competitiveness (TIN2008-03023, TIN2011-28308-C03-02 and IPT-430000-2010-054) and the Autonomous Government of Castilla y Leon (VA293A11-2 and VA301B11-2). Authors also thank David A. Velasco-Villanueva and Javier Aragón for their development work.
** This demonstration paper complements the paper [1], also published in the EC-TEL 2012.

A. Ravenscroft et al. (Eds.): EC-TEL 2012, LNCS 7563, pp. 465–470, 2012.
© Springer-Verlag Berlin Heidelberg 2012

Nevertheless, the number of available external tools and VLEs is expected to increase with the contributions of external developers, thus enriching the kinds of learning activities that practitioners might instantiate and enact.

The aim of this demonstration paper is to describe the processes of instantiating and enacting collaborative activities through GLUE!. A representative blended collaborative learning situation, which has been motivated and evaluated in [1], serve for this purpose. This and other examples of instantiation and enactment of collaborative activities with the GLUE! architecture within different VLEs can be seen at http://youtube.com/user/gsicemic.

2   A Collaborative Learning Situation

Educators delivering Advanced Networking at the University of Valladolid designed a blended collaborative learning situation [1] aimed at helping third-year students of Telecommunication Engineering reflect about the design of a message server, before starting its development. This situation includes a sequence of five collaborative activities. First, students draw the sequence diagram and the flowchart of a message server in pairs, justifying also the main decisions taken and the problems found in a different activity. Then, they review the work of two to three different pairs, being arranged in bigger groups (supergroups), for the remaining activities. Finally, supergroups agree on the final sequence diagram and flowchart, making also a presentation to explain the results to the rest of the class. This collaborative learning situation follows the *pyramid* collaborative pattern [3], and so, it can also be applied to other contexts where a gradual agreement among participants is intended to be reached.

Two educators and 51 students participated in the instantiation and enactment of this collaborative learning situation, which lasted one week in November 2011. Moodle was the VLE employed, being Google Documents, Dabbleboard and Google Presentation the tools that supported the collaborative activities.

3   Instantiation and Enactment in Moodle

Before starting the instantiation process, the Moodle administrator had already created a Moodle course for Advanced Networking, in which educators and students were enrolled. Taking this precondition, educators defined and populated, in a first step, the groups: 28 pairs (1-2 students) and 8 supergroups (6-8 students). Moodle just allows the definition of a single structure of *groups* in each course, although it recently featured a *groupings* option that supports the combination of multiple groups in a grouping. Educators thus matched the pairs with the Moodle groups and the supergroups with the Moodle groupings.

In the first activity educators wanted their students to work in pairs using the Dabbleboard external tool. To instantiate Dabbleboard within Moodle they added a new Moodle activity to the course, as usual, and selected the *gluelet* option (this is a new kind of Moodle activity that allows the integration of external tools through the GLUE! architecture) from the *add an activity* dropdown menu in the course homepage. Then, educators configured the activity as they were used to, indicating, for instance, a name, a description, and the group mode. In this screen a new drop-down menu appeared with the list of available external tools, so that educators could select the actual external tool: *Dabbleboard*. Interestingly, they selected *separate groups* in the group mode, because a new instance of the external tool had to be created and assigned to each of the groups (pairs) defined in the Moodle course. After that, a new screen showed up allowing the configuration of each of the instances that were going to be created. This screen was similar to that showed in Figure 1, although in the case of Dabbleboard no initial configurations were supported. Educators clicked on *apply to all*, and 28 Dabbleboard instances were automatically created and assigned to the Moodle users in about 1.5 minutes. Similar steps were followed to add the Google Documents external tool in the second activity. However, in this case, some parameters could be initially set for this tool: a title and a file. Educators included a generic title and uploaded a file with a template to help students carry out this activity. Then, they clicked on *apply to all*, and the 28 Google Documents instances were created and configured in about 1.5 minutes too.

The third activity was also supported by Google Documents. Here, educators selected *visible groups* in the group mode, allowing other students to see and review their partners' work. Nevertheless, only supergroup members were intended to see that work. So, educators also marked the *available for group members only* option and selected one of the groupings (e.g. *Supergroup1A*) from another drop-down menu; both features are natively available in any Moodle activity, but thanks to GLUE! they were applied to restrict the access to certain external tool instances. Besides, in the next screen they reused the previously created instances rather than creating new ones, as depicted in Figure 1. Instances from the second activity were manually assigned here (each group received the same instance employed during the second activity). It is noteworthy that the reuse of instances, of great utility in the design of complex collaborative learning flows, is a feature of the GLUE! architecture, being unsupported by Moodle built-in tools. After setting the instance each group had to reuse, this whole process was repeated for the remaining seven supergroups, thus requiring the creation of eight Moodle activities in total.

Dabbleboard was the tool supporting the fourth activity. In this case all the supergroup members had to share the same instance. Therefore, educators chose the *no groups* mode, but keeping one of the supergroups selected and the *available for group members only* option marked. One new instance of Dabbleboard was created for each supergroup, thus requiring eight different Moodle activities again. The same occurred with Google Presentations, in the fifth activity, although in this case educators could set an initial title and an initial file for each instance, as it also happened with Google Documents.

As a consequence of the whole instantiation process, educators created 36 instances of Dabbleboard, 28 instances of Google Documents and 8 instances of Google Presentations in 26 different Moodle activities. Significantly, all these actions were performed within the Moodle interface, automatizing the elements of GLUE! some of them, such as the creation of multiple instances within the

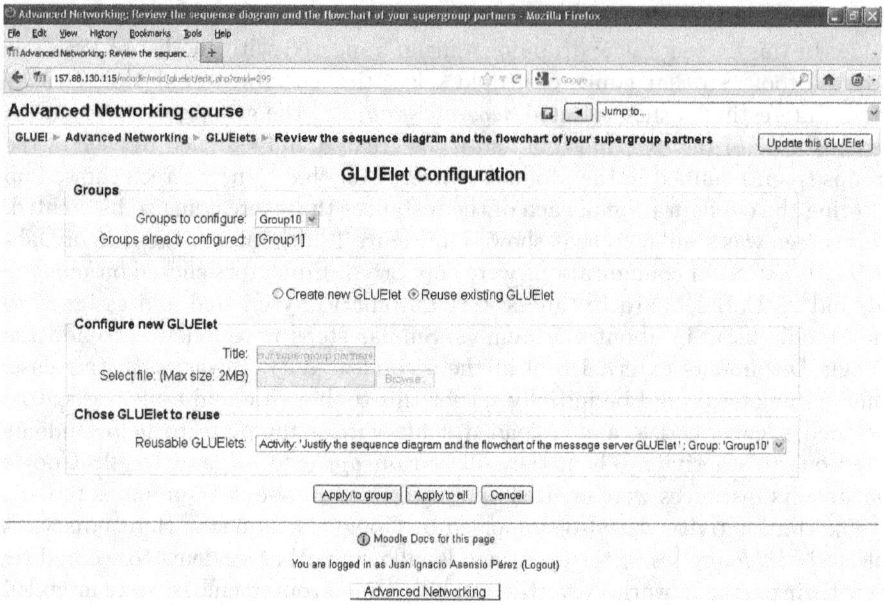

Fig. 1. Screenshot of Moodle showing the new screen where external tool instances are configured (before being created) or reused. Both options may be individually applied to each of the groups defined in Moodle. This screenshot was taken during the instantiation of the third activity, and so, some previously created instances of Google Documents were available for reuse.

same Moodle activity, and their assignment to VLE users. Despite the significant number of users, groups, tools and instances, educators only invested about 7.5 minutes in the creation of all the external tool instances [2]. For comparison, a similar course structure was created in Moodle without GLUE! (creating and configuring the instances through each tool web interface and then copying and pasting the URLs back in Moodle); the effort demanded by this manual process was about 42.5 minutes, almost 5 times more than using GLUE!.

Students enacted the five collaborative activities without any problems. As expected, in those activities where *no groups* or *separate groups* modes had been selected, they could only see the instances created for their group. Besides, in the *visible group* activity, they could also see the instances of their supergroup partners. In those activities in which several Moodle activities had been created in order to support different groupings, participants could only see the Moodle activity that belonged to their supergroup. Figure 2 shows an example with the visualization of Google Presentations within Moodle. The elements of GLUE! tried to keep the same look and feel of both the VLE and the external tool.

Interestingly, educators needed to reorganize some Moodle groups and groupings at enactment time due to some students' absences. GLUE! supported this reorganization by automatically and transparently updating the Moodle users that could access to those external tool instances affected.

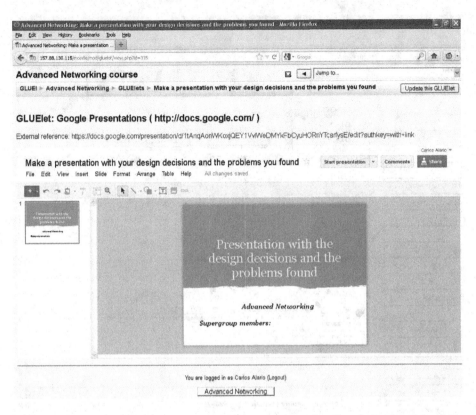

Fig. 2. Screenshot of Moodle showing the integration of a Google Presentations instance. This example corresponds to the student's view in the fifth collaborative activity of the situation presented in this paper.

4   Instantiation and Enactment in Other VLEs

This collaborative learning situation can also be instantiated and enacted in other VLEs (e.g. LAMS or MediaWiki), provided that an element called VLE adapter, which enables the interoperability between the VLE and GLUE!, is available. These VLE adapters are responsible for matching the VLE features regarding the management of groups, roles and activities to the functionality provided by the GLUE! architecture. As an example, Figure 3 illustrates the instantiation of the Advanced Networking situation in LAMS. In this VLE, external tool instances are separately configured in the *LAMS authoring environment*, being later automatically created when deploying the lesson in a course with particularized students in the *LAMS monitoring environment*; this process also happens with LAMS built-in tools. Other LAMS distinctive features like the use of branchings (in order to distinguish which pair has to review each document) can also be employed. Further details about the integration of external tools in LAMS with GLUE! can be found in [4].

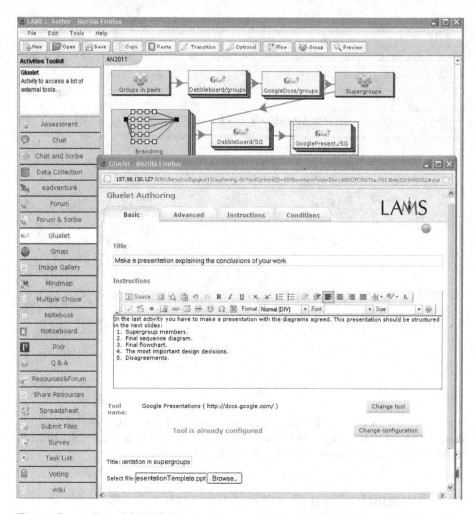

Fig. 3. Screenshot of LAMS showing the sequence created for Advanced Networking within the LAMS authoring environment: educator's view

References

1. Alario-Hoyos, C., et al.: Integration of External Tools in VLEs with the GLUE! Architecture: A case Study. In: Ravenscroft, A., Lindstaedt, S., Delgado Kloos, C., Hernández-Leo, D. (eds.) EC-TEL 2012. LNCS, vol. 7563, pp. 371–376. Springer, Heidelberg (2012)
2. Alario-Hoyos, C., et al.: GLUE!: An Architecture for the Integration of External Tools in Virtual Learning Environments. In: Computers and Education (submitted, 2012)
3. Hernández-Leo, D., et al.: CSCL Scripting Patterns: Hierarchical Relationships and Applicability. In: Proceedings of the 6th IEEE International Conference on Advanced Learning Technologies, Kerkrade, The Netherlands, pp. 388–392 (2006)
4. Alario-Hoyos, C., et al.: Integration of external tools with GLUE! in LAMS: requirements, implementation and a case study. In: Proceedings of the 6th International LAMS and Learning Design Conference, Sydney, Australia, pp. 40–48 (2011)

Energy Awareness Displays

Prototype for Personalised Energy Consumption Feedback

Dirk Börner, Jeroen Storm, Marco Kalz, and Marcus Specht

Centre for Learning Sciences and Technologies, Open Universiteit, Heerlen, The Netherlands
{dirk.boerner,jeroen.storm,marco.kalz,marcus.specht}@ou.nl

Abstract. The paper presents the "Energy Awareness Displays" project that makes hidden energy consumption data visible and accessible for people working in office buildings. Besides raising awareness on the topic and introducing relevant conservation strategies, the main goal is to provide dynamic situated feedback when taking individual consumption actions at the workplace. Therefore a supporting infrastructure as well as two example applications to access and explore the consumption information have been implemented and evaluated. The paper presents and discusses the approach, the developed infrastructure and applications, as well as the evaluation results.

Keywords: Energy Conservation, Ubiquitous Learning, Situational Awareness, Feedback.

1   Introduction

Modern office buildings are usually equipped with building automation systems that provide among others central energy management and monitoring services. Data from such systems is often gathered through proprietary software and made available only to a selected audience of engineers or facility managers. Typically, the level of detail of the gathered data does not go beyond a breakdown for the whole building, floor, or department. The main idea of the presented project is to make this data and thus the information that is hidden deep within the office building's infrastructure visible and accessible for the people working in the building - right up to a personal level of detail. In doing so the project sets up to change the energy consumption behaviour as well as the attitudes towards energy conservation of employees.

Besides raising employees' awareness on the topic and introduce relevant conservation strategies, the main goal was to provide dynamic situated feedback when taking actions. The underlying assumption is that the raised awareness on the actual consumption fosters a change in behaviour among employees and thus leads to reduced total energy consumption for the employing organisation. The idea was to reach the goal by the means of so-called eco-visualizations [1], a novel approach to display (real time) consumption data for the goal of promoting ecological literacy. On the long-term this visual, situated, real-time feedback on electricity consumption and

A. Ravenscroft et al. (Eds.): EC-TEL 2012, LNCS 7563, pp. 471–476, 2012.

respective conservation opportunities should facilitate environmental learning and behavioural change. The theoretical foundation and implications have been elaborated in [2].

2   Approach and Implementation

The presented project elaborated and developed an infrastructure that supports the concept of "Energy Awareness Displays" in office buildings with the following functionality:

- Inclusion of individual energy consumption information (device specific or personal level of detail).
- Aggregation of available information extending and enriching the overall energy consumption picture.
- Sensoring and logging to measure the effectiveness in terms of energy conservation and enable the prototypical evaluation.

Based on the supporting infrastructure respective display prototypes have been developed upon the following characteristics: (a) public interactive representation of the overall and individual energy consumption in several levels of detail, (b) explorative comparison of the consumption information in relation to fellow employees, departments, and/or floors, and (c) motivating and persuading conservation facilitation patterns based on the presented information, such as visual incentives.

The described approach required accessing and using external services offering the needed functionality. For the inclusion of individual energy consumption information the Plugwise[1] system was chosen. The system provides the needed sensor hardware to manage appliances and get access to energy consumption details. Furthermore the included software allows configuring the informational access via web services. The result is a wireless smart meter plugs network that can be accessed using the bundled software. The system was set up in such a way that individual appliance, room, and group information could be accessed. A basic application programming interface (API) can be used to access this information. The existing API was slightly adapted and enhanced to deliver all needed information in the right format. All changes are implemented based on the existing Plugwise Source[2] software template engine.

For the aggregation of available information respectively the logging of sensor data the Pachube[3] system was used. The system offers a free real-time open data web services that allows to aggregate, store, and access all kinds of sensor data, e.g. energy, home automation, and weather data can be aggregated, enriched, and accessed utilising different means. The system was set up to aggregate all the available sensor data for each room, i.e. (daily) total power usage and additionally the occupation.

---

[1] http://www.plugwise.com/
[2] http://www.plugwise.com/idplugtype-f/source
[3] https://pachube.com/

2.1   Infrastructure

The developed software infrastructure supporting the intended end-user applications is conceptually based upon the architectural framework Robotlegs[4], implementing a Model-View-Controller+Service (MVC+S) design utilising the Dependency Injection (DI) pattern. The framework is implemented in Actionscript 3. Based on the open-source Flex SDK 4.5.1 the infrastructure has been implemented using the Adobe Flash Builder[5] development environment.

Following a shared library approach the infrastructure is comprised of a library that bundles all necessary functionality for applications developed on top of it. Based on Robotlegs this library bundles model, command, event, and service components. The applications then consist of views and respective mediators that handle their functionality. Each application simply incorporates the shared libraries' functionality.

2.2   Applications

On top of the outlined infrastructure a mobile and a web/desktop end-user application have been developed using the Adobe Flash Builder development environment. Based on the open-source Flex SDK 4.5.1 the environment supports the development of mobile, web, and desktop applications. The applications visualise the gathered information within the infrastructure. Thus the information can be accessed and explored online or with existing institutional or personal devices, including desktop computers, tablets, smartphones, and so on.

Mobile Application. The developed mobile application consists of a title and navigation bar as well as a content area. When launched the application shows an overview of available rooms. The list items are rendered in such a way that each item presents at a glance its title, the current power usage, and the daily total usage. The list is sorted on the daily total usage in descending order. The coloured circles indicate visually the current power consumption (green = 0W, yellow <= 10W, red > 10W). When selecting items detailed information for the room is shown. When navigating to the groups section the application switches to the overview of available groups, providing the same functionality as for rooms.

Web/Desktop Application. When launched the developed web/desktop application shows a simple dashboard. The lists provide an overview of available rooms/groups and their appliance(s). The lists are sorted on the daily total usage in descending order. Thereby the appliance items are rendered in such a way that each item presents at a glance its title and the current power usage. The coloured circles again indicate visually the current power consumption (green = 0W, yellow <= 10W, red > 10W). When selecting items in the lists detailed information for the room, group, or appliance is shown. In addition to that users can explore, relate, or compare the item's consumption.

---

[4] http://www.robotlegs.org/
[5] http://www.adobe.com/products/flash-builder.html

3   Evaluation and Results

As part of the design cycle the developed display prototypes and used visualisation techniques have been evaluated in user-studies to reveal which are most effective in communicating energy consumption data and motivating energy conservation. Furthermore surveys have been conducted to assess whether dynamic visual feedback and the provided facilitation patterns can promote the conservation of electricity at the workplace and measure the increased awareness on the topic as well as changed attitudes and/or changes in behaviour. Furthermore the user acceptance and interest have been measured. The methodology and detailed evaluation results are presented in [3].

In an informative study university employees have been asked about their opinion on energy consumption and conservation at the workplace. The respondents (N=190) had to rate several statements on a 7-Likert-Scale describing their awareness or willingness ranging from not at all up to completely. The median results show that the respondents want to be more aware about their own energy usage and would like to receive more information on how to save energy at the workplace. Most likely they would reduce their individual consumption accordingly. Furthermore the results show that they would like to compare their consumption with colleagues, although they are not profoundly convinced.

In a comparative study university employees working in the office building where the prototype was intended to be deployed have then been asked about their awareness, concern, and attitude regarding energy consumption and conservation at the workplace. The respondents (N=58) had to rate several statements on a 5- respectively 7-Likert-Scale describing their awareness, concern, and attitude ranging from not at all up to completely. After deploying the prototype the study has been repeated among the employees who actually used the prototype (N=14). Both results were then compared. Comparing the median results reveals that the respondents' self-assessed ability to estimate their own energy consumption increased, while still staying relatively low. Furthermore the respondents' concern about their own energy consumption increased after deploying the prototype. Interestingly their concern about personal efforts and the attitude to take more conservation actions is consistent.

To clarify this the respondents were furthermore asked to indicate their actual energy conservation behaviour as well as motivating/demotivating reasons. Comparing the results highlights that in total 5 actions with high conservation potential (e.g., disconnect power supply units when not in use, deactivate screen savers) are not performed more often. The reasons can be manifold and need to be explored in further research. Either the questioned actions have already become part of daily practice and are thus not performed explicitly or participants really need more information on what actions to take in which situation. On the other hand 6 conservation actions are performed equally or even more often (e.g., switch off lighting when leaving a room, use appliance built-in energy saving options) then before.

To evaluate the prototype the participants who used the web/desktop application have been asked to give some feedback. To do so the participants (N=14) had to rate the statements presented in Table 1 on a 7-Likert-Scale ranging from not at all up to completely.

Table 1. Prototype evaluation: rated statements and means

Statement	Median
Did you make use of the energy dashboard?	2
Have you been aware what kind of information was visualized?	4
Did you understand the information given?	5
Was the used information visualization appealing to you?	4
Was the information presented useful and relevant for you?	3
Were you satisfied with the amount of information presented?	4
Were you satisfied with the granularity of the information presented?	4

The results show that although not all participants made extensive use of the display, the information visualized was perceived and understood. Furthermore the actual visualization was rated appealing, useful, and relevant. Thereby the amount of information presented as well as the information granularity satisfied their needs.

The sensoring and logging to measure the effectiveness in terms of energy conservation and enable the prototypical evaluation has been done using the introduced Pachube system. For each room and group a respective feed has been created. Each feed aggregates the total power usage and the daily power usage of the room or group. For rooms with shared workplaces, additionally the total power usage and daily power usage of each workplace are aggregated. On the short term several effects have been observed. Among others the most interesting one is that participants were especially interested in investigating and adapting their consumption patterns, e.g. switch off their appliances over the weekend instead of leaving the appliances in stand-by.

4   Discussion and Conclusions

The evaluation results presented above show the general interest in the topic and indicate the effectiveness of the introduced means towards the conservation of energy. On the long term the sustainability of these effects as well as the actual conservation potential of the deployed infrastructure of course needs to be examined and validated.

In the context of the conducted user studies and when presenting the project ideas several helpful comments as well as critical issues were raised that reflect some major points of discussion. Although the prototype was well received by the participants the actual daily usage was not as high or frequent as expected. As suggested by one participant the tool should maybe be promoted more or possibly it's use should even be enforced. Another solution would be to promote the information itself, trying to put it even more in context and thus prevailing daily practice and working routines.

Some participants also raised general concerns about the energy saving potential at the workplace and thus the usefulness of the prototype. Especially the usefulness and legitimacy of comparing the energy consumption among colleagues, departments, or buildings was questioned. The opinions drift apart widely at that point, which indicates the need for further research and discussions.

Regarding it's instructional capabilities and the application within the described learning context the prototype goes beyond the mere level of information perception. Instead the addressed situational awareness demands at least the comprehension of the available informational cues. In order to make use of the prototype efficiently and thus eventually conserve energy, even demands to forecast and estimate the implications of the personal consumption behaviour. In the terms of the used feedback characteristics the prototype provides simple verification feedback that can be more elaborated on demand. Thereby the timing can be described as immediate, although the delivery of information is not happening in real-time due to technical restrictions. The feedback intends to convey at best relational rules as learning outcome, while not going beyond the confirmatory analysis of errors.

Besides measuring the effectiveness of the prototype, an informative study, a comparative study, and a user evaluation of the prototype were conducted. The results indicate the general interest in the topic as well as the usefulness of the prototype. Nevertheless further work needs to be invested especially in the long-term sustainability of the behavioural change, design implications and improvements, as well as the way of embedding the prototypes into daily practice.

References

1. Holmes, T.G.: Eco-visualization. In: Proceedings of the 6th ACM SIGCHI Conference on Creativity & Cognition, pp. 153–162. ACM Press, New York (2007)
2. Börner, D., Kalz, M., Specht, M.: Beyond The Channel: A Literature Review On Ambient Displays For Learning. In: Computers & Education. Elsevier, Amsterdam (in Press)
3. Börner, D., Storm, J., Kalz, M., Specht, M.: Energy Awareness Displays: Motivating conservation at the workplace through feedback. International Journal of Mobile Learning and Organisation (in Press)

I-Collaboration 3.0: A Model to Support the Creation of Virtual Learning Spaces

Eduardo A. Oliveira[1,2], Patricia Tedesco[1], and Thun Pin T.F. Chiu[1]

[1] Informatic Centre, Federal University of Pernambuco (UFPE)
[2] Recife Center for Advanced Studies and Systems (CESAR)
Recife – PE – Brazil
{eao,pcart,tptfc}@cin.ufpe.br

Abstract. The growth of Web 2.0 and the effective technological convergence of mobile devices, social networks and blogs increased the potential for collective work of global nomads in digital environments. In modern digital society, nomadism changes significantly the way that users, relate, organize themselves and communicate in many different digital environments. With respect to technology-enhanced learning, the consequences of this state of affairs are that students are much more comfortable using their own social tools, and thus are not happy to spend time and effort using particular virtual learning environment (VLE). Thus, one way of keeping students motivated and exploiting their online time is to take advantage of the social tools they already use. In this direction, this article presents i-collaboration 3.0, a system that aims to create distributed and personalized virtual learning spaces on web-based tools (e.g. Twitter, Facebook). The system supports learning in distributed environments that students already know and considers their needs and preferences to provide contents.

Keywords: ubiquitous learning, social recommendations, personalized learning.

1 Introduction

Although there is a large offer of virtual learning environments (VLE) in the market, little has been done to motivate students to use them [1]. In fact, available VLE have presented the same basic functionalities over the years, which has led to a general feeling of isolation, demotivation and high student evasion rates [2]. This situation gets worse when one considers that learners are used to interacting with various other tools, that get them information anytime, anywhere, and, thus, are not interested in learner yet another software just for studying. Although current VLE are multiplatform (web, mobile devices, digital TV), and take usability issues into consideration to improve the learners' experience, merely adding new collaborative tools to VLE, is not enough to cater for the learner's individual needs. To further complicate matters, traditional VLE centralize information, making themselves the only gateway to courses, when perhaps a more natural way would be to make information accessible in

A. Ravenscroft et al. (Eds.): EC-TEL 2012, LNCS 7563, pp. 477–482, 2012.

a distributed way – thus instructors could explore whichever tools the students are most comfortable with. Hence, we argue that to provide for more effective learning, information should be distributed over the platforms the learner uses, instead of only in a VLE. The information must also be personalized.

As a contribution to the challenges (extensibility, interoperability, contextualization and data reuse and integration) found in the virtual learning environments, which we also believe that contributes to reducing the number of problems currently found in distance learning, this article presents i-collaboration 3.0 model. The model has as main objective to contribute to the creation of virtual learning spaces distributed, collaborative and personalized through the web (social networks and sites, for example).

This article is organized in 4 sections. Section 2 presents a background and related works. Section 3 discusses the proposed i-collaboration 3.0 system. Finally, Section 4 presents the conclusions and further work.

2   Background

In the context of technology-enhanced learning, system designers have tried to systematically exploit the modeling potential of computers and develop systems that support learners through adaptive or intelligent operation. Adaptive and Intelligent systems are model-based systems although they have different purposes in supporting learning. These learning strategies aim to address the new needs of the new digital users, dynamic and nomadic.

Teaching and learning are increasingly supported by mobile tools or occur in an environment where there is a wide availability of mobile devices[3], such as PDAs, smartphones, tablets and notebooks. The area concerned with learning with mobile devices is known as m-learning (mobile learning) [4,5]. M-learning is characterized by learning to be supported by many mobile devices (cell phones, notebooks, netbooks, tablets, PDAs, ... [4]. Some authors even disagree about the size of mobile devices used in learning, by categorizing m-learning in accordance with these devices. For some, m-learning should be restricted to small-size devices like cell phones [6]. Like Georgiev and colleagues [7] and Trifonova [8], we believe that m-learning is an extension of e-learning through mobile devices. In the student perspective, the m-learning happens when there is learning without been confined to a particular space or when the access to knowledge is available through any mobile device [9].

Another concept associated with the teaching and learning is the p-Learning (pervasive learning), which relates to the use of small devices (sensors, PDAs, etc.). These devices, often embedded in mobile devices, can get information about the context of the learning environment through the use of environmental models available on a dedicated computer, or through the dynamic construction of this learning strategy.

Finally, ubiquitous learning (u-Learning) integrates m-Learning and p-Learning. While the student is moving with their mobile device, the system dynamically supports their learning, proactively serving the needs of users, acting in a transparent way. Thus, the educational process may occur continuously, comprehensively and transparently [3].

The context, associated with concepts of mobile learning, ubiquitous learning, among others, supports the adaptive systems in content selection, adaptation to support navigation and presentation adapted. And the mobility and ubiquitous environments, in turn, support the student distributed learning network. The learning content can be directed to better meet the needs of each student.

To meet this challenge, researchers in the field of adaptive systems try to overcome the shortcomings of traditional approaches, which deal with all users in the same way (*one-size-fits-all*), exploring ways in which they can adapt their behavior to the goals, tasks, interests and other characteristics of interested students [10]. In educational contexts, while the definitions of "adaptive systems" differ in the literature [10], many of the interpretations converge along the lines of the system's ability to adjust itself to suit individual learners' characteristics and needs.

An Adaptive Educational System (AES) is a system that aims at adapting some of its key functionalities (for example, content presentation and/or navigation support) to the learner needs and preferences [1]). Thus, an adaptive system operates differently for different learners, taking into account information accumulated in the individual or group learner models. Respectively, an Intelligent Tutoring System (ITS) aims to provide learner-tailored support, similarly to what a human tutor would do [1]. To achieve this, ITS designers apply techniques from the broader field of Artificial Intelligence (AI) and implement extensive modeling of the problem-solving process in the specific domain of application. Both AES and ITS seek primarily to meet the individual needs of each student in an intelligent (autonomous) way. The main difference between AES and ITSs relates to their overall goals. While AES focus on adapting content and interfaces, ITS directly focus on supporting the learning of each student, simulating behavior of a virtual tutor (communication). In this article we assume that both are intelligent and autonomous systems.

To promote the adaptation and personalization of learning contents in a distributed manner in the Internet (creation of virtual learning spaces – contents available into twitter, facebook, msn, gtalk, ...), based on each student profile and needs, we present here the i-collaboration 3.0 model. During this research we did not find any related work with the creation of adaptive and distributed virtual learning spaces.

3   I-Collaboration 3.0

To contribute to the minimization of the challenges found in VLEs (communication difficulties, centralized access, interoperability and data integration, ...), that we believe also contribute to minimizing the various problems currently found in the distance education (motivation and isolation feeling), this work has as main objective to contribute to the creation of distributed virtual learning spaces. The support will be provided through i-collaboration 3.0 model, an extension of i-collaboration (1.0) model [1].

I-Collaboration 3.0 tries to ensure decentralized (distributed) access to learning contents available in different Web 2.0 tools (Twitter, MSN, Blogs, ...) and social networks (Facebook, Orkut, ...). The system also integrates students' data to personalize the learning contents (the students' are distributed in the Internet – the same student can learn using MSN, Facebook and Twitter, at the same time, for example), based on the particular tastes and needs of each student (identified through de student behavior in Twitter, MSN, ...). With the virtual learning spaces support, the students will be able to study through the Web, using platforms and environments that they already meet and frequently use.

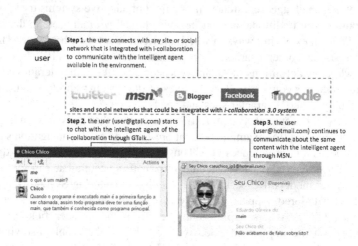

Fig. 1. I-collaboration 3.0 example of use

As shown in the scenario presented in Figure 1, we assume that Twitter, MSN, a blog (Blogger site), Facebook and Moodle (VLE) are integrated with the i-collaboration 3.0 model. To assume that the system is integrated with these environments is to assume that these environments are using i-collaboration 3.0. In the presented scenario, a single instance of an intelligent agent, which is provided by the i-collaboration 3.0 model, is available in each of these environments (such as a contact on MSN, as a user in Twitter, and as a chatterbot in Sites and VLEs, ...). Despite the fact that the intelligent agent appears in many different environments, the model provides a single agent to these environments, this ensures that the same intelligent agent will be used in all environments. The student talks across different environments, with the same bot. If a computer science student starts communicating with the intelligent agent in Gtalk, asking him about the main function of a program: "what is a main function?" he will get an answer about the main function, as requested. A few minutes later, the student goes to the MSN and asks the same thing to the intelligent agent: "main" (because he is still with doubts). At this time, the intelligent agent recognizes the student (that has communicated with him through Gtalk) asking him about the same thing (and in a few interval of time – context sensitive [5]). The intelligent agent infers about students interactions with him, such as student question, student environment, studied contents, student exams scores, student profile [6] and answers him with new questions: "We do not talk about it?" "You need more help with this issue?". If student needs more help, the intelligent agent must suggest to this student related contents based on his doubts in programming introduction.

The advantage to provide a single intelligent agent in the system is in the fact that with only one agent, we can also have a single integrated database in the model (based on students' interaction with the agent in distributed environments). With a single database, the student, which communicates with the agent in Gtalk and in MSN, can now be identified on these and in any other environment that the intelligent agent is presented. If a student interacts with the intelligent agent through Facebook, the agent will know, referring to the historical database of the student that he has already communicated with him through Twitter and MSN, and that he demonstrated interest in studying programming concepts before. I-Collaboration 3.0 extends the model of i-collaboration 1.0 [1,2] in order to make it accessible to all students of VLEs with decentralized, integrated and adaptive features.

A big challenge in developing the virtual learning spaces, as well as interoperability of distributed data on the Web, is due to the personalization of these distributed contents, so that each student has their needs met in the environments that they use to acquire new knowledge. According to Vieira and colleagues [9], the quality of context-aware services is directly related to the quality of the information collected. Context can support i-collaboration 3.0 to improve how contents are provided to each student, adapting them based on the students' individual profiles, based their own needs (and on their favorite environments).

To order exams and to suggest other logins on other Web tools 2.0/social networks, students use special commands such as "#exam", to do an exam, and "#addEnvironment Gtalk mylogin@gmail.com" to set a new login to the student (student is in MSN adding a Gtalk login, for example – teaching the bot his others logins distributed in the Web). These metadata are monitored through Drools inference engine (rule-based reasoning). Drools is responsible for integrating students distributed data and for considering context while students are making questions (repeated questions, in a small period of time, about the same subject, means that the student is finding difficulties and needs help, for example).

As a way of enabling the various customers running separately and accessing the same database of the proposed model, all in a distributed way (even on different servers) and custom, we chose to work with RMI architecture. In this architecture, different customers of different environments (MSN, Twitter, Web, ...) can work in a distributed computing (multiple JVMs). The client (company or institution that wishes to have an instance of the i-collaboration 3.0 available) download the model API and implements a method for the environment that he wants and get an instance of the system. The main class has the following method signature: public String getResponse (String 'questionText, String userId, EnvironmentType environmentType). Any customer interested in using the i-collaboration 3.0 must use this method, stating the text of the student, the student ID and the environment that the student is communicating with the intelligent agent.

This work is prototyped and implemented, and was tested during two months by two different developers (proof of concept) to check the integration of Twitter, MSN and Gtalk with i-collaboration 3.0. The tests also checked the personalization of the learning contents (based on students' profiles and on students' interaction environments) and the server stability and performance. The tests showed promising results (results obtained from logs analysis). The intelligent agent has personalized the contents based on students' needs and profiles, considering his exams scores, studied contents, frequently asked questions, and his environments (text adaptation – 140 chars for Twitter). A greater experiment is being prepared for new tests with students.

4   Conclusions and Further Work

In modern digital society, nomadism changes significantly the way that users, relate, organize themselves and communicate in many different digital environments. With respect to technology-enhanced learning, the consequences of this state of affairs are that students are much more comfortable using their own social tools, and thus are not happy to spend time and effort using particular VLE.

Through the development of i-collaboration 3.0 concept and model, this paper presented a contribution to mitigate the problems that have made difficult the use of distributed personalized learning. The model seeks to deal with each student in a unique way, in the environment that the student feels better, thus motivating these students to learn. I-collaboration 3.0 supports the creation of virtual learning spaces.

The system is also being expanded and integrated with new platforms (Gmail and Facebook). In the future, the results of a new experiment will be published and the model will be available under a software license for the scientific community use.

References

[1] Oliveira, E.A., Tedesco, P.: Putting the Intelligent Collaboration Model in practice within the Cleverpal Environment. In: 2009 International Conference of Soft Computing and Pattern Recognition (IEEE), Malacca, Malaysia, pp. 687–690 (2009), doi:10.1109/SoCPaR.2009.13

[2] Oliveira, E.A., Tedesco, P.: i-collaboration: Um modelo de colaboração inteligente personalizada para ambientes de EAD. Revista Brasileira de Informática na Educação 18, 17–31 (2010)

[3] Roschelle, J., Pea, R.: A walk on the WILD side: How wireless handhelds may change computer-supported collaborative learning. In: International Conference on Computer-Support Collaborative Learning, CSCL 2002, Boulder, Colorado (2002)

[4] Traxler, J.: Current State of Mobile Learning. In: Ally, M. (ed.) Mobile Learning: Transforming the Delivery of Education and Training, pp. 9–24. AU Press, Athabasca University (2009)

[5] Woodill, G.: The mobile learning edge. McGraw Hill, New York (2011)

[6] Keegan, D.: Mobile Learning: The next generation of learning. Disponivel em (2005), http://learning.ericsson.net/mlearning2/files/workpackage5/book.doc

[7] Georgiev, T., Georgieva, E., Smrikarov, A.: M-Learning - a New Stage of E-Learning. In: International Conference on Computer Systems and Technologies - CompSysTech 2004 (2004), doi:citeulike-article-id:1318542

[8] Trifonova, A.: Mobile Learning: Review of the literature. Department of Information and Communication Technology. University of Trento (2003)

[9] Vavoula, G.N., Lefrere, P., O'Malley, C., Sharples, M., Taylor, J.: Producing Guidelines for Learning, Teaching and Tutoring in a Mobile Environment. Paper Presented at the Proceedings of the 2nd IEEE International Workshop on Wireless and Mobile Technologies in Education, WMTE 2004 (2004)

[10] Brusilovsky, P., Nejdl, W.: Adaptive Hypermedia and Adaptive Web. Practical Handbook of Internet Computing. CRC Press LLC (2004)

Learning to Learn Together
through Planning, Discussion and Reflection
on Microworld-Based Challenges

Manolis Mavrikis[1], Toby Dragon[2], Rotem Abdu[4], Andreas Harrer[3],
Reuma De Groot[4], and Bruce M. McLaren[2,5]

[1] London Knowledge Lab, Institute of Education, London, UK
[2] Center for E-Learning Technology (CeLTech), Saarland University, Germany
[3] Catholic University Eichsttt-Ingolstadt, Bavaria, Germany
[4] Hebrew University of Jerusalem, Israel
[5] Carnegie Mellon University, USA

Abstract. This demonstration will highlight the pedagogy and functionality of
the Metafora system as developed by the end of the second year of the EU-funded
(ICT-257872) project. The Metafora system expands the teaching focus beyond
domain-specific learning to enable the development of 21st century collaborative competencies necessary to learn in today's complex, fast-paced environment. These competencies — termed collectively as "Learning to Learn together"
(L2L2) — include: distributed leadership, planning / organizing the learning
process, mutual engagement, seeking and providing help amongst peers, and reflection on the learning process. We summarise here the Metafora system, its
learning innovation and our plan for the demonstration and interaction session
during which participants will be introduced to L2L2 and Metafora through
hands-on experience.

Keywords: CSCL, learning to learn together, planning, discussion, microworlds.

1 Introduction

The EU funded Metafora project (ICT-257872), launched in July 2010, is focusing on
the development of a Computer-Supported Collaborative Learning (CSCL) system to
scaffold a process referred to as "Learning to Learn Together" (L2L2). Recognising
that collaborative work and training of meta-cognitive skills are better practiced and
learned in environments where students face serious and difficult challenges, Metafora's
pedagogical designers organize each Metafora classroom scenario around one of several
lengthy or real-world challenges. The challenges encourage students to interact with
microworlds (including simulators and games) where they either build digital artifacts
(models) that allow them to engage in collaborative problem solving or simply test
hypotheses or theories related to the challenges. The pedagogy behind the Metafora
project and the activities that can be undertaken have been described in detail in project
deliverables[1] and other publications [1–4]. As a brief summary, we first acknowledge
that L2L2 is a complex competency, not easily decomposed into a clear-cut division

---

[1] See http://www.metafora-project.org

A. Ravenscroft et al. (Eds.): EC-TEL 2012, LNCS 7563, pp. 483–488, 2012.
© Springer-Verlag Berlin Heidelberg 2012

of independent underlying skills. However, the Metafora project has identified several key skills that are necessary to any process in which students are learning together, and on a higher level, learning how to become better group learners. These competencies include: distributed leadership, mutual engagement, help seeking/giving and reflection on the group learning process.

We present in Section 2 how the Metafora platform and tools are designed to support the L2L2 process while Section 3 summarises our learning innovation. Section 4 offers our plan for the demonstration and interaction session.

2   The System

2.1   The Platform

The Metafora platform (shown in Fig. 1) serves both as a toolbox of various learning tools and as a communication architecture to support cross-tool interoperability. The tool-box facet of the system provides a graphical container framework in which the diverse learning tools can be launched and used in similar ways as their stand-alone usage. Basic functionalities that are globally available are user management (login / logout, and group membership for both local groups of students sitting at one computer as well as remote, collaborative groups), a chat system to discuss and organize work between group members, and a help request function that is present across the entire platform. We now describe the various tools that reside within the platform container.

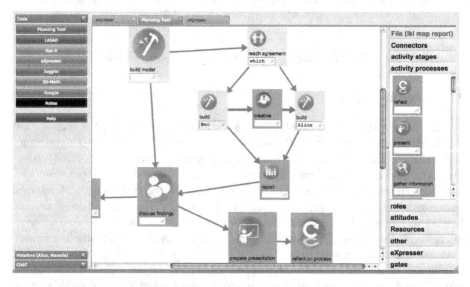

Fig. 1. Screenshot of the Metafora platform with several learning tools opened (see tabs on the upper border). The current focus is on the planning tool (started activities are marked yellow, finished activities in green). In this example the teams of Ben and Alice are each building their own model in a microworld.

2.2   The Planning / Reflection Tool

The planning/reflection tool (see Fig. 1) is a web-based application offering a visual language for planning, enacting, and reflecting on Metafora learning activities. Even though it is built as a stand-alone web application it is central to Metafora as it acts as an entry gate and a pivot to the other tools. Students can create or modify plans for facing a challenge. Their plan then provides a method for students to enact their planned steps, offering an automatic login to the various tools for their planned activities, and providing the work context needed to tackle specific tasks within the challenge. The tool acts as a shared space where students mark activities as started and finished, thereby making the plan also a visual representation of their achievements and current status. In that sense it also acts as a shared artifact for reflection on students' L2L2 process.

2.3   Discussion Tools and Referable Objects

Metafora provides discussion tools to allow general communication and collaboration for teams, but also aimed specifically to support the L2L2 process by allowing discussion and argumentation spaces to integrate artifacts created in other tools. Two discussion tools serve different purposes. First, the chat tool offers a quick and ever-present space for students to gain each other's attention and share informal thoughts in situ as they are working with any of the Metafora tools. Second, LASAD [5] offers a more structured approach to discussion through argumentation graphs (see Fig. 2) which has been shown to improve discussion and argumentation skills [6].

Both the chat functionality and the LASAD system are customized to display and offer links to *referable objects* from other tools. These referable objects are artifacts shared from other tools that can be viewed (text or thumbnail images) as components of the

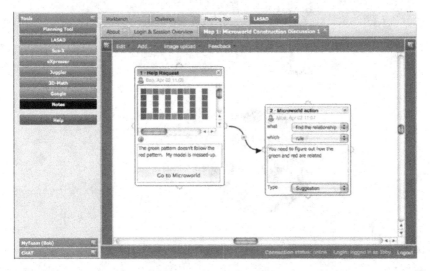

Fig. 2. A sample discussion map in LASAD. A referable object from a microworld (eXpresser) is embedded as a thumbnail within a Help Request box.

discussion, but can also be accessed in the context of the original creator tool through return links (see an example in Fig. 2).

The requirement behind referable objects emerged from early experimentation with the system and was supported by previous related research e.g. [7]. By using referable objects, students can include planning cards or microworld objects in their discussion without the need of anaphoric or deictic language. This allows continuous dialog that is explicitly linked with and contextualized by the students work in other tools. This kind of dialog promotes L2L2 activities such as offering help to one another, and reflecting on ideas in an ongoing processes of negotiation of new meaning for the referenced artifacts.

2.4   Analysis and Visualization

As each tool stands as an independent learning application, these systems offer their own analysis of student work. This automated analysis ranges from low-level activity indicators (such as indicating the creation or modification of artifacts) to high-level analyses (such as identification of whether a student is struggling). The intelligent components of the tools that create these various analyses report them to a centralized analysis communication channel for the entire Metafora platform. A central analysis agent can then monitor this channel, and offer higher-level analysis. The theory behind this work and first implementation steps can be seen in more detail in [4]. This analysis information is used to offer both direct feedback to students (through a notification system) and useful summary information to students and teachers (through visualization tools that filter and aggregate information). Defining and creating these high-level analyses and the specifics of what information should be displayed, to whom, and when is an on-going effort based on prototypes and Wizard of Oz experimentation (see [8]).

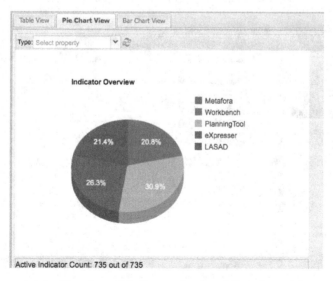

Fig. 3. An example of a filtered set of indicators showing different types of activity (creation, modification, etc.) for the discussion tool

3   Learning Innovation

Successive versions of the Metafora prototype have been used in several pilot experimentations in 4 countries by the various project teams including: the Hebrew University of Jerusalem, Israel; the London Knowledge Lab, the Institute of Education, UK; University of Exeter, UK; and the Educational Technology Lab, National and Kapodistrian University of Athens, Greece. The results of these pilots have been used to improve Metafora in an iterative fashion and also to refine the underlying L2L2 pedagogy and theory. This design-based approach has helped us pinpoint L2L2 behaviours that are enabled and encouraged by the Metafora platform and tools while students are undertaking challenges.

For example, the availability of referable objects (text or thumbnail images) from other tools to the discussion space meets the key requirements of mutual engagement as it allows students to bring individual work into a collaborative space. This functionality also offers the ability for students to seek and offer help to one another, allowing them to share individual artifacts and to exemplify problems or concepts that need to be mutually understood in order to offer support. Lastly, it also offers opportunity for reflection, on both learning activities (giving students a space to compare and discuss the artifacts they have created) and on group dynamics (providing opportunities for students to discuss their overall workflow, contributions, etc.). In short, the availability of the discussion space and its enhancement with referable objects promotes group meaning making [7]. Similarly, students' interaction with the visual language, in the shared space provided by the planning tool, encourages both the orchestration of activities but also fruitful meta-level discussions. When students collaboratively reflect upon the work undertaken to solve a challenge (i.e. task assignments, leadership distribution) they engage in co-construction as well as self- and other-directed explaining — three key mechanisms responsible for learning from collaborative problem solving (c.f. [9]).

4   Demonstration Plan

During the demonstration and interaction session the conference participants will be first and foremost introduced to the theory of learning to learn together through hands-on experience with the Metafora platform. As they interact, we will share our insights on how students and teachers experience and practice these higher-level learning skills through use of the system in various activities in science and mathematics.

The Prototype-SLAM presentation will focus on key technical and conceptual innovations of the system such as the visual language for orchestrating collaborative activities and the referable objects. Following the presentation, we will make available handouts and leaflets that describe the Metafora system, the breadth of learning topics covered by our current scenarios, and our results from the second year of the project. We will also have a dedicated laptop to display videos from actual use in school studies.

In the provided booth we will have 2-3 connected laptops to simulate a collaborative session. Members from the Metafora team will demonstrate the system and act as guides for the participants who will be able to make and modify plans, experiment with referable objects, and see the summaries of their work and types of feedback offered. We will scaffold users to interact with particular challenges that requires them to use microworlds such as the eXpresser microworld [10]. The collaborative task will encourage them to share and discuss their work with others. Through this interaction, participants will be able to appreciate how the integrated tools of Metafora create novel opportunities for collaboration and peer tutoring by allowing students to easily share and discuss their work. Throughout this experience, the system as a whole and the provided feedback will demonstrate the meaning of "Learning to Learn together", and how the system monitors and scaffolds this L2L2 process.

Acknowledgements. We would like to thank the European Union under the Information and Communication Technologies (ICT) theme of the 7th Framework Pro-gramme for R&D (FP7) for funding this research. The contents of this paper do not represent the opinion of the EU, which is not responsible for any use that might be made of them. We are also grateful to the teachers and students who are contributing to the project and to all the other members of the Metafora team for their collaboration work and stimulating discussions.

References

1. Wegerif, R., Yang, Y., De Laat, M., Pifarre, M., Yiannoutsou, N., Moustaki, F., Smyrnaiou, Z., Daskolia, M., Mavrikis, M., Geraniou, E., Abdu, R.: Developing a Planning and Reflection tool to Support Learning to Learn Together. In: Proceedings of IST-Africa 2012 (2012)
2. Smyrnaiou, Z., Moustaki, F., Kynigos, C.: Students' constructionist game modelling activities as part of inquiry learning processes. Electronic Journal of e-Learning. Special issue on Games-Based Learning (2012)
3. Abdu, R., DeGroot, R., Drachman, R.: Teacher's Role in Computer Supported Collaborative Learning. In: CHAIS Conference, pp. 1–6 (2012)
4. Dragon, T., McLaren, B.M., Mavrikis, M., Geraniou, E.: Scaffolding Collaborative Learning Opportunities: Integrating Microworld Use and Argumentation. In: Ardissono, L., Kuflik, T. (eds.) UMAP Workshops 2011. LNCS, vol. 7138, pp. 18–30. Springer, Heidelberg (2012)
5. Loll, F., Pinkwart, N., Scheuer, O., McLaren, B.M.: In: How Tough Should It Be? In: Simplifying the Development of Argumentation Systems using a Configurable Platform. Bentham Science Publishers (2012)
6. Scheuer, O., Loll, F., Pinkwart, N., McLaren, B.: Computer-supported argumentation: A review of the state of the art. International Journal of Computer-Supported Collaborative Learning 5(1), 43–102 (2010)
7. Stahl, G.: Group Cognition: Computer Support for Building Collaborative Knowledge (Acting with Technology). The MIT Press (2006)
8. Mavrikis, M., Dragon, T., McLaren, B.M.: The Design of Wizard-of-Oz Studies to Support Students Learning to Learn Together. In: Proceedings of the International Workshop on Intelligent Support in Exploratory Environments: Exploring, Collaborating, and Learning Together (2012)
9. Hausmann, R.G., Chi, M.T.H., Roy, M.: Learning from Collaborative Problem Solving: An Analysis of Three Hypothesized Mechanisms. In: 26th Annual Conference of the Cognitive Science Society, pp. 547–552. Lawrence Erlbaum (2004)
10. Mavrikis, M., Noss, R., Hoyles, C., Geraniou, E.: Sowing the seeds of algebraic generalization: designing epistemic affordances for an intelligent microworld. Journal of Computer Assisted Learning (in press)

Making Learning Designs Happen in Distributed Learning Environments with GLUE!-PS

Luis Pablo Prieto[*], Juan Alberto Muñoz-Cristóbal, Juan Ignacio Asensio-Pérez,
and Yannis Dimitriadis

University of Valladolid, Paseo Belén 15, 47011 Valladolid, Spain
{lprisan,juanmunoz}@gsic.uva.es, {juaase,yannis}@tel.uva.es
http://gsic.uva.es

Abstract. There exist few virtual learning environments (VLEs) which allow teachers to make learning design decisions explicit and reusable in other environments. Sadly, those few VLEs that do so, are not available to most teachers, due to institutional decisions and other contextual constraints. This panorama is even grimmer if a teacher wants to use not only the tools offered by the institutional VLE, but also other web 2.0 tools (in a broader, so-called "Distributed Learning Environment"). By using the GLUE!-PS architecture and data model, teachers are now able to design learning activities using a variety of learning design tools, and to deploy them automatically in several different distributed learning environments. The demonstrator will show two authentic learning designs with different pedagogical approaches, and how GLUE!-PS helps set up the ICT infrastructure for both of them into two different distributed learning environments (one based on Moodle, the other on wikis).

Keywords: learning design, deployment, virtual learning environments, web 2.0 tools, distributed learning environments.

1 Introduction: Designing Learning, and Then... What?

The discipline of learning design (LD, [1]), and other sibling and ancestor fields (instructional design, etc) have now a long and established position in educational research. However, we still see relatively low penetration of such learning design practices in our schools and universities, especially where ICT tools are involved. The failure of educational standards (e.g. IMS-LD, [2]) to achieve widespread adoption is a much-discussed topic in the field of TEL, which exceeds the scope of this paper.

However, as discussed in [3], some researchers believe that a big part of this limited adoption of (technology-supported) learning design comes mainly from the fact that few solutions exist that allow a teacher to use any of the wide array of existing LD authoring tools, and translate those designs into the ICT infrastructure needed to enact the design ideas in the (physical or virtual) classroom. Those few solutions that exist, are either not usable by teachers without a technological background, or

---

[*] Corresponding author.

A. Ravenscroft et al. (Eds.): EC-TEL 2012, LNCS 7563, pp. 489–494, 2012.

simply are not widespread in educational institutions, and thus are not a viable option for a great majority of teachers who "work with what they got". The problem is even worse in the (increasingly common) case of teachers wanting to go beyond the walls of the VLE-included tools, and use external web 2.0 tools for their learning designs. In this case, the effort and time needed to set up and orchestrate most non-trivial learning designs in such a distributed learning environment is out of the reach of all but a few teachers. This is what we call the **"deployment gap"**.

This paper presents a demonstrator for GLUE!-PS, a service-oriented architecture and data model designed to bridge this deployment gap. The GLUE!-PS proposal is described in the next section, which is followed by two examples of GLUE!-PS usage that follow the learning design from the original teachers' ideas to their implementation in distributed learning environments (DLEs) that integrate mainstream VLEs (such as Moodle[1]) and external web 2.0 tools (such as wikis or shared web apps). The paper closes with remarks about this proposal's relevance and applicability.

2   GLUE!-PS: An Architecture to Deploy Learning Designs in Distributed Learning Environments

The aforementioned "deployment gap" problem is common to many authentic TEL environments, where ad-hoc enactment solutions exist (e.g. for CSCL activities) on one side, while most institutions adopt general-purpose learning environments like Moodle or Blackboard[2]. However, this problem is even more insidious in the increasingly common case of "distributed learning environments" [4], where a central VLE or personal learning environment (PLE) is used along other learning tools, especially web 2.0 tools like wikis, blogs, shared apps or social media.

In this demonstration we present the Group Learning Unified Environment - Pedagogical Scripting (GLUE!-PS), a service-oriented architecture that aims at allowing a non-technology-expert teacher to deploy learning designs, **authored with a variety of learning design tools and languages**, into **distributed learning environments** comprising one of multiple virtual learning environments (VLEs), plus multiple other external learning tools (such as web 2.0 tools). It is important to note that this last part (the integrated use of external learning tools) is provided through the usage of the GLUE! architecture [5].

As described in [3], the GLUE!-PS architecture is based on adapters (see Fig. 1), which translate the original learning designs to a common data model, which is then translated to the models and concepts of the different target VLEs (also creating and linking the needed external resources, such as web 2.0 tools). This central data model (also described in [3]) was developed to include the most common traits of existing learning design languages, that are *deployable* into current mainstream VLEs. Although these two translations forcefully introduce a certain loss of information from the original designs, analytical evaluations and experiments with teachers show initial evidence that the final result represents the original designs well enough to be used in real situations [6].

---

[1] http://moodle.org (Last visit: 30/03/12)

[2] http://www.blackboard.com (Last visit: 30/03/12)

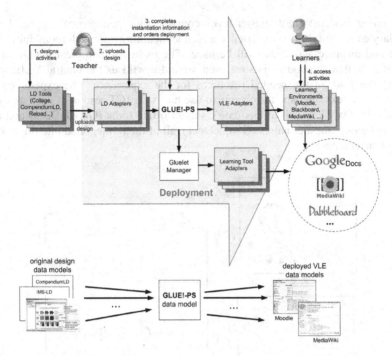

Fig. 1. Simplified GLUE!-PS architecture and data model (adapted from [3])

3   The Demonstrator

In the "Prototype-SLAM" session, a **fully functional prototype** of the GLUE!-PS architecture reference implementation will be shown. In order to demonstrate how the whole design lifecycle can be followed from learning design idea to an enactment-ready ICT-supported course in a distributed learning environment, we will use two real examples of learning designs and their deployment. These learning designs have been taken from real educational experiences in higher education, which are being enacted during this academic year. The designs and deployments were made by two different teachers with very different approaches and needs, as they went from their learning design ideas to the ICT infrastructure that embodied those ideas in two very different distributed learning environments comprising a VLE (Moodle vs. a Media-Wiki[3]-based wiki) and external web 2.0 tools (GoogleDocs[4], Dabbleboard[5]).

In our first example, a university teacher has an idea of proposing a complex colla-borative learning experience, following the **Jigsaw pattern[6]** (a very common

---

[3] http://www.mediawiki.org (Last visit: 30/03/12)

[4] https://docs.google.com (Last visit: 30/03/12)

[5] http://www.dabbleboard.com (Last visit: 30/03/12)

[6] A jigsaw implies the subdivision of a problem into parts, which are first studied separately by "experts". Later, a global solution is proposed by "jigsaw groups" that comprise experts in every sub-problem.

collaborative pattern) in a master-level course about pedagogical approaches in secondary education. In this experience a blended learning approach (combining face-to-face and distance activities) will be used. The main technological feature of this experience is that the course is structured around a **wiki** as the central VLE, where students can find all the needed resources for the experience. The experience will span several weeks, and she also wants to use non-wiki ICT resources such as shared whiteboards (Dabbleboard, in this case), shared office tools (GoogleDocs) and individual and group questionnaires (GoogleDocs), all of them integrated into the wiki for student convenience.

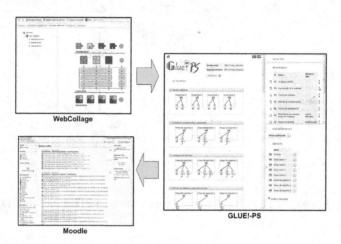

Fig. 2. Screenshots of the different applications involved in one example from learning design to implementation: WebCollage (top-left), GLUE!-PS (right), Moodle (bottom-left)

In order to make a computer-interpretable learning design with this idea, and given the pattern-based nature of the collaborative activities, the teacher chooses **WebCollage**[7] [7] as her LD authoring tool. WebCollage allows non-expert teachers to design collaborative learning activities based on collaborative learning flow patterns (such as Jigsaw, Pyramid, etc). Moreover, unlike many other LD tools, WebCollage also allows teachers to particularize their learning designs for a concrete classroom, setting the number and composition of groups on each phase. Fig. 2 shows a screenshot of the WebCollage LD tool where we can see the three activities that conform the "experts phase" of the design.

In the case teacher would like to use a different VLE (Moodle) to enact this very same learning design (e.g. because her institution enforces the usage of an institutional environment, or because she has shared her design with another fellow teacher who prefers to use Moodle), WebCollage and GLUE!-PS would allow her to do the particularization and deployment to this new target VLE, as long as GLUE!-PS has an LE adapter for it. Fig. 2 shows a real deployment of the same design into a Moodle-centric distributed learning environment.

---

[7] http://gsic.uva.es/wic2 (Last visit: 30/03/12)

As a short follow-up to this full-cycle demonstration, we will show briefly how a different (but equally authentic) blended learning experience was designed (in a few minutes minutes) by a teacher in a hurry, who wanted to deploy a blended learning activity based around a role playing situation. Since she found that the WebCollage tool did not suit her needs (her conception of the role-playing clashed with WebCollage's), she chose to use a different, simple learning design tool, the **Pedagogical Pattern Collector** (PPC, developed by the London Knowledge Lab, see [8]) and GLUE!-PS to create a Moodle course (she uses Moodle for the whole course). In this case, due to the lack of particularization information in PPC, the teacher used the GLUE!-PS graphical interface to set up the participants, groups and web tools for the lesson, and finally deployed it to her **Moodle** course. This very same design could have been deployed, e.g. to MediaWiki as the first design was, if the teacher had used such a VLE to centralize her course. These two mini-cases show how it is *the teacher* who chooses the best technological solution (combination of LD tool and VLE) to suit her needs and contextual constraints, and not the other way around.

All in all, this demonstration will show the ability of the GLUE!-PS to cover **four different conversions** from learning design to enactable course in a distributed learning environment (two learning design tools, by two virtual learning environments). The fact that those distributed learning environments include not just mainstream VLE tools (e.g. Moodle), but also external web 2.0 tools like GoogleDocs or Dabbleboard (thanks to the GLUE! architecture), only highlights further the myriad of possible learning situations for which GLUE!-PS is a relevant orchestration help. It is also important to note that, due to the adaptor-based architecture of GLUE!-PS, and the relative simplicity of its data model, such variety of combinations has been attained at a comparably low cost in development efforts. Moreover, the approach is fully extensible to adapt also other LD tools and VLEs that may emerge in the future.

4   Conclusions

The current GLUE!-PS prototype has already undergone several iterations of design and usability testing with teachers. Moreover, this prototype, along with the WebCollage authoring tool, has already been tested in the deployment of 37 learning designs made by non-technology expert teachers (see [6] for more details). Also, it has been used to deploy and enact collaborative learning designs in several authentic situations (mostly collaborative blended learning experiences in higher education): two professional development workshops about learning to design collaborative activities, and two master-level courses where complex collaborative flows with VLEs and external web tools were needed. More experiences, using different learning design tools and VLEs are also being conducted in the upcoming weeks.

Even at this early stage of development, GLUE!-PS has shown the potential for this kind of system to unload part of the (considerable) orchestration burden that enacting blended collaborative learning in a distributed learning environment can impose on the shoulders of teachers. We believe that the opening of GLUE!-PS (and the code of its reference implementation) to the public will help a wide array of teachers and institutions in overcoming one of the main current barriers for wide-spread adoption of learning design in TEL: the deployment gap. This opening will also enable multiple implementations of the GLUE!-PS adaptors for different learning design approaches, institutional VLEs and sets of learning tools. Such adaptors ecosystem may allow teachers to choose the learning design approach that best fits their needs, and then convert those ideas into ICTs ready to be used in her authentic setting.

Acknowledgements. The authors would like to thank Beatriz and Alejandra, the teachers who kindly provided the example learning designs and authentic contexts. This work has been partially funded by the Spanish Ministry of Economy and Competitiveness (TIN2008-0323, TIN2011-28308-C03-02 and IPT-430000-2010-054) and the Autonomous Government of Castilla y Leon, Spain (VA293A11-2 and VA301B11-2).

References

1. Koper, R., Tattersall, C.: Learning Design: a Handbook on Modelling and Delivering Networked Education and Training. Springer (2005)
2. IMS Global Learning Consortium: IMS Learning Design v1.0 Final Specification (2003), http://www.imsglobal.org/learningdesign
3. Prieto, L.P., Asensio-Pérez, J.I., Dimitriadis, Y., Gómez-Sánchez, E., Munoz-Cristóbal, J.A.: GLUE!-PS: A Multi-language Architecture and Data Model to Deploy TEL Designs to Multiple Learning Environments. In: Kloos, C.D., Gillet, D., Crespo García, R.M., Wild, F., Wolpers, M. (eds.) EC-TEL 2011. LNCS, vol. 6964, pp. 285–298. Springer, Heidelberg (2011)
4. MacNeill, S., Kraan, W.: Distributed Learning Environments: A Briefing Paper. Technical Report, JISC Center for Educational Technology and Interoperability Standards, CETIS (2010)
5. Alario-Hoyos, C., Wilson, S.: Comparison of the main alternatives to the integration of external tools in different platforms. In: Proceedings of the International Conference on Education, Research and Innovation (ICERI) (2010)
6. Muñoz-Cristóbal, J.A., Prieto, L.P., Asensio-Pérez, J.I., Jorrín-Abellán, I.M., Dimitriadis, Y.: Lost in Translation from Abstract Learning Design to ICT Implementation: a Study Using Moodle for CSCL. Accepted at to the European Conference on Technology-Enhanced Learning, EC-TEL (2012)
7. Villasclaras-Fernández, E.D., Hernández-Gonzalo, J.A., Hernández-Leo, D., Asensio-Pérez, J.I., Dimitriadis, Y., Martínez-Monés, A.: InstanceCollage: A Tool for the Particularization of Collaborative IMS-LD Scripts. Educational Technology & Society 12(4), 56–70 (2009)
8. Laurillard, D., Charlton, P., Craft, B., Dimakopoulos, D., Ljubojevic, D., Magoulas, G., Masterman, E., Pujadas, R., Whitley, E., Whittlestone, K.: A constructionist learning environment for teachers to model learning designs. Journal of Computer Assisted Learning (2011), doi:10.1111/j.1365-2729.2011.00458.x

Math-Bridge: Adaptive Platform
for Multilingual Mathematics Courses

Sergey Sosnovsky[*], Michael Dietrich[*], Eric Andrès[*], George Goguadze[†],
and Stefan Winterstein[*]

[*] DFKI, Centre for e-Learning Technology, Saarbrücken, Germany
[†] Saarland University, Department of Computer Science, Saarbrücken, Germany

Abstract. Math-Bridge is an e-Learning platform for online courses in mathematics. It has a number of unique features: it provides access to the largest in the World collection of multilingual, semantically annotated mathematical learning objects; it models students' knowledge and applies several adaptation techniques to support more effective learning, including personalized course generation, intelligent problem solving support and adaptive link annotation; it facilities a direct access to learning objects by means of semantic and multilingual search. All this student interface functionality is complemented by the teacher interface that allows managing students, groups and courses, as well as tracing students' progress with the reporting tool. Overall, Math-Bridge offers a complete solution for organizing technology-enhanced learning of mathematics on individual-, course- and/or university level.

Keywords: e-learning, mathematics education, adaptation, course generation, adaptive navigation support, e-learning platform, multilingual and multicultural aspects.

1 Introduction

The Math-Bridge[1] platform has been developed as a joint effort of educators and computer scientists from nine universities and seven countries in order to take a significant step towards improving European educational practices in the field of Mathematics. This project has targeted one of the most urgent problems that the majority of European countries are facing now: insufficient and inconsistent mathematical competencies of a large number of school graduates and first-year university students studying science, engineering and technology disciplines. Math-Bridge implements a solution to this problem, which is based on achieving several operational objectives:

---

[1] The work described here has been supported by the EU *eContentplus* program under the grant "Math-Bridge: European Remedial Content for Mathematics" (ECP-2008-EDU-428046).

A. Ravenscroft et al. (Eds.): EC-TEL 2012, LNCS 7563, pp. 495–500, 2012.

1. To collect and harmonize high-quality remedial content developed by experts in bridging-level mathematics and make this content broadly accessible on the Web;
2. To enable cross-cultural and multi-lingual presentation of this content, thus promoting its reuse across the borders;
3. To motivate the technological reuse of the content by implementing it in a shareable format and enriching it with metadata based on open standards;
4. To offer different types of personalized access to the content, thus supporting multiple usage scenarios of the platform: from individual exploratory e-Learning to classroom-based knowledge training and testing;
5. To foster the adoption of the platform by increasing its usability not only for students, but also for other stakeholders, including teachers and university officials.

After the 3 years of project development, these goals have been attained. The Math-Bridge platform has been developed and successfully evaluated in real bridging courses. This paper presents the implementation and the design of Math-Bridge focusing on its most important characteristics[2].

2   Content and Knowledge Base

2.1   Mathematical Content Collections

The Math-Bridge content base consists of several collections of learning material covering the topics of secondary and high school mathematics. These collections were originally developed by mathematics educators for teaching real bridging courses. All content has been sliced into individual learning objects, transformed into the OMDoc[3] format for mathematical documents and provided with metadata. The total number of learning objects in the Math-Bridge content base is almost 11000, including 5000 interactive exercises, 1500 learning examples, 2100 instructional texts, 1000 concept definitions, etc.

2.2   Metadata Schema

Math-Bridge employs very rich metadata schema for annotating individual learning objects and entire collections. The metadata elements can be divided into the following three categories:

- descriptive metadata used for administrative, cataloguing and licensing purposes; represented mainly using the Dublin Core standard[4].

---

[2] To get more information about Math-Bridge, try its demo and download the full version of the platform including the entire content collection, visit: http://www.math-bridge.org

[3] http://www.omdoc.org

[4] http://www.dublincore.org

- pedagogical metadata helping authors to specify multiple educational properties of learning objects (e.g. difficulty, learning context, field of study or cognitive processes involved).
- semantic metadata connecting different learning objects to one another (e.g. specifying a definition for a concept, or a prerequisite concept for an exercise).

Overall, Math-Bridge metadata plays the core role in the overall architecture of the platform. It enables learning objects discovery, course composition, students' knowledge tracing and subsequent adaptation of the learning content.

2.3   Math-Bridge Ontology

An ontology for the target subset of mathematical knowledge has been created. It serves as a reference point for all content collections and provides the source of the most abstract semantic metadata. The ontology is used by the system logic for modeling students' knowledge, and adaptive course generation. The ontology defines more than 600 concepts. It is available in OMDoc and OWL.[5]

2.4   Multilingual/Multi-cultural Aspect and Mathematical Notation Census

Math-Bridge content is available in seven languages: English, German, French, Spanish, Finnish, Dutch and Hungarian. The user can specify the language, in which she would like to read the content. To support multilingual students, individual learning objects can be translated on the fly. It is important to mention, that Math-Bridge translates not only the text but also the presentation of formulae. Although mathematics is often called a "universal language", this is not fully true. In many countries, the same mathematical concepts use very different symbols [5]. In order to address this challenge, Math-Bridge separates the semantic and the presentation layer of math symbols. Inside the content, symbols are encoded using unambiguous entities, and when presented to the user, a correct notation is chosen based on the current language. A public "notation census" has been conducted to document different notations of all symbols in all languages[6].

3   Technology-Enhanced Learning of Mathematics

Figure 1 presents the main layout of the student interface of Math-Bridge. It consists of three panels: the left panel is used for navigation; the central panel – for reading learning content and interacting with exercises; and the right panel provides access to learning objects details, as well as some additional features, such as semantic search and social feedback toolbox.

---

[5] http://www.math-bridge.org/content/mathbridge.owl
[6] http://wiki.math-bridge.org/display/ntns/Home

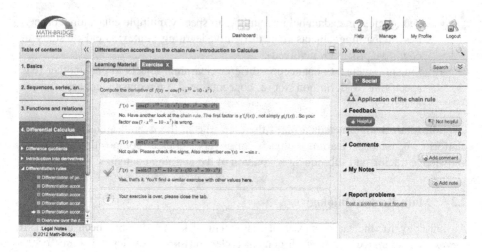

Fig. 1. Math-Bridge Student Interface

3.1   Tracing Students' Progress and Modeling Their Knowledge

Math-Bridge logs every student interaction with learning content. Loading a page, answering an exercise, accessing an individual learning object through the search tool will be stored in the student's log file and will help tracing her progress.

Interactions with exercises are used by the student-modeling component of Math-Bridge to produce a meaningful view on the student's progress. Every exercise in Math-Bridge is associated with one or several concepts and/or theoretical learning objects (such as theorems and definitions). A correct answer to the exercise is interpreted by the system as an evidence that the student knows associated knowledge and will result in the increase of knowledge levels for the corresponding concepts in this student's student model.

3.2   Personalized Courses

The course generator component of Math-Bridge allows students to automatically assemble a course tailored according to their needs and adapted based on their current knowledge state. To generate a course, students need to select the target topics and a learning scenario. Several scenarios are available within Math-Bridge: a student can choose to explore a new topic, train a particular competency, prepare for an exam, master a previous topic or a assemble a course that will focus on the current gaps in student's knowledge. Each course type is generated based on a set of pedagogical rules defining the top-level structure of the course and the learning goals. The generation tool queries the student model and the metadata storage in order to assemble a didactically valid sequence of learning objects. Pedagogical metadata (such as exercise difficulty) and semantic metadata (such as prerequisite-outcome relations) play the central role in this process.

3.3   Adaptive Navigation Support

The amount of content available within Math-Bridge is massive. Some of the pre-defined bridging courses consist of thousands of learning objects. In order to help students find the right page to read and/or the right exercise to attempt, Math-Bridge implements a popular adaptive navigation technique – adaptive annotation. The annotation icons show to the student how much progress she has achieved for the part of learning material behind the icon. Math-Bridge computes annotations on several levels: each course in the Math-Bridge dashboard, each topic within a course table of contents and each content page under a topic are provided with individual progress indicators aggregating student's learning activity on the corresponding level.

3.4   Interactive Exercises and Problem Solving Support

Interactive exercises play two important roles in Math-Bridge. First of all, they maintain constant assessment of students' knowledge thus providing the input for the student-modeling component. Second they give students the opportunity to train mathematical competencies and apply in practice theoretical knowledge acquired by reading the rest of the content.

The exercise subsystem of Math-Bridge can serve multi-step exercises with various types of interactive elements and rich diagnostic capabilities. At each step, Math-Bridge exercises can provide students with instructional feedback ranging from flag messages to adaptive hints and explanations.

Math-Bridge can automatically generate interactive exercises powered by external domain reasoner services. Currently, Math-Bridge uses a collection of IDEAS domain reasoners[7] that provide stepwise diagnosis of students' actions and help generating advanced feedback and hints on every step of the solution.

The Math-Bridge platform also implements functionality for integrating third-party exercise services that maintain the full cycle of student-exercise interaction. As a result, students can access within Math-Bridge both, native Math-Bridge exercises and exercise served by remote systems. The integration is seamless for the student (Math-Bridge makes no difference in how native and external exercises are launched) and fully functional (Math-Bridge makes no difference in how students' interactions with native and external exercises are logged and interpreted by its modeling components). Currently Math-Bridge integrates two external exercise systems: STACK[8] and MatheOnline[9].

3.5   Semantic Search of Learning Objects

In addition to navigating through the course topics, students have a more direct way to find learning objects of their interest – by using the Math-Bridge search tool. They

---

[7] http://ideas.cs.uu.nl/www/
[8] http://www.stack.bham.ac.uk/
[9] http://www.mathe-online.at

can use default search based on simple string matching, advanced search that allows more precise specification of general search parameters (exact or practical matching, lexical or phonetic matching) and semantic search. The semantic search mode fully utilizes the advanced metadata schema of Math-Bridge. Students can specify the type of the desired learning object (e.g. only exercises), its difficulty (e.g. only easy exercises), its target field of study (e.g. only easy exercises designed for physics students), etc.

4   Technology-Enhanced Teaching of Mathematics

Math-Bridge offers teachers and university IT specialists a complete arsenal of tools necessary to setup, administer and teach online courses.

4.1   Content and Course Management

Teachers can create their courses from scratch or reuse one of the existing tables of contents. They can design assessment tests, exams and questionnaires, and author individual learning objects and collections of new material.

4.2   User and Group Management

There are three categories of users in the system. Students can access learning content individually or as a part of a course. Teachers can manage their courses, including content visibility and student roster. Administrators have access to all aspects of Math-Bridge user management. They can change user parameters, and rights, modify group membership, assign a teacher to course. Naturally, administrators can also do everything that other users can.

4.3   Course Monitoring with the Reporting Tool

It is easy for teachers to monitor students' progress within Math-Bridge: a dedicated reporting tool allows them to trace individual student's performance or results of the entire course. The reporting tool can also help in discovering potentially problematic learning objects (e.g. an exercise that nobody has solved correctly).

5   Conclusion

Math-Bridge is a full-fledged e-Learning platform developed to help individual learners, classes of students, as well as entire schools and universities to achieve their real-life educational goals. Math-Bridge implements a number of advanced technologies to support adaptive and semantic access to learning content. Fostering the adoption of these technologies by the general public is the primary goal of Math-Bridge.

MEMO – Situated Learning Services for e-Mobility

Holger Diener[1], Katharina Freitag[2], Tobias Häfner[3], Antje Heinitz[3], Markus Schäfer[4],
Andreas Schlemminger[5], Mareike Schmidt[2], and Uta Schwertel[2]

[1] Fraunhofer-Institut für Graphische Datenverarbeitung, Joachim-Jungius-Str.11, 18059 Rostock
[2] imc AG, Altenkesseler Str. 17/D3, 66115 Saarbrücken
[3] Copendia GmbH & Co KG, Friedrich-Barnewitz-Str. 8, 19061 Rostock-Warnemünde
[4] Berufsbildungszentrum der KH Märkischer Kreis e. V., Handwerkerstr. 2, 58638 Iserlohn
[5] PLANET IC GmbH, Hagenower Str. 73, 18059 Schwerin
holger.diener@igd-r.fraunhofer.de,
{katharina.freitag,mareike.schmidt,uta.schwertel}@im-c.de,
{haefner,heinitz}@copendia.de,
m.schaefer@kh-mk.de,
schlemminger@planet-ic.de

Abstract. The expected large-scale introduction of electric vehicles creates a
need for up-to-date and just-in-time available learning materials and tools. In
this paper we demonstrate how a set of web-based and mobile learning and collaboration services support situated learning for e-Mobility and similar
domains. The approach is currently developed within the MEMO project.

Keywords: electric mobility, situated learning, learning and collaboration
services, Web 2.0, mobile learning, eBooks, learning apps.

1   Introduction

Electric mobility (e-Mobility) is seen as a central technology to ensure a sustainable individual mobility in the future. In 2009, the German government launched the National
Development Plan for Electric Mobility[1] with the aim of seeing one million electric vehicles on Germany's roads by 2020. This expected large-scale introduction of electric cars
requires building up know-how and skills within various target groups: 450.000 specialists
in 38.000 car garages [1] need to learn how to maintain and repair electric vehicles; emer-
gencyservices have to know about dangers related to high-voltage components; education
providers have to adapt their curricula – as a recent study[2] from VDE shows, this topic has
not yet found its way into curricula of impacted courses of studies or occupations. Last but
not least, customers need advice and help. Since technologies in the e-Mobility domain are
heterogeneous (from battery technology to communication technology within the car) and
develop rapidly, traditional forms of face-to-face training with static predefined training
materials will not suffice to provide the necessary up-to-date knowledge quickly.

---

[1] http://www.bmu.de/english/mobility/doc/44799.php
[2] http://www.vde.com/de/Verband/Pressecenter/Pressemeldungen/
Fach-und-Wirtschaftspresse/Seiten/2010-73.aspx

A. Ravenscroft et al. (Eds.): EC-TEL 2012, LNCS 7563, pp. 501–506, 2012.

In this paper we therefore suggest to enhance the existing professional system of education and training in the automotive crafts domain with more dynamic and flexible methods for learning and collaboration. The approach is currently developed within the MEMO project (Media Supported Learning- and Collaboration Services for Electric Mobility)[3]. MEMO implements web and mobile based learning services supporting knowledge transfer, collaboration and situated learning on different end-user devices.

In section 2 we present specific requirements and challenges of the car mechanics and similar domains, in section 3 we outline the architecture of MEMO and in section 4 we present three examples of learning services provided by MEMO. Section 5 concludes and outlines potentials for a sustainable use of the services.

2   Requirements from the Car Mechanics Domain

Car mechanics gain many experiences through concrete practical actions within the mechatronic systems: learning by doing [2]. In the context of education, however, an isolated action is not sufficient. At its best, it can create routines and the consolidation of tradition. For learning and developing professional competences, practical actions and their effects have to be cognitively reflected, connected and embedded into a wider context [3]. Furthermore, vocational education in Germany is characterized by many dualities which result in disruptions: spatial (workplace vs. classroom), temporal (learning, working vs. leisure time) and regarding media (tools and machines vs. educational materials in vocational schools).

Behaviourist e-Learning approaches tried to overcome these dualities. Despite some success – also in the mechanics domain [4] – there remain open questions in particular as to situated learning: How is learning to be integrated into the practical actions during the work process, where a mechanic is especially sensitized to learn?

Hence, we derive two central didactic requirements for technology-enhanced learning in this area: Firstly, the learning materials have to be precisely tailored to the specific needs and have to be available just-in-time. Secondly, the learning device (desktop computer, laptop, tablet or smart phone) should be flexibly selectable. Didactic scenarios meeting these requirements have long been utopian for technical reasons. With the emergence of web 2.0 technologies, easy access to IT technology including powerful mobile devices and the availability of cloud-based services, new opportunities arose. Content can be easily generated by any user and can be widely distributed and socialized [5] – at any time and in any place. Initial experiments[4] in the domain of car mechanics where apprentices produce just-in-time available, free learning content for peers showed that situated learning can be achieved. MEMO builds on these findings.

---

[3] The MEMO project and the research described in this paper are co-funded by the German Federal Ministry of Economics and Technology (cf. http://www.memo-apps.de).
[4] http://www.youtube.com/user/kfz4metube

3   The MEMO Approach – Concept and Architecture

The MEMO project aims at offering a cloud of learning services supporting different end-users in the area of e-Mobility. MEMO offers *content services* for up-to-date information and learning materials, *collaboration services* for sharing and discussion, *simulation services* for situated component-specific training, *testing services* for teaching, formal and informal learning, and other services. Content for the services can either be user-generated, stem from external certified sources or it can be authored or curated by the MEMO expert team ensuring quality and trust. All services are accessible at the MEMO reference portal[5].

The MEMO architecture (see Fig. 1) is designed to be open and inter-operable. Certain MEMO services can be integrated into external portals to extend already existing offerings or can be used directly on mobile devices. Vice versa, MEMO allows the integration of learning services from external providers.

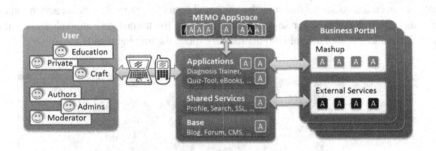

Fig. 1. MEMO Architecture

To realize this combination (mash-up) of reusable services MEMO builds on the open source Liferay Portal Community Edition[6] as underlying platform technology. The *base layer* offers standard web 2.0 functionalities for portal and content management, collaboration and social features (e.g. forums, blogs). Liferay contains these functionalities as built-in applications (portlets[7]), which can be customized for end-user needs. On the *shared services* level, we customized portlets, e.g. for user and access rights management, search (semantic capabilities), and created an interface for external services. On the *applications* layer, MEMO provides new learning services, either as portlets deployed on the Liferay server (e.g. Diagnosis Trainer), or as standalone applications (e.g. the interactive eBooks for iPads or the Quiz-Authoring-Tool).

All learning services are made available through the so-called *MEMO AppSpace*, a concept familiar from marketplaces for mobile app providers. Here users can search, filter and select services, and can access them for the respective supported devices.

---

[5] http://www.e-auto-dienste.de

[6] http://www.liferay.com,   version 6.0.6, Mashups in Liferay:
http://www.liferay.com/web/alberto.montero/blog/-/blogs/
gadgets-and-widgets:-using-liferay-as-a-mashup-platform

[7] http://www.sigs.de/publications/js/2005/03/kussmaul_JS_03_05.pdf

The current implementation of MEMO provides already 13 learning and collaboration services including the Diagnosis Trainer, interactive eBooks and the Quiz Tool.

4   Learning Services for e-Mobility – Examples

Diagnosis Trainer. Within a common training method in automotive mechatronics, trainers manipulate cars by installing different defects in order to train a strategic defect diagnosis. Trainees get these manipulated cars together with an error description and have to find the causes of failure by examining the cars as efficiently as possible using their standard diagnosis devices.

Based on this training method, we developed the MEMO service Diagnosis Trainer (DT) as a portlet for the MEMO reference portal using the web application framework Vaadin[8]. Users can examine a car by clicking on navigation and action points and selecting a specific action, e.g. checking for current passage (see Fig. 2). During a training course, trainers typically ask comprehension questions in order to check if a trainee just guesses. Within DT authors can create questions that must be answered correctly after selecting an action; otherwise the action will fail. After selecting the diagnosis, the trainee gets an evaluation and a prewritten description of the solution in order to explain the training results.

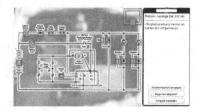

Fig. 2. Diagnosis Training Service

The Diagnosis Trainer can be neatly rendered on mobile devices, and can be used to enhance hands-on-training by providing just-in-time information on a tablet, as trainees will work in the training garage. For trainers the DT offers an author mode, where they can create tasks for their trainees. This includes adding pictures of the car, and creating navigation and action points. For each action point, trainers can integrate measurement actions and their appropriate results as well as comprehension questions. Since trainers usually manipulate fully functional cars, a fully functional basic model of the car is built in the DT. Afterwards trainers copy this basic model and change some measurement results to represent the failure behavior of the car.

A first time author is able to create a basic model with four screens, fifteen action points and five additional questions in about two hours. Manipulating this model to create different tasks takes about five minutes each.

Interactive eBooks. Modern forms of digital media can be one way to tackle the current lack of up-to-date and just-in-time available teaching materials for e-Mobility. Therefore MEMO produces several interactive eBooks on central topics of e-Mobility. The first prototype of this series, "Range of Electric Vehicles" (see Fig. 3), was created with the Freeware 'Apple iBooks

---

[8] https://vaadin.com/home

Author' and is available for the iPad in the Apple iBook Store[9]. In close cooperation with vocational trainers and craftsmen, the MEMO project will create further interactive eBooks – also for other platforms – of high educational value in the upcoming months and distribute them among trainers for the use in teaching.

Fig. 3. Interactive eBook

The additional use of media, like instructional videos or slides, is a widespread method in educational settings to make lessons more interesting. But in this traditional approach, media is not directly connected to the learner's textbook or even integrated in her/his book. So learners have to switch constantly between different files and formats. This media disruption can have a negative effect on the learning process. With interactive multimedia textbooks learners find all relevant information materials in one document: short instructional videos, audio files or multiple choice tests are incorporated at the corresponding locations in the book. Furthermore, it enables the learner to interact with the eBook, rather than just consuming its content. Interactive eBooks can thus help to bridge the initially mentioned dualities. Moreover, they can be easily authored, updated, are cheap and can be used just-in-time on mobile devices.

Quiz-Tool. The MEMO Quiz-Tool supports users to learn and understand technical terms related to e-Mobility (e.g. "fuel cell"). Users can access the tool when they need it and can apply the acquired knowledge immediately (e.g. preparing for an exam, looking-up technical terms at the workplace). The tool integrates three main types of learning content: test questions and exercises, a glossary and micro-learning-units.

The initial prototype of the Quiz-Tool was realized as a Moodle[10] course using built-in modules like Quiz or Glossary and server-side extensions for mobile support.

The Quiz-Tool consists of a *Quiz-Learning-Tool*, an environment to display the content, the *Quiz-Package* as a learning bundle with 5-10 items per content type and the *Quiz-Authoring-Tool* for the creation of Quiz-Packages. The user learns term definitions using small exercises. A simple algorithm repeats still unknown terms of a learning bundle until the user knows all of them correctly.

The Quiz-Learning-Tool can be used on different end-user devices: on tablets and PCs (as HTML application with full multimedia functionality) and on smart phones (as Android-App restricted to images and audio files). The Quiz-Tool supports different types of exercises and test

Fig. 4. Quiz Tool

---

[9] http://itunes.apple.com/us/book/reichweite-von-elektromobilen/id514673534?ls=1

[10] http://moodle.org

questions (drag'n drop, multiple choice etc.) and a glossary display supporting multimedia files. The micro-learning-units present edited learning contents with multimedia assets. An integrated display of the learning progress serves as user guidance.

The Authoring-Tool is only available for the PC and can be used on- and offline from within the preferred browser. It enables the template-based intuitive creation of Quiz-Packages, preview and export functionality.

Integrating the Quiz-Tool as learning and teaching application into the workplace has several advantages: New knowledge can be quickly integrated into the tool and tailored to the different needs. Collaborative learning and problem solving is supported, since both experts and learners can contribute content. The tool can be used just-in-time on different end-user-devices and hence ideally supports situated learning.

5   Conclusion

In this paper we demonstrated how a set of learning services can support situated learning for e-Mobility and similar domains. The approach supports user-generated as well as professionally produced content and will allow end-users to access the content also from mobile devices through "Learning Apps". MEMO currently streamlines the existing services, creates and adds diverse learning content for the services, and develops new services (e.g. Help Forum, Charging-App, games etc.). Moreover, the learning services will be evaluated by end-user-groups in realistic learning scenarios.

The distribution of services through the MEMO AppSpace generates various possibilities for commercial use: Individual services can be sold for a small fee or can be offered as free light-versions supported by advertisements. Alternatively, the services can remain without charge but the community contributes free learning contents in return.

References

1. Deutsches Kraftfahrzeuggewerbe Zentralverband (ZDK). Jahresbericht 2010/2011. Bonn: Wirtschaftsgesellschaft des Kraftfahrzeuggewerbes mbH Bonn (2011)
2. Dewey, J., Oelkers, J.: Demokratie und Erziehung: eine Einleitung in die philosophische Pädagogik. Translated by Erich Hylla. Beltz, Weinheim, Basel (2010)
3. Lisop, I., Huisinga, R.: Arbeitsorientierte Exemplarik. Subjektbildung - Kompetenz - Professionalität. Frankfurt am Main. G.A.F.B.-Verlag (2004)
4. Schäfer, M.: Netzbasiertes Lernen in produktverarbeitenden Handwerksbetrieben mit hohem Beratungsaufwand, realisiert durch Bildungsstätten des Handwerks. Schlussbericht. Förderkennzeichen: BMBF 01MD137. Hannover: Technische Informationsbibliothek u. Universitätsbibliothek Hannover (2004)
5. Dittmann, D., Schäfer, M., Zielke, T.: Audio-visuelle Lernbausteine als Ausdruck digital vergesellschafteter Produktionsformen von Wissen: Ein subjektorientierter Blick auf Chancen und Grenzen im Authoring-Prozess. In: Hambach, S., Martens, A., Urban, B. (eds.) Proceedings of the 4th International eLBa Conference, eLearning Baltics 2011. Fraunhofer Verlag, Stuttgart (2011)

PINGO: Peer Instruction for Very Large Groups

Wolfgang Reinhardt[1], Michael Sievers[1], Johannes Magenheim[1],
Dennis Kundisch[2], Philipp Herrmann[2], Marc Beutner[3], and Andrea Zoyke[4]

[1] University of Paderborn
Department of Computer Science
Computer Science Education Group
Fuerstenallee 11, 33102 Paderborn, Germany
{wolle,msievers,jsm}@uni-paderborn.de
[2] University of Paderborn
Faculty of Business Administration and Economics
Information Management & E-Finance
Warburger Straße 100, 33098 Paderborn, Germany
{dennis.kundisch,philipp.herrmann}@wiwi.uni-paderborn.de
[3] University of Paderborn
Faculty of Business Administration and Economics
Business and Human Resource Education II
Technologiepark 9, 33100 Paderborn, Germany
marc.beutner@wiwi.uni-paderborn.de
[4] University of Paderborn
Faculty of Business Administration and Economics
Business and Human Resource Education
Technologiepark 9, 33100 Paderborn, Germany
andrea.zoyke@notes.uni-paderborn.de

Abstract. In this research, we introduce a new web-based solution that
enables the transfer of the widely established Peer Instruction method
to lectures with far more than 100 participants. The proposed solution
avoids several existing flaws that hinder the widespread adoption of PI
in lectures with larger groups. We test our new solution in a series of
lectures with more than 500 participants and evaluate our prototype
using the technology acceptance model. The evaluation results as well as
qualitative feedback of course participants indicate that our new solution
is a useful artifact to transfer the PI method to large groups.

Keywords: peer instruction, classroom response systems, student
activation, interaction, mobile learning, open teaching concepts, cloud
computing.

1 Introduction

Peer Instruction is a cooperative teaching-learning method that is well suited to
involve students even in large auditoriums. Similar to the *ask-the-audience* lifeline in *Who Wants to Be a Millionaire?*, students can be involved in the lecture
through so-called clickers, which contributed to substantially enhanced learning

A. Ravenscroft et al. (Eds.): EC-TEL 2012, LNCS 7563, pp. 507–512, 2012.
© Springer-Verlag Berlin Heidelberg 2012

success. Peer Instruction is a common method in many science courses where there are even some standardized catalogs of questions. The general approach of involving students in the course is often also realized using so-called classroom response systems.

Not least because of the expensive infrastructure and complex software, the method has not yet been used on large scale, in spite of the many educational benefits. Given the widespread proliferation of smartphones and laptops for students, we developed a web-based Peer Instruction application in the PINGO project. The PINGO project also designs not yet explored Peer Instruction concepts for very large lectures in economics and evaluates the approach in a multi-perspective manner.

In this paper we describe the general idea of Peer Instruction, the hurdles that prevent a more widespread adoption of the method and how the PINGO project tries to overcome those issues. Moreover, we introduce the first prototype of the system, report on early results from the evaluation of the prototype and give a brief outlook to the future usage of the PINGO infrastructure.

2 The Peer Instruction Concept

Similar to the situation when the audience of *Who Wants to Be a Millionaire?* becomes an active part of the show when the candidate uses his *ask-the-audience* lifeline, the Peer Instruction (PI) methods enables lecturers to activate their students in a lecture. Using PI in a lecture helps to overcome the rather passive role of students and to make them active participants of the lecture.

The development of PI is mainly attributed to the Harvard-teaching physicist Eric Mazur, who uses the method for more than 20 years especially in larger introductory courses in science. He further elaborated the method as part of the so-called Galileo project (see [3,4,14]). Peer Instruction has since been established in many American and British universities and colleges as common teaching-learning method. This was accompanied by an intensive scientific discussion of the method in relevant journals (e.g. see [6,10,12,13,15,16]).

Through this cooperative teaching-learning method, students are able to better reflect new course content, interpret and link the content with existing knowledge. Furthermore, their problem-solving capabilities are stimulated. Unlike in traditional head-on lectures, the focus in courses using the PI method is not on the pure transfer of content. Instead, students are already preparing for the lecture in beforehand and it is the clarification of ambiguities and questions, the discussion of difficult to understand concepts and the linking of the content with prior knowledge that is in the foreground. Typically, a brief introductory lecture is followed by a multiple-choice question. Depending on the distribution of responses, either the introductory lecture will be repeated and intensified or peer discussions are conducted with the students sitting next to oneself, which is followed by a repeated voting on the same question. In the case of a high number of correct answers, the remaining ambiguities are explained in a plenary discussion and then a new topic is discussed.

2.1 Peer Instruction and ICT

Since the mid-1990s, the implementation of PI in lectures is supported by appropriate information technology applications. The lecturer projects multiple-choice questions using a video projector. Answering the questions by the students takes place via handy *clickers* having the size of a mobile device or a remote control. A *Personal Response System* (PRS) consisting of a receiving device and an appropriate software collects the responses from the registered clickers and immediately displays the results on the projector.

It is undisputed that the appropriate use of the PI method leads to significant learning success among students (see [2,7,8,17]). Nevertheless, a variety of reasons causes that this method is still not widespreadly used (also see [9]):

1. *High one-off expenditure:* An existing course design must be adapted to the use of the PI method.
2. *Catalog of questions:* Except for some areas of natural science there are no existing and freely available or purchasable catalogs of questions that can be used with the PI method[1].
3. *Clickers:* Distributing and collecting the clickers in courses with about 100 students is very complex; at present it is impossible in courses with over 250 students. The risk that clickers are forgotten, lost or taken is rising with increasing group size.
4. *Costs:* The clickers and especially the receiving device are very expensive.
5. *Installation:* First, an extensive installation of software on the laptop of the lecturer is required. In addition, the software installation is not possible for all operating systems and the integration of the application in Microsoft PowerPoint does not work in all existing versions of Microsoft Office.
6. *Configuration:* The preparation of the very detailed configurable software for the usage in a lecture causes high setup costs for lecturers. Moreover, the usage of the software cannot be adapted dynamically during the lecture.
7. *Exclusive use:* The use is exclusive, i.e. when the clicker are being used in a lecture, they cannot be used in another lecture or course.

The widespread adoption of the PI method is mainly hindered by the technical, organizational and financial hurdles (reasons 3 to 7). Within the PINGO project we developed a scalable system that reduces these obstacles by providing a web-based interface for both lecturers and students. Our approach assumes that most of the students have access to the Internet during a lecture using their netbooks, laptops or smart devices.

3   The PINGO Infrastructure

Based on the above elaborations, we designed the PINGO infrastructure as scalable service that is able to handle some thousands of users and responses per second without hassle.

---

[1] Catalogs of questions are mostly available for introductory courses in chemistry, physics, astronomy and partly in computer science [1].

3.1   The Backend Architecture

The current version of the PINGO backend is designed to run on a computing cloud that automatically starts new server instances when needed. Therefore, we monitor the actual load on the servers and scale up computation power when clients would have to wait to long or if the server could struggle to handle all requests. The actual implementation is realized in Ruby on Rails and the NoSQL database mongoDB. For pushing events from the server to all connected clients as well as for the synchronization of timers we use Socket.IO.

3.2   Web and Mobile Clients

The web-based administrative interface supports lecturers in the setup of the lectures in which they use PINGO. Once the lecture is created, PINGO supports lecturers to instantly setup new polls. They just have to specify the question type, the number of possible answers and the duration of the poll. Polls do not need to be prepared in beforehand but can rather be added within seconds. There is no need for providing the poll's question or naming the answer options as this is typically done in the medium used for presenting the lectures' content.

Students can participate in the polls using any WWW-enabled device. They only have to select the poll channel of the lecture and any new poll will automatically pushed to their device. If they access the poll using a mobile phone or smart device, an optimized interface for such devices is shown. Participation in the poll is only possible during the lecturer-defined timespan.

3.3   Preliminary Evaluation Results

A first prototype of the PINGO system was tested in an introductory course business information systems. This course was an ideal candidate to evaluate this prototype: There were more than 1,000 enrolled students and 95% of these students possess one or more web-enabled devices. The first tests showed very satisfying results on the technological level (stability, scalability to large groups etc.). However, the most important challenge when introducing a new technology is to achieve user acceptance. In their seminal work, Davis et al. (1989) proposed the revised Technology Acceptance Model (TAM) to predict user acceptance of information technology [5]. According to this model, the main predictors of technology acceptance are the constructs *perceived ease of use* and *perceived usefulness, attitude towards using* and the *behavioral intention to use* the technology. Until today, the TAM was validated in many studies and applied in numerous empirical settings (see [11,18]. Therefore, we use these well-established constructs as a first indicator for the expected, general acceptance of the PINGO system.

Participation in the test during the lecture was voluntary. After the test, we handed out a paper-based questionnaire to the course participants. The questionnaire included four items for the *perceived ease of use*, six items for the *perceived usefulness*, five items for the *attitude towards using*, and three items for the *behavioral intention to use* as well as some demographic questions. All of the items where measured on a seven point Likert scale.

438 of the approximately 600 course participants who showed up at the day of the lecture completed the questionnaire. Respondents were on average at the beginning of their second year of studies and gender was distributed almost equally among them (52% woman, 48% men). These demographics echo the demographics of the whole population of enrolled students in this course. We take this as a weak indicator that our results are not biased through non-response of some groups of students. The average *perceived ease of use* among respondents was evaluated with 6.22 (7 indicating the maximum for all scales), the average *perceived usefulness* with 4.93, the average *attitude towards using* with 5.49 and the average *behavioral intention to use* with 5.70. All of these results are a positive indicator for the future usage of the system. Reassuringly, we also received very positive personal feedback regarding the usage of the PINGO system from students enrolled in the course. These positive first evaluation results encourage us to further develop the PINGO system and to make it publicly available in the near future.

4   Conclusion and Outlook

In this paper we described *Peer Instruction* as a well-studied and very promising teaching-learning method with the potential to engage students to become more active in large lectures. We discussed seven reasons that prevent the widespread adoption of the method in large courses and described how the PINGO project tries to overcome them by providing a scalable web-based IT support for Peer Instruction. The PINGO project allows the application of PI in any kind of lectures and only requires WWW-enabled devices for setting up polls, participating in them and analyzing the results.

The developed architecture and interfaces are free to use for lectures at the University of Paderborn. All of them are subject to extensive evaluation in the summer term 2012. Here, several lecturers from business sciences are using PINGO in their courses with 30 to 1.200 students. The results from the evaluation will be incorporated in the further development of PINGO. Once the infrastructure reaches a mature state, we will design a viable business model for PINGO that will allow licensing for other universities and business organizations.

References

1. Chase, J.D., Okie, E.G.: Combining cooperative learning and peer instruction in introductory computer science. SIGCSE Bulletin 32(1), 372–376 (2000)
2. Cortright, R.N., Collins, H.L., DiCarlo, S.E.: Peer instruction enhanced meaningful learning: ability to solve novel problems. Advances in Physiology Education 29, 107–111 (2005)
3. Crouch, C.H.: Peer instruction: An interactive approach for large classes. Optics & Photonics News 9(9), 37–41 (1998)

4. Crouch, C.H., Mazur, E.: Peer instruction: Ten years of experience and results. American Journal of Physics 69, 970–977 (2001)
5. Davis, F.D., Bagozzi, R.P., Warshaw, P.R.: User acceptance of computer technology: A comparison of two theoretical models. Management Science 35(8), 982–1003 (1989)
6. Fagen, A.P., Crouch, C.H., Mazur, E.: Peer instruction: Results from a range of classrooms. Physics Teacher, 206–209 (2002)
7. Fels, G.: Die Publikumsfrage in der Chemievorlesung. Chemkon 16, 197–201 (2009)
8. Giuliodori, M.J., Lujan, H.L., DiCarlo, S.E.: Peer instruction enhanced student performance on qualitative problemsolving questions. Advances in Physiology Education 30, 168–173 (2006)
9. Kay, R.H., LeSage, A.: Examining the benefits and challenges of using audience response systems: A review of the literature. Computers & Education 53(3), 819–827 (2009)
10. Kay, R.H., LeSage, A.: A strategic assessment of audience response systems used in higher education. Australasian Journal of Educational Technology 25(2), 235–249 (2009)
11. Koufaris, M.: Applying the technology acceptance model and flow theory to online consumer behavior. Information Systems Research 13(2), 205–223 (2002)
12. Lantz, M.E.: The use of 'clickers' in the classroom: Teaching innovation or merely an amusing novelty? Computers in Human Behavior 26(4), 556–561 (2010)
13. MacArthur, J.R., Jones, L.L.: A review of literature reports of clickers applicable to college chemistry classrooms. Chemistry Education Research and Practice 9(3), 187–195 (2008)
14. Mazur, E.: Peer Instruction: A User's Manual. Prentice Hall (1997)
15. Moss, K., Crowley, M.: Effective learning in science: The use of personal response systems with a wide range of audiences. Computers & Education 56(1), 36–43 (2011)
16. Pilzer, S.: Peer instruction in physics and mathematics. Primus 11(2), 185–192 (2001)
17. Smith, M.K., Wood, W.B., Adams, W.K., Wieman, C., Knight, J.K., Guild, N., Su, T.T.: Why peer discussion improves student performance on inclass concept questions. Science 323, 122–124 (2009)
18. Szajna, B.: Empirical evaluation of the revised technology acceptance model. Management Science 42(1), 85–92 (1996)

Proportion: Learning Proportional Reasoning Together

Jochen Rick[1], Alexander Bejan, Christina Roche, and Armin Weinberger[1]

[1]Department of Educational Technology, Saarland University
{j.rick,a.weinberger}@mx.uni-saarland.de

Abstract. Proportional reasoning is a broadly applicable skill that is fundamental to mathematical understanding. While the cognitive development of proportional reasoning is well understood, traditional learning methods are often ineffective. They provide neither real-time feedback nor sophisticated tools to scaffold learning. Learners often cannot connect embodied notions (this glass is half full) to their symbolic representations ($\frac{1}{2}$). In this paper, we introduce Proportion—a tablet applications for two co-located learners to work together to solve a series of increasingly difficult ratio / proportion problems. We motivate the work in previous research on proportional reasoning, detail the design and outline the questions this design-based research aims to address.

1 Proportional Reasoning

Ratios and proportions play a critical role in a student's mathematical development [1]. It is a broad topic, ranging from elementary concepts of dividing a whole into halves to being able to manipulate fractions to solve algebraic equalities. Because of its importance and depth, the topic is covered repeatedly and in increasing sophistication in several grade levels. The cognitive development around ratios and proportions is well documented and moves through relatively distinct stages [2]. Proportional reasoning is realized through multiple strategies, where one strategy will be appropriate for one set of problems but inappropriate for another set. For instance, in cases where the denominators are the same, the ratio of two fractions is the same as the ratio of the respective numerators ($\frac{3}{8} : \frac{5}{8} = 3 : 5$); if the denominators are different, this strategy fails ($\frac{3}{7} : \frac{5}{4} \neq 3 : 5$). Gaining competence requires both acquiring such strategies and understanding when and how to apply them [3]. Even students who show clear competence in applying a strategy successfully to one problem often fail to realize that the same strategy applies to another problem.

Proportional reasoning is notably difficult to teach [1]. One issue is that the topic is usually taught and tested through word problems (e.g., if 300 grams of cherries are used to produce 400 milliliters of jam, how many milliliters of jam can be produced from 600 grams of cherries?). These problems attempt to connect embodied concepts (cherries, jam) to symbolic counterparts (300, 400). Embodied proportional reasoning, relying on rules-of-thumb (larger denominator means smaller amount) and estimation (9 is about twice as much as 4), is

A. Ravenscroft et al. (Eds.): EC-TEL 2012, LNCS 7563, pp. 513–518, 2012.

particularly important for learners to relate their everyday experiences to mathematical concepts [4]. Unfortunately, the embodied connection in word problems is weak: Learners cannot actually manipulate quantities of cherries to produce jam. Furthermore, they receive little feedback and support by working the problem with pen and paper. Hence, a learner might employ the wrong strategy (one that worked previously, but does not apply to that problem) over an entire sequence of problems without realizing their misconception. *Real-time feedback* on task progress can allow students to more quickly realize which of their current strategies to employ or when to generate new ones [5]. Consequently, physical manipulatives that give some level of real-time feedback (e.g., two $\frac{1}{4}$ blocks can be stacked together to form one $\frac{1}{2}$ block) have been shown to be a particularly useful technique for learning proportional reasoning [3]. Digital manipulatives can further enhance the experience by providing more sophisticated feedback and bridge the gap between an embodied experience and its corresponding symbolic representation [4, 6].

Another useful technique for supporting learning is to provide *tools* that highlight specific elements of a problem. First, the tool can provide real-time feedback. A balance beam will only balance if the ratios are correct. Second, students can gain competence in using the tool to solve problems. Tool competence can be important in applying concepts in the real world. Using a tablespoon to keep adding increments of flour and sugar to a recipe while keeping their ratio intact is a practical cooking skill. Third, learners can apply strategies learned with the tool even when the tool is gone. Students might learn to use a measuring stick to precisely solve a problem and later be able to use step lengths to estimate the solution to a similar problem.

2   The Proportion Tablet Application

The Proportion iPad application was designed to foster proportional reasoning by connecting symbolic representation (whole numbers and fractions) to embodied notions (in this case, the visual height of a column). Learners, aged 9–10, solve a series of ratio / proportion problems by positioning two columns (one on the left and one on the right) in proportion to the numerical values assigned to the respective column heights. As two novices working together with a *reflective tool* (i.e., one that provides feedback on the underlying concepts) can help each other converge on a more accurate understanding [7], we intend for Proportion to be used by two co-located learners. The tablet is positioned vertically on a table in front of the two learners (Figure 1). We have developed Proportion to make use of a variety of *tools* throughout a wide-ranging *curriculum*.

2.1   Tools

Proportion provides *tools* implemented in four interfaces (Figure 2). Without any support (Figure 2a), learners must estimate the ratios. With a fixed 10-position grid (2b), learners have precise places that they can target, thereby using their

Fig. 1. Two children working together with Proportion

mathematical understanding of the task to solve problems. One effective strategy is for users to select the grid line that corresponds to their respective numbers. This works well for simple ratios, such as 4 : 9. For the common-factor problem shown in Figure 2b, that strategy does not work. As illustrated, the children tried a novel strategy of positioning the columns based on the last digit of the number. Of course, this did not work and they were able to realize that this was not a viable strategy. With relative lines (2c) that expand based on the position of the columns, learners can use counting to help them solve the problem. They can also learn more embodied strategies, such as maximizing the size of the larger column to make it easier to correctly position the smaller column. When the lines are labeled (2d), other strategies can be supported. For instance, in the fraction-based problem shown, a viable strategy is to arrange columns so that whole numbers (e.g., 1) are at the same level.

Proportion provides two levels of *real-time feedback* using an owl avatar. If the ratio of the two columns is close to the correct answer, the owl announces "close" (2a). If the ratio is within a very small zone, then it is pronounced as "correct" and the application moves on to the next problem. When designing this feedback, it was important that learners not just solve the problem based on the feedback without strategically engaging the problem. Hence, the close feedback was designed to give no information about which direction the correct answer lies. Concurrently, learners need enough feedback to make progress when they are

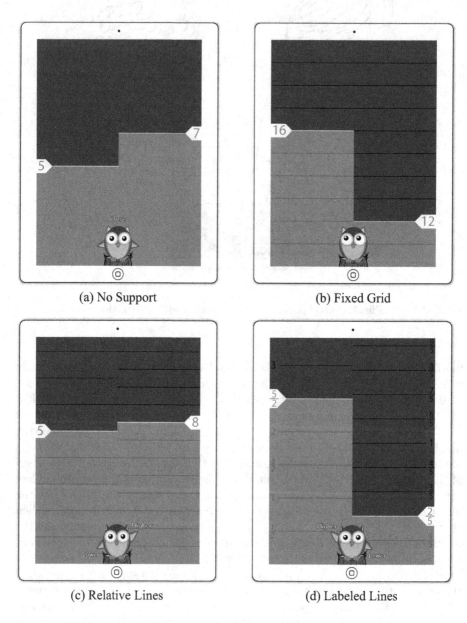

(a) No Support   (b) Fixed Grid

(c) Relative Lines   (d) Labeled Lines

Fig. 2. Four interfaces for supporting learners in solving problems

testing out or discovering a new strategy. To better support this, the sensitivity of the zones is adjusted for the problems. The first time a new strategy is needed, the zones are relatively large, allowing learners to more easily stumble upon the solution. As the sequence progresses, the zones become smaller, making it uncomfortable for learners to simply employ a stumble-upon strategy. The zones are larger for estimation tasks (e.g., 2a) where precision is difficult even when learners employ a correct strategy. Conversely, the zones are smaller when the interface should support precision, thereby coaxing learners to take advantage of those tools.

In rare cases, a problem can be too difficult for learners to work out with the existing support. To ensure that learners do not require external help (e.g., from a teacher), directional feedback appears after one minute (2c & 2d). With directional feedback, learners can size the columns correctly without understanding the underlying principle; however, the curriculum requires students to employ the same strategy multiple times. One minute was chosen as short enough to avoid damaging frustration but long enough so that it becomes uncomfortable for learners to use it to avoid learning a strategy.

2.2 Curriculum

Based on established proportional reasoning strategies, we created a curriculum and refined it through two rounds of user testing. That curriculum contains 215 problems split into 21 levels. Each level targets a different proportional reasoning strategy, from comparing simple whole numbers (1 : 5) to complex fractions ($\frac{11}{2} : \frac{4}{3}$).

3 Design-Based Research

At an average of 25 seconds per problem, learners would be able to finish the entire problem sequence in about 90 minutes; however, that is not how Proportion will be used. As a research application, it is intended to be used to compare multiple conditions, such as one without verbal prompting versus one with verbal prompting. As time on task is a dominant factor in learning success, this work aims to control for that variable. All groups will work for an hour. Even high performing groups are unlikely to finish as the problems go well beyond the targeted grade level.

The research with Proportion aims to shed light on two broad research topics. First, it will investigate *how children communicate to collaborate*. Previous work has demonstrated that children readily use their interactions with the interactive surface to communicate with their partners [8]. This work aims to tease apart the role of verbal and gestural communication. In particular, it will investigate how scripting the collaboration to encourage verbalization [9, 10] impacts learning and task performance.

Second, it will investigate issues of *equity of collaboration* for tablet-based collaboration. On tabletops, it becomes difficult for users to access all parts of the surface; therefore, users tend to concentrate their interactions in areas closer to their position at the tabletop [11]. Such separation is not possible for a tablet: Every user has good access to all parts of the interactive surface. Proportion was designed to have an interface split across the users. Learners usually assume ownership of the column on their side. What happens when the convention breaks down? Such physical conflict has been shown to highlight cognitive conflict and thus lead to conceptual change [12].

Acknowledgements. We would like to thank Michael Gros of the Saarland LPM (Landesinstitut für Pädagogik und Medien) for facilitating our access to schools and those schools for supporting our development and research efforts.

References

[1] Lamon, S.J.: Ratio and proportion: Connecting content and children's thinking. Journal for Research in Mathematics Education 24(1), 41–61 (1993)

[2] Noelting, G.: The development of proportional reasoning and the ratio concept: Part I — differentiation of stages. Educational Studies in Mathematics 11, 217–253 (1980)

[3] Tourniaire, F., Pulos, S.: Proportional reasoning: A review of the literature. Educational Studies in Mathematics 16, 181–204 (1985)

[4] Abrahamson, D., Trimic, D.: Toward an embodied-interaction design framework for mathematical concepts. In: Proceedings of IDC 2011, pp. 1–10. ACM Press, New York (2011)

[5] Vosniadou, S., Ioannides, C., Dimitrakopoulou, A., Papademetriou, E.: Designing learning environments to promote conceptual change in science. Learning and Instruction 11(4&5), 381–419 (2001)

[6] Leong, Z.A., Horn, M.S.: Representing equality: A tangible balance beam for early algebra education. In: Proceedings of IDC 2011, pp. 173–176. ACM Press, New York (2011)

[7] Roschelle, J.: Learning by collaborating: Convergent conceptual change. The Journal of the Learning Sciences 2(3), 235–276 (1992)

[8] Rick, J., Marshall, P., Yuill, N.: Beyond one-size-fits-all: How interactive tabletops support collaborative learning. In: Proceedings of IDC 2011, pp. 109–117. ACM Press, New York (2011)

[9] Chi, M.T.H., Leeuw, N., Chiu, M.H., Lavancher, C.: Eliciting self-explanations improves understanding. Cognitive Science 18(3), 439–477 (1994)

[10] Lambiotte, J.G., Dansereau, D.F., Angela, M., O'Donnell, M.D.Y., Skaggs, L.P., Hall, R.H.: Effects of cooperative script manipulations on initial learning and transfer. Cognition and Instruction 5(2), 103–121 (1988)

[11] Rick, J., Harris, A., Marshall, P., Fleck, R., Yuill, N., Rogers, Y.: Children designing together on a multi-touch tabletop: An analysis of spatial orientation and user interactions. In: Proceedings of IDC 2009, pp. 106–114. ACM Press, New York (2009)

[12] Pontual Falcão, T., Price, S.: What have you done! The role of 'interference' in tangible environments for supporting collaborative learning. In: Proceedings of CSCL 2009, ISLS, pp. 325–334 (2009)

Supporting Goal Formation, Sharing and Learning of Knowledge Workers

Colin Milligan, Anoush Margaryan, and Allison Littlejohn

Caledonian Academy, Glasgow Caledonian University, Glasgow, Scotland, United Kingdom
(colin.milligan,anoush.margaryan,allison.littlejohn)@gcu.ac.uk

Abstract. This paper describes prototype tools to support goal formation and sharing to assist knowledge workers in participating and managing their participation in learning networks. We describe the concept of Charting as the process whereby an individual monitors and optimises their interaction with the people and resources who contribute to their learning and development and how tools to support learning goal articulation and sharing can provide an integrative function for a Personal Learning Environment. The paper then describes the development of prototype tools which support goal articulation and sharing and discusses how such tools might integrate with existing learning and development practices, concluding with some questions for further research.

Keywords: informal learning, knowledge sharing, goal formation, workplace.

1   Introduction

Contemporary workplaces are undergoing profound shifts [1] that leave knowledge workers facing a number of difficult challenges. First, they are expected to manage and self-regulate their own learning alongside accomplishing their work tasks [2], a challenge further complicated by the fact that, in the workplace the individual is left to balance learning to address short-term work challenges with learning for their future career. Second, the types of problems addressed by knowledge organisations are growing in complexity. Solving complex problems requires that knowledge workers recognize the limits of their own expertise, and develop skills to cooperate within interdisciplinary teams, and develop extensive learning networks both within and outside their organization [3]. Complex work processes afford important contexts for learning in the workplace. However, since work and learning may be in conflict, powerful learning opportunities can be lost as focus is on achieving the goals of the organisation, rather than those of the individual. Third, in the current technological landscape, tools which support knowledge work have traditionally been developed independently of those which seek to facilitate learning. Further, these tools tend to be polarised between those which support individual production and group collaboration, rather than permitting the wide spectrum of interactions which individuals and teams engage in and which are an essential component of complex knowledge work.

A. Ravenscroft et al. (Eds.): EC-TEL 2012, LNCS 7563, pp. 519–524, 2012.
© Springer-Verlag Berlin Heidelberg 2012

Traditionally, learning and development and knowledge management in organisations have been addressed in isolation, from different disciplinary standpoints and within different organizational functions. Recent EU projects such as APOSDLE (*http://www.aposdle.tugraz.at*) and IntelLEO (*http://www.intelleo.eu/*) have developed systems which are more reactive to the needs of learners, but these systems have tended to focus on matching learning needs to pre-existing learning content, whereas a key feature of knowledge worker learning may be that learning content does not pre-exist in one place, but rather, the learner must learn through identifying, collecting and leveraging knowledge resources (people and content) to create new learning opportunities. To support knowledge workers effectively, organisations must provide their workers with technical systems which address the three issues identified above together. Such a Personal Work and Learning Environment (PWLE – supporting work and learning, though in the remainder of this paper we will use the more widely understood term Personal Learning Environment, PLE) should seek to balance support for the needs of the organisations with that of the individual, and should extend and integrate existing tools, without disrupting the individual workers current practice.

This paper describes a mechanism by which knowledge workers (individually and collectively) can be supported in focusing on their learning and development whilst they work. We present the concept of **Charting** to describe the group of processes through which an individual manages and participates in the learning and development and knowledge management essential to effective learning in the workplace. In previous studies [6-8], we have described how, by examining the learning practices of knowledge workers in a multinational organisation, we were able to identify four key knowledge behaviours representing different ways in which an individual interacts with the people and resources in their learning network. Our findings indicate that individuals **consume** knowledge created by others. Additionally, they **connect** with other people and resources relevant to their own learning goals. They **create** new knowledge and knowledge structures and they **contribute** this knowledge back to the collective for others to benefit from. Other studies have identified similar sets of behaviours. For instance Dorsey [9] outlined a set of seven distinct actions related to the use of knowledge: retrieving, evaluating, securing and organizing information, as well as analyzing, collaborating around and re-presenting knowledge. Karrer [10] conducted an ethnographic study examining employees' knowledge management practices and identified the following behaviours: scanning and finding information; networking and collaborating with other people; organising and improving information.

The four groups of behaviours described above (*consume, connect, create, and contribute: the 4c's*) represent the key ways in which a learner interacts with his or her learning network. But how does an individual *manage* their learning? One key mechanism is to set learning goals as a means to identify gaps in knowledge, understanding or skills, and as a starting point for planning to achieve those goals. An individual's learning goals are a central component of Charting, the organising principle for planning and managing learning. Learning goals, rather than a formal curriculum, predefined content, or organisational competencies, provide a focus for knowledge worker's sensemaking process, as they relate the new knowledge they are generating for work to the wider context of the knowledge they already have. Learning goals are individually set, but influenced heavily by others in the workplace and may be shared with co-workers or with colleagues outside the organisation. Furthermore, learning goals provide a purpose for interaction with other people and resources when learning. In other words, learning goals serve as a "social object" around which people interact [11-12].

2   Supporting the Processes of Charting

For a knowledge worker, learning is inseparable from work, and therefore tools to support learning processes must integrate closely with the normal tools which an individual uses to conduct their work. Many of these tools exist already, and indeed form the basis of many existing Personal Learning Environments. Indeed the four knowledge behaviours outlined above might be better thought of as a way of grouping the functions of an archetypal PLE. A PLE should contain tools to allow individuals to:

- Discover and **Consume** knowledge and resources created by others, leveraging value from the collective (these tools might include: *dashboards, RSS readers, search tools* etc.).
- **Connect** with others who share interests or goals to develop ideas, share experience, provide peer-support, or work collaboratively to achieve shared goals (*email and chat, video conferencing, social networks, microblogging tools*).
- **Create** new knowledge (and knowledge structures) by combining and extending sources (people and resources and personal reflections) to create a dynamic, faithful and individually focused view of the knowledge and understanding they possess about a given topic, and how different topics inter-relate within their personal world-view. This sense-making process is continual, and ensures that the knowledge space evolves with the ideas of the individual, their network and beyond (*blogs, collaboration tools, knowledge visualisation and structuring tools*).
- **Contribute** new knowledge back to the network formally (as reports, publications, and other standalone artefacts) and informally (as reflections, ideas, ratings and other context-dependent content) for the benefit of the individual, their local group and the wider community (*microblogging tools, blogs and activity stream tools*).

In addition to these core components of a PLE, a Charting toolset must provide some mechanism to bring all these functions together to promote the ongoing interaction of the individual with the resources and people they interact with rather than just the organisation and discovery of it. Tools for goal articulation and sharing can perform this role, if we use learning goals as the social object around which communities of learners can coalesce to help each other learn. Previous work such as that carried out by TenCompetence (http://www.tencompetence.org/) and current work being undertaken by IntelLEO [14] suggests that a focus on goals is appropriate.

3   Goal Articulation and Sharing Tools

A central principle underpinning the concept of a Personal Learning Environment is that rather than attempting to create a single tool which does everything, it is preferable to allow each learner to construct their own ecosystem of learning tools [14]. While most of the functions needed for a powerful PLE already exist as generic web 2.0 services, we felt that appropriate tools to support goal articulation do not already exist (or at least do not exist with the correct function set). To further explore the potential of using goals as the social object around which learning and knowledge sharing can occur, a set of prototype tools have been designed (*http://charting.gcu.ac.uk/*). These *Charting tools* are designed to be used by anyone dealing with new knowledge who needs to learn (structure and manipulate that knowledge) as part of their work practice. By articulating and sharing goals, learners create opportunities for interaction with others in their network (and beyond) who may share their goals. The more learners in the system, the more effective goal discovery is likely to be, although closed system setups (for instance for workers in the same organisation) are also possible. In principle the tools could also be used in formal education settings to support peer learning.

When the learner first joins the system, they follow a simple registration process, install a bookmarklet for Chrome or Firefox and articulate a series of goals. These goals are then published publicly in the system. As the learner goes about their daily practice, they can use the bookmarklet to associate resources with a specific goal. This process is similar to the Delicious social bookmarking service (*http://delicious.com/*) where notes can be made by the user as the resource is saved. In addition to web based resources, snippets of text from word processor documents can also be marked and associated with a specific goal. Notes may also be added via the bookmarklet, without any associated web resource. Providing simple ways to create and contribute new knowledge to the system emphasises that a user structures new knowledge by making their own connections between disparate resources, filling in the gaps and something more than a collection of resources. Of course in addition to notes and reflections stored in this way, the user may also choose to keep a blog or create content independently, and then attach that content to a specific goal through the normal procedure. Over time, the user develops a set of resources and notes which constitutes the knowledge and understanding they possess for that goal. The tools therefore offer the individual a natural and efficient way of collecting and structuring their knowledge for that learning goal. The social element of the tools is that any learner can adopt any public goals contributed by other users in the system. Once adopted, this learner then gains access to all the public resources and notes created by the original user. This affords two usage scenarios. First, users can share goals which they are collaborating on (for example two (or more) co-workers who are on the same project and need to develop a joint understanding of a new area). Second, learners can search and discover goals and associated artefacts of other learners who are unknown to them, and gain an insight into how these learners achieved the goal they set (in this respect, the design of the system mimics that of the 43things web service: *http://43things.com/*). A lightweight commenting system allows any resource or note in the system to serve as the locus of a communication between all subscribers to a given goal.

These prototype Charting tools are written in the Ruby programming language and will shortly be released under an Open Source license. A demonstration site allowing users to test the Charting tools can be found at: *http://charting.gcu.ac.uk/* This site also provides user instructions and links to further materials.

4   Conclusion and Future Questions

This paper has described the concept of Charting as the set of processes that enable an individual to manage and interact with the people, tools and resources which constitute their learning network, and describes a set of prototype Charting tools that support goal articulation, sharing and interaction. These prototype tools have been designed as a lightweight open source service which can be customised to integrate with existing tools that support four key knowledge behaviours (consume, connect, create and contribute) we observed in previous studies [6-8]. A key design principle of Charting tools do not seek to replicate functionality provided by existing tools.

There is still much work to be done determining whether goals can in fact be an effective social object. We are currently trialling the tools in a small closed learning community to further understand whether learning goals provide an effective mechanism for organising and structuring knowledge. Questions which must be further investigated include: *Do knowledge workers articulate learning goals?* (as opposed to work goals) *How personal is the language used to articulate goals?* (Can learning goals articulated by one knowledge worker be recognized as relevant by another worker in a different context but with similar learning needs?), *At what granularity are goals shared most effectively?, Does the act of sharing goals provide any benefit to the individual, and the community as a whole?* And *Can activity towards learning goals be linked back to formal development planning for performance appraisal?* The prototype tools presented here can help us explore these and other questions.

Creating tools for articulating, managing and sharing goals is one key component of a Personal Learning Environment, but visualizing knowledge structures and learning networks presents a further challenge to creating an effective PLE. Once knowledge is structured and recorded within a Charting system, it is imperative that it can be visualised and accessed as a whole in order that the learner (and others in their learning network) can explore the knowledge structure they have created. Our tool development has focused on goal articulation and sharing, but we recognise that visualisation of knowledge structures is a vital component of any mature Charting system.

Much can be learned from social software trends outside education and learning. The social bookmarking service Delicious (*http://delicious.com/*) influenced the design of the original prototype tools, with a simple bookmarklet to add new resources and the use of tags to permit simple organisation of resources. Now second generation social bookmarking tools such as Pinterest (*http://pinterest.com*) and social curation tools such as scoopit (*http://scoop.it/*) have emerged. In these, we move further from the original idea of bookmarking to emphasise personal and group curation (the action of bringing a group of resources together and contributing it back to the community). Similarly, tools such as Evernote (*http://evernote.com/*), demonstrate how software vendors have recognised the importance of bringing content from different sources together, and of providing strong tools for content creation alongside tools for content consumption, and structuring and organising knowledge. Such evolution of social software points the way to the quality and clarity of user experience which users come to expect for all the tools they use.

References

1. Ellström, P.-E.: Integrating Learning and Work: Problems and prospects. Human Resource Development Quarterly 12(4), 421–435 (2001)
2. IBM The IBM Global Human Capital Study (2009), http://www-935.ibm.com/services/us/gbs/bus/html/2008ghcs.html
3. Dron, J., Anderson, T.: Collectives, Networks and Groups in Social Software for e-Learning. In: Proceedings of World Conference on E-Learning in Corporate, Government, Healthcare, and Higher Education 2007, pp. 2460–2467. AACE, Chesapeake (2007)
4. Milligan, C.D., Beauvoir, P., Johnson, M.W., Sharples, P., Wilson, S., Liber, O.: Developing a Reference Model to Describe the Personal Learning Environment. In: Nejdl, W., Tochtermann, K. (eds.) EC-TEL 2006. LNCS, vol. 4227, pp. 506–511. Springer, Heidelberg (2006)
5. Wilson, S., Liber, O., Johnson, M.W., Beauvoir, P., Sharples, P., Milligan, C.: Personal Learning Environments: Challenging the Dominant Design of Educational Systems. Journal of e-Learning, Knowledge and Society 3(2), 26–38 (2007)
6. Margaryan, A., Milligan, C., Littlejohn, A., Hendrix, D., Graeb-Koenneker, S.: Self-regulated learning in the workplace: Enhancing Knowledge Flow Between Novices and Experts. In: 4th International Conference on Organizational Learning, Knowledge and Capabilities (OLKC), Amsterdam, April 26-28 (2009)
7. Littlejohn, A., Milligan, C., Margaryan, A.: Charting Collective Knowledge: Supporting Self-Regulated Learning in the Workplace. J. Workplace Learning 24(3), 226–238 (2012)
8. Littlejohn, A., Milligan, C., Margaryan, A.: Collective Learning in the Workplace: Important Knowledge Sharing Behaviours. Int. J. of Adv. Corp. Learning 4(4), 26–31 (2011)
9. Dorsey, P.A.: Personal Knowledge Management Education Framework for Global Business. In: 17th Turkish National Information Systems Congress, Istanbul (2000)
10. Karrer, T.: Initial Knowledge Work Framework (2008), http://www.workliteracy.com/pages/knowledge-work-framework/
11. Engeström, J.: Why Some Social Networks Work, and Others Don't (2005), http://www.zengestrom.com/blog/2005/04/why-some-social-network-services-work-and-others-dont-or-the-case-for-object-centered-sociality.html
12. Knorr-Cetina, K.: Objectual Practice. In: Knorr-Cetina, K., von Savigny, E., Schatzki, T. (eds.) The Practice Turn in Contemporary Theory, 1st edn., Routledge, New York, NY, pp. 175–188 (2001)
13. Siadaty, M., Jovanovic, J., Pata, K., Holocher-Ertl, T., Gasevic, D., Milikic, N.: A Semantic Web-enabled Tool for Self-Regulated Learning in the Workplace. In: Proceedings of ICALT 2011, pp. 66–70 (2011)
14. Zubrinic, K., Kalpic, D.: The Web as Personal Learning Environment. Int J. Emerging Technologies in Learning 3, 45–58 (2008)

U-Seek: Searching Educational Tools
in the Web of Data

Guillermo Vega-Gorgojo, Adolfo Ruiz-Calleja,
Juan Ignacio Asensio-Pérez, and Iván M. Jorrín-Abellán

University of Valladolid. Paseo de Belén 15, 47011 Valladolid, Spain
{guiveg@tel,adolfo@gsic,juaase@tel,ivanjo@pdg}.uva.es
http://www.gsic.uva.es/

Abstract. SEEK-AT-WD is an open Linked Data-based registry of educational tools that crawls the Web of Data to obtain tool metadata, thus significantly reducing the overall effort of data generation and maintenance. Since SEEK-AT-WD is an infrastructure, there is a need of end-user applications that can consume these data and provide additional value to the educational domain. In this regard, we present here U-Seek, an interactive searcher of educational tools that uses SEEK-AT-WD as the back end. U-Seek has been specially designed for educators, supporting the formulation of semantic searches by multiple criteria through a direct manipulation graphical user interface. Traditional keyword-based searches are also supported and can be combined with aforementioned semantic searches, thus resulting in enhanced flexibility and response accuracy.

1 Introduction

SEEK-AT-WD[1] (Support for Educational External Knowledge About Tools in the Web of Data) [1] is a Linked Data-based [2] registry of educational tools. The key idea of SEEK-AT-WD is to obtain tool metadata already available on structured Web data sources (e.g. Dbpedia[2], a view of the Wikipedia), automatically generate tool descriptions according to an educational vocabulary [3], and openly publish such descriptions to be consumed by other educational applications. Following this approach, SEEK-AT-WD offers a substantial collection of tools (3508 as of March 2012) that is automatically updated, thus dramatically reducing the associated data generation and maintenance costs.

The tool dataset of SEEK-AT-WD can be publicly accessed at http://seek.rkbexplorer.com/, so it is possible even for a third-party institution to build applications that consume this dataset. Perhaps the most obvious use of SEEK-AT-WD is to develop a searcher for educators, allowing them to discover suitable tools for their practice. Such an application should effectively communicate the

---

[1] A research paper about SEEK-AT-WD has been submitted to the EC-TEL; you can gain more insight at http://www.gsic.uva.es/seek/

[2] http://dbpedia.org/

A. Ravenscroft et al. (Eds.): EC-TEL 2012, LNCS 7563, pp. 525–530, 2012.

tool information structure to educators, making easy the formulation of queries and obtaining accurate responses for their needs. With this aim we present here U-Seek, an interactive searcher of educational tools specially devised for educators following a participatory design strategy [4] with both final users and software developers.

Beyond traditional keyword-based searches, U-Seek also supports semantic searches referred to a number of features annotated in the SEEK-AT-WD dataset; namely, tool types (e.g. a whiteboard), supported tasks (e.g. synchronous communication), mediating artifacts (e.g. a video clip) and technical capabilities (e.g. iOS compatible). With the aim of facilitating query formulation, semantic searches are visually constructed through a direct manipulation graphical user interface. Indeed, U-Seek can be seen as an evolution of Ontoolsearch [5], a former semantic searcher of educational tools, though limited due to the use of an isolated and difficult to evolve tool dataset.

Therefore, we can summarize the following benefits of U-Seek: 1) search of educational tools obtained from a huge, open, collaboratively created and automatically updated dataset, 2) makes the user conscious of the underlying information structure by using visual representations, and 3) fine-grained queries and accurate responses by supporting semantic searches (besides keyword-based searches). A deployment of U-Seek is available for testing purposes at http://www.gsic.uva.es/seek/useek/. In this paper we first describe the functionality provided by showcasing the user interface. Then, the application architecture is briefly depicted and some implementation details are given.

2   Searching Educational Tools with U-Seek

To illustrate the functioning of U-Seek, we are going to employ an authentic learning situation corresponding to an undergraduate software engineering course. Students enrolled in this course are engaged in a software project and given various assignments; specifically, they have to model various UML diagrams and to write a number of reports in groups. In the remaining of this section we depict how an educator can find suitable tools to support this situation with U-Seek.

An educator that faces U-Seek can compose her query by multiple criteria. For instance, she can browse the taxonomy of tool types (see (1) in Figure 1), arranged in a manipulable graph with different controls available: the graph can be panned and zoomed by scrolling and dragging, node positions can be changed by dragging the nodes around, star-shaped nodes can be expanded with a double-click and circle-shaped nodes can be collapsed. The educator may eventually find a suitable tool type such as UML editor (2) and click on this node to add it to her query. The ongoing query is shown in (3) as a list of restrictions; in the example shown, the query tool type: UML editor is very close to the intended target (a tool for modeling UML diagrams), so the educator can just press the *Search tools!* button (4) and evaluate the 20 results (5) that were obtained.

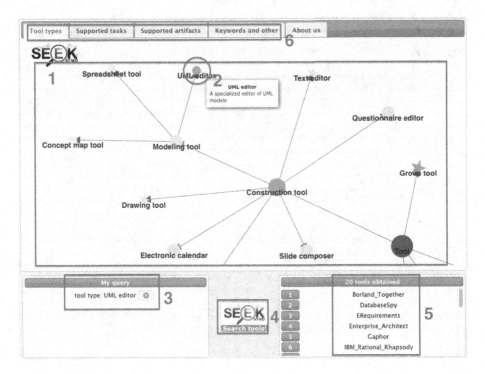

Fig. 1. Snapshot of U-Seek during the formulation of a query in which the user wants to find UML editors

In addition, educators can search by other criteria using the tab controls of the user interface (6). Similar to tool types, manipulable graphs of supported tasks (see Figure 2) and mediating artifacts are provided. Note that these views are complementary, so an educator can choose a path or another depending on her personal preferences, and different types of restrictions can be combined in a query. Indeed, due to the design of the underlying educational vocabulary [3], alternative query formulations may produce the same set of tool instances, e.g. `task type: Modeling + artifact type: UML Model` is equivalent to the precedent query. As a result, query formulation is very flexible.

Taking up the learning situation again, a possible query for discovering collaborative writing tools can be the following: `task type: Writing + tool type: Group tool`. This is shown in Figure 2, along with the list of results obtained for this query in the bottom right frame. When an item of this list is selected, a description of the tool is displayed along with the tool categories and additional information about the developer, licenses, supported operating systems and websites, if available (see the dialog box in Figure 2). In this case, the information about Google Docs was obtained from Dbpedia (including the most specific tool types), while SEEK-AT-WD deduced additional metadata, specially supported

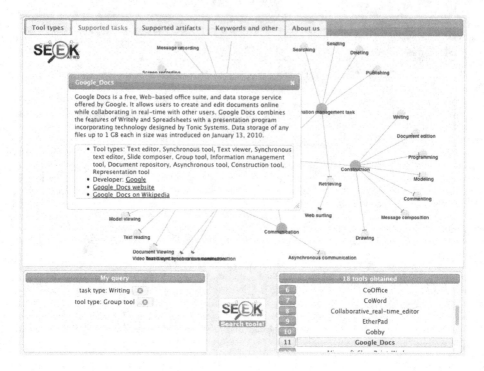

Fig. 2. Snapshot of U-Seek showing the results obtained for collaborative writing tools and details of one instance

task types and mediating artifacts; this way, it is possible to provide alternative search paths for tool discovery. Furthermore, SEEK-AT-WD periodically crawls origin data sources, so modifications (e.g. a new feature of Google Docs annotated in Dbpedia) and new tool additions are propagated to U-Seek without supplementary intervention.

To conclude this section, we give some remarks about additional search options that can be accessed through the *Keywords and other* tab. As in a conventional searcher, an educator can include keywords to her query. Note, however, that it can be done in combination with other restrictions such as tool types, thus serving to further refine a semantic search. For example, looking for a drawing tool with support for the SVG standard can be done with the query `task type: Drawing + keyword: svg` (obtaining 11 out of 303 drawing tools, see Figure 3). Finally, it is possible to restrict for licensing information, e.g. freeware, and supported operating system, e.g. Linux. This is specially interesting to comply with budget or platform restrictions that may determine the eventual selection of a tool. Licensing and operating system information is directly obtained from Dbpedia and, since there are so many choices available, we have included autocomplete input fields in order to quickly find and select an existing value (see Figure 3).

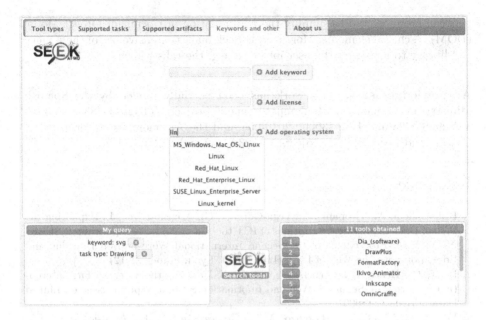

Fig. 3. Snapshot of U-Seek showing the additional search options included in the *Keywords and other* tab and the autocomplete input field

3   Architecture and Implementation Details

U-Seek is a web application that relies on SEEK-AT-WD as the back end. Architecturally, U-Seek consists of a user interface component and a connector that manages the communication with SEEK-AT-WD. By interacting with the user interface, an educator can scan the information structure and acknowledge available query options. Furthermore, U-Seek keeps track of the ongoing query, adding or removing restrictions in response to user actions. Upon a search request, the connector submits the query to SEEK-AT-WD and waits for a response. Finally, obtained results are formatted and displayed through the user interface.

When communicating with SEEK-AT-WD, U-Seek queries its SPARQL endpoint. While this is a common method to access Linked Data sources [2], it should be noted that queries have to be expressed in the SPARQL query language [6]. As a result, U-Seek performs the necessary query conversions in its exchanges with SEEK-AT-WD.

U-Seek is coded in Javascript, the programming language of the Web [7, p. 1]. Therefore, it should work on a wide range of devices with a modern browser[3]. U-Seek uses the Infovis[4] toolkit for the manipulation and visualization of graphs. In addition, the popular jQuery[5] library is employed to lessen incompatibilities

---

[3] We have tested U-Seek with Firefox 11, Google Chrome 17 and Safari 5.
[4] http://thejit.org/
[5] http://jquery.com/

between browsers and to simplify the manipulation of Document Object Model (DOM) [7, ch. 15] elements. Moreover, U-Seek makes extensive use of the jQuery UI[6] library to construct the user interface, e.g. the tabs widget.

Acknowledgements. This work has been partially funded by the Spanish Ministry of Economy and Competitiveness projects TIN2011-28308-C03-02, TIN2008-0323 and IPT-430000-2010-054, and the Autonomous Government of Castilla and León, Spain, projects VA293A11-2 and VA301B11-2.

References

1. Ruiz-Calleja, A., Tiropanis, T., Vega-Gorgojo, G., Asensio-Pérez, J.: Exploiting the Web of Data to provide descriptions of ICT tools: a preliminary report about SEEK-AT-WD. In: Proceedings of the Second International Workshop on Learning and Education with the Web of Data (LiLe 2012), Lyon, France (2012)
2. Bizer, C., Heath, T., Berners-Lee, T.: Linked data – the story so far. International Journal on Semantic Web and Information Systems, Special Issue on Linked Data 5(3), 1–22 (2009)
3. Vega-Gorgojo, G., Bote-Lorenzo, M.L., Gómez-Sánchez, E., Asensio-Pérez, J.I., Dimitriadis, Y.A., Jorrín-Abellán, I.M.: Ontoolcole: Supporting educators in the semantic search of CSCL tools. Journal of Universal Computer Science (JUCS) 14(1), 27–58 (2008)
4. Muller, M.J., Kuhn, S.: Participatory design. Communications of the ACM 36(6), 24–28 (1993)
5. Vega-Gorgojo, G., Bote-Lorenzo, M.L., Asensio-Pérez, J.I., Gómez-Sánchez, E., Dimitriadis, Y.A., Jorrín-Abellán, I.M.: Semantic search of tools for collaborative learning with the ontoolsearch system. Computers & Education 54(4), 835–848 (2010)
6. Prud'hommeaux, E., Seaborne, A. (eds.): SPARQL Query Language for RDF. W3C recommendation (2008)
7. Flanagan, D.: JavaScript: The Definitive Guide. O'Reilly Media, Sebastopol (2011)

---

[6] http://jqueryui.com/

XESOP:
A Content-Adaptive M-Learning Environment

Ivan Madjarov and Omar Boucelma

Aix-Marseille Univ, LSIS, Domaine Universitaire de Saint-Jérôme,
Avenue Escadrille Normandie-Niemen, 13397 Marseille, France
{ivan.madjarov,omar.boucelma}@lsis.org

Abstract. This demo paper illustrates content adaptation for mobile devices. Adaptation considers the context of the client and also the environment, where the client request is received. A device independent model is demonstrated in order to achieve automatic adaptation of a content based on its semantic and the capabilities of the target device. A Web Services-based Framework is presented for adapting, displaying and manipulating learning objects on small handheld devices. A speech solution allows learners to turn written text into natural speech files, in using standard voices. This demo paper also illustrates the main features of XESOP system: authoring of heterogeneous data, integration by means of Web services' invocation in a Learning Course Management System.

Keywords: Web Services, Data integration, Content adaptation, m-Learning.

1 Introduction

The challenge of mobile networking is the context-aware content adaptation. Usage of multimedia services and especially the presentation of multimedia content are more challenging in a mobile environment than on stationary devices as a result of the diversity of mobile devices and their parameters.

This demo paper highlights personalization in m-Learning following an adaptive approach. An integrated Web-based learning and m-Learning environment has been implemented. The demo will also highlight a framework that utilizes the hierarchical displaying of multimedia units with index extraction and content summarization. Web services technology is used to provide flexible integration in which all the learning components and applications are autonomous and loosely coupled. The realization combines textual content adaptation with audio transcoding to better fulfil student needs. The contribution of the work to be demonstrated is as follows:

- An extended learning object model, based on LOM[1] definition, for coping with scientific text processing;

---

[1] http://ltsc.ieee.org/wg12/files/LOM_1484_12_1_v1_Final_Draft.pdf

A. Ravenscroft et al. (Eds.): EC-TEL 2012, LNCS 7563, pp. 531–536, 2012.

- Extended authoring capabilities which defines the XML structure and semantic content of a multimedia pedagogical document [3];
- Web service-based data and application integration with standard LMS [2];
- Adaptive Web services based content adaptation tools for a plethora of mobile devices [1].

2   Demonstration Software Environment

To supply device-independent Web-accessible information that can be browsed in a readable way on different devices and software platforms we adopted methods for effective mobile device recognition, and for mobile Web browsers functionalities identification. For an effective mobile device recognition method we use the header field in the HTTP protocol. To prove the LOs (*Learning Objects*) portability on mobile browsers we conducted a series of tests. Each page contains a test element, e.g., styled text, tables, scripting, DOM and Ajax, MathML, SVG, multimedia embedded objects, image, sound, XML with XSLT. Analysis of the test results showed that a multimedia pedagogical content is suitable for a set of mobile browser. So we apply a suitable page-adaptation technique that analyses XML course structure and generated pages into smaller, logically related units that can fit into a mobile devices browser [1].

For encoding textual information and content assembly, an XML semantic editor suite is used (Fig. 1) and a tree structure of a generic learning document is generated [2]. Depending on course specificity the author can represent texts, diagrams, mathematical formulas or data in tables. To meet author's needs we developed: (1) a MathML editor for mathematical expressions, (2) a SVG editor for vector graphics creation, (3) a QTI[2] editor for student's progression evaluation, (4) a schema for table generation and (5) a chart editor for data presentation. Binary data of multimedia content is embedded directly into XML course content. If an author inserts an image or any binary data, the semantic editor will encode it using the Base64 encoding method. This single XML collection can be managed easily by providing proper XSLT transformation files. Each collection is validated by an XML Schema-based grammar to define all the elements and attributes that are allowed within a Xesop pedagogical document. The Xesop language consists of three XML "*core*" schemas named *cours.xsd*, *math.xsd* and *svg.xsd* completed with the project-specific *manifest*, which allows the validation of the whole pedagogical collection [2].

A collaborative authoring system imposes the storage of learning collections in an appropriate database. The authors have chosen an NXDB (*native XML database*) which allows the storage of XML documents in their native format. This choice, in opposition to that of a relational database, is explained by the nature of learning documents which are in general of narrative types, i.e., document-centric and not data-centric. Formatted XHTML and PDF versions of extracted learning content can be published in a LCMS via Web services [2].

---

[2] Question and Test Interoperability: www.imsglobal.org/question/

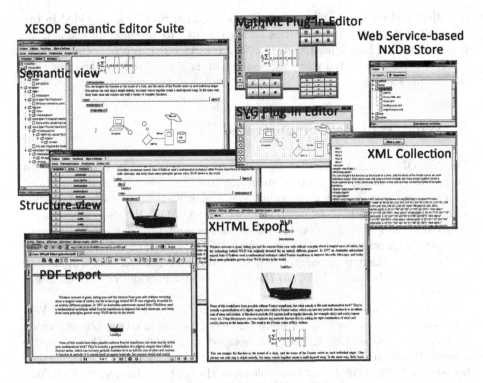

Fig. 1. Xesop Semantic Editor Suite

M-Learning pedagogical content can be provided in the form of a visual presentation as text, pictures or tables. Optionally, it may also be delivered as sound data in the form of an acoustic presentation in an audio format. We have extended the functionality of the Web service-based OSES suite with four additional services. The first one (XICT) is able to create a hypertext index on the basis of the course tree structure. The second service represents an XML content adaptation tool (XCAT) that uses profiles (*XML metadata files*) for automatic content adaptation displayed on the mobile browser. Profiles are adjusted in the function of detection: (1) mobile device profile issue from the WURFL[3] (*Wireless Universal Resource File*), this is supported by the fourth service, and (2) of mobile browser profile [1]. The third service is based on Mbrola[4] speech synthesizer free library to produce speech output from a text paragraph. The XML speech adaptation tool (XSAT) converts the associated text to index item content to an audio output.

Course content adaptation process is an overall index of hyperlinks. Each link points to a node in the hierarchical structure of a created course in XML format. On a *"click"*, the corresponding content is first adapted, then downloaded

---

[3] urfl.sourceforge.net

[4] tcts.fpms.ac.be/synthesis/mbrola.html

and displayed on the mobile screen. The navigation process is provided in two dimensions: top level index entries and hyperlinks to the next/previous page. If a text item is highlighted then the XCAT service is executed, otherwise the XSAT service is executed when the sound icon is highlighted for the same item (for details see Fig. 2).

3   What Will Be Demonstrated

The XESOP system that is demonstrated consists of set of Web services (that we developed) including: content authoring, content managing and publishing, remote exercising, service discovering, client identification, content adaptation and contextualization, data integration, etc. In XESOP, Web services application interaction is as follows: (1) the requester and provider entities become known to each other via the implemented jUDDI service; (2) the requester and provider entities agree on the service description and semantics that will manage the interaction between them; (3) the requester and provider entities exchange messages.

Fig.2 below illustrates the course content adaptation process for mobile Web browsers. It is composed of six main screen shots described as follows: screen 1 shows the course tree structure developed in compliance with the course schema definition [3]. Tree elements are labeled at their creation time; hence they become easily identifiable and locatable along the depth of the tree, which defines their hierarchical position in the index. Screen 2 presents the Java Web-based semantic editor for defining any pedagogical component, while screen 3 is an optional view of the course content in native XML format. Screen 4 shows the results of summarization in the form of indexes corresponding to each node of the hierarchical structure of the course. This summarized content is sent to the mobile Web browser. Screen 5 shows a possible learner interaction by choosing items from index and receiving corresponding adapted content, while screen 6 depicts an audio file played on the clients side player. If the audio icon is selected from screen 4 instead of text-link the associated text content is processed in audio output. If a binary content is chosen a standard audio message is sent.

We deployed the system in a real setting consisting of a Moodle[5]-based e-Learning system called eCUME[6] that has been adopted in our university. A preview of the demo is accessible at ivmad.free.fr/xesop/.

For system deployment we integrated the PHP-based LMS interface via Web services. An implementation of the LMS interface via Web services offers a high degree of flexibility and ease of use, in particular as SOAP libraries for PHP already exist, which leads to an easily extensible PHP and MySQL-based LMS. For services registration Apache Axis[7] as SOAP engine was employed. This tool

---

[5] moodle.org

[6] ecume.univmed.fr

[7] ws.apache.org/axis

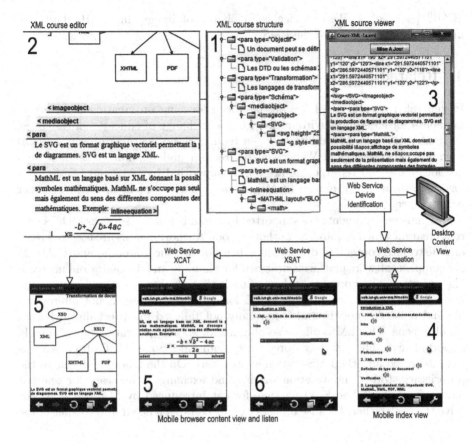

Fig. 2. Content Adaptation Process

facilitates the deployment of Web services, and it offers functionality to auto-matically generate a WSDL description of a service. For storing and managing LOs eXist[8], a Java-based open source native XML database was used. It is run-ning in the Apache Tomcat Servlet engine as Web application. For searching and updating data, eXist server supports XQuery, XPath, and XUpdate XML-based processing technologies. This database can also be invoked via XML-RPC, a REST-style Web services API, and SOAP message-based protocol. To integrate with other e-Learning and m-Learning systems we implemented a jUDDI[9] Web-services directory. The interconnection with a Web-based LMS is carried out by a WSMS (*Web Services Management System*). Thus, many Web service-based external applications can be integrated with a LMS. The publication of an XML collection created by OSES in the learning space of a LMS is achieved by the creation of a SCORM conformant *imsmanifest.xml* file and optional zipped

---

[8] `exist.sourceforge.net`

[9] `juddi.apache.org`

SCO[10] package. A Web service is in charge of integrating XML data into internal data structures of the LMS.

4   Related Work

In general, research in e-Learning has been focusing on creating a variety of metadata formats and environments for the exchange of educational resources. Currently, the interoperability between repositories of learning objects remains a challenge. Pedagogical Web resources remain underexploited, as their connection, reuse and repurposing are barely supported by LMS and LCMS platforms. Both an LMS and an LCMS manage course content and track learner performance.

Learning environments are supported by a number of key services such as content creation which requires an authoring tool. Authoring tools are used to create and distribute content in diverse domains. For instance, Moodle and Dokeos[11] are complex software platforms designed for planning and managing online learning activities. They provide authoring tools that use hypertext and multimedia features for content creation in HTML format. This content is not interoperable and reusable enough, and is normally not tailored for direct displaying on small screens. This LMSs offer also plug-ins for speech generation of a manually imported text. The result is stored in an audio file. So, the TTS framework is not integrated into the LMS-system architecture. On the other hand, due to the proprietary storage, information sharing and exchange is not easy for selected parts of a course. The SCO package format introduced by SCORM does not solve the trick either, because objects cannot be broken down into smaller units.

References

1. Madjarov, I., Boucelma, O.: Learning Content Adaptation for m-Learning Systems: A Multimodality Approach. In: Luo, X., Spaniol, M., Wang, L., Li, Q., Nejdl, W., Zhang, W. (eds.) ICWL 2010. LNCS, vol. 6483, pp. 190–199. Springer, Heidelberg (2010)
2. Madjarov, I., Boucelma, O.: Data and Application Integration in Learning Content Management Systems: A Web Services Approach. In: Nejdl, W., Tochtermann, K. (eds.) EC-TEL 2006. LNCS, vol. 4227, pp. 272–286. Springer, Heidelberg (2006)
3. Madjarov, I., Boucelma, O.: Multimodality and Context-adaptation for Mobile Learning. In: Bebo, W., Irwin, K., Philip, T. (eds.) Social Media Tools and Platforms in Learning Environments, 1st edn. LNCS, pp. 257–276 (2011)

---

[10] www.scorm.com
[11] www.dokeos.com

Part V
Poster Paper

A Collaboration Based Community to Track Idea Diffusion Amongst Novice Programmers

Reilly Butler, Greg Edelston, Jazmin Gonzalez-Rivero,
Derek Redfern, Brendan Ritter, Orion Taylor, and Ursula Wolz

Franklin W. Olin College of Engineering
Olin Way, Needham, MA 02492

Computer science is a collaborative effort, as evidenced by the proliferation of open source software and code-collaboration websites. Computer science education should introduce students to programming in an environment where collaboration is encouraged. We prototyped an integrated development environment (IDE) with connectivity to a remote database to encourage students to engage in what Etienne Wenger calls a "community of practice" [1]. This approach is based on insights from Monroy-Hernández' work on remixing in Scratch [2] with needs identified from a range of existing open source community models [3,4,5].

Existing collaboration systems allow forking and sharing entire projects, but do not track the exchange of short segments (snippets) of code. Our IDE incorporates a system to help students attribute snippets to their original authors. The intent of this mechanism is to increase students' comfort with sharing code. When a student copies a snippet from a project in the community and pastes it into another project, our IDE offers to add a citation of the exchange to both projects. The IDE gently prompts users to cite code from outside sources as well.

To introduce students to revision control without the steep learning curve of existing systems, we developed a graphical revision control system, borrowing the concept of "save points" from video games. The IDE stores save points in a remote database so students and instructors can view projects' development histories. Additionally, the database synchronizes working copies across devices.

An important goal of the project is to give computer science educators insight into how students exchange ideas. Instructors can make effective use of the revision control and snippet citation tools to observe their students' thought processes and the flow of ideas through a classroom. This prototype suggests directions in which computer science educators can emphasize collaboration.

References

1. Wenger, E.: Communities of practice: A brief introduction (2006), http://wenger-trayner.com/Intro-to-CoPs/
2. Monroy-Hernández, A., Hill, B.M., González-Rivero, J., Boyd, D.: Computers can't give credit: How automatic attribution falls short in an online remixing community. In: Conference on Human Factors in Computing Systems, CHI, May 2011. ACM (May 2011)
3. GitHub Inc.: GitHub features (2012), https://github.com/features
4. National Science Digital Library: Ensemble computing portal: Connecting computing educators (2012), http://www.computingportal.org/
5. Kölling, M.: Greenroom sharing (2010), http://greenroom.greenfoot.org/sharing

A. Ravenscroft et al. (Eds.): EC-TEL 2012, LNCS 7563, p. 539, 2012.
© Springer-Verlag Berlin Heidelberg 2012

Argument Diagrams in Facebook: Facilitating the Formation of Scientifically Sound Opinions

Dimitra Tsovaltzi, Armin Weinberger, Oliver Scheuer, Toby Dragon,
and Bruce M. McLaren

Saarland University, Saarbrücken, Germany, P.O. Box 151150
Dimitra.tsovalzi@mx.uni-saarland.de

Students use Facebook to organize their classroom experiences [1], but hardly to share and form opinions on subject matters. We explore the benefits of argument diagrams for the formation of scientific opinion on behaviorism in Facebook. We aim at raising awareness of opinion conflict and structuring the argumentation with scripts [2]. A lab study with University students (ten dyads per condition) compared the influence of argument structuring (students built individual argument diagrams before discussing in Facebook) vs. no argument structuring (only Facebook discussion) on opinion formation, measured through opinion change. The argumentation script was implemented in the web-based system LASAD to support sound argumentation [3].

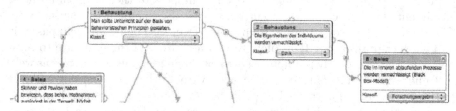

Fig. 1. View of LASAD diagram

Facebook discussions and conflict awareness led students of both conditions to change their opinions, $t(39)=8.84$, $p<.001$. Evidence suggests a connection between opinion change and the number of conflicts in a discussion. Together with a high correlation for no argument structuring between opinion change and knowledge gains, $r(20)=.54$, $p<.05$, the results suggest benefits of raising awareness of opinion conflicts in Facebook to facilitate scientific opinion formation and change.

References

1. Lampe, C., Wohn, D., Vitak, J., Ellison, N., Wash, R.: Student use of Facebook for organizing collaborative classroom activities. International Journal of Computer-Supported Collaborative Learning 6(3), 329–347 (2011)
2. Weinberger, A., Stegmann, K., Fischer, F.: Learning to argue online: Scripted groups surpass individuals (unscripted groups do not). Computers in Human Behavior 26(4), 506–515 (2010)
3. Loll, F., Pinkwart, N., Scheuer, O., McLaren, B.M.: How Tough Should It Be? Simplifying the Development of Argumentation Systems using a Configurable Platform. In: Pinkwart, N., McLaren, B.M. (eds.) Educational Technologies for Teaching Argumentation Skills, Bentham Science Publishers (to appear) (in press)

A. Ravenscroft et al. (Eds.): EC-TEL 2012, LNCS 7563, p. 540, 2012.
© Springer-Verlag Berlin Heidelberg 2012

Authoring of Adaptive Serious Games

Maurice Hendrix[1], Evgeny Knutov[2], Laurent Auneau[3], Aristidis Protopsaltis[1],
Sylvester Arnab[1], Ian Dunwell[1], Panagiotis Petridis[1], and Sara de Freitas[1]

[1] Serious Games Institute, Coventry University, Coventry, UK
{MHendrix,AProtopsaltis,SArnab,IDunwell,PPetridis,
SFreitas}@cad.coventry.ac.uk
[2] Department of Computer Science, Eindhoven University of Technology, NL
e.knutov@tue.nl
[3] Succubus Interactive, Nantes, France
laurent.auneau@succubus.fr

Abstract. Game-based approaches to learning are increasingly being recognized as having the potential to stimulate intrinsic motivation amongst learners. Whilst a range of examples of effective serious games exist, creating the high-fidelity content with which to populate a serious game is resource-intensive task. To reduce this resource requirement, research is increasingly exploring means to reuse and repurpose existing games and relevant sources of content. Education has proven a popular application area for Adaptive Hypermedia, as adaptation can offer enriched learning experiences to students. Whilst content to-date has mainly been in the form of rich text, various efforts have been made to integrate Serious Games into Adaptive Hypermedia via run-time adaptation engines. However, there is little in the way of effective integrated authoring and user modeling support for these efforts. This paper explores avenues for effectively integrating serious games into adaptive hypermedia. In particular, we consider authoring and user modeling aspects in addition to integration into run-time adaptation engines, thereby enabling authors to create Adaptive Hypermedia that includes an adaptive game, thus going beyond mere selection of a suitable game and towards an approach with the capability to adapt and respond to the needs of learners and educators.

Keywords: Adaptive Hypermedia, Adaptation, Serious Games, Educational Games, Education, Personalization.

Acknowledgements. This work has been fully supported by the mEducator project, funded by the eContentPlus programme by the European Commission. The authors thank all mEducator partners. Work has also been part supported by the European Commission under the Collaborative Project ALICE "Adaptive Learning via Intuitive/Interactive, Collaborative and Emotional Systems", VII Framework Programme, Theme ICT-2009.4.2 (Technology-Enhanced Learning), Grant Agreement n. 257639.

A. Ravenscroft et al. (Eds.): EC-TEL 2012, LNCS 7563, p. 541, 2012.

Collaborative Learning and Knowledge Maturing from Two Perspectives

Uwe V. Riss[1] and Wolfgang Reinhardt[2]

[1] SAP Research
Vincenz-Priessnitz-Str. 1
D-76131 Karlsruhe, Germany
uwe.riss@sap.com
[2] University of Paderborn
Institute for Computer Science
33102 Paderborn, Germany
wolle@upb.de

Abstract. We discuss the similarities and differences between the Co-evolution Model of Cognitive and Social Knowledge and Sociofact Theory as two new theories of collaborative learning and knowledge maturing.

Keywords: knowledge maturing, sociofact theory, coevolution.

Summary

Recently, Cress and Kimmerle have suggested the Co-evolution Model of Cognitive and Social Systems based on Luhmann's theory of communication. It allows better understanding collaborative knowledge building in wikis and other social media [1]. Alternatively, Riss and Magenheim have suggested a Sociofact Theory of social knowledge that is based on Symbolic Interactionism [2]. It emphasizes the role of deviations in common understanding of information artefacts. Although different at first glance, it is shown that both theories are based on similar ideas. Both models emphasize the crucial role of communication and the importance of cognitive conflict for social learning and point at its role for knowledge maturing in social media. Their difference mainly refers to the concept of aggregation in Sociofact Theory. Aggregation describes a social process of mutual recognition of corresponding and deviating understanding of information artefacts as precondition for knowledge maturing. It is shown that aggregation can be found in various forms, in tag clouds as well as in wiki articles.

This work has been supported by the European Union IST fund through the EU FP7 MATURE Integrating Project (Grant No. 216356).

References

1. Cress, U., Kimmerle, J.: A systemic and cognitive view on collaborative knowledge building with wikis. IJCSCL 3, 105–122 (2008)
2. Riss, U.V., Magenheim, J.: Sociofact theory the social dimension of knowledge maturing (accepted for publication in IJKBO)

A. Ravenscroft et al. (Eds.): EC-TEL 2012, LNCS 7563, p. 542, 2012.
© Springer-Verlag Berlin Heidelberg 2012

Computer Supported Intercultural Collaborative Learning: A Study on Challenges as Perceived by Students

Vitaliy Popov, Omid Noroozi, Harm J.A. Biemans, and Martin Mulder

Chair Group of Education and Competence Studies, Wageningen University, The Netherlands
{vitaliy.popov,omid.noroozi,harm.biemans,martin.mulder}@wur.nl

Abstract. This study examines challenges that are inherent in computer supported intercultural collaborative learning (CSICL) in higher education. For this purpose, a 22-item survey was completed by students (N=98) who worked collaboratively in culturally diverse pairs on an online learning task focused on the field of life sciences. Students were required to rate on a Likert scale the importance of a certain challenge in CSICL. Descriptive statistics were used to determine what challenges are perceived to be the most important by students in CSICL. The results suggest that 'a collaborative partner is not communicating properly', 'a low level of motivation' and 'insufficient English language skills' were perceived by all study participants to be the most important challenges in CSICL.

Keywords: computer supported collaborative learning, cultural diversity, challenges.

1 Purpose

This paper presents the results of a study that was conducted in a Dutch university aiming at better understanding the cross-cultural cooperation while working in culturally heterogeneous groups in CSCL environments. This study has a dual purpose: (1) to identify challenges that are inherent to CSICL in higher education based on previous research studies, and (2) to examine the extent to which culturally diverse students perceive different challenges to be important in this setting.

2 Design/Methodology

All participants (N=98) were assigned to dyads based on their disciplinary backgrounds, such that every dyad had complimentary expertise (one learner with water management disciplinary background and one learner with international development background). It resulted in 10 culturally homogeneous and 39 heterogeneous dyads. These 10 culturally homogeneous dyads were omitted from the further analysis because they did not meet the requirements of this research study.

A. Ravenscroft et al. (Eds.): EC-TEL 2012, LNCS 7563, pp. 543–544, 2012.

Participants were asked to collaborate, discuss, and argue with their assigned partner to develop possible solutions for the task (i.e. as task required students to develop a plan for fostering sustainable behavior among wheat farmers in a province of Iran and to ultimately reach an agreement about that solution). All interaction between the dyad partners was conducted online, using the chat window of the CSCL environment. A total of 26 countries were represented by our study's international participants. All students were interacting with the study personnel and with each other in English. After the experiment, participants filled in a questionnaire about the challenges in CSCL environment. The list of challenges included in the questionnaire used for this study was derived from earlier research on online collaborative learning.

3   Findings

The results of the descriptive analysis showed that almost all challenges were considered to be at least of some importance by all participants of this study (scores higher than 3 within 5-point Likert-type scale). Second, according to the students, 'a collaborative partner is not communicating properly' (M=4.17, SD=0.74), 'a low level of motivation' (M=4.13, SD=0.89) and 'insufficient English language skills' (M=4.07, SD=0.85) were the most challenging issues in CSCL for culturally diverse dyads.

Table 1. Means and Standard Deviations for the most important challenges (here presented only 10 challenges out of 22 due to space constraints)

Challenges	Mean	SD
a collaborative partner is not communicating properly	4.17	0.74
a low level of motivation	4.13	0.89
insufficient English language skills	4.07	0.85
free-riding	4.00	1.02
technical problems	3.91	0.91
attitudinal problems such as dislike, mistrust and lack of cohesion	3.89	0.90
insufficient social presence	3.74	1.01
conflicts in a collaborative pair	3.64	0.99
dominating collaborative partner	3.56	1.15
the pressure to defend group decisions whilst not agreeing with them	3.45	0.90

4   Research Limitations/Implications

Such challenges have to be considered by many different groups and collaboration forms (dyads was the only form tested in this study) in order to draw more general conclusions.

Just4me: Functional Requirements to Support Informal Self-directed Learning in a Personal Ubiquitous Environment

Ingrid Noguera, Iolanda Garcia, Begoña Gros, Xavier Mas, and Teresa Sancho

eLearn Center, Universitat Oberta de Catalunya, Barcelona, Spain
{inoguerafr,igarciago,bgros,xmas,tsancho}@uoc.edu

Abstract. The aim of this project is to design, implement and analyze the use of a personal ubiquitous learning environment in order to develop a prototype that can be commercially exploited. Initial results have shown specific requirements in terms of personalization, integration of different environments, tools and resources and features to structure and plan the knowledge.

Keywords: Self-directed learning, personal learning environments, informal learning, lifelong learning, ubiquitous learning, online platforms.

The purpose of the study is to develop a platform that provides a ubiquitous personal learning environment for the lifelong learner and integrates tools that might be of help to self plan and self structure learning pathways. In this paper we tackle the issue of functional requirements that might support informal self-directed learning [1] taking also into account mobility factors (related with ubiquitous learning). With this purpose we have collected information through three actions: a) review of the main current research on the conceptualization and implementation of personal learning environments, b) a questionnaire with seven multiple-choice questions addressed to medical professionals (N=26), and c) review of the prominent projects regarding the development and implementation of PLEs in professional contexts, higher education institutions and open environments for lifelong learning. Initial results have shown specific requirements in terms of personalization that can make the environment adaptable to the users' different levels of digital competence, learning style and needs. The platform should be close to everyday technologies, and in turn, be able to integrate and operate with other environments, tools and resources. It should also incorporate specific features and tools specially conceived to support learning, and to structure and plan the knowledge that learners acquire along and across their academic, social and professional pathways. Finally it should recommend relevant information to learners on the basis of their fields of interest.

Reference

1. Knowles, M.: Self-Directed Learning: A Guide for Learners and Teachers. Association Press, New York (1975)

A. Ravenscroft et al. (Eds.): EC-TEL 2012, LNCS 7563, p. 545, 2012.

Observations Models to Track Learners' Activity during Training on a Nuclear Power Plant Full-Scope Simulator

Olivier Champalle[1], Karim Sehaba[2], and Alain Mille[1]

Université de Lyon, CNRS
[1] Université de Lyon 1, LIRIS, UMR5205, F-69622, France
[2] Université de Lyon 2, LIRIS, UMR5205, F-69676, France
{olivier.champalle,karim.sehaba,alain.mille}@liris.cnrs.fr

Keywords: Training, skills, Observation, Analyze, Trace, Modeled Traces, Transformations, Exploration, Trace-Based Systems Framework, Full Scope Simulator, Nuclear Power Plant.

In the context of the professional training of operators of Nuclear Power Plants (NPP) on Full-Scope Simulators (FSS), the objective of our work is to propose models and tools to help trainers observe and analyze trainees' activities during preparation and debriefing. For that purpose, our approach consists in representing the actions of the operators and the simulation data in the form of modeled trace. These modeled traces are then transformed in order to extract higher information level. Trainers can visualized the different levels of trace to analyze the reasons, collective or individual, of successes or failure of trainees during the simulation.

This work is part of a research project conducted in partnership with UFPI (Training Unit Production Engineering) of EDF group. The main activity of UFPI (over 700 trainers, 3 million hours of training per year) is to provide professional training courses for staff working in the domains of power generation (nuclear, fossil-fired, hydraulic). Among its formations, the UFPI trains operators to drive nuclear power plants. For this, the trainers of the UFPI organize simulation sessions on full-scale simulators. This project aims to facilitate the debriefing and analysis phases of simulation sessions by providing tools that allow trainers to analyze the traces of trainees.

In order to validate our approach, we have developed the prototype D3KODE based on the trace model and transformation that we proposed. This prototype was then evaluated according to a protocol based on a comparative method in the context of several experiment conducted with a team of experts, trainers and trainees from EDF Group.

Acknowledgment. The ANRT and the UFPI of EDF Group finance this research work. The authors thank all the Trainers of the UFPI for their help.

A. Ravenscroft et al. (Eds.): EC-TEL 2012, LNCS 7563, p. 546, 2012.
© Springer-Verlag Berlin Heidelberg 2012

Practical Issues
in e-Learning Multi-Agent Systems

Alberto González Palomo

Sentido-Labs.com
Im Flürchen 39a, 66133 Saarbrücken, Germany
http://sentido-labs.com

Multi-Agent Systems (MAS) seem, in theory, an excellent way to integrate the various component systems used in an e-Learning platform.

Why are they not used more broadly?

I suspect the main reason is that using a multi-agent architecture requires solving many issues from the beginning that can be ignored at the prototype stage in other approaches, like fine-grained security and fault tolerance.

To solve that I am developing the e-Learning Multi-Agent System (eLMAS) that will be used as integration platform in the European project Allegro[1].

In eLMAS there are two kinds of agents: "liaison" agents that connect existing e-Learning systems to the MAS, and learner agents that represent the user, storing the learner model and planning the learning activities.

1 Component Integration

The liaison agents for the components are written in Java using JADE[2], and they can do anything a regular Java program can. Security is provided through user password authentication, code signing, and agent message signing and encryption. However these measures do not suffice if we allow agent mobility.

2 Secure Mobile Learner Agents

Learner agents need to be mobile to interact more efficiently with the different components running at each institution. To make sure they can not cause trouble in the systems they visit, they are implemented in an interpreted domain-oriented language, made simply not to have the low-level functionality needed for breaking out of the security scheme.

I chose a language called AgentSpeak invented specifically for Multi-Agent Systems[3] that implements a Beliefs-Desires-Intentions (BDI) architecture, where agents are defined by a set of initial beliefs and a set of plans. Beliefs are changed by the environment or by actions in the executed plans, and which plan to execute for a given goal is decided by a logic formula resolution engine.

The agent beliefs about the learner's knowledge are the learner model, and the planner engine provides the adaptivity.

---

[1] Interreg IV-A project Allegro: http://allegro-project.eu
[2] Java Agent DEvelopment Framework: http://jade.tilab.com/
[3] The AgentSpeak implementation is Jason: http://jason.sourceforge.net

A. Ravenscroft et al. (Eds.): EC-TEL 2012, LNCS 7563, p. 547, 2012.

Students' Usage and Access to Multimedia Learning Resources in an Online Course with Respect to Individual Learning Styles as Identified by the VARK Model

Tomislava Lauc, Sanja Kišiček, and Petra Bago

Department of Information and Communication Sciences, Faculty of Humanities and Social Sciences, Zagreb, Croatia
{tlauc,smatic,pbago}@ffzg.hr

Poster Summary

We present a research on students' learning styles and their learning activity with respect to multimedia learning resources in a virtual learning environment within a Moodle online course. We investigate the relation between learning styles based on sensory modality and their learning activity regarding different types of multimedia resources.

VARK is a sensory model developed by Neil Fleming. It is an acronym for Visual (V), Aural (A), Read/Write (R), and Kinesthetic (K).

Considering the VARK questionnaire results, we came to a conclusion that two out of three students have multimodal learning styles and prefer combining different types of resources. The most used type of resource is pictorial accompanied by text (72%). Considering log file data analysis it is shown that students with higher visual learning style scores obtained by the VARK questionnaire have lower tendency of accessing pictorial resources accompanied by text. Furthermore, at the end of each lesson we conducted a survey allowing multiple answers asking students what type of resources they have been using.

Results show that the usage of video resources is highly correlated with the access to the same resources. Correspondingly the usage of textual resources is highly correlated with the access to the same resources. Moreover, a negative correlation exists between the usage of pictorial resources and the access to textual resources, meaning that the students who preferred pictorial resources accompanied by text, accessed to textual resources less.

However, the usage of pictorial resources accompanied by text has low correlation with the access to those resources. Considering this fact, it could be possible that students spent more time studying the pictorial resources accompanied by text, therefore accessing those resources less.

Keywords: learning styles, multimedia learning, online learning, log file data analysis.

A. Ravenscroft et al. (Eds.): EC-TEL 2012, LNCS 7563, p. 548, 2012.
© Springer-Verlag Berlin Heidelberg 2012

Technology-Enhanced Replays of Expert Gaze Promote Students' Visual Learning in Medical Training

Marko Seppänen[1] and Andreas Gegenfurtner[2]

[1] Turku PET Centre, Turku University Hospital, Turku, Finland
marko.seppanen@tyks.fi
[2] TUM School of Education, Technical University of Munich, Munich, Germany
andreas.gegenfurtner@tum.de

Abstract. Based on diagnostic performance and eye tracking data, the present study demonstrates that technology-enhanced replays of expert gaze can promote the visual learning of students in clinical visualization-based training.

Keywords: Eye tracking, gaze replay, visual expertise, medical visualizations.

Diagnosing medical visualizations is difficult for students to learn [1-3]. The purpose of the study is to test whether technology-enhanced replays of expert gaze (TEREG) promote visual learning of medical students with realistic, three-dimensional stimuli. Medical students' diagnostic performance and eye movements were compared before and after exposure to TEREG. Participants were 18 undergraduate students of medicine (10 women, 8 men) with no self-reported prior knowledge in interpreting PET/CT. Results of the performance and eye movement measures indicate significant improvement in diagnostic performance and eye movements. Given the importance of motivation for transfer in technology-enhanced training environments [3-5], future studies can test if students apply their novel visual skills to related clinical domains.

References

1. Gegenfurtner, A., Lehtinen, E., Säljö, R.: Expertise Differences in the Comprehension of Visualizations: A Meta-Analysis of Eye-Tracking Research in Professional Domains. Educ. Psychol. Rev. 23, 523–552 (2011)
2. Gegenfurtner, A., Siewiorek, A., Lehtinen, E., Säljö, R.: Assessing the Quality of Expertise Differences in the Comprehension of Medical Visualizations. Vocat. Learn. 6 (2013)
3. Helle, L., Nivala, M., Kronqvist, P., Gegenfurtner, A., Björk, P., Säljö, R.: Traditional Microscopy Instruction versus Process-Oriented Virtual Microscopy Instruction: A Naturalistic Experiment with Control Group. Diagn. Pathol. 6, S8 (2011)
4. Gegenfurtner, A., Vauras, M.: Age-related Differences in the Relation between Motivation to Learn and Transfer of Training in Adult Continuing Education. Contemp. Educ. Psychol. 37, 33–46 (2012)
5. Gegenfurtner, A., Festner, D., Gallenberger, W., Lehtinen, E., Gruber, H.: Predicting Autonomous and Controlled Motivation to Transfer Training. Int. J. Train. Dev. 13, 124–138 (2009)

A. Ravenscroft et al. (Eds.): EC-TEL 2012, LNCS 7563, p. 549, 2012.
© Springer-Verlag Berlin Heidelberg 2012

Towards Guidelines for Educational Adventure Games Creation (EAGC)

Gudrun Kellner[1], Paul Sommeregger[1], and Marcel Berthold[2]

[1] Vienna University of Technology, Institute of Software Technology and Interactive Systems, Favoritenstr. 9-11/188, 1040 Vienna, Austria
{gudrun.kellner,paul.sommeregger}@tuwien.ac.at
[2] Graz University of Technology, Knowledge Management Institute, Brückenkopfg. 1, 8020 Graz, Austria
marcel.berthold@tugraz.at

Keywords: educational adventure games, DGBL, game design, e-learning, game development.

Many recent studies have shown that educational games are effective tools for learning [1]. Despite the recent popularity of game-based learning and some first general guidelines for the creation of such educational games [2], there is a lack of useful practical guidelines for specific game types that address all relevant aspects of design, implementation and testing. This might be explained with the possible lack of experience of instructional designers with computer games and game designers with education [3], but it influences the focus and the approach chosen for the game design and implementation process. The goal of our work is to develop guidelines for the creation of educational adventure games that help not to forget any aspect. In order to do so, we refer to both existing guidelines for the design of entertainment games and existing frameworks for the design of educational games. We suggest a structure of five main game development phases (conceptual design and game design, implementation, testing and validation) and also take project management into account, to not only guide through the creation of the game itself but also to support the organization of the game development process [4]. Our next steps are to put the EAGC guidelines into practice and to show their applicability in a concrete example project, an educational adventure game on electricity, which is currently under development.

References

1. Prensky, M.: Computer Games and Learning: Digital Game-Based Learning. In: Raessens, J., Goldstein, J. (eds.) Handbook of computer game studies, pp. 97--122. Mass., MIT Press (2005)
2. Hirumi, A., Appelman, B., Rieber, L., Eck, R. Van: Preparing Instructional Designers for Game-Based Learning: Part 1. TechTrends 54/3, pp. 27--37 (2010)
3. Linehan, C. et al.: Practical, appropriate, empirically-validated guidelines for designing educational games. In: CHI 2011, pp. 1979--1988. ACM Press, New York (2011)
4. Sommeregger, P., Kellner, G.: Brief Guidelines for Educational Adventure Games Creation (EAGC). In: IEEE DIGITEL 2012, pp. 120--122. IEEE Press, New York (2012)

A. Ravenscroft et al. (Eds.): EC-TEL 2012, LNCS 7563, p. 550, 2012.
© Springer-Verlag Berlin Heidelberg 2012

Author Index